Ecological Studies, Vol. 150

Analysis and Synthesis

Springer

Berlin
Heidelberg
New York
Hong Kong
London
Milan
Paris
Tokyo

Soil crust community on terra rossa soils dominated by the whitish (when dry) lichen *Diploschistes diacapsis* in a Mediterranean shrub community, Cyprus. Insert: The moist lichen with black apothecia (photo is close to actual lichen size). (Photo O.L. Lange)

Jayne Belnap Otto L. Lange (Eds.)

Biological Soil Crusts: Structure, Function, and Management

1st Edition 2001, Revised 2nd Printing 2003

With 141 Figures, 64 in Color, and 30 Tables

Springer

Prof. Dr. Jayne Belnap
US Geological Survey
Forest and Rangeland Ecosystem Science Center
2290 S. Resource Blvd.
Moab, Utah 84532
USA

Prof. Dr. Drs. h. c. Otto L. Lange
Universität Würzburg
Julius-von-Sachs-Institut für Biowissenschaften
Lehrstuhl für Botanik II
Julius-von-Sachs-Platz 3
97082 Würzburg
Germany

Cover illustration: Schematic block diagram of a biological soil crust with typical colonizers (for details see Fig. 1.1 in Chap. 1). Illustration by Renate Klein-Rödder

ISSN 0070-8356
ISBN 3-540-43757-6 Springer-Verlag Berlin Heidelberg New York

Originally published as hardcover edition in 2001 (ISBN 3-540-41075-9)

Library of Congress Cataloging-in-Publication Data applied for

Die Deutsche Bibliothek – CIP-Einheitsaufnahme

Biological soil crusts : structure, function, and management ; with 30
tables / Jayne Belnp ; Otto L. Lange (ed.). - 1. ed., rev. 2. print..-
Berlin ; Heidelberg ; New York ; Barcelona ; Hong Kong ; London ; Milan ;
Paris ; Tokyo : Springer, 2003
(Ecological studies ; Vol. 150)
ISBN 3-540-43757-6

Springer-Verlag Berlin Heidelberg New York
a company of BertelsmannSpringer Science+Business Media GmbH

http://www.springer.de

Cover design: *design & production* GmbH, Heidelberg
Typesetting: Friedmut Kröner, Heidelberg

SPIN 10882381 31/3130 YK – 5 4 3 2 1 0 – Printed on acid free paper

Preface

In arid and semiarid lands throughout the world, where the cover of vegetation is sparse or absent, the open spaces between the higher plants are generally not bare of autotrophic life, but covered by a community of highly specialized organisms. This soil-surface community consists of cyanobacteria, algae, lichens, mosses, microfungi, and other bacteria in differing proportions. Cyanobacterial and microfungal filaments, rhizinae and rhizomorphs of lichens, and the rhizinae and protonemata of bryophytes weave throughout the top few millimeters of soil, gluing loose soil particles together. This forms a crust up to a few centimeters thick that stabilizes and protects soil surfaces from erosive forces (Cameron and Blank 1966; Friedmann and Galun 1974; Belnap and Gardner 1993). These crusts occur in all hot, cool, and cold-arid and semiarid regions of the world, and may constitute up to or more than 70 % of the living cover. Biological soil crusts have only recently been recognized as having a major influence on terrestrial ecosystems.

Research on these soil-surface communities began in the 1950s. The first efforts were mostly descriptive, consisting of taxonomic studies of crusts from different environments. Limited research in the 1960s documented the ability of crustal components to fix nitrogen and influence hydrologic processes such as water runoff and infiltration rates. As the importance of these crusts in ecosystem functioning has become apparent, research efforts have intensified, with well over 3000 publications now available on the biology, ecology, and ecophysiology of soil-crust communities and their components, as well as on applied aspects such as landscape-level hydrology and management.

Globally, this consortium of soil biota has many similarities in species composition, structure, and function, in spite of occurring in unconnected and seemingly dissimilar environments. Crusts are found in an astonishing variety of higher plant communities throughout the world: in all types of desert and semidesert plant communities, ranging from shrub deserts to succulents and to open woodlands (the latter, for instance, in the United States and in Australia). Crusts are also found in steppe formations in both the Northern and Southern Hemisphere; in the gaps of evergreen shrubs and forests in the mediterranean-type climate; and on open ground or between the tundra vegetation in cold climates. On a small scale, soil-crust communities are even found in the more open types of vegetation in temperate regions, such as xerothermic local steppe formations in central Europe and in the pine barrens of the United States. Many of the dominant cyanobacteria and soil lichens have a cosmopolitan distribution, and cryptogamic communities of many crusts have almost identical growth forms. For instance, the same soil lichens dominate soil crusts of both the Sonoran and Great Basin Deserts of the western United States, while vascular plant species and climate varies greatly between these two regions. There are similarities in species composition of biological soil crusts even between arctic and hot desert environments. The physical structure of crusts is very similar in hot deserts of the world, such as the Atacama, Sonoran, Chihuahuan, African, and Australian Deserts, while very different from the physical structure of crusts in cool and cold deserts such as the Great Basin and the Arctic.

Recently, ecophysiologists have begun to study carbon gain and nitrogen fixation of soil-crust biota under controlled conditions in the laboratory and in the field. These are the first steps towards understanding the soil crusts' functioning and their importance in different ecosystems. Researchers have found the autotrophic components of soil crusts deliver assimilates to the often carbon-limited desert soils. Cyanobacteria and cyanobacteria-containing lichens can be an important source of nitrogen for plants and soils in many desert ecosystems. In this way, crusts can improve soil fertility which, in turn, influences vascular plant nutrition. Soil crusts can also influence the germination and establishment of vascular plants. Biological soil crusts reduce the susceptibility of soils to both wind and water erosion. Polysaccharides extruded by the cyanobacteria and green algae, along with the rhizinae of lichens and mosses, entrap and bind soil particles together, hardening the soil surface. Wind and water are major erosive forces, especially in deserts, where there is little protection of soil surfaces by organic matter or higher-plant cover. Thus, the presence of biological soil crusts in water-limited areas can be crucial in reducing soil loss from plant interspaces and in maintaining vascular vegetative productivity.

Extensive livestock grazing, foot traffic, and off-road vehicles break through the protective soil-crust cover. This decreases the resistance of the soil surfaces to wind and water erosion. Subsequent invasion of exotic annual plants into native perennial communities can cause dramatic changes in soil-crust flora. Such invasions, along with disturbance, can lead to substantial alterations in carbon and nitrogen inputs. Thus, land managers in arid and semiarid areas need to understand the role of crusts, and the impact of different land uses, in different ecosystems.

There have been several reviews of the literature on biological soil crusts, beginning with Cameron and Blank in 1966 and Friedmann and Galun in

1974. Harper and Marble reviewed the literature again in 1988, emphasizing studies in the United States, and West (1990) published a review that emphasized European and Israeli literature. This was followed by a brief overview by Johansen in 1993 and a review by Eldridge and Greene in 1994. Most recently, Evans and Johansen (1999) summarized our knowledge on biological crusts and ecosystem processes. Eldridge and Tozer (1997) published a practical guide for identification of soil lichens and bryophytes of Australia's dry country. However, apparently contradictory information on the role of these crusts has often been published. Moreover, many papers are difficult to find, making it impossible for many people to obtain an overview on the state of knowledge about biological soil crusts. For these reasons, we felt a general synthesis volume on biological crusts was needed as a central reference for scientists and land managers. To achieve this goal, we invited the world's leading specialists in soil-crust research to summarize known research on the different aspects of biological soil crusts. These contributions summarize our present knowledge about the role of soil crusts in different ecosystems, and point out known gaps in our current understanding. They also explore worldwide and regional patterns to help explain apparent differences in the role and function of these crusts from disparate environments.

The taxonomy of the highly specialized soil-crust organisms is a difficult field, as several of the systematic groups are insufficiently known for correct identification of all representatives down to the species level. Nevertheless, taxonomic composition of most of the crust types can be clearly characterized. In *Part I* of our Volume, characteristic soil-crust organisms and typical crust habitats throughout the world are depicted by photographs (Chap. 1). The communities of organisms forming soil crusts and the appearance of different soil-crust types are described for the different geographical regions. Discussion is centered on how species composition and dominance of crustal organisms varies with precipitation, temperature, soil physical and chemical properties, elevation, vascular plant vegetation, and other site factors. A synthesis chapter (Chap. 12) compares and discusses observed patterns on a continental and global scale. Biological soil crusts regularly serve as an environment also for heterotrophic soil organisms such as invertebrates and microfungi. *Part II* reports what little we know about this topic. Structure and fine structure of soil crusts from their different habitats, using light and electron microscopy, are analyzed in the first chapter (Chap. 15) of *Part III,* and environmental controls on crust composition and structure are discussed on various scales in the subsequent chapters. Present knowledge of crust photosynthesis and nitrogen fixation is summarized in ecophysiological chapters of *Part IV.* The influence of soil crusts on ecosystem nitrogen and carbon cycles and the effects of soil crust on vascular plant germination, survival, and nutrition is evaluated. The following parts of the

Volume are devoted to larger-scale aspects of soil-crust performance. Research on the effects of soil crusts on landscape hydrology and erosion has been mainly done in Israel, Australia, and the United States. In *Part V*, eminent case studies are presented on soil stability and soil-water relations which include water runoff, water infiltration, evaporation and sediment production, and how these factors vary with climate, soil chemistry, soil texture and structure, as well as with crustal composition and disturbance history. Since there are seemingly contradictory findings on the role that crusts play in different soil types, geographical regions, and environments, a synthesis chapter (Chap. 26) explores regional and global patterns to explain these results. Disturbance of soil crusts can impact ecosystems in many ways, including increasing susceptibility to soil erosion and decreased carbon and nitrogen inputs. Resistance and resilience of crusts to disturbance varies widely with differences in species composition and soil types. In *Part VI*, impact of native and non-native ungulates, recreation, fire, and other soil-surface disturbances on species composition, biomass, and physiological functioning of soil crusts are reported. Ways to hasten recovery of disturbed areas are examined. *Part VII* looks at strategies to manage soil crusts, considering human impact such as off-road driving and livestock grazing. How to monitor and integrate soil crusts into ecosystem management on a land-scape scale is outlined, and use of remote sensing for crust monitoring is represented. The possible impacts of global climate changes and how this might influence land management decisions are also discussed. Finally, *Part VIII* summarizes the material presented in this Volume. Resistance and resilience of soil-crust systems to disturbance are discussed, applying an evolutionary perspective. Future research needs are outlined, with management and scientific implications of the research summarized.

As evidenced by the large and diverse audiences present at biological soil-crust workshops and seminars that have taken place in the past few years (e.g., St. Clair and Johansen 1993), land management agencies, private citizens, and the scientific community are beginning to realize the importance and exciting scientific opportunities to be found in work on biological soil crusts. As worldwide needs and pressures upon arid landscapes increase, it is also incumbent upon the scientific community to better understand the role these crusts play in different, and often fragile, arid ecosystems, and to inform land-management agencies on the impacts of different types and intensities of land use on this vital ecosystem component. We sincerely hope that the present Volume will help to procure for soil crusts the attention and appreciation they deserve for the benefit of protecting arid land from long-lasting anthropogenic impacts.

Due to space restrictions, only a fraction of literature relevant to soil-crust biology is referenced in the present Volume. One of us (J.B.) maintains a web-based databank of publications concerning all aspects of biological soil crusts (www.soilcrust.org). Currently (March 2001), this web site contains over 3000 references. This literature is recommended to the readers of this Volume for more detailed information regarding biological soil crusts.

Acknowledgments. Dr. Burkhard Büdel (Kaiserslautern) is thanked for his continuous support and advice during the realization of this Volume. Soil crust research of O.L.L. was funded under the Sonderforschungsbereich 251 der Universität Würzburg. Grateful acknowledgment is made of Beth Coker-Roy, Sue Phillips, Wilma Samfaß, Michelle Schmid, and last, but not least, Dr. Andrea Schlitzberger, who assisted in all aspects of this Volume. Friedmut Kröner provided superb production advise and expertise. Finally, we thank Rose and Steve for their patience and understanding with our absorbed state during this project.

Jayne Belnap, Moab, Utah, U.S.A.
March 2001 *Otto L. Lange*, Würzburg, Germany

References

Belnap J, Gardner JS (1993) Soil microstructure of the Colorado Plateau: the role of the cyanobacterium *Microcoleus vaginatus*. Great Basin Nat 53:40–47

Cameron RE, Blank GB (1966) Desert algae: soil crusts and diaphanous substrata as algal habitats. Jet Propulsion Lab, Calif Inst Techn, Pasadena, Techn Rep 32-971:1–41

Eldridge DJ, Greene RSB (1994) Microbiotic soil crusts: a review of their roles in soils and ecological processes in the rangelands of Australia. Aust J Soil Res 32: 389–415

Eldridge DJ, Tozer ME (1997) A practical guide to soil lichens and bryophytes of Australia's dry country. Dep Land Water Cons, Sydney

Evans RD, Johansen JR (1999) Microbiotic crusts and ecosystem processes. Crit Rev Plant Sci 18:183–225

Friedmann EI, Galun M (1974) Desert algae, lichens and fungi. In: Brown GW (ed) Desert biology. Academic Press, New York, pp 165–212

Harper KT, Marble JR (1988) A role for nonvascular plants in management of arid and semiarid rangeland. In: Tueller PT (ed) Vegetation science applications for rangeland analysis and management. Kluwer Academic Publishers, Dordrecht, pp 135–169

Johansen JR (1993) Cryptogamic crusts of semiarid and arid lands of North America. J Phycol 29:140–147

St Clair LL, Johansen JF (1993) Introduction to the symposium on soil crust communities. Great Basin Nat 53:1–4

West NE (1990) Structure and function of soil microphytic crusts in wildland ecosystems of arid and semi-arid regions. Adv Ecol Res 20:179–223

Contents

Contributors

Belnap, Jayne

US Geological Survey. 2290 S. West Resource Blvd., Moab, Utah 84532, USA

Broady, Paul A.

University of Canterbury, Plant Microbial Sciences Department. Private Bag 4800, Christchurch, New Zealand

Büdel, Burkhard

Universität Kaiserslautern, FB. Biol., Allgemeine Botanik. Gebäude 13/2; Postfach 3049, 67653 Kaiserslautern, Germany

Christensen, Martha

University of Wyoming, Department of Botany. Laramie, Wyoming 82071, USA

Clark, Roger N.

US Geological Survey. PO Box 25046, MS 964, Denver, Colorado 80225, USA

Eldridge, David J.

University of New South Wales, School of Geography, Department of Land and Water Conservation. Sydney NSW 2052, Australia

Evans, R. Dave

University of Arkansas, Department of Biological Sciences. 601 Science-Engineering Bldg. Fayetteville, Arkansas 72701, USA

Galun, Margalith

Tel-Aviv-University, George S. Wise Faculty of Life Science, Department of Plant Sciences. University Campus, Ramat Aviv, Tel-Aviv 69978, Israel

Garcia-Pichel, Ferran

Arizona State University, Department of Microbiology. Tempe, Arizona 85287-2701, USA

Gärtner, Georg

Universität Innsbruck, Institut für Botanik. Sternwartestraße 15, 6020 Innsbruck, Austria

Garty, Jacob

Tel-Aviv-University, George S. Wise Faculty of Life Science, Department of Plant Sciences. University Campus, Ramat Aviv, Tel-Aviv 69978, Israel

Green, T. G. Allan

The University of Waikato, Biological Sciences Department. Private Bag 3105, Hamilton, New Zealand

Hansen, Eric Steen

The Botanical Museum. 130 Gothersgade, 1123 Copenhagen, Denmark

Harper, Kimball T.

Utah Valley State College, Department of Biology. Orem, Utah 84058, USA

Johansen, Jeffrey R.

John Carroll University, Department of Biology. 2070 N. Park Blvd., University Heights, Ohio 44118, USA

Kaltenecker, Julie H.

Idaho State Office, Bureau of Land Managment, 1387 Vinnell Way, Boise, Idaho 83709, USA

Karnieli, Arnon

Ben Gurion University of the Negev, J. Blaustein Institute for Desert Research, The Remote Sensing Laboratory. Sede Boker Campus 84990, Israel

Kinter, Cecilia Lynn

Washington State University, School of Biological Sciences. P.O. Box 644238, Pullman, Washington 99164-4238, USA

Kokaly, Raymond

US Geological Survey. P.O. Box 25046 MS 973, Denver, Colorado 80225, USA

Lange, Otto L.

Universität Würzburg, Julius-von-Sachs-Institut für Biowissenschaften, Lehrstuhl für Botanik II. Julius-von-Sachs-Platz 3, 97082 Würzburg, Germany

Phillips, Susan L.

US Geological Survey. 2290 S. West Resource Blvd., Moab, Utah 84532, USA

Prasse, Rüdiger

Universität Hohenheim, Institut für Landschafts- und Pflanzenökologie. Postfach 700562, 70593 Stuttgart, Germany

Rosentreter, Roger

Idaho State Office, Bureau of Land Managment. 1387 Vinnell Way, Boise, Idaho 83709, USA

States, Jack S.

2 Canyon Shadows Rd., Lander, Wyoming 82520, USA

Türk, Roman

Universität Salzburg, Institut für Pflanzenphysiologie, Ökophysiologie der Pflanzen. Hellbrunnerstraße 34, 5020 Salzburg, Austria

Ullmann, Isolde

Universität Würzburg, Julius-von-Sachs-Institut für Biowissenschaften, Lehrstuhl für Botanik II. Julius-von-Sachs-Platz 3, 97082 Würzburg, Germany

Warren, Steven D.

Colorado State University, Center for Ecological Management of Military Lands, Department of Forest Sciences. Fort Collins, Colorado 80523-1500, USA

West, Neil

Utah State University, Department of Natural and Renewable Resources. Logan, Utah 84322, USA

Yair, Aaron

The Hebrew University of Jerusalem, Department of Geography.
Mount Scopus Campus, Jerusalem 91905, Israel

Part I

Taxonomic Composition, Ecology, and Biogeography of Soil-Crust Communities

1 Biological Soil Crusts: Characteristics and Distribution

J. BELNAP, B. BÜDEL, and O. L. LANGE

1.1 Definition of Biological Soil Crusts

Biological soil crusts result from an intimate association between soil particles and cyanobacteria, algae, microfungi, lichens, and bryophytes (in different proportions) which live within, or immediately on top of, the uppermost millimeters of soil. Soil particles are aggregated through the presence and activity of these biota, and the resultant living crust covers the surface of the ground as a coherent layer (Fig. 1.1). This definition does not include communities where soil particles are not aggregated by these organisms (e.g., cyanobacterial/algal horizons in littoral sand and mudflats), where organisms are not in close contact with the soil surface (e.g., thick moss-lichen mats growing on top of decaying organic material, as in boreal regions), nor where the majority of the biomass is above the soil surface (e.g., large club-moss mats found in North American grasslands or dense stands of fruticose lichens, such as *Niebla* and *Teloschistes* species from the coastal fog deserts of California and of Namibia, respectively). However, the boundaries between the latter communities and biological soil crusts are fluid. In a similar fashion, there is no strict dividing line between the cyanobacterial, green algal, and fungal species that occur in soil-crust communities, yet are also found in a multitude of additional habitats (e.g., intertidal mats, tree trunks and leaves, rock faces).

Multiple names have been applied to biological soil crusts. These communities have been referred to as cryptogamic, cryptobiotic, microbiotic, microfloral, microphytic, or organogenic soil crusts (see Harper and Marble 1988; West 1990). Evans and Johansen (1999) discuss the implication of the different expressions. In this Volume, we use the term biological soil crusts because it is the broadest term which clearly states that these crusts are dependent on the activity of living organisms, in contrast to physical or chemical crusts (see Chap. 15). In addition, the term biological soil crust lacks taxonomic implications, and thus is broadly applicable to all soil crusts, regardless of their species composition.

Ecological Studies, Vol. 150
J. Belnap and O.L. Lange (eds.) Biological Soil Crusts: Structure, Function, and Management

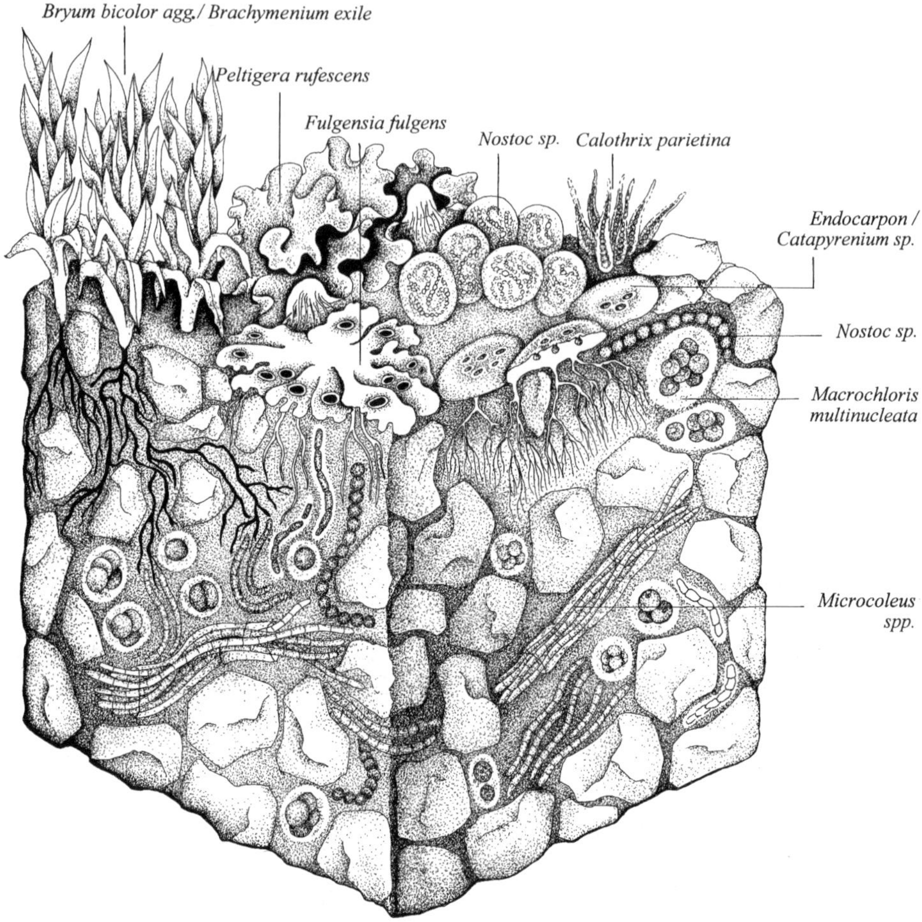

Fig. 1.1. Schematic block diagram of a biological soil crust with typical colonizers. Thickness of the layer about 3 mm, organisms not drawn to scale. (Illustration Renate Klein-Rödder)

1.2 Taxonomic Components of Soil-Crust Communities (Photos 1–28)

Cyanobacteria, and in particular species of the genus *Microcoleus* (Photos 1, 2), provide most of the cohesive quality of the biological soil crusts in arid and semiarid lands. As discussed in detail in Chapter 15, bundles of *Microcoleus* filaments are surrounded by extracellular sheaths which wind throughout the uppermost soil layers. When wet, the filaments glide out of their sheaths, and in a phototactic reaction, move up towards the soil surface. Upon drying, the filaments leave the surface, and the exposed filaments secrete new sheaths. The result of this frequent movement is copious sheath material dispersed

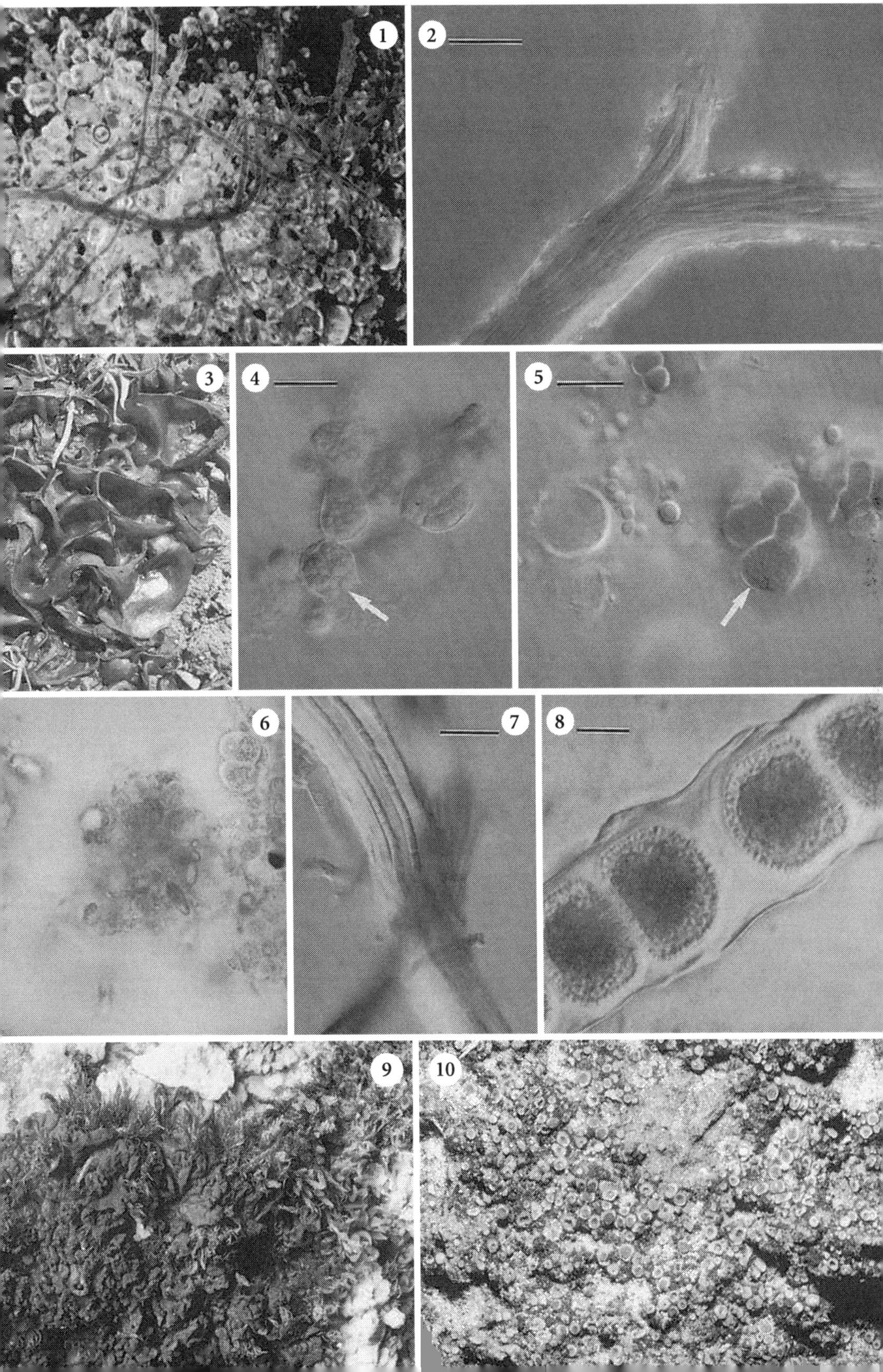
1
2
3
4
5
6
7
8
9
10

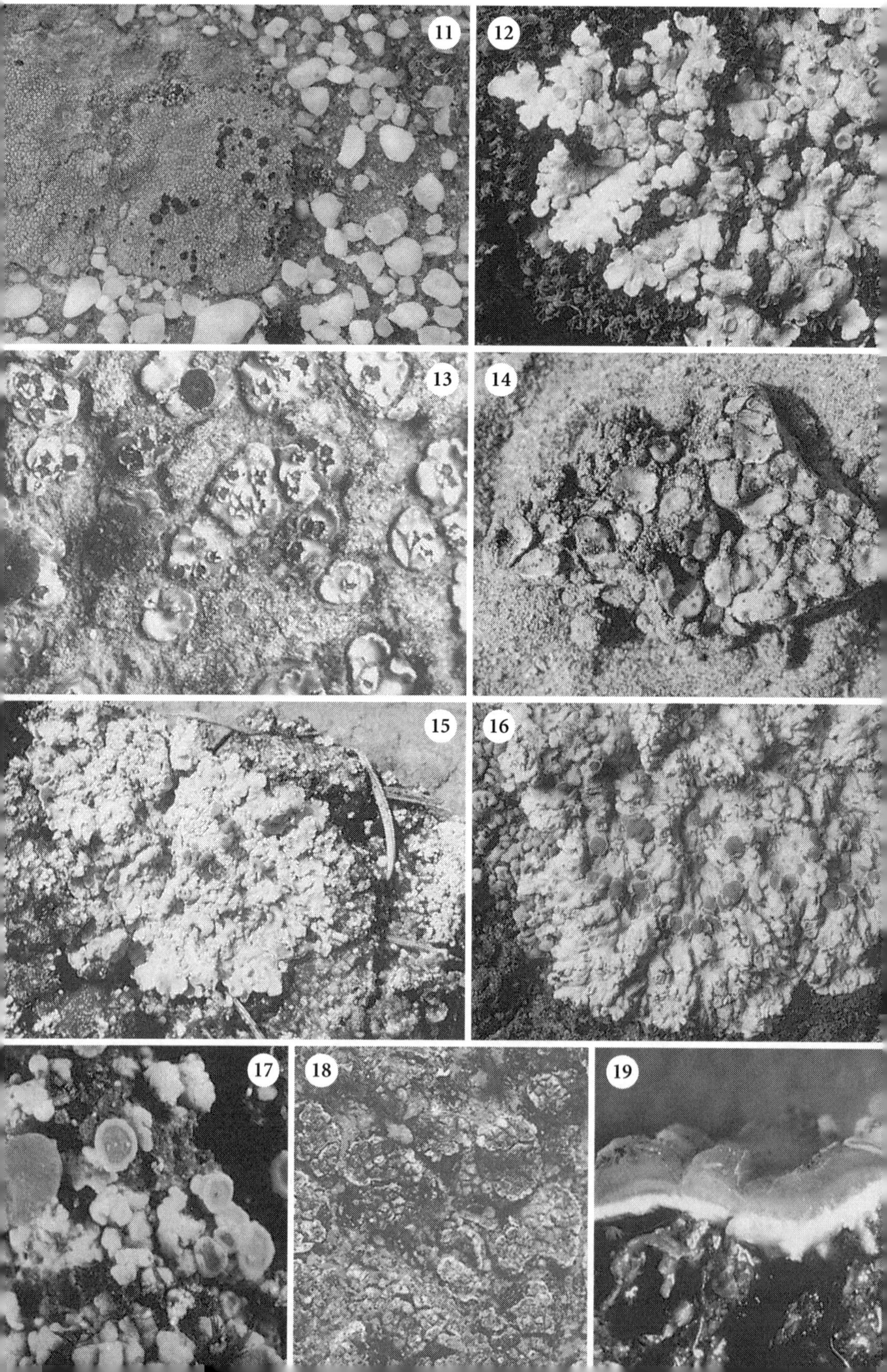
11
12
13
14
15
16
17
18
19

20
21
22
23
24
25
26
27
28

29
30
31
32
33
34
35
36
37

38

39

40

41

42

43

44

45

46
47
48
49
50
51
52
53
54

55
56
57
58
59
60
61

Soil Crust Forming Organisms

Photo 1. Green filaments of *Microcoleus vaginatus* (Cyanobacteria) between quartz grains. Soil crust from Utah. *Bar* 0.1 mm. (Photo J.Belnap)

Photo 2. *Microcoleus paludosus* (Cyanobacteria) with numerous trichomes united in a common sheath. Soil crust from Tunisia. *Bar* 25 µm. (Photo B.Büdel)

Photo 3. *Nostoc commune* (Cyanobacteria). Young, wet thalli on soil, Germany. *Bar* 20 mm. (Photo O.L.Lange)

Photo 4. *Chroococcidiopsis* sp. (Cyanobacteria) showing baeocytes (arrow). Hypolithic habitat, Namib Desert, Namibia. *Bar* 20 µm. (Photo B.Büdel)

Photo 5. *Pleurocapsa minor* (Cyanobacteria) pseudofilaments with baeocyte formation in one cell (*arrow*). Hypolithic habitat, Namib Desert, Namibia. *Bar* 20 µm. (Photo B.Büdel)

Photo 6. *Macrochloris multinucleata* (Chlorophyta). Photo shows dry cells a few minutes after wetting. Soil crust from the Negev Desert, Israel. *Bar* 20 µm. (Photo B.Büdel)

Photo 7. *Schizothrix telephoroides* (Cyanobacteria) trichomes in a typically reddish colored, common sheath. Soil crust from the humid savanna along the Orinoco, Venezuela. *Bar* 20 µm. (Photo B.Büdel)

Photo 8. *Zygogonium ericetorum* (Chlorophyta) cells with characteristic plastids and cell wall. Soil crust from a forest margin, Germany. *Bar* 10 µm. (Photo B.Büdel)

Photo 9. *Collema tenax* (gelatinous cyanolichen) hydrated thallus on mosses. Soil crust in local steppe formation, Germany. (Photo H.Ullrich)

Photo 10. *Peltula patellata* (peltate cyanolichen) with whitish thallus and red apothecia. Sonoran Desert (Baja California, Mexico). (Photo B.Büdel)

Photo 11. *Lecidella crystallina* (crustose-areolated phycolichen). Namib Desert, Namibia. (Photo H.Zellner)

Photo 12. *Squamarina lentigera* (squamulose phycolichen) and *Tortella* cf. *inclinata* (Musci). Local steppe formation, Germany. (Photo H.Ullrich)

Photo 13. *Buellia subcoronata* (placoid phycolichen). Semiarid Australia. (Photo D.J.Eldridge)

Photo 14. *Psora crenata* (placoid phycolichen) with laminal pycnidia. Sonoran Desert (Baja California, Mexico). (Photo B.Büdel)

Photo 15. *Baeomyces placophyllus* (phycolichen with crustose-squamulose thallus and upright podetia). Clearing in a *Fagus sylvatica* forest, Germany. (Photo B.Büdel)

Photo 16. Phycolichens *Fulgensia fulgens* (yellow-whitish, crustose-squamulose) and *Toninia sedifolia* (bluish, bullate; *left corner*). Local steppe formation, Germany. (Photo H.Ullrich)

Photo 17. *Fulgensia bracteata* var. *alpina* (crustose-granular phycolichen). Alpine steppe formation, Austria. (Photo R.Türk)

Photo 18. *Eremastrella crystallifera* (placoid phycolichen; Fensterflechte or window lichen). Semiarid Australia. (Photo D.J.Eldridge)

Photo 19. *Eremastrella crystallifera* cross section, showing the transparent upper cortex above the algal layer, and with the underlying medulla. South Africa. *Bar* 0.5 mm. (Photo B.Büdel)

Photo 20. *Xanthomaculina convoluta*, a vagrant, foliose phycolichen in the dry, rolled (*black*) and in the wet, unrolled (*green*) state. Namib Desert, Namibia. (Photo O.L.Lange)

Photo 21. *Catapyrenium squamulosum* (placoid phycolichen) and *Riccia* sp. (Hepaticae; notice wet, unfolded leaflets). Dry savanna in North Transvaal, South Africa. (Photo B.Büdel)

Photo 22. *Cladonia convoluta* (foliose phycolichen). Mediterranean shrub, Cyprus, dry. (Photo O.L.Lange)

Photo 23. The same specimen as Photo 22, after moistening. (Photo O.L.Lange)

Photo 24. *Peltigera rufescens* (lobated cyanolichen). Local steppe formation, Germany. (Photo B.Büdel)

Photo 25. *Exormotheca holstii* (Hepaticae, Marchantiales) with air-chambers. Namib Desert, Namibia. (Photo B.Büdel)

Photo 26. *Campylopus introflexus* (Musci). Soil crusts on the margins of higher rainfall areas, east Australia. (Photo D.J.Eldridge)

Photo 27. *Gigaspermum repens* (Musci), small male and larger female individuals. Soil crusts throughout dry Australia. (Photo D.J.Eldridge)

Photo 28. *Bryum campylothecium* (Musci). Southern Australia, predominantly in woodlands and low shrublands. (Photo D.J.Eldridge)

Landscapes and Habitats of Soil-Crust Occurrence

The Americas

Photo 29. Eastern Oregon: *Artemisia tridentata* shrub-steppe with extensive soil crusts in interspaces. (Photo J.Belnap)

Photo 30. Same location as Photo 29: rolling soil crust with mosses and lichens. (Photo J.Belnap)

Photo 31. Colorado Plateau, Utah: pinnacled soil crusts in pinyon-juniper woodland. (Photo O.L.Lange)

Photo 32. Similar location as Photo 31: pinnacled soil crust with white *Diploschistes muscorum* and light green *Squamarina lentigera* (phycolichens). (Photo O.L.Lange)

Photo 33. Same location as Photo 31: pinnacled soil crust with *Psora cerebriformis* (white phycolichen). (Photo O.L.Lange)

Photo 34. Sonoran Desert flood plain (Baja California, Mexico): soil crusts in succulent bush. (Photo B.Büdel)

Photo 35. Same location as Photo 34: rugose soil crust with *Peltula richardsii* (dark green cyanolichen with maroon apothecia). (Photo B.Büdel)

Photo 36. Venezuela: humid savanna with *Trachypogon* grass. (Photo B.Büdel)

Photo 37. Same location as Photo 36: cyanobacterial *Schizothrix*-dominated soil crust. (Photo B.Büdel)

Europe and Asia

Photo 38. Alps, Austria, Großglockner area: alpine mat vegetation with soil crusts in interspaces, 2700 m a.s.l. (Photo R.Türk)

Photo 39. Alps, Austria, Hochtor: alpine cold desert with soil crusts 2630 m a.s.l. (Photo R.Türk)

Photo 40. Same location as Photo 39: alpine soil crust with *Protoblastenia terricola,* whitish phycolichen with brown apothecia. (Photo R.Türk)

Photo 41. Local steppe formation with soil crusts in a typical German landscape, near Goslar. (Photo H.Ullrich)

Photo 42. Similar location as Photo 41: lichen soil crust with phycolichens *Buellia epigaea* (bluish white) and yellow *Fulgensia bracteata*, gypsic soil. (Photo H.Ullrich)

Photo 43. Germany, clearing in a *Fagus sylvatica* forest: successional soil crust dominated by the green alga *Zygogonium ericetorum.* (Photo B.Büdel)

Photo 44. China, Ganzu Province: dwarf shrub-steppe (*Artemisia* and *Haloxylon*) with extensive rugose crusts formed by cyanobacteria and lichens. (Photo O.L.Lange)

Photo 45. Same location as Photo 44: soil crusts composed of the cyanobacteria *Nostoc flagelliforme* (*arrow*), phycolichens *Psora* sp. (*red*) and *Diploschistes* (*white*), and the black cyanolichen *Collema* sp. (Photo O.L.Lange)

Photo 46. A special Chinese meal prepared with black cyanobacterium *Nostoc flagelliforme* (Facai) in a baked, sliced eggroll. (Photo O.L.Lange)

Middle East and Africa

Photo 47. Jordan, Wadi Arava: gravel desert with extensive moss crust. (Photo W.Frey)

Photo 48. Same location as Photo 47: moss soil crust formed by Crossidio-laevipili-Toruletum-atrovirentis association. (Photo W.Frey)

Photo 49. Israel (Negev Desert) Nizzana: smooth soil crust on dunes between *Cornulaca monacantha.* (Photo R.Bornkamm)

Photo 50. Similar location as Photo 50: smooth soil crust, dominated by the cyanobacterium *Microcoleus sociatus*, (Photo G.Kidron)

Photo 51. Africa, Namib Desert: extensive areas covered by rugose lichen-dominated soil crusts. (Photo O.L.Lange)

Photo 52. Similar location as Photo 51: rugose soil crusts with phycolichens *Lecidella crystallina* (*whitish crust*), *Xanthoparmelia walteri* (*green, foliose*) and *Xanthomaculina hottentotta* (*blackish, foliose*). (Photo O.L.Lange)

Photo 53. South Africa, Knersvlakte: shrub desert paved with quartz pebbles and lichen cyanobacteria-dominated soil crusts. (Photo B.Büdel)

Photo 54. Same location as Photo 53: soil crust with the red phycolichen *Psora decipiens* and quartz pebbles. (Photo B.Büdel)

Photo 55. Zimbabwe *Brachystegia* woodland with rugose soil crusts in interspaces. (Photo J.Belnap)

Photo 56. Same location as Photo 55: cyanobacterial (*Porphyrosiphon notarisii*)-dominated soil crust with some mosses. (Photo J.Belnap)

Australia, Antarctica

Photo 57. South Australia, eucalypt-mallee woodland with extensive cover and diversity of soil crust organisms. (Photo D.J.Eldridge)

Photo 58. Similar location as Photo 55: soil crust rich in phycolichens (e.g., *Diploschistes*, *Endocarpon* spp.) and mosses (e.g., *Didymodon torquatus*, *Desmatodon convolutus*). (Photo D.J.Eldridge)

Photo 59. South Australia, *Atriplex vesicaria* dominated shrubland: recovering soil crusts after topsoil removal more than 35 years ago. (Photo D.J.Eldridge)

Photo 60. Antarctica, Victoria Land, Hallett Station: flush of moss and algae in front of a penguin rookery. (Photo O.L.Lange)

Photo 61. Same location as Photo 60: soil crust composed of the moss *Bryum subrotundifolium* and the green alga *Prasiola* sp. (*dark green*). (Photo O.L.Lange)

---

throughout the uppermost soil layers. As the sheath material is sticky, it adheres to, and aggregates, soil particles. The cohesive quality of these crusts can be easily seen when the surface of a desert soil is broken. Rather than breaking into individual soil particles, the soil breaks into larger pieces. If these pieces are held aloft, small soil clumps can be see dangling down, held in place by cyanobacterial sheaths.

Taxonomy of the crust-forming *Microcoleus* species is still in the state of flux. *Microcoleus chtonoplastes* Thuret and *M. vaginatus* (Vauch.) Gom. are the two most frequent soil-crust species. While *M. chtonoplastes* prefers habitats with higher salinity (see Chap. 17), *M. vaginatus* is the main biomass contributor in many desert soil crusts. Recent studies using 16S rDNA sequencing (F. Garcia-Pichel, A. Lòpez-Cortéz, and U. Nübel, unpubl.) revealed that all samples of *M. vaginatus* coming from different localities of the world (Utah, California, Spain, and Israel) are virtually identical. They represent a 'good species', forming a monophyletic, deeply branched taxonomic cluster. However, this *M. vaginatus* cluster is more related to the filamentous cyanobacterium *Trichodesmium* than to the *M. chtonoplastes* cluster. The strains of the third soil-crust species, *M. sociatus* W. et G.S.West from Israel and Utah, are also identical. They form yet another distinct cluster that is not closely related to either *M. vaginatus* or *M. chtonoplastes*. Thus, the present genus *Microcoleus* seems to be polyphylous in nature and may be split into several genera in the future. However, further research is needed for such decisions, and in the present Volume the classic taxonomy will be maintained.

In addition to *Microcoleus*, cyanobacteria with single filaments also occur in biological soil crusts. *Nostoc* species are common in crusts, growing both within and on top of the soil. Sheets of *Nostoc commune* (Photo 3) and the horse hair-like, vagrant threads of *Nostoc flagelliforme* (Photo 45) spend a substantial period of their life cycle lying loose on soil surfaces in almost all types of habitats, including the Arctic, tropics, and deserts. These species may be the only soil-crust organisms used for human food (Photo 46). *Scytonema* and *Calothrix* spp., with falsely branching trichomes, are frequent. Unicellular species are represented by the coenobia-forming (many cells in a common slime matrix) genus *Gloeocapsa*, and by single-celled Chroococcales (Photo 4). While all soil-crust cyanobacteria contribute carbon to soils, many also contribute nitrogen as well (Chap. 18–20).

Eukaryotic algae are represented mainly by coccoid species (e.g., of the genera *Chlorococcum, Macrochloris,* and *Stichococcus*, Photos 5, 6). In addition to these Chlorophyta, Euglenophyta and Chromophyta of the classes Tribophyceae and Diatomophyceae are also reported (see Johansen 1993). As a whole, hundreds of different species of cyanobacteria and eukaryotic algae are associated with biological soil crusts (Evans and Johansen 1999). Due to their limited biomass, the majority of these species, although widespread in distribution, are usually of secondary importance in the formation of crusts. The same is true for most of the bacteria which have been isolated from soil crusts. However, while little is known about fungal species found in crusts, they often contribute significantly to soil aggregate formation (Chap. 13).

Unlike the endedaphic (subterranean) soil cyanobacteria and algae, the epedaphic lichens and bryophytes have almost all of their photosynthetic tissue on or above the soil surface (Fig. 1.1). Both phycolichens (with green algal photobionts) and cyanolichens (with cyanobacterial photobionts) with different growth forms (as defined by Jahns 1988) are found. Crustose lichens cover the soil with an appressed, more or less even but discontinuous, layer of thalli. More or less isolated crustose thallus scales occur in placoid genera like *Psora, Buellia* or *Trapelia* spp. (Photos 13, 14, 18, 21, 33, 54), and shield-like scales can form peltate thalli which are attached by a central holdfast, as in the genera *Endocarpon* or *Peltula* (Photos 10, 35). When thalli are more continuous, as in *Diploschistes* or *Lecidella,* the thallus surface is usually divided in small areoles (Photos 11, 40, 42, see frontispiece). Squamulose species such as *Squamarina* represent a transition to the foliose lichens; here, the margins of the individual thallus lobes are raised above the substrate (Photos 12, 15, 16). Foliose soil-crust lichens are exemplified by species of *Peltigera* (Photo 24). The transition to the fruticose form is represented by *Peltula* and *Toninia* spp. (Photos 10, 16), both with inflated thallus lobes, while examples of soil crust fruticose species include *Cladonia* (Photos 22, 23) or *Cladia* species. A special growth form associated with soil-crust communities in steppes and deserts are erratic and vagrant lichens (Wanderflechten) that have no attachment to the ground (Elenkin 1901; Rosentreter 1993), such as *Xanthoparmelia* spp. in North America and in the Asian steppes, *Chondropsis semiviridis* in Australia and New Zealand, *Xanthomaculina convoluta* in Namibia (Photo 20), and the different types of manna lichens (Crum 1993; see Chap. 9). Most of the lichens mentioned so far are heteromerous in structure, i.e., their thallus is stratified in definite layers. There are also several important soil-crust cyanolichens with unstratified (homoiomerous) thalli and a gelatinous consistency. Examples include the genera *Collema* (Photo 9), *Gloeoheppia, Peccania,* and *Gonohymenia.*

Soil lichens have soil-anchoring structures (such as rhizoptae, rhizinae, and rhizomorphs, see Poelt and Baumgärtner 1964; Sanders 1994), which penetrate the uppermost soil layers. One type of these appendages is shown in Fig. 1.2 for *Buellia epigaea* (see Photo 42). Thin, corticated, and branched rhizomorphs (Rhizinenstränge) originating from the lower side of the thallus squamules extend for 4 to 5 mm into the soil. These strands are densely covered with a fine tomentum of rhizinae and single hyphae (Hyphenfilz) which fill the soil matrix. Thicker rhizomorphs of other species (e.g., *Toninia sedifolia,* Fig. 1.3; see Photo 16) connect the epedaphic thallus squamules with each other. They form a dense, subterranean network of fungal material, intimately connected to the soil particles. In *Psora cerebriformis* (see Photo 33) this system of rhizomorphs and rhizinae can penetrate soil down to 14 mm (O.L. Lange, unpubl.). While these structures are not known to transport water or nutrients, they firmly fasten the lichens to the soil, and thus enable them to withstand the erosive forces of wind and water. In addition, their subterranean structures contribute decisively to soil stability.

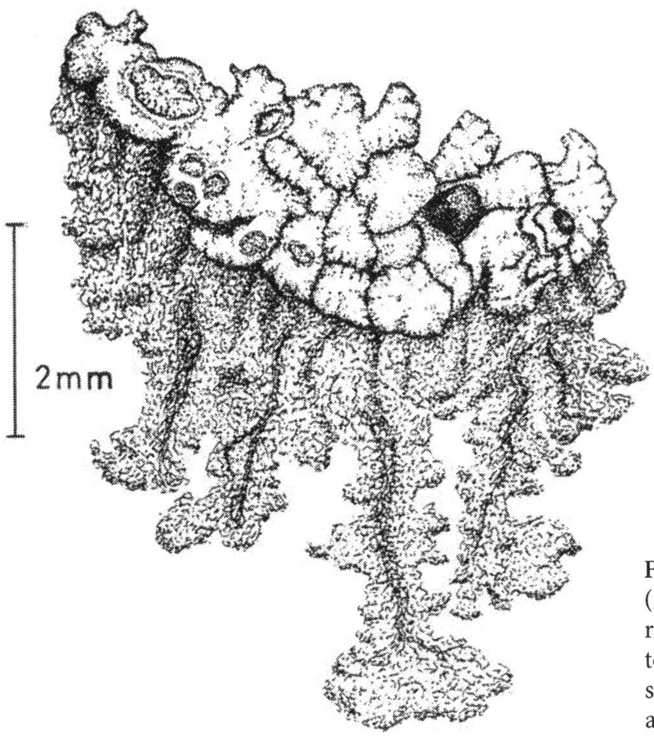

Fig. 1.2. *Buellia epigaea* (phycolichen) with thin rhizomorphs covered with tomentum of rhizinae and single hyphae. (After Poelt and Baumgärtner 1964)

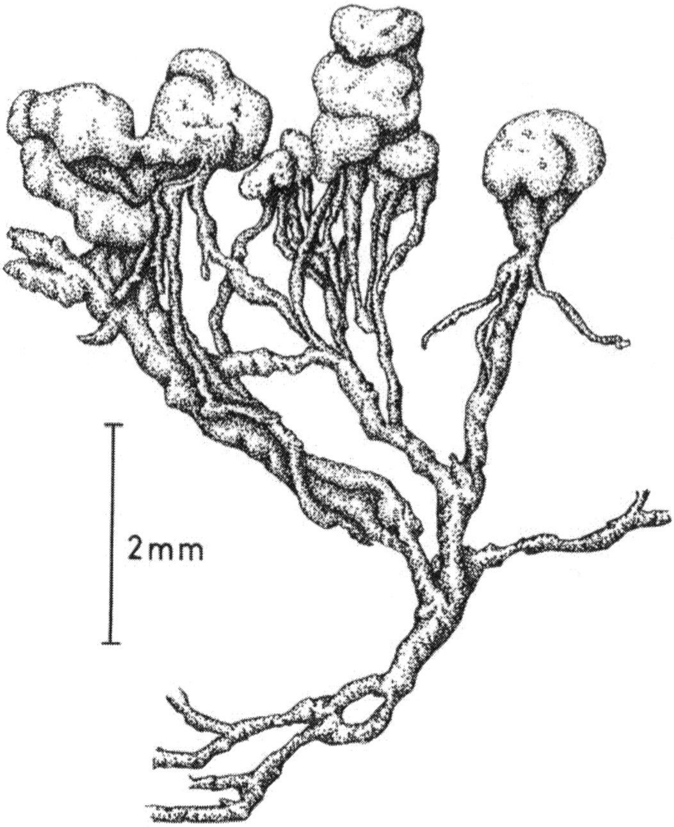

Fig. 1.3. *Toninia sedifolia* (phycolichen) with system of branched rhizomorphs, which develop a tomentum of rhizinae at their ends, which is not shown. (After Poelt and Baumgärtner 1964)

The presence of bryophytes usually characterizes the more moist habitats of biological soil crusts. Both annual and perennial representatives of the Musci (true mosses; Photos 26, 27, 28, 48, 61) and the Hepaticae (thallose and leafy liverworts; Photos 21, 25) occur in soil-crust communities. Eldridge and Tozer (1996) report 56 bryophyte species (41 mosses, 15 liverworts) associated with biological soil crusts in semiarid and arid eastern Australia (see also Eldridge and Tozer 1997). However, the majority of publications dealing with arid-land bryophytes do not report their affiliation to soil-crust communities. Thus, a general overview about typical growth forms involved in soil-crust formation is not yet possible. The Pottiaceae seem to be the most abundant family associated with soil-crust habitats, and are represented by genera like *Aloina, Barbula, Crossidium, Pottia,* and *Tortula. Riccia* (Photo 21) seems to be the most important liverwort genus present in soil-crust communities. Frey and Kürschner (1988, 1991a) conducted extensive studies on ecology and life strategies of the terrestrial bryophyte vegetation of the Arabian Peninsula and in the Judean Desert (Israel), and described soil-crust habitats for many communities.

Terrestrial bryophytes are attached to the soil by rhizoids that often grow down more than 10 mm. Thallose liverworts such as *Riccia* (Photo 21) are closely appressed to the ground by a dense mat of unicellular rhizoids which form aggregates with the soil particles. In most of the mosses, the lower parts of the stems, together with their rhizoids, are involved in the anchoring process. Species such as *Gigaspermum mouretii* (Photo 27) have subterranean, rhizome-like, persistent main axes creeping through the soil, and under favorable conditions, above-ground lateral branches with stems and leaves (Frey and Kürschner 1991b). In addition, multibranched, subterranean moss protonemata are interspersed throughout the soil-crust matrix. These structures become intimately bound with the soil particles and contribute to the stability of the uppermost soil layers. In contrast to most soil-crust lichens, some mosses are able to grow up through small amounts of newly deposited sand or dust (see Fig. 15.2b).

1.3 Habitats and Distribution of Soil-Crust Communities (Photos 29–61)

As soil-crust organisms have a limited ability to grow upwards from the soil surface, they are generally unable to compete with phanerogamous plants for light. Consequently, crust development is limited where climate conditions permit development of a closed vascular plant canopy or thick litter layers (Chap. 21). Thus, in arctic, boreal, temperate, and tropical regions, crusts occur only where the overstory cover is sparse or disappears due to limited water or local disturbance. Locally limited water facilitates crust development in dunes, in local steppe formations of Europe and Asia, in open woodlands (e.g., pine barrens of the southeast US), and within the discontinuous, evergreen mediterranean vegetation found worldwide. Following disturbance such as treefall or volcanic eruptions, crust organisms are generally the first to colonize.

Crust organisms have low moisture requirements, and their ability to utilize small rainfall events, snow melt, fog, and/or dew as water sources enables them to exist where moisture deficit limits vascular plant cover and productivity. In these low-productivity environments, the biological crusts colonize gaps left open by higher vegetation. This is true on a global, regional, landscape, and microsite scale (Chaps. 12, 16, 17); thus, crust communities occur in a large variety of phanerogamic formations in many different vegetation zones worldwide. Well-developed crusts occur in winter-cold steppes and grassland regions throughout the world. However, the most conspicuous development of crusts occurs in hot, cool, and cold semiarid and arid areas where plants are widely spaced. Vegetational communities in these regions range from evergreen and deciduous woodlands, salt-bush communities, grassland, shrub, and succulent formations to areas with fixed dunes or regions where vascular plants are restricted to water-collecting depressions. Soil-crust communities also colonize the spaces between vascular plants in alpine areas and polar regions (see Part I). A selection of typical soil-crust habitats in different vegetation formations around the world are depicted in Photos 29–45 and 47–81.

Since all soil-crust biota are only metabolically active when wet, moister habitats (e.g., under plant canopies, under thin plant litter, or on northern exposures) in dry regions are favored by most crust species. Mosses and liverworts are often only found in such microhabitats, due to higher moisture requirements than lichens and cyanobacteria. These latter two groups are found both in the moister microhabitats as well as in more exposed, drier microhabitats. Soil-crust communities in areas with higher effective precipitation or those under vascular plants are often dominated by mosses, while in drier habitats they are dominated by lichens, cyanobacteria, and green algae (West 1990). Desert habitats with fog and dew (as the Namib, see Chap. 9, and the Central Negev Desert) favor phycolichens, whereas lack of dew, less rain, and higher temperatures (as in the Arava Valley, Dead Sea area) favor cyanolichens (Galun et al. 1982; see Chap. 18). Under still more extreme conditions, soil-crust communities consist almost exclusively of endedaphic cyanobacteria.

Successional stages of crust communities can be distinguished (Chap. 27). Large filamentous cyanobacteria generally colonize first, followed by smaller cyanobacteria and green algae. After soil surfaces have stabilized, lichens and mosses appear. In temperate and tropical regions, these organisms influence soil formation as the vascular vegetation becomes re-established, a process which can last for decades. Under arid and semiarid conditions, where water limits vascular plant cover, crusts remain as the final stage of succession in plant interspaces (see Chaps. 21, 27).

Biological soil crusts are found on almost all soil types, as they can colonize soils with high to low levels of nutrients and soil salts, as well as a wide variety of soil textures. They are, however, limited on clay soils with a high shrink-swell coefficient. Soil pH and chemistry influence the composition of the crust communities. Green algae are favored on more acidic soils and/or in areas with higher effective precipitation. Areas with alkaline soils, high salt content, and/or lower effective precipitation favor cyanobacteria (see Chap. 17). Lichens grow across the pH gradient, although species composition may change (Chap. 2). Extensive moss and lichen cover is found on highly stable soils, such as gypsum and calcite, which also have high water-holding capacity and high levels of phosphorus and sulfur. Nitrogen-fixing species have been found to prefer sites with low soil N (Starks et al. 1981). The highly mobile and larger cyanobacteria (such as *Microcoleus* spp.) are generally dominant where sediments are less stable, such as coarse sandy soils or in recently disturbed areas (Belnap 1993). Because the smaller cyanobacteria, green algae, lichens, and some mosses cannot grow up through large amounts of deposited sediment, they are generally found on either more inherently stable surfaces (e.g., alluvial or gypsiferous soils), or on soil surfaces initially stabilized by subterranean cyanobacteria (Chap. 27).

A frequent phenomenon in arid regions is the often almost luxurious hypolithic growth of cyanobacteria and microalgae on the lower surface of translucent quartz pebbles on the desert surface (Fensteralgen or window algae; Vogel 1955). These organisms profit from dew condensation, which accumulates below the stones, and the stones protect them from solar radiation. Diatoms (Rumrich et al. 1989) and lichens (Schubert 1982) also colonize this microhabitat. These stones are tightly glued to the soil surface by these organisms, and thus contribute to soil surface stability.

1.4 Functional Adaptations of Soil-Crust Organisms

Given the wide range of vegetation types in which soil crusts occur, and given the different types of biological soil crusts (see Sect. 1.5), the similarity of the general appearance of soil-crust communities and the combination of growth forms involved worldwide is impressive. Many comparable structures and functions occur amongst nonrelated taxa both locally and globally, indicating that adaptation to the soil surface environment has produced a parallel response among these divergent taxa and implying convergent evolutionary trends. As an example, Vogel (1955) describes three placoid soil-crust lichens from South Africa with almost identical morphological features, very similar anatomical thallus structures, and similar functional adaptations, despite belonging to quite different taxonomic groups (species of the genera *Lecidea, Endocarpon,* and *Toninia;* Fig. 1.4).

Another striking feature is the taxonomic similarity of soil-crust communities worldwide (Chap. 12). There are many genera, and even some species, of cyanobacteria, algae, lichens, and bryophytes which occur on almost all continents. For example, the taxonomically well-defined lichen *Psora decipiens* (Hedwig) Hoffm. (Schneider 1979; Timdal 1986) is found on soil crusts in North America, Mexico, Africa, around the Mediterranean Sea, in Europe, throughout the Middle East, in India, in many areas of Asia, in Australia, New Zealand, and even in the high Arctic and the nival belt of the Alps. The species *Collema tenax*, *C. coccophorum,* and *Catapyrenium squamulosum* have a similarly widespread distribution (see Part I). Thus, on a worldwide basis, the soil-crust communities not only resemble each other with respect to convergent features, but also contain a significant portion of common taxa on both the

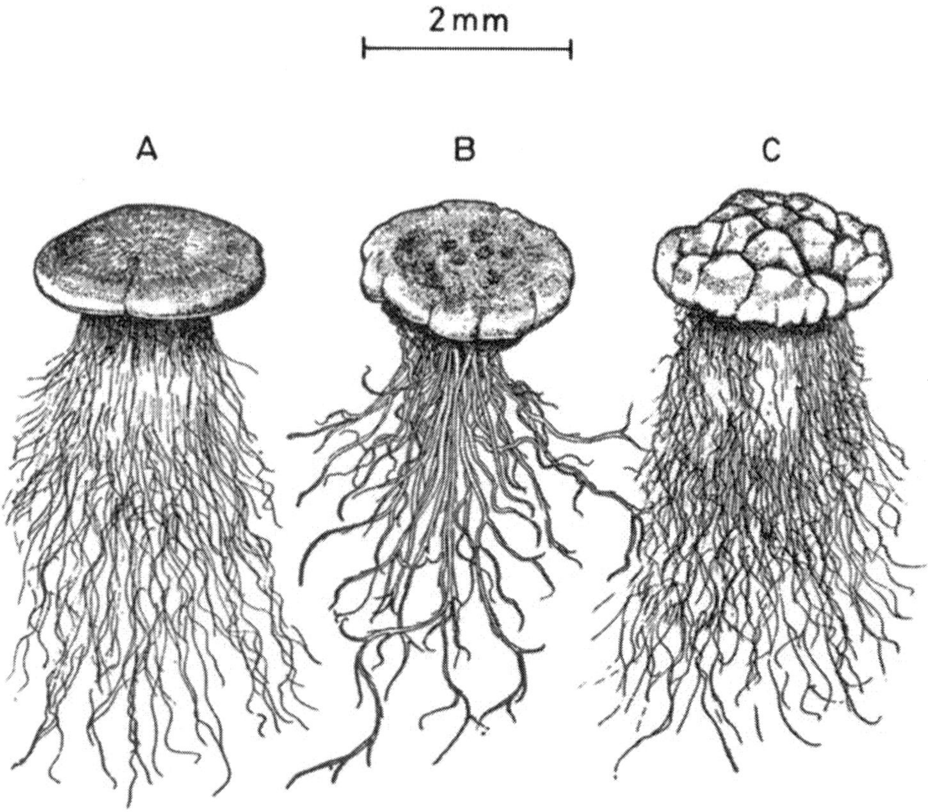

Fig. 1.4A–C. Convergent growth forms of soil lichens in South Africa: thalli of *Lecidea* (**A**), *Endocarpon* (**B**), and *Toninia* (**C**) sp. with rhizinae and hyphae. (After Vogel 1955)

generic and species level. This high degree of taxonomic similarity occurs in spite of crusts occurring in climatic zones ranging from hot deserts to polar regions, indicating that many crust taxa possess a broad ecological amplitude, while being highly specialized for the habitat at the soil surface.

The microenvironment in which the soil-crust biota are found (the uppermost millimeters of soil in relatively unshaded sites) is one of the most extreme habitats for autotrophic organisms on earth. The soil surface is a place of radiant energy conversion, and is where the highest and lowest temperatures occur within the soil-atmosphere profile. Temperatures up to almost 70 °C have been measured in terrestrial lichen thalli in a local steppe formation under temperate climatic conditions (southwest Germany; Lange 1953), and freezing temperatures can fall below −20 °C at the same site. Thus, for soil-crust organisms, the ability to tolerate extreme temperatures is one of the necessary requirements for existence. Indeed, when dehydrated, their distribution is not limited by even the most extreme environmental temperatures. In a study by Lange (1953), terrestrial species from European temperate steppe formations belonged to the most resistant of all lichens tested. Viability of lichen photobionts was not impaired after 30 min of experimental exposure to 90 °C (*Squamarina cartilaginea, S. lentigera*) or even 100 °C (*Peltigera rufescens, Cladonia pocillum*). Desiccated soil-crust mosses showed similar heat-resistance limits (Lange 1955): from 100 °C (*Tortella nitida, Ceratodon purpureus*) to 110 °C (*Barbula gracilis*). Volk (1979) reported that soil-crust liverworts (*Riccia* spp. and *Exormotheca* sp.) from Namibia survived a treatment of several hours at 100 °C. In contrast to desiccated organisms, the heat resistance of hydrated lichens and mosses is much lower, highlighting the importance of evaporative cooling in avoiding heat damage when direct sun radiation warms up a hydrated thallus.

In accordance with their habitat requirements, the photosynthetic metabolism of soil-crust lichens is adjusted to relatively high temperatures. This is especially true for the homoiomerous cyanolichen *Collema tenax* from Utah (Lange et al. 1998). In addition, net photosynthesis of phycolichens can also occur at extremely low temperatures. A *Diploschistes* species and *Cladonia convoluta* from a Mediterranean soil-crust community maintained low, but measurable, net photosynthesis down to a thallus temperature of −11.5 and −22 °C respectively (Lange 1965; see Chap. 18). This is in agreement with the seemingly unlimited cold resistance of many lichens (Kappen and Lange 1972).

Poikilohydric soil-crust colonizers are often exposed to long periods of desiccation between infrequent moistening events. Dry weight-related water content of lichen thalli can reach 5% or less, terminating all metabolic processes. Many cyanobacteria, algae, lichens, and mosses tolerate strong experimental desiccation for extended periods (Bewley and Krochko 1982). *Cladonia convoluta* from a soil-crust site in southwest Germany was not impaired after 56 weeks of experimental drying (Lange 1953). *Nostoc flagelliforme* and *N. commune* from terrestrial habitats in China recovered after a drought period of 2 years with respiration, photosynthesis, and nitrogen fixation reactivating in that order (Scherer et al. 1984). Volk (1984) reported that *Riccia canescens* survived 7 years of desiccation. The stress hormone ABA is involved in induction of desiccation tolerance with bryophytes (Hartung and Gimmler 1994). This was extensively studied for the extremely xerophilic soil-crust liverwort *Exormotheca holstii* (Marchantiales, see Photo 25) from the Namib Desert by Hellwege et al. (1994).

Any anabolic metabolism of poikilohydric soil-crust organisms needs sufficient hydration. The periods of time under moistened, active conditions are short; often, lack of water is the limiting factor for existence. Many physiological and morphological features of poikilohydric organisms are viewed as adaptive strategies to prolong active periods and/or to make growth and production more efficient (see general reviews by Kappen 1973, 1988 and

Rogers 1977 for lichens; Proctor 1981 for bryophytes). Examples of such adaptations include the high water-holding capacity of some lichen thalli ("succulence" of gelatinous lichens, Galun 1963), and the special water-conducting and storing structures, glass hairs, and revolute leaf margins (to reduce transpirational water loss) of desert mosses (Frey and Kürschner 1991b, 1998). Phycolichens are capable of utilizing extremely small amounts of water for photosynthetic activity, and are able to use water vapor for reactivation (Chap. 18). This is not the case with soil-crust mosses (Rundel and Lange 1980), thus accounting for their absence or rarity in hyperarid environments.

Epedaphic crust organisms are also exposed to high levels of irradiance, further stressing these organisms. In fact, photoinhibition was observed in the Antarctic when soil-crust mosses were exposed to cycles of freezing and thawing (Lovelock et al. 1995; T.G.A. Green, pers. comm.). Soil-crust organisms have developed a variety of strategies to protect themselves against damaging effects of excessive radiation. Desiccation presents one effective photoprotective mechanism. Desiccated thalli of shade-adapted *Peltigera rufescens* from a local steppe formation near Würzburg survived experimental treatments of high light without any impairment, in contrast to hydrated specimens (Demmig-Adams et al. 1990a). Some low moisture content and/or long extended periods of exposure to high light can decrease the susceptibility of lichens to excessive light (Solhaug and Gauslaa 1996; Gauslaa and Solhaug 1999).

The photobionts of soil-crust lichens are covered by an upper fungal cortex and epinecral layer which, through their structure and pigmentation, reflect and absorb radiation. This may help the sensitive autotrophs to avoid excessive radiation (Rikkinen 1995). Indeed, only a fraction of incident light is transmitted through the cortex of many lichen thalli. For the soil-crust species *Psora crenata* and *P. decipiens* from the Sonoran Desert, only about 50 % of photosynthetically active radiation reaches the upper part of the photobiont layer (Dietz et al. 2000). Much lower degrees of transmittance are reported from *Peltula* species from South Africa and Mexico, where 89 to 93 % of incident radiation is absorbed by the fungal cortex, including the upper part of the photobiont layer (Büdel and Lange 1994). Thickness, structure, and pigmentation of the upper cortex of soil-crust lichens clearly show adaptive features to the light environment of their habitat (Vogel 1955; Kappen 1973; Dietz et al. 2000). Hygrochasic thallus movements, often found in lobate soil-crust lichens such as *Chondropsis viridis, Xanthomaculina convoluta,* or *Cladonia convoluta,* are interpreted as photoprotective (see Rogers 1977; Büdel and Wessels 1986). The flat thallus lobes curl when dry, so that the surface area exposed to insolation is reduced and the now upside-down photobiont layer is protected. Similar curling features are known for desert bryophytes and liverworts (Frey and Kürschner 1991b).

Free-living terrestrial cyanobacteria cells also protect themselves by synthesizing UV-screening pigments (Garcia-Pichel and Castenholz 1991,1993; see Chap. 30). Seasonal time courses of pigment production are established in *Nostoc commune* colonies in terrestrial habitats (Bilger et al. 1997). The pigmentation in cyanolichens from high light-intensity habitats (including *Collema* cf. *coccophorum* from a soil crust in southern Utah) protects the photobiont from damaging UV rays (Büdel et al. 1997).

If excessive photosynthetic radiation is absorbed by hydrated lichens, green algal photobionts are able to effectively dissipate this energy. There is much evidence that the carotenoid zeaxanthin plays a decisive role in this process (Demmig-Adams et al. 1990b). Cyanobacteria lack the xanthophyll cycle needed to generate this carotenoid, and it is hypothesized that they instead use canthaxanthin for photoprotection (Leisner et al. 1994). Thallus content of canthaxanthin is positively correlated with light levels in the habitats of the studied lichens, and it is especially high in the soil-crust lichens *Gloeoheppia, Gonohymenia, Heppia,* and *Peccania* spp. of the Judean Desert (on the slopes of the Dead Sea, Leisner et al. 1993). The seasonal cycle of canthaxanthin production is also clearly correlated to the light received by *Peltigera rufescens* in a temperate terrestrial habitat (Lange et al. 1999).

1.5 Classification of Biological Soil Crusts

Biological soil crusts can be classified into several broad types based on habitat conditions, taxonomic composition, physical appearance, and function. The species composition of soil crusts is still not adequate for a large-scale floristic, phytosociological grouping of soil-crust communities. There have been several suggestions for crust classification using the growth forms of dominant species, soil substrate, and development processes as criteria (see West 1990). Soil algal growth types were defined by Komáromy (1976). Following Johansen (1993), Evans and Johansen (1999) typify crust formations encountered in the western United States. For the rangelands of Australia, Eldridge and Greene (1994) identify three forms of biological crusts: hypermorphs (above ground), perimorphs (at ground) and cryptomorphs (hidden below ground). Based on these earlier definitions, we propose a topographic classification of crust types that facilitates a worldwide comparison of crust morphology (Chap. 15). Four crust categories are used: smooth, rugose, pinnacled, and rolling. Smooth crusts are found in hot deserts where soils do not freeze, and are almost exclusively endedaphic cyanobacteria, algae and fungi (Photos 49, 50, 55, 56). Chemical crusting is also common in this crust type. The other three crust categories generally have epedaphic colonizers (lichens and mosses) in addition to endedaphic autotrophs and fungi. Rugose crusts have low surface roughness (<2 cm), generally from scattered lichen and/or moss clumps (Photos 47, 48, 51, 52, 57, 58, 59). Under moister conditions in temperate regions, rugose crusts can also be dominated by an epedaphic layer of filamentous algae (such as *Zygogonium*), which can penetrate 4 mm deep into the soil (Photos 8, 43). Pinnacled and rolling crusts occur only where soils freeze in winter. Pinnacled crusts are dominated by cyanobacteria, but locally can have up to 40 % moss-lichen cover. Pinnacles range up to 15 cm high (Photos 31, 32, 33). Rolling crusts occur in regions where high precipitation results in heavy lichen-moss cover, and frost-heaving of these surfaces results in a gently rolling surface about 5 cm high (Photos 29, 30, 40, 41, 42, 44, 45). This classification, discussed in detail in Chapter 15, is a highly generalized scheme, and the four categories are connected by intermediate crust types.

1.6 Problems with Species Identification and Sampling

Sampling and describing soil-crust communities presents some fundamental difficulties. First, species identification for many crust components is problematic. For instance, few cyanobacterial, green algal, bacterial, or fungal genera have received the attention necessary to establish clear taxonomic relationships or to provide definitive identification criteria for the many forms a species can assume. Isolated cells or filaments are often found that cannot be determined with the commonly used identification techniques. In addition, many species may be present in insufficient numbers for identification. In these cases, culturing is necessary. However, even if firm taxonomic identification is possible, culturing cannot provide frequency or biomass estimates. Thus, characterization of the microbial components of soil crusts is difficult.

Second, lichens and bryophytes have received greater taxonomic attention than the microbial components. Nevertheless, many species can only be determined through microscopic examination, or by specialists. This often hinders fieldwork aiming at statistically sound descriptions of species composition of crust communities. Some attempts have been made to facilitate recognition of diversity of soil-crust biota by nonspecialists. Often, dominant genera or easily recognizable indicator species can be identified, allowing a general characterization of crust communities. Morphological groups of soil-crust organisms can also be defined, which is being successfully used for rangeland monitoring (Chap. 32).

Third, species composition and abundance can also vary drastically with season. Although perennial lichen and moss components generally show little change from season to season, ephemeral mosses and especially cyanobacterial and algal populations can show large seasonal fluctuations (Johansen and Rushforth 1986; see Chap. 28). Therefore, representative assessments of the ecological role of soil-crust components and spatial comparisons often require measurements be made in all seasons, and this is often difficult.

Lastly, it is rare that the scientist involved in characterizing the species composition of a soil crust is a specialist in all of the taxonomic groups occurring in a given community. This often results in varying levels of emphasis for different crust groups, depending on the investigator, and complicates comparisons between described crust communities. In addition, descriptions of the different components or functions of a given crust community are often published in journals of widely-separated disciplines (lichenology, phycology, bryology, ecology, geology, etc.), making collation of this information difficult.

1.7 Heterogeneous State of Knowledge in Different Geographic Areas

The intensity of study on the structure, function, and management of biological soil crusts has varied widely between different geographic areas. For example, there is a large literature on crusts in the western United States, Australia, and Israel, and yet very little on crusts in South America or Asia. It might be concluded that research has been concentrated in certain areas because crusts are more important or more apparent in those ecosystems. This is definitely not the case. The amount known about crusts in a given area reflects more the location of interested scientists, rather than the importance of the crust to a given ecosystem. Many remote regions have never even been surveyed for the presence of soil crusts. In addition, there are serious limitations for obtaining and evaluating local literature from large areas of the world, especially when results are published in languages like Russian or Chinese.

In the following chapters of Part I of this Volume, the different geographic areas are treated separately by appropriate specialists. The chapters summarize the actual state of available information. The disparity of information regarding biological soil-crust communities on different continents is immediately apparent. Further research on the role of soil crusts in understudied habitats is needed, especially in areas where human activities are accelerating.

Acknowledgments. The authors are extremely grateful to the following colleagues for their help with pictures: Reinhard Bornkamm (Berlin), David J. Eldridge (Sydney), Wolfgang Frey (Berlin), Giora Kidron (Jerusalem), Roman Türk (Salzburg), Hans Ullrich (Goslar), and Hans Zellner (Würzburg).

References

Belnap J (1993) Recovery rates of cryptobiotic soil crusts: inoculant use and assessment methods. Great Basin Nat 53(1):89–95

Bewley JD, Krochko JE (1982) Desiccation-tolerance. In: Lange OL, Nobel PS, Osmond CB, Ziegler H (eds) Physiological plant ecology II. Water relations and carbon assimilation. Encyclopedia of plant physiology 12B. Springer, Berlin Heidelberg New York, pp 325–378

Bilger W, Bohuschke M, Ehling-Schulz M (1997) Annual time course of the contents of carotenoids and UV-protective pigments in the cyanobacterium *Nostoc commune*. Bibl Lichenol 67:223–234

Büdel B, Lange OL (1994) The role of cortical and epinecral layers in the lichen genus *Peltula*. Cryptogam Bot 4:262–269

Büdel B, Wessels DCJ (1986) *Parmelia hueana* Gyeln., a vagrant lichen from the Namib Desert, SWA/Namibia. I. Anatomical and reproductive adaptations. Dinteria 18:3–15

Büdel B, Karsten U, Garcia-Pichel F (1997) Ultraviolet-absorbing scytenemin and mycosporine-like amino acid derivatives in exposed, rock-inhabiting cyanobacterial lichens. Oecologia 112:165–172

Crum H (1993) A lichenologists view of lichen manna. Contrib Univ Mich Herb 19:293–306

Demmig-Adams A, Máguas C, Adams III WW, Meyer A, Kilian E, Lange OL (1990a) Effect of high light on the efficiency of photochemical energy conversion in a variety of lichen species with green and blue-green phycobionts. Planta 180:400–409

Demmig-Adams B, Adams III WW, Czygan F-C, Schreiber U, Lange OL (1990b) Differences in the capacity for radiationless energy dissipation in the photochemical apparatus of green and blue-green algal lichens associated with differences in carotenoid composition. Planta 180:582–589

Dietz S, Büdel B, Lange OL, Bilger W (2000) Transmittance of light through the cortex of lichens from contrasting habitats. Bibl Lichenol 75:171–182

Eldridge DJ, Greene RSB (1994) Microbiotic soil crusts: a review of their roles in soil and ecological processes in the rangelands of Australia. Aust J Soil Res 32:389–415

Eldridge DJ, Tozer ME (1996) Distribution and floristics of bryophytes in soil crusts in semi-arid and arid eastern Australia. Aust J Bot 44:223–247

Eldridge DJ, Tozer ME (1997) A practical guide to soil lichens and bryophytes of Australia's dry country. Dep Land and Water Cons, Sydney

Elenkin AA (1901) Wanderflechten der Steppen und Wüsten. Bull Jard Imp Bot St Pétersbourg 1:1–45

Evans RD, Johansen JR (1999) Microbiotic crusts and ecosystem processes. Crit Rev Plant Sci 18:183–225

Frey W, Kürschner H (1988) Bryophytes of the Arabian Peninsula and Socotra. Floristics, phytogeography and definition of the Xerthermic Pangaean element. Nova Hedwigia 46:37–120

Frey W, Kürschner H (1991a) Lebensstrategien von terrestrischen Bryophyten in der Judäischen Wüste. Bot Acta 104:172–182

Frey W, Kürschner H (1991b) Das Fossombronio-Gigaspermetum mouretii in der Judäischen Wüste. 2. Ökosoziologie und Lebensstrategien. Cryptogam Bot 2/3:73–84

Frey W, Kürschner H (1991 c) Morphologische und anatomische Anpassungen der Arten in terrestrischen Bryophytengesellschaften entlang eines ökologischen Gradienten in der Judäischen Wüste. Bot Jahrb Syst 112:529–552

Frey W, Kürschner H (1998) Wüstenmoose: Anpassungen und Überlebensstrategien im täglichen Kampf mit der Sonne. Biol Unserer Zeit 28:231–240

Galun M (1963) Autecological and synecological observations on lichens of the Negev, Israel. Isr J Bot 12:179–186

Galun M, Bubrick P, Garty J (1982) Structural and metabolic diversity of two desert-lichen populations. J Hattori Bot Lab 53:321–324

Garcia-Pichel F, Castenholz RW (1991) Characterization and biological implications of scytonemin, a cyanobacterial sheath pigment. J Phycol 27: 395–409

Garcia-Pichel F, Castenholz RW (1993) Occurrence of UV-absorbing, mycosporine-like compounds among cyanobacterial isolates and an estimate for their screening capacity. Appl Environ Microbiol 59:163–169

Gauslaa Y, Solhaug KA (1999) High-light damage in air-dry thalli of the old forest lichen *Lobaria pulmonaria* – interactions of irradiance, exposure duration and high temperature. J Exp Bot 50:697–705

Harper KT, Marble JR (1988) A role for nonvascular plants in management of arid and semiarid rangeland. In: Tueller PT (ed) Vegetation science applications for rangeland analysis and management. Kluwer Academic Publishers, Dordrecht, pp 135–169

Hartung W, Gimmler H (1994) A stress-physiological role for abscisic acid (ABA) in lower plants. Prog Bot 55:157–173

Hellwege EM, Dietz K-J, Volk OH, Hartung W (1994) Abscisic acid and the induction of desiccation tolerance in the extremely xerophilic liverwort *Exormotheca holstii.* Planta 194:525–531

Jahns HM (1988) The lichen thallus. In: Galun M (ed) Handbook of lichenology, 1. CRC Press, Boca Raton, pp 95–143

Johansen JR (1993) Cryptogamic crusts of semiarid and arid lands of North America. J Phycol 29:140–147

Johansen JR, Rushforth SR (1986) Cryptogamic soil crusts: seasonal variation. Great Basin Nat 46:632–640

Kappen L (1973) Response to extreme environments. In: Ahmadjian V, Hale ME (ed) The lichens. Academic Press, New York, pp 311–380

Kappen L (1988) Ecophysiological relationships in different climatic regions. In: Galun M (ed) Handbook of lichenology, 2. CRC Press, Boca Raton, pp 37–100

Kappen L, Lange OL (1972) Die Kälteresistenz einiger Makrolichenen. Flora 161:1–29

Komáromy ZP (1976) Soil algal growth types as edaphic adaptation in Hungarian forest and grass steppe ecosystems. Acta Bot Acad Sci Hung 22:373–379

Lange OL (1953) Hitze- und Trockenresistenz der Flechten in Beziehung zu ihrer Verbreitung. Flora 140:39–97

Lange OL (1955) Untersuchungen über die Hitzeresistenz der Moose in Beziehung zu ihrer Verbreitung. I. Die Resistenz stark ausgetrockneter Moose. Flora 142:381–399

Lange OL (1965) Der CO_2-Gaswechsel von Flechten bei tiefen Temperaturen. Planta 64:1–19

Lange OL, Belnap J, Reichenberger H (1998) Photosynthesis of the cyanobacterial soil crust lichen *Collema tenax* for arid lands in southern Utah, USA: role of water content on light and temperature response of CO_2 exchange. Funct Ecol 12:195–202

Lange OL, Leisner MR, Bilger W (1999) Chlorophyll fluorescence characteristics of the cyanobacterial lichen *Peltigera rufescens* under field conditions II. Diel and annual distribution of metabolic activity and possible mechanisms to avoid photoinhibition. Flora 194:413–430

Leisner MR, Bilger W, Czygan F-C, Lange OL (1993) Lipophilous carotenoids of cyanobacterial lichens from different habitats, including an extreme desert site. Cryptogam Bot 4:74–82

Leisner MR, Bilger W, Czygan F-C, Lange OL (1994) Light exposure and the composition of lipophilous carotenoids in cyanobacterial lichens. J Plant Physiol 143:514–519

Lovelock CE, Osmond CB, Seppelt RD (1995) Photoinhibition in the Antarctic moss *Grimmia antarctici* Card. when exposed to cycles of freezing and thawing. Plant Cell Environ 18:1395–1402
Poelt J, Baumgärtner H (1964) Über Rhizinenstränge bei placodialen Flechten. Österr Bot Z 111:1–18
Proctor MCF (1981) Physiological ecology of bryophytes. In: Schulze-Motel W (ed) Advances in bryology 1. Cramer, Vaduz, pp 79–166
Rikkinen J (1995) What's behind the pretty colours? A study on the photobiology of lichens. Bryobrothera 4:1–239
Rogers RW (1977) Lichens of hot arid and semi-arid lands. In: Seaward MRD (ed) Lichen ecology. Academic Press, London, pp 211–252
Rosentreter R (1993) Vagrant lichens in North America. Bryologist 96:333–338
Rumrich U, Rumrich M, Lange-Bertalot H (1989) Diatmeen als "Fensteralgen" in der Namib-Wüste und anderen ariden Gebieten von SWA/Namibia. Dinteria 20:23–29
Rundel PW, Lange OL (1980) Water relations and photosynthetic response of a desert moss. Flora 169:329–335
Sanders WB (1994) Role of lichen rhizomorphs in thallus propagation and substrate colonization. Cryptogam Bot 4:283–289
Scherer S, Ernst A, Chen T-W, Böger P (1984) Rewetting of drought-resistant blue-green algae: time course of water uptake and reappearance of respiration, photosynthesis and nitrogen fixation. Oecologia 62:418–423
Schneider G (1979) Die Flechtengattung *Psora* sensu Zahlbruckner. Bibl Lichenol 13: 1–291
Schubert R (1982) Lichens of central Asia. J Hattori Bot Lab 53:341–343
Solhaug KA, Gauslaa Y (1996) Parietin, a photoprotective secondary product of the lichen *Xanthoria parietina*. Oecologia 108:412–418
Starks TL, Shubert LE, Trainor FR (1981) Ecology of soil algae: a review. Phycologia 20:65–80
Timdal E (1986) A revision of *Psora* (Lecideaceae) in North America. Bryology 89:253–275
Vogel S (1955) Niedere "Fensterpflanzen" in der südafrikanischen Wüste. Eine ökologische Schilderung. Beitr Biol Pflanz 31:45–135
Volk OH (1979) Beiträge zur Kenntnis der Lebermoose (Marchantiales) aus Südwestafrika (Namibia) I. Mitt Bot Staatssamml Münch 15:223–242
Volk OH (1984) Beiträge zur Kenntnis der Marchantiales in Südwest-Afrika/Namibia IV. Nova Hedwigia 39:117–143
West NE (1990) Structure and function of soil microphytic crusts in wildland ecosystems of arid and semi-arid regions. Adv Ecol Res 20:179–223

2 Biological Soil Crusts of North America

R. Rosentreter and J. Belnap

2.1 Introduction

Biological soil crusts in North America are diverse and found in many different habitats. On a broad scale, there are several different vegetation zones or ecoregions in western North America that contain biological soil crusts as major components (Fig. 2.1, Table 2.1). These include hot deserts (Mojave, Chihuahuan, Sonoran; see Chap. 1, Photos 34, 35), cool deserts (Great Basin, Colorado Plateau; also often denoted as semideserts, Photos 29–33), the coastal woodlands, chaparral and islands of California; and the subhumid grasslands of the Great Plains. These habitats span a large range of annual precipitation (75–740 mm) and annual mean temperatures (4–24°C). Soil crusts are also found on a smaller scale in shallow lithic sites, infertile forest soils, alpine habitats, and more mesic ecosystems as an early successional vegetation type.

Most research on North American crusts has occurred in the western United States. This area lies between two major mountain masses, the Rocky Mountains on the east and the Sierra Nevada on the west. While over 200 mountain ranges exceeding 4000 m in elevation occur in this region, most of the area consists of large basins at –70 m to 300 m elevation. The northern cool deserts (Great Basin, Colorado Plateau) receive up to half their precipitation as snow, and soils freeze annually, while the hot deserts to the south (Mojave, Sonoran, Chihuahuan) receive almost exclusively rain, and soil freezing is absent. Potential evaporation in these deserts ranges from 1500–4000 mm annually (Nash and Moser 1982; MacMahon 1987).

Although there are species differences in crusts of the different deserts of North America and although some crust species are local or regional endemics, many species are found in all regions with soil crusts. Ubiquitous cyanobacterial and green algal species include *Calothrix parietina* (Naeg.) Thur., *Hantzschia amphioxys* (Ehr.) Grunow, *Microcoleus vaginatus* (Vauch.) Gom. (Photos 1,2), *Nostoc commune* Vauch. (Photo 3), *Pinnularia borealis* (Ehr.) Rabh., *Schizothrix calcicola* (Ag.) Gom., *Scytonema hofmannii* (Ag.), and *Tolypothrix tenuis* (Khtz.) Schmidt. Lichens found in all regions include *Catapyrenium squamulosum* (Ach.) Breuss (Photo 21), *Collema tenax* (Sw.)

Ecological Studies, Vol. 150
J. Belnap and O.L. Lange (eds.) Biological Soil Crusts: Structure, Function, and Management

Table. 2.1. Soil, climatic and vegetative characteristics for low and mid-elevations in North American arid, semiarid, and subhumid ecoregions. (After Bailey 1994, MacMahon 1987)

Ecoregion	Annual precipi-tation (mm)	Moisture Season/Form	Mean annual temperature Range (°C)	Vegetation type
Sonoran Desert	75–300	Summer/rain Fall/rain	15–24	Mixed thorn scrub
Mojave (low elevation)	45–200	Winter/rain	10–24	Shrubland
Mojave (high elevation)	200–610	Winter/rain	3–10	Shrubland
Chihuahuan Desert	200–425	Summer/rain	10–18	Shrubland
Colorado Plateau	205–510	Winter/snow Spring/rain Summer/rain	4–13	Shrubland woodland
Great Basin	130–490	Winter/snow Spring/rain	4–13	Sagebrush steppe Salt desert shrub
Columbia Basin	260–635	Winter/rain Spring/rain	4–14	Perennial grassland
California Chaparral	175–635	Winter/rain	15–18	Chaparral grassland
Great Plains	250–900	Spring/rain summer/rain	4–18	Prairie grassland

Fig. 2.1. Arid and semiarid ecoregions of western North America. Areas *shaded in black* are cool deserts, while *gray-shaded areas* are hot deserts

Ach. (Photo 9), *C. coccophorum* Tuck., *Heppia lutosa* (Ach.) Nyl., *Psora decipiens* (Hedw.) Ach., and *Toninia sedifolia* (Scop.) Timdal, while cosmopolitan mosses include *Bryum argenteum* Hedw. and *Pterygoneurum ovatum* (Hedw.) Dix.

2.2 Factors Influencing Distribution and Composition of North American Soil Crusts

Precipitation and Temperature. In lower elevation areas, total crust cover is generally inversely related to vascular plant cover, as less plant cover leaves more soil surfaces available for crust colonization, and is positively correlated with increasing precipitation. These relationships hold until precipitation is high enough for vascular plant cover to preclude the presence of soil crusts. In drier regions, almost all soil surfaces not otherwise occupied by rocks, plants or litter are colonized to some degree by some crustal species. Exceptions to this include actively disturbed areas such as wash bottoms, moving sand dunes, or places where recurring anthropogenic activities disturb the soil surface. North and east slopes favor crustal development in lower elevation desert regions of North America. Slope angle does not generally affect crust cover or species richness, except where the slope or soils are unstable (Rosentreter 1986; Kaltenecker et al. 1997).

The relative species composition of crusts is also related to potential evapotranspiration (PET) of a site. Cyanobacteria heavily dominate crusts of hot desert sites (Sonoran, Mojave and Chihuahuan) where PET is high. As PET decreases northward (from the Colorado Plateau, across the Great Basin and the Columbia Plateau into Canada), relative lichen and moss diversity and cover increases as well. In addition, while lichen cover is greater than moss cover in hot deserts, there is a greater abundance of mosses than lichens in the more northerly areas (e.g., the Columbia Plateau; Ponzettii et al. 1998). The lichen and moss flora also changes along this precipitation and temperature gradient (Fig. 2.2).

Season of rainfall may also play a part in determining crust cover and composition. Crust species differ in how fast they can reach net carbon gains after moistening, and also in how long they can continue this gain (Lange et al. 1997, 1998; Lange 2000; see Chaps. 18–20). Because soils dry so quickly in summer, rainfall events can result in greater carbon loss for some species compared to others (Jeffries and Klopatek 1987). For instance, nighttime rain often will benefit only those lichens with high water-holding capacity (e.g., *Collema* spp.) which can stay moist long enough during early morning hours to achieve net carbon gain (Lange 2000). Ecoregions that receive summer rainfall and lower overall rainfall (i.e., Sonoran Desert) tend to have greater numbers

Figure **2.2.** Distribution of common soil lichens as influenced by temperature and precipitation. Note that some lichens occur across many climatic zones, while others are very climate-specific

of heterocystic cyanobacteria genera (e.g., *Lyngbya, Calothrix, Schizothrix* and *Nostoc*), and lower numbers of nonheterocystic genera (e.g., *Microcoleus*). This is in contrast to regions with winter rainfall and/or greater rainfall that are heavily dominated by *Microcoleus*. Soil lichen diversity is lower in summer-rainfall regions than in regions without summer rainfall. For instance, sandy soils on the Colorado Plateau (with 35–40 % summer rainfall) can have 40 % lichen cover with six to ten species, while just 100 km south in the Sonoran Desert with 55 % summer rainfall, sandy soils generally have <10 % lichen cover and only two to four species. However, PET is also higher in the Sonoran Desert than on the Colorado Plateau, and the relative role of total precipitation versus precipitation timing has not been well investigated.

Soils. Soil texture and chemistry strongly influence the species composition of North American crust communities (and worldwide; see Chaps. 2–12, 15, 16). More stable, fine-textured soils (such as gypsiferous and silty loam soils) support greater cover and more varied populations of cyanobacteria, lichens, and mosses than less stable, coarse-textured soils (Kleiner and Harper 1977), which often support only highly mobile, large filamentous cyanobacteria (such as *Microcoleus* spp.). Embedded rocks near or at the soil surface can increase crust cover by providing additional water and protection from disturbances. Shallow soils often support a wide variety of cyanobacteria, lichens, and mosses (regardless of soil texture), relative to adjacent deep soils. Physical and chemical crusts can also enhance cover of biological crusts by increasing water pooling times through reduced infiltration and increasing soil surface stability (see Chap. 27 for a more detailed discussion).

Soil chemistry can also influence crust cover and composition. Cyanobacteria dominate alkaline soils, while green algal species dominate acidic soils (Grondin and Johansen 1995). Some soil lichens are indicative of soils with low and high levels of calcium carbonate (Table 2.2). Overall, crust cover and lichen diversity is increased on highly calcareous soils. In North American deserts (and deserts worldwide), highly calcareous soils derived from limestone or gypsum can support up to 80 % lichen cover, while adjoining, less calcareous substrates may have less than 10 % cover (J. Belnap, unpubl. data).

Vascular Plant Community Structure. The distribution, shape, and height of vascular plants can either increase or decrease biological soil crust cover and influence crust species composition. Vascular vegetation reduces the overall soil surface available for colonization (especially in regions with lower PET). At the same time, shading provided by plant cover can enhance crust development through increased soil moisture, especially in the hotter, drier

Table 2.2. Great Basin soil-crust lichens that are calcium carbonate indicators. (After McCune and Rosentreter 1995)

Low calcium carbonate	High calcium carbonate
Acarospora schleicheri	*Aspicilia fruticulosa*
Arthonia glebosa	*Aspicilia hispida*
Aspicilia reptans	*Buellia elegans*
Aspicilia filiformis	*Caloplaca tominii*
Cladonia borealis	*Collema tenax*
Diploschistes muscorum	*Fulgensia bracteata, F. desertorum*
Leptochidium albociliatum	*Heppia lutosa*
Megaspora verrucosa	*Phaeorrhiza nimbosa*
Ochrolechia upsaliensis	*Psora cerebriformis*
Placynthiella uliginosa	*Psora decipiens*
Psora nipponica	*Psora tuckermanii*
Xanthoparmelia wyomingica	*Toninia sedifolia*

deserts. Soils under shrubs tend to be dominated by mosses in the cooler deserts (Colorado Plateau and Great Basin), while interspaces tend to be dominated by cyanobacteria and lichens (St. Clair et al. 1993). Invasive exotic plants generally decrease the structural diversity of native vascular plant communities and often fill previously biologically crusted interspaces. In most ecosystems of the Great Basin, this has resulted in less cover and lower species richness of soil crusts (Rosentreter 1994; Kaltenecker 1997).

Morphology. The morphology of biological crusts varies between regions of North America. Rugose crusts are present in the hot deserts (e.g., Sonoran, Chihuahuan, and Mojave), pinnacled crusts occur on the Colorado Plateau and southern Great Basin, and rolling crusts occur in the northern Great Basin northwards (for a detailed discussion on crust types, see Chap. 15). Cyanobacterial communities in these deserts are often layered in the soil. Genera with heavy UV-protective pigmentation (e.g., species of *Schizothrix*, *Calothrix*, *Nostoc*) are generally within the top 1 mm of soil, while less pigmented genera (e.g., species of *Microcoleus*, *Oscillatoria*) are found below this depth (for a more detailed discussion, see Chap. 30).

2.3 Species Composition of Soil Crusts in North American Ecoregions

2.3.1 Hot deserts

There are three major hot deserts in North America: Sonoran, Mojave, and Chihuahuan (Fig. 2.1). These deserts are found within a few degrees of 30°N latitude. Summers are hot, with daytime temperatures frequently over 30 °C. Winters are mild, with a mean January temperature of 10 °C. Snow falls very rarely in these deserts and soils seldom freeze. Crusts at lower elevations are either smooth or rugose, while pinnacled crusts occur at higher elevations.

2.3.1.1 Sonoran

The Sonoran Desert extends north from Sonora, Mexico (roughly 23°N latitude) to southern Arizona (35°N latitude). This region is mostly underlain by volcanic rocks, and is dominated by depositional environments consisting of large, sweeping valleys and deep soils with well-defined calcrete layers. Most elevations are between 0 and 1000 m, and annual rainfall is 75–300 mm, of which 55 % occurs during the summer. High air temperatures result in high

PET. Soils do not freeze in this desert. In general, this is the hottest US desert. Vegetation ranges from upland communities of large, columnar cactus or perennial grasses to lowland communities dominated by *Larrea tridentata* or cacti-succulents. Most plants are widely spaced, as is typical of the other hot deserts (Nash and Moser 1982; MacMahon 1987).

Biological soil crusts are rugose in the Sonoran Desert, and heavily dominated by cyanobacteria. In Baja California, Mexico, the most commonly encountered cyanobacteria are *Nostoc commune* and *Schizothrix calcicola*, followed by *Microcoleus vaginatus*, *Nostoc piscinale* Kütz., and *Plectonema tomasinianum* var. *gracile* Hansg. (Flechtner et al. 1998). Further north in Arizona, commonly occurring species include *Calothrix parietina*, *Microcoleus vaginatus*, *M. paludosus* (Khtz.) Gom., *M. sociatus* (W. et G.S.West), *Nostoc commune*, *N. microscopicum* (Carm.) Harv. et Hook., *N. muscorum* Ag., *Schizothrix calcicola*, and *Scytonema hofmannii*. Other common species include *Lyngbya aestuari* (Mert.) Liebm., *Plectonema nostocorum* Born., *Phorimdium* sp., *Porphyrosiphon fuscus* Gom., *Oscillatoria* sp., and *Tolypothrix tenuis* (Cameron 1962, 1972). The diverse and fairly even mix of cyanobacterial species found in the Sonoran Desert differs from other US desert areas (including the similarly hot Mojave), whose cyanobacterial floras are heavily dominated by *Microcoleus vaginatus*.

The most common soil green algal species in Baja California, Mexico, are the chlorophytes *Myrmecia astigmatica* Vinatzer, *M. biatorellae* (Tsch. et Plessl.) Boye-Pet., and *Bracteacoccus minor* (Chodat) Petrova, and the diatoms *Hantzschia amphioxys*, *Hantzschia amphioxys* f. *capitata* O.Müller, *Luticola cohnii* (Hilse) Mann, *Luticola mutica* (Kütz.) Mann and *Pinnularia borealis* var. *scalaris* (Ehr.) Rabh. Chrysophyte cysts are also very common (Flechtner et al. 1998). Further north in Arizona, the most common coccoid green algae include *Chlorella vulgaris* Beij., *Navicula* sp., *Nitzschia* sp., *Palmogloea protuberans* (Sm. et Sow.) Khtz., *Pinnularia* sp., *Protococcus grevillei* (Ag.) Crouan, *Stichococcus subtilis* (Khtz.) Klerck, and *Trochiscia hirta* (Lagerh.) Hansg. (Cameron 1960, 1972).

Cyanobacteria and green algal species are common under diaphanous rocks in all hot deserts, including the Sonoran, and these organisms glue the rocks firmly to the soil surface. In these microhabitats, water is available longer, UV light is decreased, and temperatures are more favorable for microfloral growth. Hypolithic species include *Anacystis* spp., *Bracteacoccus* spp., *Calothrix parietina*, *Chlorococcum* spp., *Coccochloris* spp., *Nostoc* spp., *Protococcus grevillei*, *Protosiphon cinnamomeus*, *Scytonema hofmannii*, and *Trochiscia hirta* (Cameron 1960; Cameron and Blank 1967).

Soil lichens occur as sparse patches in the Sonoran Desert. The most common lichens include *Catapyrenium squamulosum*, *Collema coccophorum* and *C. tenax*. Other less common lichens include *Acarospora strigata* (Nyl.) Jatta, *Heppia lutosa*, *H. adglutinata* (Kremph.) A. Massal., *H. despreauxii*

(Mont.) Tuck., *Peltula richardsii* (Herre) Wetm., and *P. patellata* (Bagl.) Swinscow et Krog. In coastal fog deserts of Baja California, rugose lichen crusts are replaced by extensive carpets of the fruticose lichen *Desmazieria* spp. Common soil crust mosses include *Bryum argenteum*, *B. caespiticium* Hedw., *Crossidium aberrans* Holz. et Bartr., *Tortula plinthobia* (Sull. et Lesq.) Aust., *T. euryphylla* Zand., *T. obtusifolia* Schwaegr., *Pterygoneurum ovatum*, and *Syntrichia ruralis* (Hedw.) Web. et Mohr. Large thalloid liverworts, *Athalamia hyalina* (Sommert.) Hatt and *Riccia sorocarpa* Bisch., are also common in this desert. As elevation increases (and PET decreases), the lichen and moss flora is more diverse, and approximates that found on the Colorado Plateau (see Sect. 2.3.2.1; Nash et al. 1974; Pearson and J. Belnap, unpubl.; L.L. Stark, unpubl.).

Much of the early ecological work on soil crusts occurred in the Sonoran Desert. Studies showed that soil crusts in this desert contribute nitrogen to soils and plants, increase soil surface stability, and influence vascular plant establishment in this desert (McIlvanie 1942; Mayland et al. 1966). For a more detailed discussion, see Chapters 19, 21, and 26.

2.3.1.2 Mojave Desert

The Mojave Desert lies in the rain shadow of the Sierra Nevada Mountains, and consists of internally-draining, wide valleys filled with mixed sediments washed from scattered sedimentary (mostly limestone) and igneous (mostly granite) mountains. Consequently, soil texture and chemistry varies widely. Elevations range from –70 m to 1300 m, with mountains up to 3500 m. Annual precipitation at low elevation is 45–200 mm, with most falling in winter. Summer temperatures average 29 °C, and winters average 13 °C. Soils do not freeze except at the higher elevations. Summer temperatures can be extremely high (54 °C), with ground temperatures exceeding 90 °C. *Larrea tridentata* communities occur in areas with <175 mm rainfall; *Coleogyne ramosissima*, *Atriplex spinosa*, and *Lycium* sp. communities occur at 175–200 mm; and *Atriplex* sp. and *Artemisia* sp. communities are found in areas with >200 mm (Nash and Moser 1982; MacMahon 1987).

The rugose biological soil crusts in the Mojave Desert are heavily dominated by cyanobacteria, with occasional lichen-moss patches. Morphologically, soil crusts in the Mojave Desert are very similar to those found in the Sonoran Desert. Unlike the Sonoran Desert, however, the Mojave cyanobacterial flora is generally dominated by *Microcoleus vaginatus*, though there are exceptions (Durrell 1962). Other common species include *M. chthonoplastes* (Fl. Dan.) Thuret et Gom., *Anabaena variabilis* (Kütz.), *Calothrix castellii* Fremy., *Nostoc commune*, *N. muscorum*, *Phormidium tenue* (Menegh.) Gom., *Schizothrix calcicola*, and *Scytonema hofmannii*. Algal species include *Chlorococcum infusionum* (Schrank) Meneghini, *Chlorella vulgaris*, *Protococcus viridis* Agardh., and *Rhizoclonium hieroglyphicum* (C.A.Ag.) Kütz.

Soil lichen species are limited, with *Catapyrenium squamulosum*, *Collema coccophorum,* and *C. tenax* being the most common. *Acarospora strigata* is important at some sites. Other common species include *Psora decipiens*, *Fulgensia desertorum* (Tomin) Poelt, *F. bracteata* Hoffm., *Heppia lutosa*, *Peltula patellata*, and *Toninia sedifolia* (Photo 16). The dominant mosses of Mojave soil crusts are *Syntrichia caninervis* Mitt., *Pterygoneurum ovatum*, *P. subsessile* (Brid.) Jur., *Tortula inermis* (Brid.) Mont., *Crossidium aberrans*, *Didymodon vinealis* (Brid.) Zand., and *D. australasiae* (Hook. et Grev.) Zand. Other significant mosses include *Bryum argenteum*, *Tortula atrovirens* (Sm.) Lindb., *Crossidium crassinerve* (De Not.) Jur., and *Microbryum starkeanum* (Hedw.) Zand. Thalloid hepatics are rare (L.L. Stark, unpubl.). As in the Sonoran, increases in elevation result in occurrence of lichen and moss species common to the Colorado Plateau (see Sect. 2.3.2.1).

2.3.1.3 Chihuahuan

The Chihuahuan Desert is on a fairly high plain (1000–2000 m), situated between the Sierra Madre Oriental and Occidentale Mountains. Over 80 % of the soils are derived from calcareous materials (mostly limestone), with the remaining soils mostly of volcanic origin. The landscape is characterized by large alluvial plains, bajadas, and scattered mountains, as is typical of the other hot deserts. Total precipitation is 200 mm to 425 mm, and falls mostly during summer. Areas with >250 mm rainfall are generally grasslands (unless converted to shrubs by grazing), while <250 mm areas are dominated by *Larrea tridentata*. Soils do not freeze, and rugose crusts are the dominant form of crust (Nash and Moser 1982; MacMahon 1987).

As in most of western North American, the cyanobacterial flora of the Chihuahuan Desert is dominated by *Microcoleus vaginatus*. Other common species include *M. paludosus*, *Nostoc commune*, *Oscillatoria* sp., *Phormidium* sp., *Plectonema nostocorum*, *Schizothrix californica* Dr., *S. lamyi* Gom., and *Scytonema hofmannii*. Common green algae include *Astasia* sp., *Chlorella* sp., *Chlorococcum* sp., *Fritschiella* sp., *Gonium* sp., *Navicula* sp., *Palmogloea protuberans*, and *Porphyrosiphon fuscus* (R.E. Cameron, unpubl.; Shields et al. 1957).

The dominance of calcareous soils results in high cover and species diversity of lichens. The lichen flora is dominated by *Collema coccophorum*, *C. tenax*, *Catapyrenium squamulosum*, *C. lacinulatum* (Ach.) Breuss, *Heppia lutosa*, *Peltula obscurans* (Nyl.) Gyelnik, and *P. richardsii*. Other common species include *Lecidea rubiformis* (Wahl.) Hook., *Psora decipiens*, *Fulgensia desertorum*, *F. bracteata*, *Heppia lutosa*, and *Toninia sedifolia*. Common mosses include species of *Bryum*, *Crossidium aberrans*, *Tortula atrovirens*, *Pterygoneurum ovatum*, *Syntrichia ruralis*, species of *Weissia*, and *Pseudocrossidium crinitum* (Schultz) Zand. Common liverworts include *Plagiochasma rupestre* (Forst.) Steph. and *Reboulia hemisphaerica* (L.) Raddi (Nash et al. 1974; L.L. Stark unpubl.).

2.3.2 Cool Deserts

The Colorado Plateau, Great Basin, and Columbia Plateau are considered cool deserts, and are located above 37°N latitude. Summer daytime temperatures seldom reach 30 °C. Low winter temperatures result in frequent soil freezing. Elevations are principally over 1300 m, and winter snow is common. Crusts in these deserts have an either pinnacled or rolling morphology.

2.3.2.1 Colorado Plateau

The Colorado Plateau is a uniformly uplifted area of predominantly sedimentary rock with deep, water-cut canyons. While most US deserts are mostly depositional landscapes dominated by basins with deep soils, the Colorado Plateau is a highly erosional environment with shallow soils. Elevations generally exceed 1500 m, with mountains reaching 4300 m. Summers average 16–26 °C and winters average 0–8 °C. Soils freeze annually. Precipitation averages 150–250 mm annually, with about 35–40 % falling as summer monsoons (Ashcroft et al. 1992). The most frequently encountered plant communities are dominated by the genera *Coleogyne*, *Atriplex*, *Artemisia*, *Pinus* and *Juniperus*. Soils in the north are mostly fine-textured soils derived from shales; the middle portion is dominated by coarse-textured, sandy soils derived from sandstones; and the southern part is dominated by coarse-textured volcanic-derived soils (Nash and Moser 1982; MacMahon 1987).

Despite this wide range of soils, the composition and morphology of crust communities is very similar in these different regions. In relatively undisturbed areas, crusts are heavily pinnacled, with a microtopography that can reach 15 cm. Soil crusts are heavily dominated by *Microcoleus* spp. (often >90 % of the biomass), with the most common species being *M. vaginatus*. In well-developed crusts, *Scytonema myochrous* and *Nostoc commune* are common as well; however, they seldom make up more than 20 % of the cyanobacterial biomass. Less common cyanobacteria include *Anabaena variabilis*, *Calothrix parietina*, *Chroococcus turgidus* (Kütz.) Naegeli, *Gloeothece linearis* G.M.Smith, *Lyngbya limnetica* Lemm., *N. paludosum* Kütz., *Oscillatoria geminata* Menegh., *O. hamelii* Fremy, *Phormidium minnesotense*

(Tild.) Drouet, *P. tenue*, *Plectonema radiosum* (Schiederm.) Gom., *Schizothrix calcicola*, and *Tolypothrix tenuis*.

Green algal diversity is fairly high (>40 species; Johansen et al. 1984), but biomass in field-collected soils is extremely low (<1 %). Common species in the Chlorophyta include *Chlorella protothecoides* Krug., *Klebsormidium flaccidum* (Kütz.) Silva, Mattox et Blackwell, *Friedmannia israeliensis* Chant. et Bold, *Lobococcus irregularis* (Boye-Pet) Reis, *Oedogonium* sp. and *Ulothrix tenerrima* Kütz. Common species in the Bacillariophyta include *Hantzschia amphioxys*, *Navicula mutica* Kütz., *N. paramutica* Bock, and *Pinnularia borealis* (Grondin and Johansen 1993; Johansen et al. 1981, 1982, 1984).

The lichen flora of sandy and silty soils on the Colorado Plateau soil crusts is dominated by *Collema tenax* and *C. coccophorum*. *Catapyrenium squamulosum* is almost always associated with *Collema*, but with less cover. Other common species include *Psora decipiens*, *Toninia sedifolia*, *Fulgensia bracteata*, and *F. desertorum*. Less common species include *Buellia elegans* Poelt, *Squamarina lentigera* G.H.Weber (Photo 12), *Diploschistes muscorum* (Ehr.) Zahlbr., *Psora cerebriformis* W.Weber, *Heppia lutosa*, *Caloplaca tominii* Sav., and *Endocarpon pusillum* Hedwig. More rarely seen species include *Acarospora schleicheri* (Ach.) Massal, *Cladonia pyxidata* (L.) Hoffm., *Lepraria membranacea* (Dicks.) Vain., *Psora globifera* Ach., *P. tuckermanii* R.Anderson ex Timdal, *Physconia muscigena* (Ach.) Poelt, and *Toninia tristis* (Th. Fr.) Th. Fr. (Nash 1974; Nebeker and St. Clair 1984; Rajvanshi et al. 1998). The lichen flora of gypsiferous soils in this region is discussed in Section 2.4. The most common mosses encountered are *Syntrichia caninervi* and *S. ruralis*. Other frequent mosses include *Bryum argenteum*, *B. caespiticium*, *B. lisae* De Not., *Ceratodon purpureus* (Hedw.) Brid., *Crossidium aberrans*, *Didymodon rigidulus* Hedw., *D. vinealis*, *Funaria hygrometrica* Hedw., *Pterygoneurum ovatum*, *P. subsessile* (Brid.) Jur., *Tortula brevipes* (Lesq.) Broth., and *T. mucronifolia* Schwaegr. Liverworts are quite uncommon (Flowers 1973; Anderson and Rushforth 1976; Johansen et al. 1984; L.L. Stark, unpubl.).

2.3.2.2 Great Basin

The Great Basin is the most northerly of the US deserts, and is characterized by internally draining, broad plains covered with deep soils. Soils are mostly derived from sandstones and limestones in the south, while northern soils are mostly of volcanic origin. Annual precipitation is 130–490 mm and elevations generally exceed 1300 m, with mountains reaching 4000 m. Summer highs average 27 °C, and winters average 2 °C. PET is low relative to other US deserts. The southern parts of the Great Basin are dominated by *Atriplex* sp. and perennial grasses. Cacti and other succulent vegetation are notably absent. The northern portions are dominated by *Artemisia* sp.

Unlike other US deserts, soil crusts are heavily dominated by lichens and mosses, often covering 80% or more of the soil surfaces (Nash and Moser 1982; MacMahon 1987). Extensive coverage of mosses and lichens in this region, and large amounts of vascular plant roots on the surface, result in a tightly bound soil surface and soil crusts generally have a rolling morphology (see Chap. 1).

The cyanobacterial flora of most Great Basin soils is dominated by *Microcoleus vaginatus*. Other common species include *Anabaena variabilis*, *A. doliolum* Bharadw., *Chroococcus minimus* (Keissl.) Lemm, *Gloeothece palea* (Khtz.) Rabenh., *Lyngbya limnetica*, *L. scotti* Fritsch, *L. subtilis* West, *Nostoc commune*, *N. paludosum*, *N. pruniforme* (L.) Ag., *Oscillatoria agardhii* Gom., *Phormidium fragile* (Menegh.) Gom., *P. tenue*, *Scytonema schmidtii* Gomont, *Tolypothrix bouteillei* (Breb. et Desm.) Forti, and *T. tenuis*.

Green algal diversity is higher in the Great Basin (with over 72 species) than in the more southern desert regions and, in a few acidic soils, green algae are more numerous than cyanobacteria (Johansen et al. 1993). Common species in the Chlorophyta include *Bracteacoccus* sp., *Chlorella protothecoides* Krueg., *Klebsormidium flaccidum*, *Chlorococcum* sp., *C. arenosum* (Archib. et Bold), *Macrochloris cohaerens* (Vinatzer) Ettl. et Gaertner, *Friedmannia israelensis*, *Leptosira* sp., *Myrmecia bisecta* Reisigl, *Deasonia ovatum* (Deason et Bold) Ettl. et Gaertner, *Palmellopsis texensis* (Grover et Bold) Ettl. et Gaertner, *Pleurastrum* sp., *Oedogonium* sp., *Protosiphon* sp., and *Ulothrix tenerrima* Kütz. Common species in the Xanthophyceae include *Ellipsoidion oocystoides* Pascher, *Xanthonema debilis* (Vischer) Silva, and *Pleurochloris commutata* Pascher. Common species in the Bacillariophyta include *Achnanthes* sp., *Hantzschia amphioxys*, *Navicula mutica*, and *Pinnularia borealis*.

The lichen flora in the southern Great Basin is very similar to that of the Colorado Plateau (see Sect. 2.3.2.1). Going northward in the Great Basin, most soil lichens found on the Colorado Plateau are still present, although some are less common, and additional species appear (Fig. 2.2). Some of these additional species include *Aspicilia desertorum* (Kremp.) Meresh., *Candelariella terrigena* Räsänen, *Placynthiella uliginosa* (Schrader) Coppins et P.James, *Lepraria neglecta* (Nyl.) Erichsen, *Leptochidium albociliatum* (Desmaz.) M.Choisy, *Leptogium lichenoides* (L.) Zahlbr., *Massalongia carnosa* (Dickson) Körber, *Ochrolechia inaequatula* (Nyl.) Zahlbr., *Physconia detersa* (Nyl.) Poelt, *Psora crenata* (Taylor) Reinke, *P. lurida* (With.) DC., *P. luridella*, *Psorotichia nigra* Magnusson, and *Peltigera rufescens* (Weiss) Humb. (Photo 24; Rosentreter 1986). Some lichens are indicators of late successional *Artemisia*-steppe. These include *Acarospora schleicheri*, *Massalongia carnosa*, *Pannaria cyanolepra* (type b, B. McCune, unpubl.), *Trapeliopsis wallrothii* (Flörke ex Sprengel) Hertel et G.Schneider, *Trapeliopsis* sp. nov. (B. McCune, unpubl.), and *Texosporium sancti-jacobi* (Tuck.) Naxv.

Mosses commonly found in the Great Basin include *Bryum argenteum*, *B. caespiticium*, *B. lisae*, *Ceratodon purpureus*, *Funaria hygrometrica*, *Pterygoneurum ovatum*, and *Syntrichia ruralis*. Common liverworts include *Athalamia hyalina*, *Riccia beyrichiana* Hampe ex Lehm., and *R. sorocarpa* Bisch.

2.3.3 California Chaparral and Channel Islands

In California, biological soil crusts are a major component of valley and foothill woodlands (including the understory grasslands), coastal chaparral, and the maritime Channel Islands. The valley and foothill woodlands are dominated by various oaks (*Quercus* sp.) with an understory perennial/annual grassland. Elevations are from sea level to 1300 m, and fog and dew are frequent. Rainfall averages 500 mm annually. The mean average temperature is 16 °C, with a mean amplitude of only 10–19 °C. Chaparral forms a nearly continuous cover of closely spaced shrubs. This community occurs from sea level to 2000 m on rocky, nutrient-poor soils. Chaparral occurs in Mediterranean climate conditions: cool, wet winters and hot, dry summers. The winter-dominant rainfall is 200–1000 mm annually, and fog and dew are common. Mean winter temperatures range from 0–10 °C, and summer temperatures often exceed 40 °C at inland sites (Barbour and Billings 1988).

Common lichens in these two habitat types include *Catapyrenium squamulosum*, *Cladonia fimbriata* (L.) Fr., *C. furcata* (Huds.) Schrad., *C. macilenta* Hoffm., *C. ochrochlora* Flk., *C. verticillata* (Hoffm.) Schaer., *Collema coccophorum*, *Diploschistes scruposus*, *Lepraria membranacea*, *Leptochidium albociliatum* (Dezmaz.) M. Choisy, *L. lichenoides*, *L. palmatum*, *Peltigera canina* (L.) Willd., *P. collina* (Ach.) Schrad., *Peltula patellata*, *Psora decipiens*, and *Xanthoparmelia taractica* (Kremph.) Hale (Hale and Cole 1988). Common mosses include *Grimmia maritima* Turn., *Ceratodon purpureus* and *Encalypta vulgaris*.

The Channel Islands, off the coast of Santa Barbara, support extensive soil crust cover. On San Nicolas Island, vascular vegetation is dominated by *Artemisia californica*, *Isocoma veneta*, *Lotus argophyllus*, and *Lupinus albifrons*. The climate is characterized by cool summers and mild, moist winters. Mean annual temperature is 15 °C. Temperatures above 32 °C are rare, and freezing temperatures have never been recorded. Average precipitation is 175 mm, with most falling during winter months. Fog occurs throughout the year, relative humidity averages 70 %, and days are generally windy.

Forty species from 15 genera of soil green algae and cyanobacteria have been recorded from San Nicolas Island. Soils are dominated by the cyanobacteria *Microcoleus vaginatus* and *Schizothrix calcicola*. Other common species include *Nostoc* spp. and *Tolypothrix* spp. (R. Fletchner et al., unpubl.). The soil lichen flora of these islands is very diverse and cover is high. Prevalent species include *Collema tenax*, *Catapyrenium squamulosum*, *Acarospora schleicheri* (Ach.) Massal, *Endocarpon lepidallum* Nyl., *Diploschistes diacapsis*, *D. scruposus*, *Fulgensia desertorum*, *Leprocaulon microscopicum* (Vill.) Gams ex D.Hawksw., *Siphula* sp., and *Toninia sedifolia*. Although moss cover is high, only *Barbula vinealis* Brid. has been identified from these crusts (J. Belnap, unpubl.).

2.3.4 Great Plains Grasslands

The Great Plain grasslands are characterized by rolling plains ranging from 750 m (in the central US) to 1500 m (at the base of the Rocky Mountains) elevation. This area is underlain by Precambrian rocks covered by deep soils derived from sandstones, limestones, and shales. This region is characterized by cold, dry winters and hot, humid summers, with an average annual temperature of 8 °C. Rainfall is concentrated in the summer. Short-grass steppe is the most westerly of these grasslands, and receives about 250 mm rainfall annually. Grasses are short and cover is sparse. Mixed grass prairie rainfall averages 350–580 mm annually, and supports a mix of short and tall grass species. The tall-grass prairie has an annual rainfall of 500–900 mm, and is dominated by closely spaced, tall grasses. Soil crusts are most abundant in the short-grass steppe, where plants are short, spaced far apart, and the accumulation of plant litter occurs mostly at the base of the grasses, and crusts are least abundant in the tall grass community. This region evolved with periodic fire and grazing of large bison herds, unlike other arid and semiarid regions of the US (Mack and Thompson 1982).

Little work has been done on Great Plains soil crusts. Common soil green algae and cyanobacteria include surface-dwelling colonies of *Nostoc commune*, as well as the ubiquitous genera *Microcoleus*, *Oscillatoria*, *Phormidium*, *Scytonema*, *Chlorococcum*, and *Ulothrix* (Schulten 1985). The dominant lichens are vagrants, which roll about within the grasslands. Their mobile nature is well adapted to occasional physical disturbance, as they reproduce by vegetative fragmentation (Rosentreter 1993). The most common vagrant lichens in the Great Plains are *Xanthoparmelia camtschadalis* (Ach.) Hale, *X. chlorochroa* (Tuck.) Hale, and *X. vagans* (Nyl.) Hale. When the grass and other vascular plants are sparse, the lichens *Aspicilia hispida* Meresch., *Cladonia cariosa* (Ach.) Spreng., *Diploschistes scruposus* Schreb., *Collema tenax*, *Endocarpon pusillum*, and *Physconia muscigena* (Ach.) Poelt can occur. When soils are calcareous, one also finds species common to many other arid regions, including *Fulgensia bracteata*, *Heppia lutosa*, *Psora decipiens*, *Calo-*

placa tominii, and *Squamarina lentigera* (Freeman 1998; R. Rosentreter, unpubl.).

Bryophytes are abundant. Common species include *Astomum muehlenbergianum* (Sw.) Grout, *Phascum cuspidatum* Hedw., *Weissia controversa* Hedw., *Ephemerum spinulosum* Bruch et Schimp., *Funaria hygrometrica*, *Homalothecium nevadense* (Lesq.) Ren. et Card., *Brachythecium albicans* (Hedw.) B.S.C., and *Syntrichia ruralis*. In disturbed areas, *Ceratodon purpureus* and *Bryum argenteum* are common (Freeman 1998).

2.4 Unique Crustal Communities in North America

Special geologic features or soils can create conditions that promote the growth of biological soil crusts where they would not otherwise occur. Some of these sites support biological soil crusts containing uncommon associations or rare species. These unique crustal communities are not common across the landscape but are at times locally abundant.

Gypsum. Gypsiferous soils are fine-textured, sometimes acidic, and with high concentrations of sulfate and calcium. Vascular plant cover at these sites is often sparse and species diversity limited. Gypsiferous soils generally support high lichen diversity compared to directly adjacent soils (Anderson and Rushforth 1976; St. Clair et al. 1993) including *Diploschistes diacapsis* (Ach.) Lumbsch, *Psora decipiens*, *Collema tenax*, *Catapyrenium squamulosum*, *Buellia elegans* Poelt, and *Squamarina lentigera*. Several lichens are restricted to gypsiferous soils, including *Acarospora nodulosa* (Dufour) Hue var. *nodulosa* (St. Clair and Warrick 1987), *Catapyrenium daedaleum* (Kremp.) Stein, *Gypsoplaca macrophylla* (Zahlbr.) Timdal (Timdal 1990), and *Lecanora gypsicola* St. Clair et Newberry (Rajvanshi et al. 1998).

Sand Barrens. Throughout the US, pockets of sandy (>90 % sand) soils support a combination of soil crust-forming species common to deserts and a lichen-moss mat common to more northerly latitudes. These patches are not a successional sequence, but are constant through time. In central Ohio, algae and cyanobacteria in acidic soils were found to contain *Oscillatoria subrevis* Scmidle, *Cyanothece aeruginosa* (Nageli) Komaek, *Lyngbya* spp., *Microcoleus vaginatus*, *Fottea pyrenoidosa* Broady, *F. stichococcoides* Hindak, and *Desmococcus olivaceus* (Pers. et Ach.) Laundon. Lichens are fruticose [e.g., *Cladonia rangiferina* (L.) Nyl., *C. cariosa* (Ach.) Sprengel, and *C. pyxidata*]. Mosses include *Ceratodon purpureus*, *Polytrichum commune* Hedw., *P. juniperinum* Hedw., *P. piliferum* Hedw., and *Steerecleus serrulatus* (Hedw.) Robins (D. Neher et al., unpubl. data). On sandy barrens in New Jersey and

Michigan, soil crusts are dominated by cyanobacteria (including *Nostoc* sp. and *Microcoleus vaginatus*) and *Collema* sp. In New Jersey, the adjoining lichen-moss mats are dominated by the lichens *Cladonia coccifera* (L.) Willd., *C. pyxidata*, *C. uncialis* (L.) Hoffm., *C. deformis* (L.) Hoffm., and *C. stellaris* (L.) Rabenh. and the mosses *Polytrichum juniperinum* Hedw., *P. piliferum* Hedw, and *Pleurozium schreberi* (Brid.) Mitt.

Glades, Lithic and Shallow Soil Sites. Forest glades are widespread across North America but are limited both locally and in overall area. In some glade areas, the lichens *Baeomyces rufus* (Huds.) Rebent. or *B. fungoides* (Swartz.) Ach. form extensive crust colonies (R. Rosentreter, pers. obs.). In western and more northern boreal forest openings, *Cladonia* spp., *Peltigera* spp., and occasionally *Multiclavula* spp. are the dominant soil crust species. These genera are also common on disturbed soil along roadways and cutbanks throughout the nonarid portions of the continent. A recent survey of northeastern Ohio showed 41 taxa, of which *Coccomyxa subglobosa* Pascher, *Navicula minima* Grunow, *Pinnularia microstauron* (Ehr.) Cleve, *P. viridis* (Nitzsch) Ehr., *Pleurochloris commutata* Pascher, *Stauroneis anceps* Ehr., and *Stichococcus dubius* Chodat were the most common (Grondin and Johansen 1995). Only two cyanobacterial species were observed: *Phormidium minnesotense* (Tilden) Drouet and *Myxosarcina apectabilis* Geitler. No common soil diatoms were observed.

Lithic and shallow soil sites are often colonized by biological crusts, since these sites are extremely droughty, which limits vascular plant growth. Biological crusts may dominate these sites even in moist climates. Some lithic sites in wet climates are leached of soil nutrients and often dominated by gelatinous lichens such as *Leptogium lichenoides* or nitrogen-fixing cyanobacteria such as *Nostoc* sp. (R. Rosentreter, pers. obs.). In eastern forests, shallow disturbed or compacted acidic soils (e.g., along roadsides) are often dominated by green algal species.

Thermal Springs. Parts of western North America contain extensive active or dormant thermal (hot) springs. *Nostoc* spp., *Chlorogloeopsis* spp., and *Tolypothrix* spp. occur in areas frequently sprayed by hot water. Soils adjacent to these springs are often clear of vegetation, due to episodic flooding by hot water, and generally clear of snow in winter. These soils can often be relatively infertile, as hot water leaches many minerals, and can contain high levels of toxic heavy metals. Here, soil crusts are dominated by *Microcoleus vaginatus*, *Nostoc commune* and *Scytonema myochrous*, resembling those of the Colorado Plateau, despite being located in much more northerly latitudes (e.g., Montana). Common lichen species include *Diploschistes diacapsis*, *Psora decipiens*, *Collema tenax*, *Catapyrenium squamulosum*, *Buellia elegans*, and *Squamarina lentigera* (F. Garcia-Pichel, pers. comm.).

Playas. Playas form where ephemeral water pools on poorly drained soils. Playas are often barren of vascular vegetation, and support *Microcoleus* spp., *Nostoc* spp., and either attached or vagrant forms of *Dermatocarpon miniatum* (L.) Mann. *Nostoc commune* found in these sites includes the common filamentous form, the colonial-eggs form, and occasionally the long hair-like unattached form (*N. flagelliforme*). Small playettes covered with fine silts can also be colonized by crustose lichens more typically occurring on rocks, such as *Lecanora muralis* (Schreber) Rabenh. (Rosentreter 1986; Rosentreter and McCune 1992).

Alpine Sod. Biological soil crusts are a common community component of many alpine and subalpine sites in North America. Alpine and subalpine sites contain some species sympatric with the arid regions of the continent. In contrast, a few lichens are restricted to alpine crusts and include *Lecidoma demissum* (Rutstr.) G.Schneider et Hertel, *Solorina* spp., *Lepraria neglecta*, *Ochrolechia frigida* (Swartz.) Lynge, and *Thamnolia subuliformis* (Ehrh.) Culb. (DeBolt and McCune 1993). The nitrogen-fixing *Stereocaulon* sp. is also present (Lamb and Ward 1974). In the subalpine zone, biological crusts are often associated with snowdrifts. Characteristic lichens include *Solorina crocea* (L.) Ach. and *Lepraria neglecta* (McCune 1998). Cyanobacterial crusts are also common in these areas and are generally dominated by species of *Nostoc* (Reisigl 1964). For further discussion of alpine areas, see Chapter 5.

Arctic Tundra. Alpine and tundra soils are often covered by crust communities in the early stages of succession, and later dominated by fruticose lichens (e.g., *Cladina mitis, Cetraria cucullata, and C. islandica*) and grasses. Areas with poor drainage can be totally covered by surface mats and soil colonies of *Nostoc commune* (Liengen 1999), similar to other portions of the Northern Hemisphere including Greenland (see Chap. 4). Soil crusts are a permanent feature in many communities, including dry steppe-like vegetation, open *Dryas* heaths, and fell-fields. Here, cyanobacterial crusts are common, and often dominated by *Nostoc* species (Hansen 1997).

2.5 Conclusions

Biological soil crusts in North America are characteristic of many of the arid and semiarid vegetation types in western North America. While the heavily pinnacled crusts of the Colorado Plateau are the most visible among ecoregions, soil crusts are a major component of most western North American habitats. Crusts from the different ecoregions of North America are distinct. In the hotter, drier regions where soils do not freeze, rugose crusts are dominated by cyanobacteria. As precipitation increases and temperatures decrease, freezing soils produce pinnacled and rolling crusts dominated by lichens and mosses. Many of the species found in soil crusts are the same or closely related to species occurring in other arid regions worldwide.

Acknowledgments. Many thanks to Larry St. Clair for his contribution on gypsiferous soils and to Burkhard Büdel, Otto L. Lange, Sue Phillips, Michelle Schmid, and Julie Kaltenecker for helpful comments.

References

Anderson DC, Rushforth SR (1976) The cryptogam flora of desert soil crusts in southern Utah, U.S.A. Nova Hedwigia 28:691–729

Ashcroft GL, Jensen DT, Brown JL (1992) Utah climate. Utah State University, Logan, UT

Bailey RG (1994) Description of the ecoregions of the US. Misc Public 1391. USDA Forest Service, Washington DC

Barbour MG, Billings WD (1988) North American terrestrial vegetation. Cambridge University Press, New York

Cameron RE (1960) Communities of soil algae occurring in the Sonoran desert in Arizona. J Ariz Acad Sci 1:85–88

Cameron RE (1962) Species of *Nostoc vaucher* occurring in the Sonoran Desert in Arizona. Trans Am Microsc Soc 81:379–384

Cameron RE (1972) Comparison of soil microbial ecosystems in hot, cold, and polar desert regions. In: Rodin LE (ed) The eco-physiological foundation of ecosystems productivity in arid zones. Nauk, Leningrad, USSR, pp 185–192

Cameron RE , Blank GB (1967) Soil studies: microflora of desert regions, VIII. Distribution and abundance of desert microflora. JPL Space Programs Summary No 37–44. Vol 4:193–201. Pasadena, CA

DeBolt A, McCune B (1993) Lichens of Glacier National Park, Montana. Bryologist 96:192–204

Durrell LW (1962) Algae of Death Valley. Trans Am Microsc Soc 81:267–273

Flechtner VR, Johansen JR, Clark WH (1998) Algal composition of microbiotic crusts from the central desert of Baja California, Mexico. Great Basin Nat 58:295–311

Flowers S (1973) Mosses – Utah and the West. Brigham Young University Press, Provo, UT

Freeman CC (1998) The flora of Konza Prairie: a historical review and contemporary patterns. In: Knapp AK, Briggs JM, Hartnett DC, Collins SL (eds) Grassland dynamics: long-term ecological research in a tallgrass prairie. Oxford University Press, Cambridge, pp 69–80

Grondin AE, Johansen JR (1993) Microbial spatial heterogeneity in microbiotic crusts in Colorado National Monument. I. Algae. Great Basin Nat 53:24–30

Grondin AE, Johansen JR (1995) Seasonal succession in a soil algal community associated with a beech-maple forest in northeastern Ohio, USA. Nova Hedwigia 60:1–12

Hale ME Jr., Cole M (1988) Lichens of California. University of California Press, Berkeley

Hansen ES (1997) Studies of the lichen flora of coastal areas in central West Greenland. Nova Hedwigia 64:505–523

Jeffries DL, Klopatek JM (1987) Effects of grazing on the vegetation of the blackbrush association. J Range Manage 40:390

Johansen JR, Rushforth SR, Brotherson JD (1981) Subaerial algae of Navajo National Monument, Arizona. Great Basin Nat 41:433–439

Johansen JR, Javakul A, Rushforth SR (1982) Effects of burning on the algal communities of a high desert soil near Wallsburg, Utah. J Range Manage 35:598–600

Johansen JR, St Clair LL, Webb BL, Nebeker GT (1984) Recovery patterns of cryptogamic soil crusts in desert rangelands following fire disturbance. Bryologist 87:238–243

Johansen JR, Ashley J, Rayburn WR (1993) Effects of rangefire on soil algal crusts in semiarid shrub-steppe of the lower Columbia Basin and their subsequent recovery. Great Basin Nat 53:73–88

Kaltenecker JH (1997) The recovery of microbiotic crusts following post-fire rehabilitation on rangelands of the western Snake River Plain. Unpubl. MS Thesis, Boise State University, ID

Kaltenecker JH, Wicklow-Howard M, Rosentreter R (1997) Microbiotic crusts of the Lemhi Resource Area and their effect on soil-water relationships in sites recovering from long-term grazing. Unpubl. Report, USDI-BLM, Salmon, ID

Kleiner EF, Harper KT (1977) Soil properties in relation to cryptogamic groundcover in Canyonlands National Park. J Range Manage 30:202–205

Lamb IM, Ward A (1974) A preliminary conspectus of the species attributed to the imperfect lichen genus *Leprocaulon* Nyl. J Hattori Bot Lab 38:499–553

Lange OL (2000) Photosynthetic performance of a gelatinous lichen under temperate habitat conditions: long-term monitoring of CO_2 exchange of *Collema cristatum*. Bibl Lichenol 75:307–332

Lange OL, Belnap J, Reichenberger H, Meyer A (1997) Photosynthesis of green algal soil crust lichens from arid lands in southern Utah, USA: role of water content on light and temperature responses of CO_2 exchange. Flora 192:1–15

Lange OL, Belnap J, Reichenberger H (1998) Photosynthesis of the cyanobacterial soil-crust lichen *Collema tenax* from arid lands in southern Utah, USA: role of water content on light and temperature responses of CO_2 exchange. Funct Ecol 12:195–202

Liengen T (1999) Conversion factor between acetylene reduction and nitrogen fixation in free-living cyanobacteria from high arctic habitats. Can J Microbiol 45:223–229

Mack RN, Thompson JN (1982) Evolution in steppe with few large, hooved mammals. Am Nat 119:757–773

McCune B, Rosentreter R (1995) Field key to soil lichens of central and eastern Oregon. Unpubl Rep. USDA Bureau of Land Management, Boise, ID

MacMahon JA (1987) Disturbed lands and ecological theory. In: Jordan W (ed) Restoration ecology: a synthetic approach to ecological research. Cambridge University Press, New York

Mayland HF, McIntosh TH, Fuller WH (1966) Fixation of isotopic nitrogen in a semi-arid soil by algal crust organisms. Soil Sci Soc Am Proc 30:56–60

McCune B (1998) Lichens of granitic peaks in the Bitterroot Range, Montana and Idaho, USA. In: Glenn MG, Harris RC, Dirig R, Dole MS (eds) Lichenographia Thomsoniana: North American lichenology in honor of John W. Thomson. Mycotaxon, Ithaca, NY, pp 281–294

McIlvanie SK (1942) Grass seedling establishment and productivity-overgrazed vs. protected range soils. Ecology 23:228–231

Nash III, TH (1974) Lichens of the Page environs as potential indicators of air pollution. J Ariz Acad Sci 9:97–101

Nash III TH, Moser TJ (1982) Vegetational and physiological patterns of lichens in North American deserts. J Hattori Bot Lab 53:331–336

Nash III TH, White S, Nash JM (1974) Composition and biomass contribution of lichen and moss communities in the hot desert ecosystems. US/IBP Desert Biome Res Memo 74–19:159–165

Nebeker GT, St Clair LL (1984) The lichen flora of Navajo National Monument, Arizona. Mycotaxon 19:413–422

Ponzettii J, Youtie B, Slazer D, Kimes T (1998) The effects of fire and herbicides on microbiotic crust dynamics in high desert ecosystems. Unpubl. Report, USGS-BRD, Forest and Rangeland Ecosystem Science Center, Oregon State University, Corvallis, OR

Rajvanshi V, St Clair LL, Webb BL, Newberry CC (1998) The terricolous lichen flora of the San Rafael Swell, Emery County, UT, USA. In: Glenn MG, Harris RC, Dirig R, Dole MS (eds) Lichenographia Thomsoniana: North American lichenology in honor of John W. Thomson. Mycotaxon, Ithaca, NY, pp 399–406

Reisigl H (1964) Zur Systematik und Ökologie alpiner Bodenalgen. Österr Bot Z 111:402–499

Rosentreter R (1986) Compositional patterns within a rabbitbrush (*Chrysothamnus*) community of the Idaho Snake River plain. In: McArthur ED, Welch BL (eds) Symposium on the Biology of *Artemisia* and *Chrysothamnus*. General Techn Rep INT GTR 200. USDA Forest Service, Intermountain Research Station, Ogden, UT, pp 273–277

Rosentreter R (1993) Vagrant lichens in North America. Bryologist 96:333–338

Rosentreter R (1994) Displacement of rare plants by exotic grasses. In: Monsen SD, Kitchen SG (eds) Proceedings: ecology and management of annual rangelands. General Technical Report INT-GTR-313. USDA Forest Service, Intermountain Research Station, Ogden, UT, pp 170–175

Rosentreter R, McCune B (1992) Vagrant *Dermatocarpon* in western North America. Bryologist 95:15–19

Schulten JA (1985) Soil aggregation by cryptogams of a sand prairie. Am J Bot 72:1657–1661

Shields LM, Mitchell C, Drouet F (1957) Alga- and lichen-stabilized surface crusts as soil nitrogen sources. Am J Bot 44:489–498

St Clair LL, Warrick RB (1987) *Ascarospora nodulosa* (Duf.) Hue. var. *nodulosa*: a new record for North America. Bryologist 90:48–49

St Clair LL, Johansen JR, Rushforth SR (1993) Lichens of soil crust communities in the intermountain area of the western United States. Great Basin Nat 53:5–12

Timdal E (1990) *Gypsoplacaceae* and *Gypsoplaca*, a new family and genus of squamiform lichens. Bibl Lichenol 38:419-42

3 Biological Soil Crusts of South America

B. BÜDEL

3.1 Introduction

In contrast to the African continent with the transcontinental Saharan Desert, desert regions in South America are restricted to a strip along the western coast of the subcontinent. In common with the African continent, the South American subcontinent has large savanna and grassland biomes. Thus, climatic and edaphic conditions of large areas seem to be well suited for the development of a rich terrestrial cryptogamic vegetation. However, only a few reports are available dealing with the presence, composition, and ecology of biological soil crusts in the different biomes of South America.

3.2 Humid Savanna, Venezuela

One of the first investigations dealing with the species composition of soil crusts of the large *Trachypogon*-savanna areas (Chap. 1, Photo 36) in the northern region of the Orinoco Llanos was recently published by San José and Bravo (1991). This reflects how long this biological phenomenon has been ignored. These authors sampled algal crusts from several localities of *Trachypogon* savannas and found them dominated by cyanobacteria (species of the genera *Lyngbya, Oscillatoria, Porphyrosiphon, Anabaena, Nodularia, Plectonema, Scytonema,* and *Stigonema*). Green alga of the Zygnemophyceae (*Cylindrocystis, Cosmarium*) were found as well. In a later study, Büdel et al. (1994) found the following species in crust flakes of the *Trachypogon* savanna around Puerto Ayacucho (Photos 36, 37): *Schizothrix telephoroides* Gomont (Photo 7), *Scytonema myochrous* (Dillwyn) Agardh ex Bornet et Flahault em. Jaag, *Stigonema ocellatum* (Dillwyn) Thuret ex Bornet et Flahault, *Tolypothrix robusta* Gardner, and the filamentous green alga *Zygogonium* cf. *ericetorum* Kützing. Lichens have not been found in these biological crusts. While mosses sometimes occur, they have not been identified. *Schizothrix telephoroides* appears to

Ecological Studies, Vol. 150
J. Belnap and O.L. Lange (eds.) Biological Soil Crusts: Structure, Function, and Management

play an important role in the early establishment of soil crusts. This can be observed in older, flat rock pools of inselbergs, where detritus accumulates and soils have begun development. *S. telephoroides* is often the first cyanobacterium to stabilize young soils in the rock pools.

In the nutrient budget of the oligotrophic savanna systems, the role of biological nitrogen fixation has been overlooked for a long time (San José and Bravo 1991; Medina 1993). The cyanobacterial crusts cover 10–12 % of the *Trachypogon* savannas, and are estimated to contribute between 0.4 and 0.7 mg N m^{-2} day^{-1} (Santaella 1985; Medina 1993). However, these rates are not enough to compensate for potential N losses through fire (Medina 1993). Examining the CO_2 exchange of the cyanobacterial crusts, San José and Bravo (1991) found a maximum of 0.207 mg CO_2 fixation m^{-2} s^{-1} in the Guanipa savanna. These crusts seemed resistant to heat stress, which resulted in optimal carbon gains over a broad temperature range (see Chap. 18).

3.3 Caatinga, Brazil

Caatinga is a type of deciduous, xerophilous vegetation with woody, thorny, small-leaved plants that drop their leaves during the dry season. Cacti, bromeliads, and seasonal herbs are also important in this community (Bastos et al. 1998). The Caatinga biome occupies 1 000 000 km^2, mainly in the northeast of Brazil. The climate is semiarid with a long dry period (7–11 months). Its cryptogamic flora has been poorly studied. In their study of caatinga bryoflora, Bastos et al. (1998) mention that it is not easy to find bryophytes in open areas, as most of them take refuge in more moist microhabitats. However, they list a number of hepatics and mosses found on soil and which are probably involved in crust formation. The species mentioned by Bastos et al. (1998) include the hepatic *Riccia vitalii* S.Jovet-Ast and the mosses *Bryum argenteum* Hedw., *Hyophiladelphus agrarius* (Hedw.) Zander, *Hyophyla involuta* (Hook.) Jaeg., and three species of *Fissidens*: *F. bryoides* Hedw., *F. gardneri* Mitt., and *F. radicans* Mont. According to the authors, acrocarpous mosses (erect, short, and turf-forming) are the most numerous in the open caatinga environment. Species of the Ricciaceae were the only hepatics found on soil in their study.

Although the cyanobacterial flora of Brazil is relatively well investigated compared to other regions of South America (Sant'Anna and de Pavia Azevedo 1995), very little is known about their involvement in biological soil crusts. In their monograph on the cyanobacterial Oscillatoriaceae, Sant'Anna and de Pavia Azevedo (1995) mention some cyanobacteria that often are actively involved in soil-crust formation. These include: *Lyngbya majuscula* Harvey ex Gomont, *L. martensiana* Meneghini ex Gomont, *Microcoleus* cf.

amplus Gardner, *M. sociatus* West et West, *M. vaginatus* Gomont, *Oscillatoria acuminata* Gomont, *O. irrigua* Kützing ex Gomont, *O.* cf. *jenensis* G.Schmid, *O. proboscidea* Gomont, *O. splendida* Greville ex Gomont, *O. tenuis* Agardh ex Gomont, *Phormidium crouani* Gomont, *P. retzii* Gomont, and *P. setchellianum* Gomont. *Nostoc* sp. is also mentioned as being often associated with other soil algae. In total, the authors conclude that there is a great similarity to the cyanobacterial flora from India and Africa, with a high percentage of cosmopolitan species.

3.4 Chaco, Argentina

In the Chaco and Formosa provinces in the north of Argentina, DeHalperin et al. (1976) investigated biological soil crusts at 12 localities within an area extending approximately 1200 km from east to west, and 900 km from north to south. This included habitats with subhumid to semiarid climates. All crusts investigated by the authors were restricted to the upper few millimeters of the soil and were dominated by cyanobacteria. Mosses and diatoms regularly occurred in the crusts, whereas unicellular or filamentous green algae were only rarely observed. However, only the cyanobacteria have been studied in detail, and the authors found 48 species belonging to the following 11 genera: "*Anacystis*" (8 species), "*Coccochloris*" (1), *Lyngbya* (5), *Microcoleus* (4, including *M. chtonoplastes* [Fl. Dan] Thuret ex Gom., *M. paludosus* [Kütz.] Gomont, and *M. vaginatus* [Vaucher] Gomont), *Oscillatoria* (11), *Phormidium* (9), *Porphyrosiphon* (2), *Schizothrix* (3, incl. *S. calcicola* [Agardh] Gomont), *Symploca* (1), *Nostoc* (8, incl. *N. commune* (L.) Vaucher ex Bornet et Flahault and *N. muscorum* Agardh ex Bornet et Flahault), and *Scytonema* (2, *S. guyanense* [Montagne] Bornet et Flahault, *S. hofmannii* Agardh ex Bornet et Flahault).

Two different types of cyanobacterial crusts were distinguished by the authors: stratified and nonstratified. Stratified crusts always had either *Nostoc commune* or other *Nostoc* species as the uppermost layer, underlain by filaments of mainly *Microcoleus chtonoplastes*, but also other species of *Microcoleus*, *Porphyrosiphon*, *Symploca*, *Lyngbya*, and *Oscillatoria*. Stratified crusts with an abundance of *Nostoc commune* and/or *Nostoc* sp. had amino acid nitrogen (N) levels 5–8 times that of underlying soils or soils of surrounding areas without crusts. The nonstratified crusts, with either no or low numbers of *Nostoc* spp., had amino acid N levels 1.2–2.8 times greater than underlying soil or surrounding crust-free soils. Nitrite and nitrate N concentrations were 1.3–2 and 1.7–3.5 times higher in stratified crusts, and 1.5 times higher to 2.5 times lower in nonstratified crusts, than those of underlying soils or the surrounding crust-free soils, respectively. In nonstratified crusts, nitrite and nitrate N values were 1.2 times greater to 1.2 times less than in surrounding crust-free soils.

3.5 Coastal Desert, Chile

In the extremely arid coastal strip of northern Chile at Caldera and near Antofagasta, Schwabe (1960) collected soil crusts consisting mainly of four hitherto undescribed cyanobacterial species. An interesting type of an endedaphic soil crust near Antofagasta, with the two species, *Calothrix desertica* Schwabe and *Schizothrix atacamensis* Schwabe, was found (Fig. 3.1). *Calothrix* occurred near the soil surface, while *Schizothrix* was found under *Calothrix.* The second crust at Caldera was formed by the two species, *Plectonema polymorphum* Schwabe and *Schizothrix adunca* Schwabe. Besides the four species mentioned above, no other species could be detected in the two soil-crust formations. All four species secrete sticky polysaccharide sheaths, and thus glue sand grains together, stabilizing the upper soil layers (Schwabe 1960). In addition, Forest and Weston (1966) investigated soils of the Atacama Desert for the presence of cyanobacteria. They found *Schizothrix calcicola* (Agardh) Gomont, and two unicellular species and isolated them. However, as they used Drouet (1959, 1981) for identification, it is not known what species were actually observed.

In an extremely arid landscape like the Atacama Desert, which receives most of its humidity from fog events, one would surely expect crustose, soil crust-forming lichens. However, there is only one short report from Follmann (1965) about such organisms. In an area extending about 500 km from north to south in the Atacama Desert, he discovered soil-crust lichens of the Fensterflechten type (window lichens; see Chaps. 1, 9). These lichens were morphologically very similar to those found in Africa by Vogel (1955). However, while the families were the same (Buelliaceae, Dermatocarpaceae, Leci-

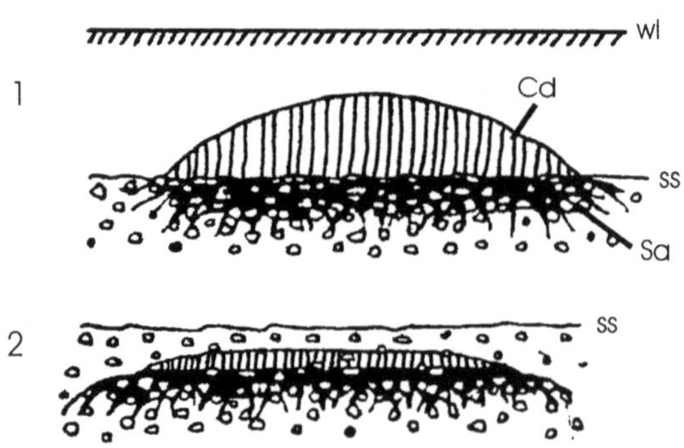

Fig. 3.1, 1, 2. Cyanobacterial soil crust consisting of *Calothrix desertica* (*Cd*) and *Schizothrix atacamensis* (*Sa*); Atacama Desert, near Antofagasta. Cyanobacterial crusts: **1** inundated in water (*wl* water level) and **2** on humid sand (*ss* soil surface). (After Schwabe 1960)

deaceae), the spectrum of these undetermined species was completely different from those from South Africa (Follmann 1965). In addition, species from the Acarosporaceae also occurred in the Atacama Desert. The author found that six lichens of the Fensterflechten type still had 44% thallus water contents at a soil surface temperature of 62 °C, 6 h after moistening by dewfall in the early morning.

References

Bastos CJP, Albertos B, Bôas SBV (1998) Bryophytes from some caatinga areas in the state of Bahia (Brazil). Trop Bryol 14:69-75

Büdel B, Lüttge U, Stelzer R, Huber O, Medina E (1994) Cyanobacteria of rocks and soils of the Orinoco lowlands and the Guyana uplands, Venezuela. Bot Acta 107:422-431

DeHalperin DR, DeMule MCZ, DeCaire GZ (1976) Biodermas algales como fuente de nitrógenos en suelos sub-húmedos y semi-áridos (Provincias de Chaco y Formosa, Argentina). Darwiniana 20:341-370

Drouet FA (1959) Myxophyceae. In: Ward HB, Whipple GC (eds) Fresh water biology, 2nd edn. WT Edmundson, New York, pp 95-114

Drouet FA (1981) Revision of the Stigonemataceae with a summary of the classification of the blue-green algae. Beih Nova Hedwigia:1-221

Follmann G (1965) Fensterflechten in der Atacamawüste. Naturwissenschaften 52: 434–435

Forest HS, Weston CR (1966) Blue-green algae from the Atacama Desert of Northern Chile. J Phycol 2:163–164

Medina E (1993) Mineral nutrition: tropical savannas. Prog Bot 54:237–253

San Jose JJ, Bravo CR (1991) CO_2 exchange in soil algal crusts occurring in the *Trachypogon* savannas of the Orinoco Llanos, Venezuela. Plant Soil 135:233–244

Sant'Anna CL, de Paiva Azevedo MT (1995) Oscillatoriaceae (Cyanophyceae) from São Paulo State, Brazil. Nova Hedwigia 60:19–58

Santaella SE (1985) Fijación biológica de nitrógeno en sabanas de *Trachypogon* spp. Tesis de Grado, Fac. De Ciencias, Esc. De Biología, UCV, Venezuela

Schwabe GH (1960) Zur autotrophen Vegetation in ariden Böden. Blaualgen und Lebensraum IV. Oesterr Bot Z 107:281–309

Vogel S (1955) Niedere "Fensterpflanzen" in der südafrikanischen Wüste. Beitr Biol Pflanz 31:45–135

4 Lichen-Rich Soil Crusts of Arctic Greenland

E.S. Hansen

4.1 Introduction

Literature on individual components of biological soil crusts such as lichens, mosses, algae, and cyanobacteria, is available from all parts of the arctic area. However, there is a lack of thorough descriptions of the taxonomic composition and structure of the arctic soil crusts as a unit. General facts on arctic soil-crust organisms are given by Aleksandrova (1988) and Longton (1988). This chapter addresses specific lichen-rich terrestrial communities of Greenland, based on works by the author from 1969 to 1998.

4.2 Soil-Lichen Communities Associated with Major Vegetation Types

Biological soil crusts are particularly common in the high arctic part of Greenland because vascular plant cover is low. However, they are also of great importance in the dry areas in more southern parts of Greenland. Below are discussed the different soil-crust communities associated with some of the major vegetation types found in Greenland.

4.2.1 Steppe-Like Vegetation Dominated by Sedges and Grasses

Different steppe-like communities are of great importance at Kangerlussuaq (67°02'N, 50°40'W) (Fig. 4.1). The area has a low-arctic, continental climate with 140 mm annual precipitation (Böcher 1954). The *Carex supina–Calamagrostis purpurascens–Kobresia myosuroides* steppe is fairly rich in crustaceous lichens. The neutral pH soils are thin and quick to desiccate after rain, and have a somewhat cracked appearance. *Caloplaca tominii* Savicz and *Physconia muscigena* (Ach.) Poelt are the two dominant lichens. *C. tominii* is often the

Ecological Studies, Vol. 150
J. Belnap and O.L. Lange (eds.) Biological Soil Crusts: Structure, Function, and Management

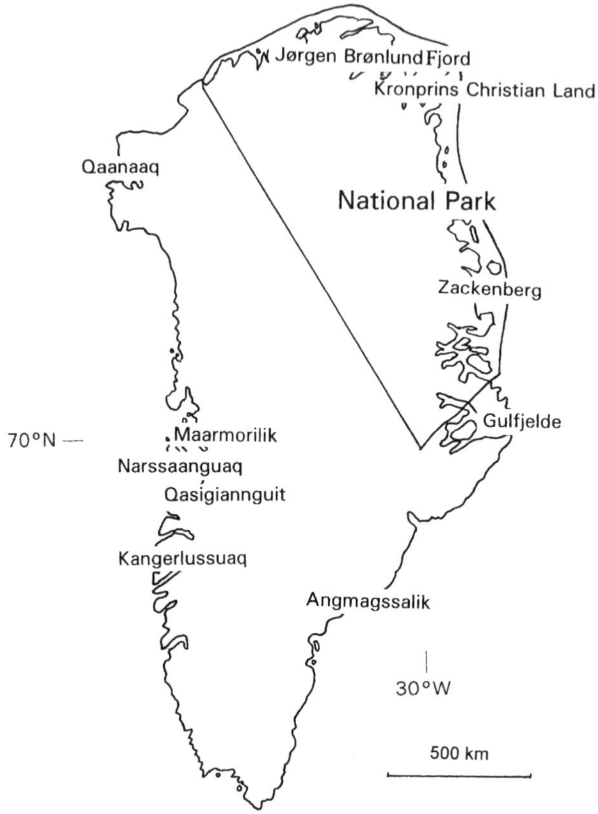

Fig. 4.1. Greenland, showing areas and localities with important soil-crust communities. The delimitation of the National Park of North and East Greenland is shown on the map

only lichen species found on loess soil between the dense *Kobresia myosuroides* stands. Other lichens include *Acarospora rhizobola* (Nyl.) Alstrup, *Fulgensia bracteata* (Hoffm.) Räsänen (Chap. 1, Photo 17), *Phaeorrhiza sareptana* (Tomin) H.Mayrhofer et Poelt v. *sphaerocarpa* (Th.Fr.) H.Mayrhofer et Poelt, *Psora vallesiaca* (Schaer.) Timdal, *Rinodina mniaraea* (Ach.) Körb., *Toninia tristis* (Th.Fr.) Th.Fr., *Arthrorhaphis alpina* (Schaer.) R.Sant., *Endocarpon pusillum* Hedw., *Catapyrenium lachneum* (Ach.) R.Sant. [syn. *Placidium l.* (Ach.) DeLesd.], *C. squamulosum* (Ach.) Breuss [syn. *Placidium s.* (Ach.) Breuss; Photo 21], and *C. michelii* (A.Massal.) R.Sant. (syn. *Placidium m.* A.Massal.) (Hansen 1986; Breuss and Hansen 1988). *Diploschistes muscorum* (Scop.) R.Sant. [infested by *Acarospora schleicheri* (Ach.) A.Massal.], or *Psora rubiformis* (Ach.) Hook. sometimes dominate on thin loess over gneiss. *Xanthoria elegans* (Link) Th.Fr. is a common epilithic lichen in this area. Apart from *Cladonia pocillum* (Ach.) Grognot, *Physconia muscigena,* and *Peltigera lepidophora* (Nyl. ex Vain.) Bitter, macrolichens are of minor importance in this community, and mosses, green algae, and cyanobacteria are rare. Occasionally, *Cladonia luteoalba* Wheldon et A.Wilson grows along tracks on the slope. *Peltigera rufescens* (Weiss) Humb. and *Stereo-*

caulon glareosum (Savicz) H.Magn. bind the sand and form sand crusts below Ravneklippen (67°00'N, 50°40'W).

At Qasigiannguit (68°49'N, 51°12'W) in Disko Bugt, *Carex rupestris, Kobresia myosuroides,* and *Artemisia borealis* form a characteristic mosaic with the following soil-crust lichens: *Peltigera lepidophora, Phaeophyscia constipata* (Norrl. et Nyl.) Moberg, *Physconia muscigena, Collema undulatum* Laurer ex Flot. var. *granulosum* Degel., *Catapyrenium lachneum, C. cinereum* (Pers.) Körb., *Rinodina mniaraea,* and *Diploschistes muscorum.* The community occurs on thin, strongly weathered amphibolite soils (Hansen 1997). Steppe-like vegetation, dominated by *Carex rupestris, Calamagrostis purpurascens,* and *Kobresia myosuroides* in the subcontinental Maarmorilik area (71°07'N, 51°17'W), has soil crusts characterized by the lichens *Catapyrenium lachneum, Fulgensia bracteata, Megaspora verrucosa* (Ach.) Hafellner et V.Wirth, *Toninia sedifolia* (Scop.) Timdal (Photo 16), *Psora vallesiaca* (Schaer.) Timdal, and *Solorina spongiosa* (Ach.) Anzi (Hansen 1991).

4.2.2 Open *Dryas* Heaths

Openings in *Dryas* heaths are sometimes very rich in soil-crust lichens. South of Gulfjelde (70°30'N, 22°05'W) in Central East Greenland, the slightly alkaline soils contain *Peltigera rufescens* (dominant), *P. lepidophora, Solorina bispora* Nyl., *Cladonia pocillum, Psora decipiens* (Hedw.) Hoffm., *P. rubiformis, Catapyrenium lachneum, C. cinereum, Caloplaca jungermanniae* (Vahl) Th.Fr., *C. tetraspora* (Nyl.) H.Olivier, *C. tiroliensis* Zahlbr., *Lecanora epibryon* (Ach.) Ach., *Rinodina roscida* (Sommerf.) Arnold, *R. mniaraea, Dacampia hookeri* (Borrer) A.Massal., *Pertusaria bryontha* (Ach.) Nyl., and *Leproloma vouauxii* (Hue) J.R.Laundon (Hansen 1995b). Abundant information on soil flora in the Scoresby Sund region has been provided by Hartz (1895), but little is known about soil crusts in this area. At Narssaanguaq (69°52'N, 54°22'W), Gelting (1954) studied a *Dryas integrifolia* community on basaltic sand and clay soils and found *Collema tenax* (Sw.) Ach. emend. Degel., *Solorina spongiosa*, and *Leciophysma finmarkicum* Th.Fr.

Similar lichen crusts occur at the shore of the two saline, alkaline lakes, Store Saltsø (66°59'N, 50°37'W) and Brayasø (66°59'N, 51°02'W). At Store Saltsø, *Fulgensia desertorum* grows together with *Physconia muscigena* and *Toninia sedifolia.* The lake is rich in Mg, Na, K, HCO_3, and Cl, with a pH of about 9 (Böcher 1949). The crusts near Brayasø contain *Physconia muscigena, Leptogium saturninum* (Dicks.) Nyl., *Fulgensia bracteata, Toninia sedifolia,* and *Ochrolechia upsaliensis.* Hansen (1978a) and Daniëls (1982) list soil-crust lichens in *Dryas integrifolia* in the Angmagssalik district in Southeast Greenland.

4.2.3 Mixed Dwarf Shrub Heaths

North and northeast Greenland have many lichen-rich soil crusts in mixed dwarf shrub heaths. On stony soil (neutral to slightly alkaline reaction) near the western shore of Romer Sø (80°59'N, 19°29'W) in Kronprins Christian Land are found *Fulgensia bracteata* (dominant), *Cladonia pocillum* (subdominant), *Physconia muscigena*, *Solorina bispora*, *S. saccata* (L.) Ach., *Peltigera lepidophora*, *P. rufescens*, *Baeomyces placophyllus* Ach. (Photo 15), *B. rufus* Rebent., *Arthrorhaphis alpina*, *Acarospora rhizobola*, *Buellia elegans* Poelt, *B. epigaea* (Pers.) Tuck., *B. papillata* (Sommerf.) Tuck., *Catapyrenium cinereum*, *Dacampia hookeri*, *Psora decipiens*, *P. rubiformis*, *P. vallesiaca*, *Diploschistes muscorum*, *Caloplaca tominii*, *Lecanora epibryon*, *L. luteovernalis* Brodo, *Lecidea tesselata*, *Lecidella bullata* Körb., *Leproloma vouauxii*, *Megaspora verrucosa*, *Mycobilimbia lobulata* (Sommerf.) Hafellner, *Ochrolechia upsaliensis* (L.) A.Massal., *Phaeorrhiza nimbosa* (Fr.) H.Mayrhofer et Poelt, *Protoblastenia terricola* (Anzi) Lynge (see Photo 40), *Rinodina mniaraea*, *R. roscida*, *Toninia arctica* Timdal, *T. sedifolia*, and *T. tristis*. Scattered shrubs of *Dryas octopetala*, *Saxifraga oppositifolia*, *Cassiope tetragona*, and *Salix arctica* occur among the lichens. *Dactylina ramulosa* (Hook.) Tuck. is the dominant lichen on some small hills in the area. Based on the number of epigaeic lichens, the area is among the richest floristically in Greenland, despite its very northern location and the fragmentary appearance of the soil crusts.

Some soil crusts occur near the head of Danmark Fjord (80°34'N, 23°57'W). They are characterized by frequent *Buellia epigaea* growing in *Dryas-Saxifraga oppositifolia-Salix arctica* heaths on relatively even ground, together with *Cladonia pocillum*, *Toninia sedifolia*, *Fulgensia bracteata*, *Psora vallesiaca*, and *Leproloma vouauxii*. The heath patches and fell-fields are exposed to strong winds and have fewer lichens than more protected areas near Romer Sø. Small crusts composed of *Collema substellatum* H.Magn. and *Fulgensia desertorum* were found on thin layers of loess on dolomite rock ledges.

Hummocky dwarf shrub heaths dominated by *Dryas* and *Cassiope tetragona* occur in many places near Græselvdal (80°03'N, 23°11'W). In protected places, lichens such as *Cladonia pocillum*, *Solorina bispora*, *Fulgensia bracteata*, and *Leproloma vouauxii* form crusts on the sides of the hummocks that are up to 30 cm high. More exposed heath patches support a fairly rich lichen vegetation consisting of *Teloschistes contortuplicatus* (Ach.) Clauzade et Rondon, *Physconia muscigena*, *Solorina bispora*, *Collema substellatum*, *Cladonia pocillum*, *Acarospora rhizobola*, *Fulgensia bracteata*, *F. desertorum*, *Caloplaca tominii*, *Diploschistes muscorum*, *Lecidea tesselata* Flörke, *Psora decipiens*, *P. rubiformis*, *P. vallesiaca*, *Protoblastenia terricola*, *Phaeorrhiza nimbosa*, *Rinodina mniaraea*, *Leproloma vouauxii*, *Gypsoplaca macrophylla* (Zahlbr.) Timdal, *Toninia arctica*, and *T. sedifolia*.

Soil crusts with a similar composition, but with fewer species due to aridity, were studied in the valley west of Centrum Sø (80°11'N, 22°49'W). Here, *Collema coccophorum* Tuck. is associated with the dominant lichen, *C. substellatum*, which forms large growths among the scattered dwarf shrubs on thin layers of loess and dolomitic rocks. *Caloplaca decipiens* (Arnold) Blomb. et Forssell is the most important component of the crusts on a hill fertilized by geese. These soil crusts are very poor in mosses, algae, and cyanobacteria.

Biological soil crusts on hummocky polygon soil are common in the Jørgen Brønlund Fjord area (82°09'N, 29°53'W) in Peary Land. This area has a high-arctic, continental climate characterized by very low summer temperatures (the mean July temperature is ca. 5 °C) and low annual precipitation (20–80 mm). Sand drifting occasionally occurs. Polygon soil is developed on the leeside of slopes and hills, where snow is accumulated during winter. In some places, *Dryas octopetala*, *Saxifraga oppositifolia*, *S. flagellaris*, and *Salix arctica* totally dominate on the top of the polygon-shaped tussocks, but often this vegetation is very open, allowing lichens to colonize the soil. Foliose macrolichens such as *Peltigera rufescens*, *Solorina bispora*, *S. saccata*, and *Physconia muscigena* bind the soil into more or less coherent crusts on the top of the tussocks (ca. 40 cm broad, up to 15 cm high). *Fulgensia bracteata* and *Leproloma vouauxii* grow on the sides of the tussocks, while *Nostoc* sp. and different mosses occur in the depressions between the tussocks. Such a habitat with *Solorina bispora* and *Fulgensia bracteata* is shown in Fig. 4.2.

Unstable, sandy soil and loess in more wind-exposed places support the following lichens: *Collema substellatum* (dominant) and *Caloplaca tominii* (subdominant), *Diploschistes muscorum*, *Fulgensia desertorum*, *Gypsoplaca macrophylla*, *Protoblastenia terricola*, *Toninia sedifolia*, *T. arctica*, *Catapyrenium squamulosum*, *Arthrorhaphis alpina*, and *Psora vallesiaca* (Timdal 1990, 1991; Hansen 1993, 1995a). This crust community, which is poorer in mosses, algae, and cyanobacteria than the hummocky patches, is influenced by copric dust rich in nitrogen and phosphate from musk oxen.

The soil crusts in the *Dryas–Cassiope–Vaccinium uliginosum* heaths in the Zackenberg area (74°31'N, 20°20'W), East Greenland (Fig. 4.1), are composed of the following lichens: *Physconia muscigena*, *Cladonia pocillum*, *Solorina bispora*, *Peltigera lepidophora*, *P. rufescens*, *Phaeorrhiza nimbosa*, *Rinodina mniaraea*, *Megaspora verrucosa*, *Caloplaca tominii*, *Protoblastenia terricola*, *Ochrolechia upsaliensis*, and *Lecanora epibryon*.

Mixed heaths in central west Greenland are comparatively poor in lichen crusts. In July and August 1999 Hansen investigated several interesting soil crust communities in Inglefield Land and Washington Land. These studies, which indicate that soil crust communities are more common in northernmost west Greenland and the western part of north Greenland, will be published separately. Hansen (1989) mentions different types of dwarf shrub

Fig. 4.2. Soil crusts on polygon soil with scattered lichen vegetation near Kap Moltke at Jørgen Brønlund Fjord, Peary Land. (Photo E.S. Hansen)

heaths occurring at Qaanaaq (77°29'N, 69°15'W). Of particular interest is the *Cassiope* heath with the lichens *Arthrorhaphis citrinella* (Ach.) Poelt, *Candelariella placodizans* (Nyl.) H.Magn., and *Lecanora geophila* (Th.Fr.) Poelt growing on soils among the dwarf shrubs. *Baeomyces placophyllus* and *Dibaeis baeomyces* (L. fil.) Rambold et Hertel also occur in the crusts. A nearby *Vaccinium uliginosum–Dryas integrifolia–Carex* community has crusts dominated by *Psora rubiformis* infested by the rare Greenland species *Caloplaca psoricida* E.S. Hansen, Poelt et Søchting (Hansen et al. 1987).

The north-facing slopes around Maarmorilik are characterized by mixed vegetation of dwarf shrubs, herbs, sedges, and lichens. *Dryas integrifolia* is the dominant dwarf shrub, but species such as *Braya purpurascens*, *Carex rupestris*, and *C. misandra* are also of great importance. The soil crusts, which are slightly alkaline due to surrounding marble rocks, are covered by lichens such as *Solorina bispora*, *S. saccata*, *Dacampia hookeri*, *Catapyrenium lachneum*, *C. cinereum*, *Fulgensia bracteata*, *Gyalecta foveolaris* (Ach.) Schaer., and *Protoblastenia terricola*. The lichens stabilize the bare soil. This lichen community also occurs on south-facing, steppe-like slopes dominated by *Carex rupestris*, *Calamagrostis purpurascens*, and *Kobresia myosuroides*.

4.2.4 Fell-fields

There is no sharp floristic limit between open dwarf shrub heath and fell-fields, but the latter community is characterized by a wider vertical distribution than the dwarf shrub heaths (Hansen 1996). The crusts in the fell-fields in southeast Greenland usually are quite thin (5–10 cm) and consist of gravelly soil, which is desiccated rather quickly by the wind. *Arthrorhaphis citrinella*, *Candelariella placodizans*, and *Lecidoma demissum* (Rutstr.) Gotth. Schneid. et Hertel are frequent crustal components (Hansen 1978a). A soil-crust community developed on neutral soil and consisting of lichens such as *Peltigera rufescens*, *Cladonia pocillum*, *Arthrorhaphis alpina*, *Fulgensia bracteata*, *Psora decipiens*, *P. rubiformis*, *Phaeorrhiza nimbosa*, *Catapyrenium cinereum*, and *Toninia sedifolia* occurs at an altitude of 250–300 m a.s.l. on Aucellabjerg (74°31'N, 20°20'W) situated in the Zackenberg area (Hansen 1996).

4.2.5 Snow Patches

Snow-patch crusts rich in lichens were previously analyzed in several places in the Angmagssalik district in 1970 (Hansen 1978a). The coastal part of this area has a low arctic and oceanic climate (annual precipitation of ca. 1000 mm). In *Salix herbacea* communities, the crusts are composed of lichens such as *Solorina crocea* (L.) Ach. (sometimes dominant), *Peltigera rufescens*, *Pertusaria oculata* (Dicks.) Th.Fr., *Lepraria neglecta* (Nyl.) Lettau, and the liverwort *Anthelia juratzkana*. During most of the summer, melt-water moistens these crusts (Hansen 1978b; see also Chap. 5).

4.3 Conclusions

Biological soil crusts are widely distributed in semiarid and arid areas in north and northeast Greenland. They also occur abundantly in continental and subcontinental areas in more southern parts of Greenland, particularly at the head of large fjords. Crust communities are smaller and scattered in oceanic parts of Greenland, including the coastal region of southeast and southwest Greenland.

Steppe-like communities dominated by *Kobresia myosuroides*, *Calamagrostis purpurascens*, and *Carex supina* are of great importance in low-arctic, continental parts of southwest Greenland. Soil crusts rich in crustaceous lichens characterize these areas. Open *Dryas* heaths and more arid communities (*Carex* and *Kobresia*) are widely distributed in Greenland and have soil crusts rich in lichens. Soil crusts also occur frequently in the fell-fields, which are characterized by very open and scattered vascular vegetation. However, the soil in the fell-fields is often more acidic than, for example, the *Dryas* heaths, and this is reflected in the composition of the soil crusts, which are usually frequent in the fell-fields. Some snow-patch types are defined by their soil crusts, as shown by a low-arctic, oceanic area in southeast Greenland.

High-arctic, steppe- and desert-like areas in north and northeast Greenland with a continental climate are particularly rich in biological soil crusts. A lichen community dominated by *Collema substellatum* and *Caloplaca tominii* occurs on thin layers of loess over rocks. It is of great importance because of its disjunct lichens, including *Fulgensia desertorum*, *Gypsoplaca macrophylla,* and *Collema substellatum*. Soil-crust lichens occurring on hummocks and polygon soil usually form very characteristic, mixed communities in arid areas in northernmost Greenland.

Much information is available about the geological and climatic conditions of the different parts of Greenland, but the physical and chemical characteristics of the soil crusts are generally poorly known. The ecophysiology of the different Greenland soil-crust communities is also in need of additional studies. Undoubtedly, future research will focus on these subjects in the whole arctic area.

References

Aleksandrova VD (1988) Vegetation of the Soviet Polar Deserts. Cambridge University Press, Cambridge

Böcher TW (1949) Climate, soil, and lakes in continental west Greenland in relation to plant life. Medd Grønl 147:1–63

Böcher TW (1954) Oceanic and continental vegetational complexes in southwest Greenland. Medd Grønl 148:1–336

Breuss O, Hansen ES (1988) The lichen genera *Catapyrenium* and *Placidiopsis* in Greenland. Plant Syst Evol 159:95–105

Daniëls FJA (1982) Vegetation of the Angmagssalik District, southeast Greenland, IV. Shrub, dwarf shrub and terricolous lichens. Medd Grønl Biosci 10:1–78

Gelting P (1954) The *Rhizocarpon* species with peltate areoles occurring in Europe and North America. Bot Tidsskr 49:295–313

Hansen ES (1978a) Notes on occurrence and distribution of lichens in south east Greenland. Medd Grønl 204:1–71

Hansen ES (1978b) Notes on vertical distribution of lichens on three mountains in the Angmagssalik District, south east Greenland. Bot Tidsskr 73:55–61

Hansen ES (1986) Søndre Strømfjord, et arktisk steppeområde. Urt 3:68-72

Hansen ES (1989) The lichen flora of Qaanaaq (Thule), northwestern Greenland. Mycotaxon 35:379–394

Hansen ES (1991) The lichen flora near a lead-zinc mine at Maarmorilik in west Greenland. Lichenologist 23:381–391

Hansen ES (1993) *Collema substellatum* and *Fulgensia desertorum*, new to Greenland. Lichenologist 25:451–458

Hansen ES (1995a) The lichen flora of the Jørgen Brønlund Fjord area, northern Greenland. Bibl Lichenol 57:187–198

Hansen ES (1995b) The lichen flora near Ittoqqortoormiit/Scoresbysund in central east Greenland. Herzogia 11:197–206

Hansen ES (1996) Vertical distribution of lichens on the mountain, Aucellabjerg, northeastern Greenland. Arct Alp Res 28:111–117

Hansen ES (1997) Studies of the lichen flora of coastal areas in central west Greenland. Nova Hedwigia 64:505–523

Hansen ES, Poelt J, Søchting U (1987) Die Flechtengattung *Caloplaca* in Grönland. Medd Grønl Biosci 25:1–52

Hartz N (1895) Østgrønlands Vegetationsforhold. Medd Grønl 18:105–314

Longton RE (1988) Biology of polar bryophytes and lichens. Cambridge University Press, Cambridge

Timdal E (1990) Gypsoplacaceae and *Gypsoplaca*, a new family and genus of squamiform lichens. Bibl Lichenol 38:419–427

Timdal E (1991) A monograph of the genus *Toninia* (Lecideaceae, Ascomycetes). Opera Bot 110:1–137

5 Biological Soil Crusts of the Subalpine, Alpine, and Nival Areas in the Alps

R. Türk and G. Gärtner

5.1 Introduction

In the Alps, a relatively large amount of landscape occurs in the subalpine, alpine, and nival zones. In these zones, orographic and climatic conditions result in many areas with poorly-developed vascular vegetation. Abundant biological soil communities colonize the gaps between the plants. The most outstanding soil-crust development is found in windswept places of the alpine and nival zones, and on fluvioglacial deposits in periglacial areas.

Information regarding biological soil-crust communities from these alpine areas is very limited. In spite of many publications existing on different groups of terrestrial organisms (e.g., lichens, mosses, and soil algae) found in these regions, the phenomenon of soil crusts as a community is more or less neglected. As with lower-elevation areas where plant cover is sparse, these crusts probably play a critical role in soil stabilization and nutrient cycling.

Two types of biological soil crusts occur in these regions: (1) ephemeral communities consisting of microorganisms such as flagellates (*Chlamydomonas*), filamentous green algae (e.g., *Stichococcus*), and Xanthophyceae (*Vaucheria*) that occur in small, fast-desiccating puddles and result in green or yellow-green mats; and (2) perennial communities of cyanobacteria, algae, mosses, and lichens that show only small seasonal changes. As the first community type does not result in a permanent incrustation of the upper soil layer, these communities are not considered biological soil crusts in the context of the present Volume.

In the past decades, knowledge has been broadened by multiple excursions and investigations in the alpine and nival areas. The result of these field studies and reports in the literature are the basis of this chapter (Klement 1955; Kalb 1970; Vogel 1983; Türk and Wittmann 1987; Hofmann et al. 1988; Mayrhofer et al. 1989; Wittmann and Türk 1990; Timdal 1991; Poelt and Gärtner 1992; Türk and Hafellner 1992; Obermayer 1993; Hafellner and Wittmann 1996; Hafellner et al. 1996; Hofmann et al. 1998; Türk and Wunder 1999). Also, reports on lichens in the alpine parts of France (Clauzade and Roux 1985) and

Ecological Studies, Vol. 150
J. Belnap and O.L. Lange (eds.) Biological Soil Crusts: Structure, Function, and Management

Italy (Nimis 1993) are taken into consideration. We did not include lichens that occur on mosses or plant litter. Results of investigations on terrestrial algae are summarized by Reisigl (1964) and Ettl and Gärtner (1995).

5.2 Crust Communities over Carbonate Soils

Biological soil crusts show remarkable species diversity on carbonate soils that occur over limestone, marble, and intermediate rocks. In these areas, the higher vegetation is dominated by sedges and grasses. The plant association Seslerio–Semperviretum is characterized by *Carex sempervirens*, *Sesleria caerulea*, *Silene acaulis*, *Salix serpyllifolia*, *Dryas octopetala*, and different *Saxifraga* species. In the higher alpine zone (up to 2800 m), the Caricetum firmae association covers the windswept soils and is represented by *Carex firma*, *Dryas octopetala*, *Silene acaulis*, *Salix retusa*, *Salix serpyllifola*, *Gentiana clusii*, and *Saxifraga caesia*. Crust-species composition depends on exposure. In windswept areas of calciferous soil in the subalpine to alpine belt (2000–2700 m; Chap. 1, Photos 38, 39), the following lichens are dominant: *Psora decipiens* (Hedw.) Hoffmann, *Protoblastenia siebenhaariana* (Körb.) J.Steiner var. *siebenhaariana* [syn. *Protoblastenia terricola* (Anzi) Lynge, Photo 40], *Fulgensia bracteata* (Hoffmann) Räsänen var. *alpina* (Th.Fr.) Räsänen (Photo 17) and ssp. *deformis* (Erichs.) Poelt, *Buellia elegans* Poelt, *Catapyrenium cinereum* (Pers.) Körb., *Polyblastia tatrana* Vezda, *Diploschistes muscorum* (Scop.) R.Sant. var. *iridatus* (A.Massal.), and *Rinodina mniaraea* (Ach.) Körb. var. *mniaraea*. Also, typical algae occur in the soil crusts between the thalli of lichens and the less abundant mosses, often forming grayish or black layers on top of, or within, the uppermost zone of open mineral soil. Dominant species include the cyanobacterium *Nostoc* sp., coccal green algae such as *Chlorella* sp. and *Coccomyxa* sp., and filamentous green algae such as *Klebsormidium flaccidum* (Kützing) Silva, Mattox et Blackwell and *Leptosira* sp. In the Xanthophyceae *Heterococcus brevicellularis* Vischer, *Xanthonema solidum* (Vischer) Silva, and *Gloeobotrys piriformis* Reisigl are frequent, among other species (Reisigl 1964).

Where soil is formed in thicker layers that have a higher water-holding capacity, or on northern exposures where the snow cover remains over a longer period, the following species of lichens and algae – in places intermingled with mosses – dominate open soil patches: *Gyalecta foveolaris* (Ach.) Schaer., *G. geoica* (Wahlenb. ex Ach.) Ach., *G. peziza* (Mont.) Anzi, *Mycobilimbia berengeriana* (A.Massal.) Hafellner et V.Wirth, *Mycobilimbia lobulata* (Sommerf.) Hafellner, *Psoroma hypnorum* (Vahl) S.Gray, *Pertusaria bryontha* (Ach.) Nyl., *Involucropyrenium waltheri* (Krempelh.) Breuss, *Placidiopsis pseudocinerea* Breuss, *Phaeorrhiza nimbosa* (Fr.) H.Mayrhofer et

Poelt, and *Phaeorrhiza sareptana* (Tom.) H.Mayrhofer et Poelt var. *sphaerocarpa* (Th.Fr.) H.Mayrhofer et Poelt. Among the cyanobacteria, *Nostoc* spp. (in gelatinous colonies; Photo 3) is accompanied by filamentous taxa like *Microcoleus* sp. (Photo 1). Green algae such as *Stichococcus bacillaris* Naegeli and *Klebsormidium flaccidum* are also present.

Southern exposures, or soils with a lower water-holding capacity, are covered by gelatinous, crustose, and small foliose lichens. Most species are closely attached to the substrate, with some of them stabilizing the upper-most soil layer with rhizohyphae, rhizoptae, and rhizinae. Here, the following lichen species are dominant: *Endocarpon pusillum* Hedw., *Buellia epigaea* (Pers.) Tuck., *B. geophila* (Flörke ex Sommerf.) Lynge, *B. papillata* (Sommerf.) Tuck., *Placidium lachneum* (Ach.) de Lesd. [syn. *Catapyrenium l.* (Ach.) de Lesd.; nomenclature cf. Breuss 1990, 1997], *P. norvegicum* (Breuss) Breuss (syn. *Catapyrenium n.* Breuss), *P. squamulosum* (Ach.) Breuss [syn. *Catapyrenium s.* (Ach.) Breuss; Photo 21)], *Heppia adglutinata* (Krempelh.) A.Massal., *Toninia physaroides* (Opiz) Zahlbr., *T. sedifolia* (Scop.) Timdal (Photo 16), *T. squalida* (Ach.) A.Massal., *Collema tenax* (Sw.) Ach. em. Degel. var. *tenax* (Photo 9), *C. crispum* (Hudson) Weber ex F.H.Wigg. var. *crispum*, *Leptogium tenuissimum* (Dickson) Körber, and, as a macrolichen, *Peltigera rufescens* (Weiss) Humb. Also cyanobacteria like *Nostoc* spp. belong to this community. On very xeric cold sites, *Caloplaca tominii* Sav. can very occasionally be found.

Soil gaps between higher plants or cushions of mosses are generally covered by macrolichens such as *Cladonia symphycarpa* (Flörke) Fr., *C. pyxidata* (L.) Hoffmann, *Solorina bispora* Nyl. var. *bispora*, *S. octospora* (Arnold) Arnold, *Peltigera lepidophora* (Nyl. ex Vain.) Bitter, *Anaptychia bryorum* Poelt, and *Leptogium lichenoides* (L.) Zahlbr. var. *lichenoides*. These are associated with the bullate taxa *Toninia diffracta* (A.Massal.) Zahlbr. and *T. sedifolia*, and the crustose cyanobacterial lichen *Fuscopannaria praetermissa* (Nyl.) Jørg. *Cladonia macroceras* (Del.) Hav. also belongs to this biocenosis. This species exhibits different growth forms on calcareous and siliceous soil. On calcareous substrates, the primary squamules of the thallus are developed, sometimes producing downwards-bending brown or dark brown-colored podetia. On siliceous soil, the typical thallus with erect podetia is formed without primary squamules. Filamentous green algae also grow in this habitat, including *Klebsormidium flaccidum*, *Stichococcus bacillaris*, and the filamentous *Xanthonema hormidioides* (Xanthophyceae).

5.3 Lichens over Soil in Crevices of Calciferous Rocks

The calciferous rocks are often occupied by the plant association Potentilletum caulescentis characterized by *Potentilla caulescens*, *Asplenium ruta-*

muraria, *Rhamnus pumila*, *Saxifraga paniculata*, *Minuartia rupestris*, and *Valeriana saxatilis* (Ellenberg 1988; Reisigl and Keller 1987). On steep to vertical surfaces of rocks, soil material is deposited in crevices by occasional water flows. Conditions found in the crevices are favorable for some photophilous, crustose lichen species including *Caloplaca aurea* (Schaer.) Zahlbr., *Lecidea lurida* Ach., *Psora rubiformis* (Ach.) Hooker, *Squamarina gypsacea* (Sm.) Poelt, *Toninia diffracta* (A.Massal.) Zahlbr., *T. opuntioides* (Vill.) H.Baumgartner, *T. rosulata* (Anzi) Oliv., *T. squalida* (Ach.) A.Massal., and *T. taurica* (Szat.) Oxner. These species not only form crusts on soils within the crevices, but they also grow onto the adjacent rock surfaces.

No special cyanobacterial and algal communities are known from this special habitat, but many coccal organisms are swept in by the episodic water flows. These include the cyanobacteria *Nostoc* spp., *Gloeocapsa* spp., *Chroococcus* spp., and the green algae *Stichococcus bacillaris* and *Coccomyxa* spp.

5.4 Crust Communities over Acid Soil

The higher vegetation is dominated by typical communities of *Carex curvula* (class Curvuletea). In lower areas they are intermingled with plants from the class Nardetea; in higher areas with plants from the class Androsacetea. Flowering plants in the carpet of the plant association Curvuletum are: *Carex curvula*, *Oreochloa disticha*, *Juncus trifidus*, *Phyteuma hemisphaericum*, *Primula minima*, *Primula hirsuta*, *Primula glutinosa*, *Lloydia serotina*, and others. On wind-exposed habitats the dwarf shrub *Loiseleuria procumbens* dominates.

Acid soils develop over siliceous rocks, peat, and plant debris. These soils are poor in nutrients and carbonate content is very low. Low nutrient availability probably results in lower diversity of all species. However, the lower cover of vasular plants permits a higher abundance of soil-crustal organisms. An exception is the number of soil algal species, including epe- and endedaphic forms, which, on average, are fewer on acid than on calcareous soils. Taxa found on acid soils include many common types which can be found elsewhere. An example is the xanthophycean genus *Heterococcus*. This genus include species with a very large ecological range reaching from the desert soils of the Sahara to alpine soils of Europe and the Himalaya (Pitschmann 1963).

On soils with a low water-holding capacity or on drier, light-exposed sites, the lichens *Cetraria muricata* (Ach.) Eckf., *Cladonia coccifera* (L.) Willd., *C. bellidiflora* (Ach.) Schaer., *C. carneola* (Fr.) Fr., *C. macroceras* (Del.) Hav., *C. macrophyllodes* Nyl., *C. pyxidata*, *C. strepsilis* (Ach.) Vain., *Placidium lachneum*, *Lepraria neglecta* Vain., *Lepraria cacuminum* (A.Massal.) Lohtander, and *Pycnothelia papillaria* Dufour can be dominant. In habitats with higher water-holding capacity and moisture, the lichens *Dibaeis baeomyces* (L. f.) Rambold et Hertel, *Baeomyces rufus* (Hudson) Rebent., *Arthrorhaphis alpina* (Schaer.) R.Sant., *A. vacillans* Th.Fr., *A. citrinella* (Ach.) Poelt, *Lecidoma demissum* (Rutstr.) G.Schneider et Hertel, *Stereocaulon glareosum* (Savicz) H.Magn., *Cladonia strepsilis* (Ach.) Vain., *Mycobilimbia hypnorum* (Lib.) Kalb et Hafellner, *Solorina crocea* (L.) Ach., *S. spongiosa* (Ach.) Anzi, *Lecidea limosa* Ach., and *Thelocarpon epibolum* Nyl. var. *epibolum* dominate the crustal lichen flora.

On bare soil between higher plants, the following species of lichens occur: *Solorina octospora* (Arnold) Arnold, *Trapeliopsis granulosa* (Hoffmann) Lumbsch, *Bryonora castanea* (Hepp ex Th.Fr.) Poelt, *Ochrolechia inaequatula* (Nyl.) Zahlbr., *Rinodina mniaraea* var. *mniaraea*, *Baeomyces placophyllus* Ach., and *Placidiopsis pseudocinerea* Breuss. Gaps between the lichen thalli are covered by grayish or black layers of cyanobacteria such as *Nostoc* spp., *Schizothrix* spp., *Scytonema* spp., *Phormidium* sp., and others. Acid soils also occur in siliceous rock crevices. An example of a crustal organism in this habitat is the lichen *Catolechia wahlenbergii* (Ach.) Flotow ex Körb.

Fluvioglacial deposits recently exposed by glacial retreat are fairly unstable, so soil crust-forming organisms must be adapted to these conditions. The higher vegetation contains successional communities ranging from the pioneer stage to the stable sward stage (i.e., the association Caricetum curvulae). Some characteristic plant species are *Saxifraga aizoides*, *Cerastium uniflorum*, *Trifolium pallescens*, *Oxyria digyna*, *Geum reptans*, and others. Typical lichen pioneers in these areas include *Stereocaulon alpinum* Laurer, *S. nanodes* Tuck., *S. incrustatum* Flörke, *Cladonia pyxidata*, *C. cervicornis* (Ach.) Flot. ssp. *verticillata* (Hoffm.) Ahti, *C. coccifera* (L.) Willd., *Pertusaria geminipara* (Th.Fr.) Knight ex Brodo, *Involucropyrenium waltheri* (Krempelh.) Breuss, and *C. bellidiflora*. Green algal organisms often form macroscopic, ephemeral crusts dominated by *Bracteacoccus minor* (Chodat) Petrová, *Chlorella vulgaris* Beijerinck, *Elliptochloris subsphaerica* (Reisigl) Ettl et Gärtner, and *Stichococcus bacillaris*. Among the Xanthopyhceae are unidentified species of the genus *Chloridella* and *Xanthonema* (Reisigl 1964).

5.5 Crust Communities in Snow Patches

Two vegetation types occur in the snow patches: one is dominated by bryophytes (the Polytrichetum norvegici association) and the other by phanerogams (e.g., *Salix herbacea*, *Leucanthemopsis alpina*, *Soldanella pusilla*, *Sibbaldia procumbens*, and *Gnaphalium supinum*; the Salicetum herbaceae association). Crust communities also occur at the edges of semi-permanent snow patches. These habitats are characterized by a very short growth period for vegetation, and occasionally the snow cover remains during the whole year, at least in the center of the snow patches. Such habitats are often dominated by bryophytes. The continuously wet conditions are not favorable for lichens, which are adapted to alternating wet-dry periods. Only a few lichen species can be found at the edges of the snow patches, such as *Lecidea limosa*, *Solorina crocea*, and *Lecidoma demissum*. Most of the crust-forming cyanobacteria found are *Nostoc* spp.

5.6 Conclusions

1. Unlike in arid and semiarid areas, very little work has been done on the ecological role of crusts in high-altitude areas. Such work is needed, as nitrogen and carbon are probably limiting to plants and soil biota, and soils are exposed to much wind and water erosion. Thus, soil crusts are expected to play an important role in these ecosystems, such as they do in desert systems.
2. There are striking similarities in lichen, algal, and cyanobacterial species between high-altitude areas and deserts (see Chaps. 4, 7, 12). This is probably due to the extremely harsh conditions found on soil surfaces, regardless of habitat, and only a few very highly adapted species can handle these extreme conditions.
3. Little work has been done to identify mosses in high-altitude areas, and much work is needed.

References

Breuss O (1990) Die Flechtengattung *Catapyrenium* (*Verrucariaceae*) in Europa. Stapfia 23:1–153

Breuss O (1997) Ein verfeinertes Gliederungskonzept für *Catapyrenium* (lichenisierte Ascomyceten, Verrucariaceae). Ann Naturhist Mus Wien 98 B Suppl:35–50

Clauzade G, Roux C (1985) Likenoj de okcidenta Europo. Bull Soc Bot Centre-Ouvest, Nouvelle Serie 7–1985

Ellenberg H (1988) Vegetation ecology of Central Europe. Cambridge University Press, Cambridge

Ettl H, Gärtner G (1995) Syllabus der Boden-, Luft- und Flechtenalgen. Fischer, Stuttgart

Hafellner J, Wittmann H (1996) IAL3. Excursion 2: Alpine lichens of the central part of the Eastern Alps. Excursion guide. Graz 1996, 24 pp

Hafellner J, Türk R, Breuss O (1996) Zur Flechtenflora des Wechsel (Österreich). Österr Z Pilzk 5:211–231

Hofmann P, Türk R, Gärtner G (1988) Beitrag zur Flechtenflora Tirols: Obergurgl (Ötztaler Alpen, Nordtirol). Ber Nat-Med Ver Innsbruck 75:7–19

Hofmann P, Berger F, Obermayer W, Wittmann H, Breuss O, Sipman H (1998) Ergänzungen zur Flechtenflora der Ötztaler Alpen (Tirol, Österreich). Ergebnisse der BLAM-Exkursion 1993. Herzogia 13:155–164

Kalb K (1970) Flechtengesellschaften der Vorderen Ötztaler Alpen. Diss Bot 9:1–118

Klement O (1955) Prodromus der mitteleuropäischen Flechtengesellschaften. Feddes Repert Spec Nov, Beih 135:1–195

Mayrhofer H, Türk R, Wittmann H (1989) Ein Beitrag zur Flechtenflora von Vorarlberg (Österreich). Ergebnisse der Feldtagung der Bryologisch-lichenologischen Arbeitsgemeinschaft für Mitteleuropa im Juli 1986. Herzogia 8:207–247

Nimis PL (1993) The lichens of Italy. An annotated catalogue. Mus Reg Sci Nat Monogr 12:1–897

Obermayer W (1993) Die Flechten der Seetaler Alpen (Steiermark, Österreich). Mitt Naturwiss Ver Steiermark 123:91–66

Pitschmann H (1963) Vorarbeiten zu einer Monographie der Gattung Heterococcus. Nova Hedwigia 5:487–531

Poelt J, Gärtner G (1992) *Gypsoplaca macrophylla*, eine Flechte winterkalter Halbwüsten in den Alpen. Herzogia 9:229–237

Reisigl H (1964) Zur Systematik und Ökologie alpiner Bodenalgen. Österr Bot Z 111: 402–499

Reisigl H, Keller R (1987) Alpenpflanzen im Lebensraum. Alpine Rasen, Schutt- und Felsvegetation. Fischer, Stuttgart

Timdal E (1991) A monograph of the genus *Toninia* (Lecideaceae, Ascomycestes). Opera Bot 110:1–137

Türk R, Hafellner J (1992) Flechten im Nationalpark Hohe Tauern – Kärntner Anteil (Österreich). Carinthia II 183/103:723–757

Türk R, Wittmann H (1987) Flechten im Bundesland Salzburg (Österreich) und im Berchtesgadener Land (Bayern, Deutschland) – die bisher beobachteten Arten und deren Verbreitung. Sauteria 3:1–313

Türk R, Wunder H (1999) Die Flechten des Nationalparks Berchtesgaden und angrenzender Gebiete. Nationalpark Berchtesgaden Forschungsber 42:1–131

Vogel E (1983) Der Glungezer und seine Umgebung (Tuxer Voralpen, Nordtirol). Ber Nat-Med Ver Innsbruck 76:51–71

Wittmann H, Türk R (1990) Die Flechten im Nationalpark Nockberge. Kärntner Nationalparkschr 4:1–112

6 Biological Soil Crusts in European Temperate and Mediterranean Regions

B. Büdel

6.1 Introduction

Biological soil crusts, dominated by either algae, lichens, or mosses are, from floristic, taxonomical, and phytosociological points of view, well investigated in Europe (e.g., Fritsch and John 1942; Wilmanns 1959; de Winder 1990; Paus 1997). However, this does not mean that their role in dynamics and ecology is well known, as the role of biological soil crusts as pioneers in disturbed soils remains unstudied. In general, an extensive literature exists from mediterranean and temperate regions, often approaching local soil-crust communities from a phytosociological viewpoint. Consequently, only a selection of a few of the most relevant publications is given here.

6.2 Mediterranean Regions

Because of its floristic and vegetational similarities with the European Mediterraneis, the mediterranean region of North Africa is included here.

6.2.1 West Mediterranean Region

The Tabernas badlands in semiarid southeast Spain belong to the driest part of Europe, with a mean annual precipitation of ca. 170 mm (Alexander et al. 1994). The summers are hot and dry and winters are mild, with most precipitation falling as intense storms in late September and October. From this area, Alexander and Calvo (1990) described four physiognomic types of lichen-dominated soil crusts. The first group contained *Toninia sedifolia* (Scop.) Timdal, a *Collema* species, and an indeterminate black crust. Together, they form patches of a thin, porous crust, likely to provide some degree of surface binding. The second group contained *Psora decipiens* (Hedw.) Ach. and *Cata-*

Ecological Studies, Vol. 150
J. Belnap and O.L. Lange (eds.) Biological Soil Crusts: Structure, Function, and Management

pyrenium lachneum (Ach.) R.Sant., growing as groups of discrete squamules of 1–2 mm in diameter, thus giving rise to a discontinuous, nonporous but permeable surface (the authors define porosity in terms of coverage of soaked lichens, i.e., porous means that the soaked lichens leave some open space where the soil is exposed; permeability is defined by the penetration of water into the soil, either through the lichens or the soil itself). Surface binding is provided by rhizinae, anchoring the squamules and penetrating to depths of 1–5 mm. The two lichens *Squamarina lentigera* (G.H.Weber) Poelt, growing as imbricate squamules, and *Fulgensia fulgens* (Sw.) Elenk. (Chap. 1, Photo 16), a crustose species with raised margins, form the third group and provide a discontinuous, permeable crust with good surface binding. The last group consists of *Diploschistes diacapsis* (Ach.) Lumbsch, a robust crustose species which provides a nonporous, slightly permeable surface that frequently develops a microtopographic relief of up to 1.7 cm. Species of small mosses (*Tortula* spp.) occurred at all sites. In their experimental studies concerning erosion processes and lichen crusts, Alexander and Calvo (1990) clearly demonstrated that soil crusts protected soils from erosion.

Soil crust-forming lichens from semiarid regions of Spain (e.g., Tabernas) were partly treated in the monograph of the lichen genera *Heppia* and *Peltula* by Egea (1989). Members of these two genera that formed soil crusts included *Heppia despreauxii* (Mont.) Tuck. (*H. gigantea* Egea et Llimona), *H. lutosa* (Ach.) Nyl., *H. solorinoides* (Nyl.) Nyl. (*H. reticulata* [Duf.] Nyl.), *Peltula crispatula* (Nyl.) Egea, and *P. patellata* (Bagl.) Swinscow et Krog (Photo 10). In their study of distribution and life strategies of soil crust-forming mosses, Frey and Kürschner (1990, 1991) mention the significant association of the moss *Gigaspermum mouretii* Corb. with the cyanolichens of the genus *Collema* in soil crusts of southern Spain, but also from Morocco and Mallorca.

6.2.2 East Mediterranean Region

Soil crusts were reported from the Greek island Euboea by Roussomoustakaki (1983). She found extensive cyanobacterial crusts on solonetz and solonchak soils (= aridisols, i.e., soils without humus layer and with high salt concentration). The most common species were *Microcoleus chtonoplastes* (Fl.Dan.) Thuret ex Gom., *Lyngbya aestuari* (Mert.) Liebm., *Anabaena variabilis* Kütz., *Nostoc punctiforme* (Kütz.) Hariot, *Oscillatoria laetevirens* (Crowan) Gom., and *Nodularia spumigena* (Mert.) ex Born. et Flah. Crusts on calcareous soils also had a rich flora of cyanobacteria, with *Nostoc commune* Vauch. (Photo 3), *Microcoleus vaginatus* (Vauch.) Gom. (Photo 1), and *Schizothrix calcicola* (Ag.) Gom. as the most frequent species. She also observed that microstratification of the cyanobacteria was deeper in sandy soils than in loamy soils.

The Aegean islands are considered to be the tips of a submersed mountain system, which was connected to the Greek mainland and Asia Minor in the geological past. Investigating the lichen flora of Paros, Antiparos, and the Santorini island group, Sipman and Raus (1999) focused on soil-inhabiting lichens. The soils were generally poor in organic matter and soil crusts, with *Psora decipiens* and *Fulgensia subbracteata* (Nyl.) Poelt being the dominant lichens. Several other lichen species occurred additionally in the soil-crust communities: *Catapyrenium virescens* (Zahlbr.) Breuss, *Collema cristatum* (L.) Wigg., *C. tenax* (Sw.) Ach., *Diploschistes diacapsis*, *Heppia solorinoides*, *Leptogium schraderi* (Ach.) Nyl., *Catapyrenium cinerascens* (Nyl.) Breuss, *C. tenellum* Breuss, *C. squamulosum* (Ach.) Breuss (Photo 21), *Squamarina cartilaginea* (With.) P.W.James, *S. concrescens* (Müll.Arg.) Poelt, *S. gypsacea* (Sm.) Poelt, *S. lentigera* (Photo 12), *S. periculosa* (Schaerer) Poelt, and *Toninia sedifolia*.

Climate and vegetation differ in various regions of the eastern Mediterranean island of Cyprus. The Akamas peninsula, where Litterski (1997) did a survey of the lichens of the Troodos mountains, has a true mediterranean climate. The climate in the Troodos mountains is characterized by relatively high annual rainfall (600–1000 mm) and elevations range from 100–900 m a.s.l. Soil crusts were found on either base-rich siliceous soils or soils originating from calcareous rocks. The following species were found: *Catapyrenium daedaleum* (Krempelh.) B.Stein, *Collema crispum* (Hudson) Weber ex Wigg., *C. tenax*, *Diploschistes diacapsis*, *Fulgensia fulgens*, *F. fulgida* (Nyl.) Szat., *F. subbracteata*, *Psora decipiens*, *Squamarina cartilaginea*, *S. concrescens*, *S. lentigera*, *Toninia sedifolia*, and *T. toepfferi* (B.Stein) Navas. Calcareous soil in open gaps of the sparse evergreen mediterranean shrub vegetation of the eastern coastal plains of Cyprus is covered with a rich assemblage of soil lichens of a similar species combination (O.L.Lange, pers. comm.). The white patches of *Diploschistes diacapsis* are often dominating (see Frontispiece, this Vol.). The described east Mediterranean soil-crust vegetation resumes continuation in the Middle East region, as analyzed in Chapter 8.

6.2.3 Mediterranean Regions of North Africa

Egea (1989) mentions the following species of *Heppia* and *Peltula* as taking part in soil-crust formation in Algeria and Morocco: *Heppia despreauxii* (*H. gigantea*), *H. solorinoides* (*H. reticulata*), *Peltula crispatula*, *P. psammophila* (Nyl.) Egea (only in Algeria), *P. michoacanensis* (B. de Lesd.) Wetmore, and *P. patellata* (only in Morocco). More crust-forming lichens from Moroccan soils are reported in Werner (1979) and Alonso and Egea (1994). Werner (1979) lists a number of terricolous species in the Cordillere Betico-Rifaine region, of which 14 are probably involved in crust formation: *Fulgensia fulgens*, *Diplo-*

schistes muscorum (*D. bryophilus* [Ehr.] Zahlbr.), *Psora decipiens, Toninia sedifolia* (*T. caeruleonigricans* [Lightf.] Th.Fr.), *Collema tenax* (Sw.) Ach. and *Dermatocarpon rufescens* Th.Fr. Alonso. Eleven terricolous and potentially crust-forming species are listed by Alanso and Egea (1994): *Catapyrenium pilosellum* Breuss, *C. squamulosum, Collema tenax, Diploschistes diacapsis, Fulgensia fulgens, Heppia solorinoides, Psora decipiens, Squamarina cartilaginea, Synalissa symphorea* (Ach.) Nyl., *Toninia aromatica* (Turn.) Massal., and *T. sedifolia.*

6.3 Xerothermic Steppe Formation of Central Europe

Specific soil-lichen communities, centered in the Mediterranean and the continental areas of the Eurasian continent, can also be occasionally found in Central Europe, where suitable growth conditions occur in open local xerothermic steppe formations (Photos 41, 42). The lichens of the community Fulgensietum fulgentis are widely known as Bunte Erdflechtengesellschaft (community of colored lichens) in Central Europe (e.g., Reimers 1950, 1951; Marstaller 1968). Further xerothermic associations of soil-crust forming lichens belong to this group, e.g., the Fulgensietum alpinum (the alpine form of the complex) and the Cladonietum symphycarpiae. These soil lichen communities are well investigated and summarized in a complex syntaxonomical system (Reimers 1940; Klement 1955; Bornkamm 1958; Wilmanns 1959; Marstaller 1971; Ritschel 1974; Paus 1997). The Fulgensietum fulgentis and Cladonietum symphycarpae are composed of about 30–40 characteristic lichens and some 40 mosses, combined with a number of cyanobacteria and, rarely, filamentous green algae. Typical lichens are: *Cladonia convoluta* (Lam.) Anders, *Fulgensia fulgens, Squamarina cartilaginea, S. lentigera, Endocarpon pusillum, Toninia sedifolia, Psora decipiens, P. lurida, Peltigera rufescens, Catapyrenium squamulosum, C. lachneum, Collema tenax* (Photo 9) and *Diploschistes muscorum.* Common mosses include: *Ceratodon purpureus* (Hedw.) Brid., *Tortella inclinata* (Hedw.f.) Limpr., *T. tortuosa* (Hedw.) Limpr., *Tortula ruralis* (Hedw.) Gärtn., Meyer et Scherb., *Rhacomitrium heterostichum* agg., *R. canescens* (Hedw.) Brid., *Bryum capillare* agg., *B. caespititium* Hedw., *Barbula fallax* Hedw., and *B. convoluta* Hedw. The most important cyanobacterium is *Nostoc commune*, forming large gelatinous thalli on top of the soil (Photo 3) or living inside the soil as hormogonia during part of its life cycle.

The Bunte Erdflechtengesellschaft (i.e., communities of the complex of Fulgensietum fulgentis and Cladonietum symphycarpae) has a wide distribution, reaching from southern Algeria to the southern Swedish Alvar region (see also Sect. 6.7) in the north (Du Rietz 1925; Pettersson 1946; Ott et al.

1996), and from the Aralo-Caspian semideserts and the Mesopotamia region in the east to the Poitou and the Eifel midlands (e.g., Müller 1965) in the west. The presence of this peculiar, arid microclimate-adapted community of organisms in Central Europe – centered in the Mediterranean and the continental areas of the Eurasian continent – may be explained as a relict of the postglacial warm period (Reimers 1940).

The ecology of the lichens of the Fulgensietum fulgentis and Cladonietum symphycarpae in a local xerothermic steppe formation near the German city Würzburg was investigated by Hahn et al. (1989) and Lange et al. (1995). Field studies (Hahn et al. 1989) showed that the lichens were dry and thus inactive during a considerable amount of time throughout the year. In winter, positive net photosynthesis often occurred during the day, but carbon gain was limited by low light intensities. In summer, net photosynthesis rates were high after a rain event, but high insolation resulted in fast desiccation. Four types of photosynthetic strategies, related to thallus water content, were found among the lichens tested [Lange et al. 1995: type A (*Cladonia convoluta*: Photos 22, 23), *Diploschistes muscorum*] showed almost no depression of CO_2 uptake during suprasaturation of the thallus with water; type B (*Squamarina lentigera*) showed only a slight depression of CO_2 uptake at the highest thallus water content; type C (*Collema tenax, Toninia sedifolia*) showed a large depression which became steadily greater as water content increased; type D lichens (*Cladonia pocillum, Fulgensia fulgens*) behaved similarly to type C; however, the response curve had an inflection so that net photosynthesis was low but almost constant. This different type of response may partially explain observed microscale and macroscale vegetation patterns (see Chaps. 18, 20).

6.4 East European Xerothermic Steppe Formation

The different growth types of algal soil crusts and their relation to edaphic factors were investigated by Komáromy (1976) in Hungarian forests and grass steppe ecosystems (Hortobágy National Park). As in other crusts of the European forest zone, green algae (e.g., *Chlorococcum infusionum* [Schrank] Meneghini, *Botrydiopsis minor* [Schmidle] Chod., *Stichococcus bacillaris* Näg., *Klebsormidium flaccidum* [Kütz.] Silva, Mattox et Blackwell, *K. dissectum* [Gay] Ettl et Gärtner, *K. crenulatum* [Kütz.] Ettl et Gärtner) and xanthophytes (e.g., *Tribonema minus* [Wille] Hanzen) were dominant. Steppe soils with high salt contents were dominated by cyanobacteria (e.g., *Microcoleus lacustris* [Rabenh.] Farl., *M. sociatus* W. et G.S.West, *Phormidium foveolarum* [Mont.] Gom., *P. molle* [Kütz.] Gom., *P. paulsenianum* Petersen, *Tolypothrix tenuis* Kütz., *Nostoc commune* Vauch., and *N. microscopicum* Carm.). The author distinguished five different types of crust formation:

1. The stratose growth type occurred on acidic, nonpodzolized brown forest soil; bare soils in clearings; and also on soils with high salt content. It is composed of filamentous green and yellow-green algae that form a dense coat on the soil surface and have only a few connections with the top of the soil.
2. The *Microcoleus* growth type was found on heavily trodden paths and other places with a very solid soil surface. The filamentous cyanobacteria (*Lyngbya, Microcoleus, Phormidium*) were thoroughly interwoven with the soil particles.
3. The ramose growth type was consistently found on half-closed *Festuca* grassland and in meadows with salty (solonetz) soils. Branches of the red-brown *Tolypothrix tenuis* filaments were observed on the soil surface, while other filaments were interwoven with the soil particles.
4. The mucose growth type was a characteristic constituent of the salt pasture (salty soils) in Hungary. It is composed of cyanobacteria and algae with sticky extracellular polysaccharide material (e.g., *Nostoc commune*, *Gloeocapsa arenaria* (Hassall) Rabenh., *Chroococcus minor* (Kütz.) Naeg., and some terricolous *Chlamydomonas* species) that adheres to the soil surface. At least part of the life cycle of these algae occurs inside the soil.
5. The glutinose growth type mainly occurred in the grass steppe ecosystem on solonetz soils with some mull-ranker influence (mull = specific form of humus with an intensive mixture of clay and Ca-humat; ranker = carbonate free soils with a humus horizon only), where they prefer the permanently wet soil. This group is composed of terricolous diatoms (*Hantzschia amphyoxis* [Ehr.] Grun., *Navicula mutica* Kütz., and *Pinnularia borealis* Ehr.), that live in the mucilage of other cyanobacterial or algal species (e.g., *Nostoc*).

Soil-crust communities rich in lichens, mosses, and cyanobacteria stretch from their outpost habitats in more or less isolated central European local steppe formations far into the extended steppe regions of east Europe and Asia. This is described in Chapter 7.

6.5 European Temperate Regions

Paus (1997) described several soil-crust communities composed of algae, lichens, and mosses occurring on soils within the European *Fagus sylvatica* forest zone in Northern Germany. Since many of these communities also occur in other parts of the European temperate region, a short description of the main communities is summarized below.

Dibaeetum baeomycetis. This community occurs on bare soil and is often a pioneer in low shrub vegetation at higher altitudes. It occurs in habitats with low to high light, but is restricted to areas receiving high precipitation. The main species include the lichens *Dibaeis baeomyces* (L.f.) Rambold et Hertel, *Baeomyces placophyllus* Ach. (Photo 15), *B. rufus* (Hudson) Rebent., the moss *Polytrichum commune* L. ex Hedw., and the green alga *Zygogonium ericetorum* Kütz. On bare acidic soils, *Zygogonium ericetorum* is often the primary pioneer (Photos 8, 43), followed by mosses (e.g., *Polytrichum piliferum* Schreb ex Brid.; B. Büdel, unpubl.).

Lecideetum uliginosae. This is a common pioneer community in heathland that needs high insolation and humic soils. Where it occurs, it often covers 80 % or more of the soil surface. It is dominated by the lichens *Placynthiella uliginosa* (Schrader) Coppins et P. James, *P. oligotropha* (Laundon) Coppins et P. James, *P. icmalea* (Ach.) Coppins et P. James, *Trapeliopsis granulosa* (Hoffm.) Lumbsch, and several *Cladonia* species. Mosses rarely occur in this community.

Cladonietum symphycarpae. This is a community characteristic of the xerothermic steppe vegetation on calcareous soils and on loess. For a more detailed description, see Section 6.3.

Cladonietum callosae. This community occurs in coastal and inland lowland heath on open soil. It can cover 40–98 % of the soil surface and is composed of the lichens *Cladonia callosa* Del. ex Harm. and other *Cladonia* species, *Baeomyces rufus*, the bryophytes *Cephaloziella* sp., *Polytrichum piliferum*, and the alga *Zygogonium ericetorum*.

Inland dunes are sometimes partly covered by combined lichen-moss communities such as the Steroacauletum condensati, dominated by the moss *Polytrichum piliferum* (35 % coverage), the lichen *Stereocaulon condensatum* Hoffm., and the green alga *Zygogonium ericetorum*.

6.5.1 Disturbed Soils of Central Europe

Lukešová and Komárek (1987) reported a distinct succession of soil algae, at least partly forming soil crusts, on coal mine spoils (covering almost 400 km^2) in North Bohemia. The spoils were first colonized by green algae and moss protonemata in the first year. Cyanobacteria occurred in larger species numbers and quantities only after 18 years. In *Pinus sylvestris* forests in the Czech Republic, the filamentous green algal genus *Klebsormidium* was observed to form dense soil crusts on bare sandy soils after fires (A. Lukešová, pers. comm.).

Soil crusts composed of the cyanobacterial genus *Nostoc* (mainly *N. microscopicum*), the liverworts *Blasia* sp, *Fossombronia* sp., *Riccia sorocarpa* Bisch., and the hornwort *Phaeoceros laevis* (L.) Prosk. are commonly found along corn field margins on siliceous soils in the German midland region (Mollenhauer 1970). In the midland region, Spessart, this peculiar crust type supports the rare endocytobiosis *Geosiphon pyriforme* (Kütz.) v. Wettstein, a symbiosis of a zygomycete and *Nostoc*.

6.6 Coastal Dunes

The coastal dune area Meijendel north of The Hague (The Netherlands) is part of the "younger dunes" and was formed between the 12th and 19th centuries. Where the vegetation of the dunes is disturbed, bare sand comes to the surface and is eroded by wind, forming "blowouts". During this process, the shape of the blowouts changes and the resulting depressions are subjected to different wind speeds. Where the surface wind energy is low, phototrophic organisms colonize the surface and form a soil crust (deWinder 1990), which may initiate vegetation succession. During periods of drought, these rigid crusts play an important role in erosion control, especially in the semiarid microclimatic zones of the dunes. Laboratory tests showed that sediment production from an uncolonized surface was 100 times higher than from an uncolonized surface (Van den Ancker et al. 1985; see also Chap. 25).

The crust system described by deWinder (1990) lies well above the groundwater level and receives water only during periods of rainfall or dew. Phototrophic organisms forming a crust were present mainly on the lee side of the blowout. The organisms were found in the uppermost millimeters of the sandy surface throughout the year and stabilized the soil. The crust positively influenced the water-holding capacity of the soil. The crusts were composed of the green algae *Klebsormidium flaccidum* and *K. mucosum*, and the cyanobacteria *Oscillatoria* sp. (2 species), *Tychonema* sp. (1), *Crinalium epipsammum* deWinder, LPP strains (2; LPP = group of cyanobacteria belonging to the *Lyngbya-Phormidium-Plectonema*-complex), *Microcoleus* sp. (3), *Synechococcus* sp. (1), and *Gloeocapsa* sp. (1). Crust succession started with *Oscillatoria* spp. and the LPP group, with *Tychonema* sp. as the most important primary colonizer. In a later stage *Crinalium epipsammum* can become dominant. Mature crusts are dominated by *Klebsormidium* spp., are very rigid, and up to 1–5 mm thick.

Along the west coast of Jütland (Denmark), dunes that extend from south to north have been investigated for vegetation dynamics by Müller (1999).

White dunes in the west are the youngest, gray dunes in the middle the next oldest, and brown dunes in the east the oldest. The author found communities of mosses and lichens growing in these dunes. Since most of the lichens were fruticose lichens, the formation of a crust is somewhat doubtful. However, because they aid in soil stabilization, and because the mosses at least penetrate the uppermost millimeters of the soil, some of the results are reported here. The "crusts" are composed of the mosses *Brachythecium albicans* (Web.) B.S.G., *Ceratodon purpureus* (Hedw.) Brid., *Polytrichum piliferum*, *Hypnum jutlandicum* Holmen et Warncke, *Pleurozium schreberi* (Brid.) Mitt., and the lichens *Coelocaulon aculeatum* (Schreber) Link, *Cladonia foliacea* (Huds.) Willd, *C. verticillata*, *C. rangiformis* Hoffm., *C. pleurota* (Flörke) Schaer., *C. furcata* (Hudson) Schrader, *C. subulata* (L.) Weber, *C. fimbriata* (L.) Fr., *C. uncialis* (L.) Wigg., *C. arbuscula* (Wallr.) Flotow, and *C. portentosa* (Dufour) Coem. On white dunes, this community was found only on the lee side of the dune, whereas on gray dunes, they had the highest diversity and occurred on all parts. On brown dunes, only a reduced number of species was found. According to the author, this community was not a successional phase, but a stable vegetation type. In the younger white dunes they existed only on the lee side, as moving sand covered them on other dune aspects. In the older brown dunes, a well-developed vascular vegetation limited their occurrence.

6.7 Northern Europe

The lichen-rich soil-crust communities of arctic Greenland are described in Chapter 4. Soil crusts consisting of lichens, mosses, algae, and cyanobacteria are reported from the northern part of Sweden in the vicinity of Abisko (Skuja 1964). The vegetation there is tundra-like, and the mean annual precipitation lies well above 1000 mm. However, Skuja's work focuses on the algal and cyanobacterial flora of this region. Mosses and lichens are mentioned only briefly, and no information on genera or species is given. The crusts are patchy, with parts dominated by lichens and mosses with cyanobacteria in between, and parts dominated by cyanobacteria and algae. The crusts were composed of cyanobacteria such as *Nostoc muscorum* Ag., *N. macrosporum* Menegh., *Scytonema mirabile* (Dillw.) Bornet, *S. myochrous* (Dillw.) Ag., *Schizothrix calcicola*, *Petalonema crustaceum* (Ag.) Kirchn., *Microcoleus sociatus*, *M. vaginatus*, *Gloeocapsa rupestris* Kütz., *G. compacta* Kütz., *Gloeothece palea* (Kütz.) Rbh., *Chroococcus tenax* (Kirchn.) Hieron., and *C. turgidus* (Kütz.) Näg.; green algae such as *Gloeocystis rupestris* (Lyngb.) Rbh., several *Cosmarium*-species, *Penium* spp., and *Euastrum* spp.; and diatoms such as *Eunotia praerupta*, *Pinnularia* spp., *Hantzschia* spp., *Caloneis alpestris*, *Navicula* spp., *Diploneis ovalis* (Hilse) Cl., and *Cymbella* spp. Granhall and

Henriksson (1969) studied several soil types in Sweden, including soils of the Abisko region, and determined their ability to fix atmospheric nitrogen. Moist soils with a pH between 6 and 8 had the greatest cyanobacterial flora (biomass) and, concomitantly, N_2 fixation. According to the authors, at least one third of the nitrogen input in undisturbed soils is contributed by cyanobacteria. The nitrogen-fixing genera determined by the authors were *Nostoc, Anabaena, Nodularia, Cylindrospermum, Tolypothrix,* and *Calothrix.*

The large lichen carpets of the taiga and tundra region in Scandinavia are heavily influenced by grazing and trampling of reindeer and, after a fire, the soil is bare for some time (Ahti and Oksanen 1990). Soils are first colonized by crust-forming lichens like *Trapeliopsis granulosa*, *Placynthiella oligotropha*, and *P. uliginosa*. Later, the bryophytes *Polytrichum piliferum* and *P. juniperinum* Hedw. become important. However, lichen soil crusts are only a succession stage (0–3 years) during the development of a lichen carpet dominated by the fruticose *Cladina* and *Cladonia* species

Pettersson (1946) reported the northernmost distribution of the soil crust-forming cyanolichen *Heppia lutosa* (Ach.) Nyl. from Öland and Gotland (Sweden). The mosses *Barbula convoluta* Hedw., *Tortula ruralis*, and other moss and liverwort species, as well as the crust-forming lichens *Toninia sedifolia, Psora decipiens, Squamarina lentigera, Catapyrenium lachneum,* and *Fulgensia bracteata* (Hoffm.) Räsänen were found growing together with *H. lutosa*. This is also the northernmost occurrence of the Bunte Erdflechtengesellschaft.

References

Ahti T, Oksanen J (1990) Epigeic lichen communities of taiga and tundra regions. Vegetatio 86:39–70

Alexander RW, Calvo A (1990) The influence of lichens on slope processes in some Spanish badlands. In: Thornes JB (ed) Vegetation and erosion. John Wiley, New York, pp 385–398

Alexander RW, Harvey AM, Calvo A, James PA, Cerda A (1994) Natural stabilization mechanisms on badland slopes: Tabernas, Almeria, Spain. In: Millington AC, Pye K (eds) Environmental changes in drylands: biogeographical and geomorphological perspectives. John Wiley, New York, pp 85–111

Alonso FL, Egea JM (1994) Líquenes calcícolas y terrícolas de algunas lacalidades costeras de Marruecos. Acta Bot Malacitana 19:51–61

Bornkamm R (1958) Die Bunte Erdflechten-Gesellschaft im südwestlichen Harzvorland. Ber Dtsch Bot Ges 71:253–270

De Winder B (1990) Ecophysiological strategies of drought-tolerant phototrophic microorganisms in dune soils. PhD Thesis, Department of Microbiology, University of Amsterdam, Amsterdam

Du Rietz GE (1925) Gotländische Vegetationsstudien. Sven Växtsociol Sällsk Handl 2: 1-65

Egea JM (1989) Los géneros *Heppia* y *Peltula* (Líquenes) en Europa Occidental y Norte de Africa. Bibl Lichenol 31:1–122

Frey W, Kürschner H (1990) Das Fossombronio-Gigaspermetum mouretii ass. nov. in der Judäischen Wüste. 1. Verbreitung und Soziologie. Cryptogam Bot 2:56–63

Frey W, Kürschner H (1991) Das Fossombronio-Gigaspermetum mouretii in der Judäischen Wüste. 2. Ökosoziologie und Lebensstrategien. Cryptogam Bot 2:73–84

Fritsch FE, John RP (1942) An ecological and taxonomic study of the algae of British soils. II. Consideration of the species observed. Ann Bot NS VI, 23:371–395

Granhall U, Henriksson E (1969) Nitrogen-fixing blue-green algae in Swedish soils. Oikos 20:175–178

Hahn S, Speer D, Meyer A, Lange OL (1989) Photosynthetische Primärproduktion von epigäischen Flechten im "Mainfränkischen Trockenrasen". I. Tagesläufe von Mikroklima, Wassergehalt und CO_2-Gaswechsel zu den verschiedenen Jahreszeiten. Flora 182:313–339

Klement O (1955) Prodromus der mitteleuropäischen Flechtengesellschaften. Feddes Repert Beih 135:5–194

Komáromy ZP (1976) Soil algal growth types as edaphic adaptations in Hungarian forest and grass steppe ecosystems. Acta Bot Acad Sci Hung 22:373–379

Lange OL, Reichenberger H, Meyer A (1995) High thallus water content and photosynthetic CO_2 exchange of lichens. Laboratory experiments with soil crust species from local xerothermic steppe formations in Franconia, Germany. In: Daniels JA, Schulz M, Peine J (eds) Contributions to lichenology in honour of Gerhard Follmann. University of Cologne, Cologne, Germany, pp 139–153

Litterski B (1997) Lichen observations on the island of Cyprus. Feddes Repert 108:463–473

Lukešová A, Komárek J (1987) Succession of soil algae on dumps from strip coal-mining in the Most region (Czechoslovakia). Folia Geobot Phytotax, Praha 22:355–362

Marstaller R (1968) Die Xerothermflora der Gipshänge bei Jena (Ostthüringen) unter besonderer Berücksichtigung der Bunten Erdflechten-Gesellschaften. Hercynica 5:352–372

Marstaller R (1971) Zur Kenntnis der Gesellschaften des Toninion-Verbandes im Unstruttal zwischen Nebra und Artern sowie im Kyffhäusergebirge. Hercynica NF 8:34–51

Mollenhauer D (1970) Botanische Notizen Nr. 1: Beobachtungen an der Flechte *Geosiphon pyriforme*. Nat Mus 100:213–223

Müller T (1965) Die Flechten der Eifel mit Berücksichtigung der angrenzenden Ardennen und der Kölner Bucht. Decheniana Beih 12:1–72

Müller J (1999) Kleinräumige Vegetationsdynamik in Küstendünen-Dauerflächen. Faun-Ökol Mitt Suppl 26:33–48

Ott S, Elders U, Jahns HM (1996) Vegetation of the rock-alvar of Gotland. I. Microhabitats and succession. Nova Hedwigia 63, 433–470

Paus SM (1997) Die Erdflechtenvegetation Nordwestdeutschlands und einiger Randgebiete. Bibl Lichenol 66:1–222

Pettersson B (1946) *Heppia lutosa* (Ach.) Nyl. i Sverige. Bot Not 1:94–102

Reimers H (1940) Bemerkenswerte Moos- und Flechtengesellschaften auf Zechstein-Gips am Südrande des Kyffhäuser. Hedwigia 79:81–174

Reimers H (1950) Beiträge zur Kenntnis der Bunten Erdflechten-Gesellschaft. I. Zur Systematik und Verbreitung der Charakterflechten der Gesellschaft besonders im Harzvorland. Ber Dtsch Bot Ges 63:148–156

Reimers H (1951) Beiträge zur Kenntnis der Bunten Erdflechten-Gesellschaft. II. Allgemeine Fragen. Ber Dtsch Bot Ges 64:36–50

Ritschel G (1974) Beitrag zur Kenntnis der Verbreitung xero- und basiphiler Erdflechten in Mainfranken. Abh Naturwiss Ver Würzburg 15:7–32

Roussomoustakaki M (1983) Cyanophyta from soils of Euboea Island (Hellas). Doctoral thesis of the Department of Biology, University of Athens, Athens (in Greek with English summary)

Sipman H, Raus T (1999) A lichenological comparison of the Paros and Santorini island groups (Aegean, Greece), with annotated checklist. Willdenowia 29:239–297

Skuja H (1964) Grundzüge der Algenflora und Algenvegetation der Fjeldgegenden um Abisko in Schwedisch Lappland. Nova Acta Regiae Soc Sci Ups Ser I, 18:1–465

Van den Ancker JAM, Jungerius PD, Mur LR (1985) The role of algae in the stabilization of coastal dune blow-outs. Earth Surf Proc Landf 10:189–192

Werner RG (1979) La flore lichenique de la Cordillere Betico-Rifaine. Collect Bot 11(17): 409–471

Wilmanns O (1959) Zur Kenntnis des Toninion coeruleonigricantis Reimers 1951 in Südwestdeutschland. Bot Jb 78:481-488

7 Biological Soil Crusts of Asia Including the Don and Volga Region

B. Büdel

7.1 Introduction

Information on biological soil crusts in Asia is extremely rare and scattered throughout the literature. Language barriers make it difficult to identify relevant publications. Although scientific literature on soil algal and cyanobacterial communities from the former USSR is fairly abundant (e.g., Bolyshev and Manucharova 1947; Gollerbakh et al. 1956; Forest 1965; Novichkova-Ivanova 1980), reports on soil cyanobacteria, algae, and lichens from other countries are usually restricted to single publications chronicling expeditions to specific biomes or regions (e.g., Cogt 1979; Huneck et al. 1984, 1987, 1992). In addition, most studies focused on the soil independent of its relationship to biological soil crusts (see Novichkova-Ivanova 1980). As a result, information on crust-forming organisms is scattered and is found partly in the phycological and lichenological literature that addresses "subaerial" algae (including cyanobacteria) and/or terricolous lichens. Because of their close phytogeographical relations to Asia, biological soil crusts of the eastern European Volga and Don region are included in this chapter as well.

7.2 Semideserts and Deserts of the Lower Courses of the Volga and Ural River

The semidesert region along the lower courses of the Volga and the Ural Rivers extends eastward to the foothills of the Altai. Within this region, crusts composed of cyanobacteria, lichens, and more rarely mosses and liverworts, cover large areas (Keller 1930). The crusts are more abundant in the northern parts of the semidesert. However, these crusts do not form compact, luxurious carpets, but rather scattered black patches when dry. After rainfall, they become greenish and prominent. The soils on which the crusts occur belong to the bright chestnut or brown-colored solonetz type. According to the

Ecological Studies, Vol. 150
J. Belnap and O.L. Lange (eds.) Biological Soil Crusts: Structure, Function, and Management

author, the main activity time of the crusts is spring and autumn. During summer, activity depends on rainfall events.

The most prominent cyanobacterium in these soil crusts is *Microcoleus vaginatus* (Vauch.) Gom. Occurring less frequently are *Scytonema ocellatum* Lyngbye ex Bornet et Flahault and scattered thalli of *Nostoc commune* (L.) Vaucher ex Bornet et Flahault. *Nostoc flagelliforme* Berk. et Curtis, with its characteristic large, hair-like thalli, occurs in some parts of the region (Keller 1930). However, both *Nostoc* species mentioned contribute to crust formation only during certain stages of their life cycles. Their large, easily visible thalli do not penetrate and are not fixed to the upper soil surface, and are thus not an integral part of the crust. Only their hormogonia and thallus primordia have an endedaphic phase.

In the semideserts of the lower Volga and Ural rivers, 44 different soil lichen taxa have been reported (Keller 1930). Only a few of them are found in soil crusts, including *Diploschistes scruposus* var. *terrestris* Pers., sometimes covering large areas; *Acarospora* spp.; *Catapyrenium squamulosum* (Ach.) Breuss [syn. *Placidium s.* (Ach.) Breuss]; *Endocarpon* spp., *Fulgensia fulgens* Elenk., *Psora decipiens* Hoffm., *P. lurida* Ach., *Squamarina lentigera* (Web.) Poelt., and *Toninia sedifolia* (Scop.) Timdal.

7.3 China

Although China has large regions in which soil-crust formation is likely, our knowledge about crusts in this area of the world is rather poor. However, in his enumeration of lichens in China, Jiang-chun (1991) lists quite a number of species known to form, or often be a part of, soil crusts. Among them are important genera containing typical soil-crust formers such as *Baeomyces, Catapyrenium, Collema, Diploschistes, Endocarpon, Fulgensia, Peltula, Psora, Psorotichia, Solorina, Squamarina,* and *Toninia.*

Extensive biological crust formation was reported by Reynaud and Lumpkin (1988) from a region east of Lanzhou in the Gansu province. The area lies on aeolian loess at 1556 m elevation and receives about 358 mm of precipitation annually, primarily in summer. Average temperatures are –14 °C in January and the average July maximum is 20 °C. Forty percent (by volume) of the quartz particles had a particle size of less than 5 µm and about 50 % (by volume) had a size between 5 and 20 µm. The pH of the soil was 8. The authors found the crust composed of 11 different organisms. The eukaryotic algae included the two unicellular forms (*Bracteacoccus* sp. and *Myrmecia* sp.) and one filamentous branched alga (*Monocilia* sp.), the latter representing the highest biomass of the crust. The cyanobacteria reported included the heterocystous N_2-fixing cyanobacterium *Tolypothrix* sp. Other cyanobacteria found in the crusts were *Oscillatoria williamsii* Mev., members of the LPP group B, *Phormidium* sp., the heterocystous forms *Nostoc* spp. (3 species), and *Anabaena* sp.

O.L. Lange (pers. comm.) reports from the *Artemisia-Haloxylon* steppe formation in the province of Gansu, close to Wuwei, extensive crust formation with several lichen species typical for associations of the Psoretea decipientis Mattick (Chap. 1, Photos 44, 45). The epedaphic, edible *Nostoc flagelliforme* cyanobacterium (Chinese: Facai, Photo 46) occurred in these crust communities and was also reported by Scherer (1994) from the province of Ningxia, at the margin of the central Asian desert. *Nostoc commune*, another epedaphic, sometimes crust-forming cyanobacterium is listed by Jiang-chun (1991) as the edible Di Er with an ear-like morphology. Soil crusts rich in crustose lichens, including *Collema* sp. and the epedaphic cyanobacterium *Nostoc commune*, are also typical for the extensive grasslands of Inner Mongolia (O.L. Lange and J. Belnap, pers. comm.). The UV-protecting substances in cyanobacteria were described for the first time in *Nostoc commune* specimens and cultures originating from the central Asian desert regions of China (Scherer et al. 1988).

7.4 Mongolia and the Gobi Desert

Cyanobacteria and lichens are known to form biological soil crusts in the Gobi Desert. Friedmann et al. (1995) reported that the unicellular cyanobacterial genera *Chroococcidiopsis, Aphanocapsa,* and the filamentous genera *Nostoc, Scytonema,* and *Phormidium* form a crust underneath the translucent stones that pave many desert areas. Biomass of these organisms ranged from 0.16 to 47.07 g dry organic matter m^{-2}.

Biological soil crusts with cyanobacteria (*Microcoleus* sp.) and lichens such as *Collema* sp. are abundant in the eastern semidesert and steppe formations of Mongolia, where precipitation ranges from 150–350 mm annually (Belnap, pers. comm.). The macroscopic thalli of *Nostoc commune* and *N. flagelliforme* can often be observed as constituents of these soil crusts. In central regions with rainfall below 150 mm, thin crusts containing only *Microcoleus* sp. were found in soil samples collected by Belnap in central Mongolia (B. Büdel unpubl.). Lichens that are extensively involved in soil-crust formation in the Gobi Desert and the adjacent steppe regions in the north are the squamulose-peltate cyanolichen *Peltula radicata* Nyl., the squamulose cyanolichen *Heppia lutosa* (Ach.) Nyl., *Peltula zabolotnoji* (Elenk.) Golubk., the semifruticose cyanolichens *Lichinella algerica* (J.Steiner) Moreno et Egea, *Lichinella stipatula* Nyl., *Peccania coralloides* (Massal.) Massal., *Collema* sp. and *Leptogium* sp. and the phycolichen genera *Aspicilia, Bacidia, Buellia, Catapy-*

renium, Diploschistes, Endocarpon, Fulgensia, Lecidea, Psora, Rinodina, Squamarina, and *Toninia* (Cogt 1979; Hunek et al. 1984, 1987, 1992). Extensive soil-crust formation, composed of the lichens *Xanthoparmelia camtschadalis* (Ach.) Hale, *Diploschistes muscorum* (Scop.) Sant. and *Acarospora schleicheri* (Ach.) Massal., was mentioned for the region between Tarnijn chijd and Cecerleg uul (Hunek et al. 1992). In summarizing the biogeographical data of all soil lichens found in Mongolia, Cogt (1979) characterized the flora as xero-meridional with a large influence of arctic-alpine and boreal species.

7.5 Central Asian Takyr

"Takyr"(playa) soils are characteristic for the deserts and semideserts of Central Asia. Takyr soils are flat, delta-like areas at the end of large valleys, where water pools after heavy rainfall. Infiltration of water into the heavy clay soils of takyrs is minimal, and as a result, the takyr soil is dry again soon after flooding. They are usually devoid of higher plant vegetation and covered by a thin, brownish, elastic, somewhat lustrous, and velvety cyanobacterial crust (Bolyshev and Evdokimova 1955, after Friedmann and Galun 1974). According to the authors' report, the crusts are primarily formed by filamentous cyanobacteria of the Oscillatoriales. The most common species of the takyr crusts is *Microcoleus vaginatus,* whose sheath material glues soil particles together. Thirty-one cyanobacterial species were found in the crusts: *Phormidium* (10 species), *Lyngbya* (9), *Nostoc* (4), *Oscillatoria* (2), *Microcoleus* (2), *Aphanothece* (1), *Nodularia* (1), *Schizothrix* (1), and *Calothrix* (1) at two different localities reported by Bolyshev (1955, after Friedmann and Galun 1974). In addition, 12 unicellular green algae, three diatoms, and moss protonemata were found. *Nostoc* species in the crusts sometimes may become lichenized, resulting in the establishment of *Collema* species. Additional cyanobacteria were found by Forest (1965), including *Symploca atlantica* Gom., *Schizothrix calcicola* (Ag.) Gom., and *Microcoleus chtonoplastes* (Mert.) Zanard. Gollerbakh et al. (1956) mention in total 147 soil algal species, of which 92 belong to the cyanobacteria (*Microcoleus vaginatus,* 32 *Phormidium* spp., and other genera). At higher elevations, where flooding does not occur but where the soils still become wet, lichen crusts develop, consisting of *Diploschistes diacapsis* (Ach.) Lumbsch, *D. scruposus* (Schreb.) Norman, *Squamarina lentigera* (Web.) Poelt, together with *Collema* spp., *Aspicilia* spp., and *Psora* spp. On average, between 48 to 91 different species of soil cyanobacteria and algae occur in takyr soil crusts, with the cyanobacteria being the dominating part (Novichkova-Ivanova 1980). In these soil crusts, dry matter is between 0.5 and 1.0 (rarely 1.4) t ha^{-1} and nitrogen content is high (4.5 % of the dry matter).

7.6 Caspian Region and Steppe Formations Along the Don River

Yabukov et al. (1961) report on the soil formation processes from the western part of north Caucasian deserts. The climate in the area is extremely continental, and characterized by air temperatures of 45 °C and surface temperature up to 75 °C in the summer. Minimum air temperature in winter is –40 °C. The annual precipitation does not exceed 200 mm. Solonchak soils in small depressions form a characteristic element of the landscape. In these depressions, mosses of the genera *Tortula* and *Bryum*, together with lichens of the genus *Parmelia*, cover sandy soils to a considerable degree (60–70 %), thus stabilizing it by the formation of a crust.

Biological crusts in the north Caspian sands are dominated by thick films of cyanobacteria (Gayel and Shtina 1974) composed of *Phormidium* (mainly *P. foveolarum* Gomont and *P. corium* Gomont), *Schizothrix, Plectonema, Microcoleus vaginatus, Microcystis* (?), and *Nostoc* [mainly *N. muscorum* Agardh ex Bornet et Flahault and *N. microscopicum* (Carmichael ex Harvey) Bornet et Flahault]. Green algae, which play an important role in the sands of the Don steppe formation (below), were completely absent here.

The sandy soils of the Don steppe region were evaluated for soil-crust formation by Gayel and Shtina (1974). In this study, the authors show that soil crusts establish in the Golubinskiye sands after phanerogams colonize the sand. These initial crusts were dominated by ubiquitous green algae such as *Chlorococcum* spp., *Coccomyxa* sp., *Chlorella vulgaris* Beijerinck, and *Desmococcus olivaceus* (Pers. ex Ach.) Laundon. In addition, the filamentous green algal species *Klebsormidium montanum* (Skuja) S.Watanabe and the filamentous cyanobacteria *Schizothrix* and *Plectonema* were observed. In the Archedino-type sands, covering 300 000 ha and associated with floodplain terraces of the Don Valley between the mouths of the Medveditsa and Ilovlya Rivers, the authors found a number of different soil-inhabiting green algae and cyanobacteria. However, crust formation was found only in some areas, where the sand-binding, filamentous cyanobacteria *Phormidium, Plectonema, Schizothrix, Nostoc commune*, and *Nostoc muscorum* were found together with the green alga *Klebsormidium flaccidum* (Kütz.) Silva, Mattox et Blackwell. The authors also observed black and silvery lichen crusts, but did not identify them. Most of the identifications were performed using culture methods. The isolated algae and cyanobacteria were regularly accompanied by the development of fungi, bacteria, actinomycetes, and shelled ameba. These organisms are certainly important in the formation of soil crusts.

7.7 Japan

Two investigations by Akiyama (1961, 1963) on terrestrial organisms of the San-in region of Honshu reports 70 species of Chlorophyta, 26 species of Cyanobacteria, 12 species of Tribophyceae, 15 species of Bacillariophyceae, and one red alga (Rhodophyta) in the soil. Although the author stresses the role of the development of gelatinous sheaths or the formation of mats as an adaptive phase, it is not clearly stated whether or not crust formation occurs. In the lichen flora of Japan, Yoshimura (1987) lists a number of lichen genera potentially capable of soil-crust formation including *Baeomyces, Collema, Diploschistes, Endocarpon,* and *Solorina.*

7.8 India and Sri Lanka

Examples of soil-crust formation in India are described from the Nagpur region by Marathe and Kushaldas (1975). They reported several types of crusts, distinguished by species composition and the dominant organism. One crust was taken from garden soil and had either *Aphanothece pallida* (Kütz.) Rabenh. or *Oscillatoria obscura* Bruhl et Biswas as the dominating cyanobacteria, with associated *Cylindrospermum muscicola* Kütz. ex Born. et Flahault, *Plectonema hansgirgi* Schmidle, *Microchaete tenera* Thuret, and a few diatoms. The other crusts all came from undisturbed soils. They were: (1) dominated by the green alga *Microspora floccosa* (Vaucher) Thuret, together with *Sirocladium maharashtrense* Randhawa, *Oscillatoria amoena* (Kütz.) Gomont, and diatoms; (2) dominated by *Oscillatoria obscura*, together with *Phormidium foveolarum* (Mont.) Gomont, and diatoms; (3) dominated by *Microcoleus vaginatus*, and sometimes associated with the unicellular green algae *Chlamydomonas* spp. or (4) most commonly, dominated by the cyanobacteria *S. maharashtrense* or *O. amoena*. The presence of the soil crusts was hypothesized to influence soil pH, nitrogen, phosphorous, and potassium contents. However, the values given in their work show only small differences between colonized and noncolonized soil.

In the comprehensive floras by Awasthi (1988, 1991) covering macro- and micro-lichens of India, Nepal, and Sri Lanka, many species of terrestrial cyanolichens and phycolichens are listed which normally are characteristic of soil-crust communities (*Catapyrenium squamulosum*, *C. pusillum* Hedw., *Collema, Endocarpon, Fulgensia, Peltula, Solorina, Squamarina,* and *Toninia sedifolia*). No detailed habitat characteristics are given, so that conclusions from these publications about formation and general distribution of Indian soil crusts are not possible. However, it is expected that biological soil crusts are abundant in arid and semiarid lands in India. The soil crust-forming cyanolichens *Heppia lutosa* (Ach.) Nyl. and *Peltula patellata* (Bagl.) Swinscow et Krog are reported from northern India (Upreti and Büdel 1990; Awashti 1991). There is one report that biological soil crusts with several species of crustose and squamulose lichens are frequent in shrub and succulent formations of the Thar Desert near Jodhpur (Rajasthan) in northwest India (O.L. Lange, pers. comm.).

Fritsch (1907) mentions the presence of algal mats on soil composed of cyanobacteria in his treatment of the subaerial algae from Sri Lanka (Ceylon). This crust had a tangled growth type and was composed mainly of the filamentous genera *Tolypothrix, Hapalosiphon, Scytonema, Stigonema,* etc. However, there is no clear evidence in the text as to whether the cyanobacterial mats are involved in soil-crust formation by cementing soil particles.

References

Akiyama M (1961) Aerial and terrestrial algae in San-in region of Honshu, Japan. Bull Shimane Univ Nat Sci 10:75–89

Akiyama M (1963) Some soil algae from Japan. Bull Shimane Univ Nat Sci 15:96–117

Awasthi DD (1988) A key to the macrolichens of India and Nepal. J Hattori Bot Lab 65:207–302

Awasthi DD (1991) A key to the microlichens of India, Nepal and Sri Lanka. Bibl Lichenol 40:1–337

Bolyshev NN, Evdokimova TIO (1955) O prirode korochek takyrov (The nature of takyr crusts). Pochvovedenie NS 7–8:345–352

Bolyshev NN, Manucharova EA (1947) The distribution of algae in the profile of some desert soils. Vestn Mosk Univ Ser Biol 8:115–130

Cogt U (1979) Bodenflechten der mongolischen Volksrepublik. Feddes Repert 90: 421–440

Forest HS (1965) The soil algal community. II. Soviet soil studies. J Phycol 1:164–171

Friedmann EI, Galun M (1974) Desert algae, lichens, and fungi. In: Brown GW (ed) Desert biology. II. Academic Press, New York, pp 165–212

Friedmann EI, Allen ME, Druk AY (1995) Cyanobacteria: important primary producers in extreme arid deserts. In: Grilli Caiola M, Albertano P (eds) 13th Inter Symp on Cyanophyte Research. Programme and Abstracts. Univers Rome, pp 34

Fritsch FE (1907) A general consideration of the subaerial and fresh-water algae flora of Ceylon. A contribution to the study of tropical algae ecology. I. Subaerial algae of the inland fresh-waters. Proc R Soc Lond Ser B 79:197–254

Gayel AG, Shtina EA (1974) Algae on the sands of arid regions and their role in soil formation. Sov Soil Sci 6:311–319 (translated: Pochvovedeniye 6:67–75)

Gollerbakh MM, Novichkova LN, Zdobnikova VV (1956) The algae of the takyry. In: Kovda VA (ed) The takyry of western Turkmenistan and the means of their agricultural utilization. Nauka, Moscow, pp 38–54 (in Russian)

Huneck S, Poelt J, Ahti T, Vitikainen O, Cogt U (1984) Zur Verbreitung und Chemie von Flechten der Mongolischen Volksrepublik. Ergebnisse der Mongolisch-Deutschen

Biologischen Expeditionen seit 1962, Nr. 128. Erforsch biol Ress MVR, Halle (Saale) 4:51–62

Huneck S, Poelt J, Ahti T, Vitikainen O, Cogt U (1987) Zur Verbreitung und Chemie von Flechten der Mongolischen Volksrepublik. II. Ergebnisse der Mongolisch-Deutschen Biologischen Expeditionen seit 1962 Nr. 177. Nova Hedwigia 44:189–213

Huneck S, Ahti T, Cogt U, Poelt J, Sipman H (1992) Zur Verbreitung und Chemie von Flechten der Mongolei. III. Ergebnisse der Mongolisch-Deutschen Biologischen Expedition seit 1962 Nr. 217. Nova Hedwigia 54:277–308

Jiang-chun W (1991) An enumeration of lichens in China. Int Academic Publishers, Beijing

Keller B (1930) Die Erdflechten und Cyanophyceen am unteren Lauf der Wolga und des Ural. In: Karsten G (ed) Vegetationsbilder, 20. Reihe, Heft 8. Fischer, Jena

Marathe K, Kushaldas C (1975) Observations on the algal crusts of Nagpur soils. J Univ Bombay 44:66–76

Novichkova-Ivanova LN (1980) Soil algae of the Sahara-Gobi Desert region. Nauka Publishing House, Leningrad (in Russian)

Reynaud PA, Lumpkin TA (1988) Microalgae of the Lanzhou (China) cryptogamic crust. Arid Soil Res Rehab 2:145–155

Scherer S (1994) Anpassungen von Cyanobakterien in Wüsten. In: Hausmann K, Kremer BP (eds) Extremophile, Mikroorganismen in ausgefallenen Lebensräumen. VCH Verlagsgesellschaft, Weinheim, pp 179–193

Scherer S, Chen TW, Böger P (1988) A new UV-A/B protecting pigment in the terrestrial cyanobacterium *Nostoc commune*. Plant Physiol 88:1055–1057

Upreti DK, Büdel B (1990) The lichen genera *Heppia* and *Peltula* in India. J Hattori Bot Lab 68:279–248

Yabukov TF, Bespalova RY, Dokuchayev VV (1961) Soil formation processes during the invasion of sands by plants in the northern deserts of the Caspian region. Sov Soil Sci 6:651–658

Yoshimura I (1987) Lichen flora of Japan in color. Hoikusha Publishing Co, Osaka (in Japanese)

8 Biological Soil Crusts of the Middle East

M. GALUN and J. GARTY

8.1 Introduction

Large parts of the Middle Eastern countries (Egypt, Israel, Lebanon, Syria, Jordan, Iraq, Saudi Arabia, Kuwait) are arid and semiarid regions, and thus are characterized by limited higher plant cover. These areas are often densely covered by communities of cyanobacteria, green algae, fungi, lichens, and mosses which form typical biological soil crusts. The information available on these crusts for the Middle East is very sparse, except for Israel, where extensive research has been conducted. The study areas in Israel are the Central Negev, the Arava Valley and Judean Desert, and the northwestern Negev Desert (Fig. 8.1), which are discussed separately below. The political borders between the countries of the Middle East, now and in the past, do not coincide with any climatic, geological, vegetational or other natural partitioning. Nevertheless, the research and the information published are presented according to the countries and thus is also the format for this chapter.

8.2 Central Negev

8.2.1 Cyanobacteria and Green Algae

The dominant organisms in biological soil crusts of the Central Negev are cyanobacteria. Identification of these organisms is very difficult, as they have different shapes under different growth conditions. Also, when isolated and cultured, they undergo dramatic morphological, and perhaps also physiological, modifications. With these limitations in mind, the recorded cyanobacteria for this area (Friedmann and Galun 1974) include *Nostoc muscorum* Ag., *Microcoleus chtonoplastes* (Mert.) Zanard., *M. vaginatus* (Vauch.) Gom. (Chap. 1, Photo 1), *Oscillatoria* sp., and *Botrydium granulatum* (L.) Grev.

Ecological Studies, Vol. 150
J. Belnap and O.L. Lange (eds.) Biological Soil Crusts: Structure, Function, and Management

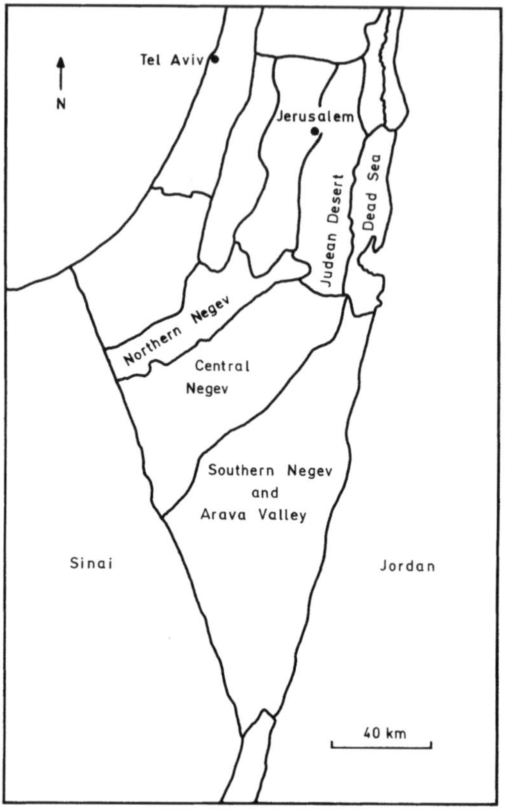

Fig. 8.1. Landscapes of south Israel

Green algae are also reported for soil crusts in the Central Negev (Shachack and Steinberger 1980), including *Protococcus grevilli* (Ag.) Crouan and *Protosiphon cinnamomens* (Menegh.) Drouet et Daily. In the Central Negev, Friedmann and Galun (1974) found *Bracteacoccus minor* (Chodat) Petrova var. *desertorum* Friedmann et Ocampo-Paus, *Chlorosarcinopsis eremi* Chantachat et Bold, *C. negevensis* f. *negevensis* and f. *ferruginea* Friedmann et Ocampo-Paus, *Protosiphon botryoides* (Kuetz.) Klebs, *Radiosphaera negevensis* f. *negevensis,* and f. *minor* Ocampo-Paus et Friedmann.

8.2.2 Lichens

The soil-crust lichens of the Central Negev listed below comprise a part of the well-studied lichen flora of the Negev in general. This list is taken from the *Lichen Flora of Israel* (Galun 1970), which was reconfirmed taxonomically and amended by Galun and Mukhtar (1996). Species include *Acarospora murorum* A.Massal., *A. reagens* Zahlbr. f. *radicans* (Nyl.) H.Magn., *Buellia epigaea* (Pers.) Tuck., *B. zoharyi* Galun in Poelt et Sulzer, *Catapyrenium squamulosum*

(Ach.) Breuss (syn. *Placidium s.* Breuss) (Photo 21), *Collema coccophorum* Tuck., *C. crispum* (Huds.) Weber ex F.H.Wigg., *C. tenax* (Sw.) Ach., *Diploschistes diacapsis* (Ach.) Lumbsch (syn. *D. steppicus* Reichert), *Fulgensia fulgens* (Sw.) Elenkin (Photo 16), *Gloeoheppia turgida* (Ach.) Gyeln., *Psora decipiens* (Hedw.) Hoffm., *Squamarina cartilaginea* (With.) P.James, *S. cartilaginea* f. *pseudocrassa* (Mattick) Poelt, *S. lentigera* (Weber) Poelt (Photo 12), *Toninia albilabra* (Dufour) H.Olivier, *T. aromatica* (Sm.) A.Massal., and *T. sedifolia* (Scop.) Timdal. Many of these lichens have thick pruinose covers, often to the extent that the original color of the thallus is concealed (Galun 1963). This is especially obvious for *Psora decipiens* growing on soil in the desert (see Chap. 1).

In 1940 Reichert described *Diploschistes diacapsis* as *D. steppicus* n. sp. I.Reichert and compared it in length with *D. albissimus.* According to Reichert "it has been found in localities in Palestine, Syria, Transjordan, Iraq, Persia and southeastern Transcaucasia, is closely associated with the vegetation of the steppe of Irano-Turanian region, and may serve as an indicator of steppe conditions". In his lichenological survey of Israel (at that time Palestine), Reichert (1937) describes a continuous cover of lichens on the soil of the Saharo-Sindian desert region of Israel and Egypt, which are regions devoid of higher plants. Reichert mentions the importance of these lichens as soil stabilizers.

8.2.3 Fungi

Rayss and Borut (1958) isolated, cultured, identified, and described 107 fungi from various soils in Israel. Among these, they found a rather rich mycoflora in the soils of the arid zones (the Negev, the Arava Valley and the Judean Desert), dominated by species of the genera *Aspergillus, Penicillium, Chaetomium,* and *Alternaria.* These genera are characterized by dark-pigmented mycelia and/or spores or conidia, by thick cell walls, and multicellular spores (Borut 1960).

8.3 Arava Valley and Judean Desert

8.3.1 Cyanobacteria and Green Algae

A dominant phenomenon in this region is the hypolithic (the lower surfaces of stones) growth of cyanobacteria and green algae. These stones are tightly glued to the soil surface by these organisms, and certainly contribute to the stability of the crust. Friedmann et al. (1967) found the following hypolithic green algae: *Bracteacoccus* sp., *Chlorosarcinopsis negevensis* Friedmann et Ocampo-Paus, *Friedmannia israeliensis* Chatanachat et Bold, *Hormidium sterile* Deason et Bold, *H. subtilissimum* (Rabenh.) Mattox et Bold, *Radiosphaera negevensis* Ocampo-Paus et Friedmann, *Stichococcus* sp., *Trebouxia* sp., *Trochiscia* sp., and *Ulothrix minuta* Mattox et Bold. They also found the following cyanobacteria: *Paracapsa* sp., *Aphanocapsa* sp., *Gloeocapsa* spp., *Lyngbya* spp., *Microcoleus chtonoplastes*, *Myxosarcina* sp., *Nostoc* spp., *Plectonema* sp., *Schizothrix calcicola* (Ag.) Gom., *Schizothrix* spp., *Scytonema* sp., and *Tolypothrix byssoidea* (Berkeley) Kirchner.

Dor and Danin (1996) recently recorded the cyanobacterial population from soil crusts in the Dead Sea Valley, and found the following species: *Aphanothece* sp., *Chroococcidiopsis* sp., *Microcoleus vaginatus* Gom., *Nostoc punctiforme* (Kuetz.) ex Hariot, *Petalonema vetulinum* (Rab. ex Born. et Flah.) Mig., *Phormidium fragile* (Menegh.) Gom., *Protococcus* sp., *Schizothrix friesii* Gom., *S. arenaria* (Bork.) Gom., and *Scytonema hofmanni* Ag. ex Born. et Plah. They found also one green algal species, *Chlorogloea microcystoides* Geit.

8.3.2 Lichens

Other important contributors of soil-crust formation in this region are terricolous cyanolichens, extensively studied by Marton and Galun (1981), including *Collema crispum*, *C. cristatum* (L.) Weber ex F.H.Wigg., *C. tenax*, *Gloeoheppia erosa* (J.Steiner) K.Marton, *G. turgida* (Ach.) Gyeln., *Gonohymenia algerica* J.Steiner, *G. collemoides* K.Marton et Galun, *G. sinaica* (Galun et K.Marton) Moreno et Egea, *G. sodomensis* K.Marton et Galun, *Heppia despreauxii* (Mont.) Tuck., *H. echinulata* K.Marton et Galun, *Peccania* sp., *Peltula bolanderi* (Tuck.) Wetm., *P. patellata* (Bagl.) Swinscow et Krog, *Psorotichia numidella* Forsell, and *Thyrea* spp. The cyanobionts of all these lichens, except for *Collema* spp., appear similar in the lichenized state, resembling members of the cyanophycean genus *Gloeocapsa*. However, after isolation from the lichens and culturing, this organism reveals a wide range of morphological diversity (Bubrick and Galun 1984).

Scattered, poorly developed populations of phycolichens are found in both the Arava Valley and the Judean Desert, including *Acarospora reagens*. f. *radicans*, *Catapyrenium squamulosum*, *C. lacinulatum* (Ach.) Breuss (syn. *Placidium l.* Breuss), *Diploschistes diacapsis*, *Fulgensia fulgens*, *Psora decipiens*, *Squamarina cartilaginea*, *S. lentigera*, and *Toninia sedifolia*.

8.3.3 Mosses

Moss rhizoids and protonemata also play an important role in soil stabilization of arid and semiarid regions (Lange et al. 1992), and aid in preventing soil erosion (Danin and Ganor 1991). In their study on life strategies of terrestrial bryophytes in the Judean Desert, Frey and Kürschner (1991) distinguished eight moss associations along a transect from the Dead Sea, through the Judean Desert, and to the Judean Mountains. The associations that characterize the phytogeographic territories this transect passes through include the Crossidio laevipili–Tortuletum atrovirentis association of the Saharo–Arabian territory, the Crossidio crassinervis–Tortuletum obtusatae, the *Grimmia mesopotamica* community, the Trichostomposo–Tortuletum obtusatae, the Fossombronio–Gigaspermetum moureti and the Crossidio crassinervis–Aloinetum aloidis associations of the Irano–Turanian territory, and the Riccio–Pseudocrossidietum hornschuchiani and the Bryo–Barbuletum acutae associations of the Mediterranean territory.

8.4 North Western Negev Desert and Sinai Peninsula (Egypt)

Karnieli and Tsoar (1995) studied biological soil crusts with the aid of space-borne sensors in the northwestern Negev and the bordering Sinai Peninsula. They showed a sharp contrast between the bright reflectance value from the Sinai side and the dark reflectance value from the Negev side due to lack of soil crusts on the overgrazed Egyptian side (see detailed discussion in Chap. 31).

8.4.1 Cyanobacteria and Green Algae

The dominant organism of the biological soil crusts in this area is the cyanobacterium *Microcoleus sociatus* W. et G.S.West (Lange et al. 1992), accompanied by *Calothrix parietina* Thuret, *Nostoc* sp., and *Gloeocapsa* sp. According to Karnieli and Tsoar (1995), these crusts also contain *Scytonema* sp., *Schizothrix* sp., *Chroococcidiopsis* sp. and *Phormidium* sp. Lange et al. (1992) found the chlorophytes *Macrochloris multinucleata* (Reisigl) Ettl et Gärtner (originally reported as *Chlorococcum* sp., O.L. Lange and B. Büdel, pers. comm.) and *Stichococcus* sp.

8.4.2 Lichens

Among the 51 lichen species recorded from the desert of the Northern and Central Sinai Peninsula, several are typical soil-crust lichens (Galun and Garty 1972) and are found mainly on soil at the foot of rocks. These include *Collema tenax* var. *vulgare* (Schaer.) Degel., *Gonohymenia sinaica*, *Psora decipiens*, *Toninia albilabra*, *T. sedifolia*, *T. aromatica*, *Diploschistes diacapsis*, *Squamarina cartilaginea*, *S. cartilaginea* f. *pseudocrassa*, *S. lentigera*, *Buellia zoharyi*, *Fulgensia desertorum* (Tomin) Poelt, and *Catapyrenium squamulosum*. Among the lichens of Egypt enumerated by Müller (1880) are two additional soil-crust species from the Egyptian desert: *Fulgensia fulgens* and *Acarospora reagens* f. *radicans*.

8.5 Distribution of Communities of Biological Soil-Crust Organisms in Israel

According to the analysis of the desert and coastal vegetation by Danin and Orshan (1999), biological soil crusts are a very frequent and broadly distributed phenomenon throughout Israel. Crusts occur from the semisteppe batha and the different shrub-steppe communities to the various types of desert and sand communities. Lichen-rich soil crusts are found also in the mediterranean belt of the country. Differential spatial distribution of mosses, lichens, and cyanobacteria on north- and south-facing hill slopes along a gradient from a mediterranean climate to an extreme arid climatic was documented by Kutiel et al. (1998). Several detailed studies describe the structure, ecological conditions, and ecosystem importance of biological soil crusts in Israel (e.g., Danin and Barbour 1982; Danin et al. 1989; Lange et al. 1992; Kidron 1995).

The Central Negev and the Arava Valley are two closely situated geographical desert regions in southern Israel (Fig. 8.1). They differ, however, in important climatic factors, with the Arava Valley being more extreme in terms of lower preciptitaion and dewfall, and higher temperature and light. As a consequence, the soil crust-forming lichen flora is completely different: phycolichens dominate in the Central Negev, whereas in the Arava Valley, the lichen flora consists almost entirely of cyanobacterial species belonging to Lichenaceae, Heppiaceae, and Collemataceae, which apparently are better suited to the extremely inhospitable conditions found there (Galun et al. 1982). In the Dead Sea Valley, even cyanolichens become rare, and soil crusts are mainly populated by endedaphic filamentous cyanobacteria, which are involved in soil polygon formation. Danin et al. (1998) describe the succesional processes in crust-biota populations colonizing alluvial terraces. Cyclic microsuccessions on loessial slopes facing the Dead Sea are correlated to microtopography and salinity within an area of a few m^2 and along an elevational gradient of 10 cm (Danin and Barbour 1982).

Predominantly smooth biological soil crusts (Photos 49, 50), consisting of endedaphic filamentous cyanobacteria and green algae, together with a few moss and lichen species, are characteristic of the sand dune systems of the western Negev (Danin et al.1989; Lange et al. 1992). Kidron (1995) has defined five different crust types. As shown by Danin et al. (1989), these crusts interact with the successional processes of higher plant vegetation, and they effectively increase sand stability. Crust formation decreases the infiltration rate of rainwater and promotes runoff generation in this region, thus impacting landscape structure (Chap. 22).

8.6 Jordan

The lichen flora of Jordan, particularly from the Azraq Oasis in the Eastern Province (El-Oqlah et al. 1986) is comparable to the composition of the lichen flora of the Central Negev, and includes soil species such as *Toninia sedifolia*, *Psora decipiens*, and *Collema tenax* (El-Oqlah and Lahham 1985; El-Oqlah 1992). Frey and Kürschner (1992, 1994) record the associations of bryophytes that they found in semiarid regions in Jordan (Photo 47, 48), including the Trichostomopso–Barbuletum trifariae, Tortuletosum iranicae, and the Tortuletum parnassicae.

8.7 Iraq

Hamdi et al. (1978) reported an abundance of *Azotobacter* and nonsymbiotic N-fixing cyanobacteria in cultivated soils of Iraq. They found fewer of these organisms in uncultivated soils. Reichert (1940) mentioned *Diploschistes steppicus* (now *D. diacapsis*) from Iraq. Schubert (1973) studied terrestrial lichen communities in the north of Mesopotamia between Baghdad and Fallfar. Typical species were *Catapyrenium squamulosum*, *Collema crispum*, *Diploschistes diacapsis*, *Fulgensia bracteata* (Hoffm.) Räsänen, *Psora decipiens*, *Squamarina lentigera*, and *Toninia sedifolia*. These communities, which he classified in the alliance of Toninion sedifoliae, occur in gaps between the vascular vegetation of *Stipa capensis* and *Artemisia herba-alba* semi-deserts, and in different kinds of steppe formations. There was a significant increase in lichen diversity and abundance with increasing rainfall from south to north. Gypsiferous soils were especially rich in soil-crust lichens.

8.8 Lebanon and Syria

In a series of lichenological studies, Werner (1954–1958) enumerated a long list of lichens from Lebanon and Syria. Most of these are saxicolous and corticolous species, well known to us from the mediterranean localities in Israel. The soil-crust species mentioned in these lists and confined to arid regions include *Diploschistes diacapsis*, *Psora decipiens*, *Squamarina lentigera*, *S. cartilaginea*, and *Toninia sedifolia*.

8.9 Saudi Arabia

8.9.1 Cyanobacteria

Chantanachat and Bold (1962) and Abu-Zinada and El Husseing (1975) identified a number of soil cyanobacteria from Saudi Arabia. However, most were found in cultivated soils, except for *Merismopedia* sp. and *Oscillatoria* sp., which they recorded from desert soils.

8.9.2 Fungi

Fathi et al. (1975) studied the fungal population of various soils in Saudi Arabia. They isolated and identified the following genera from the Rhazya desert soil: *Trichoderma*, *Penicillium*, *Mucor*, and *Sporotrichum*.

8.9.3 Lichens

Among the 67 lichen species reported from Saudi Arabia, Abu-Zinada et al. (1986) found the following soil-crust species in the Central Province (Riyadh Region): *Collema tenax*, *Gloeoheppia turgida*, *Gonohymenia* sp., *Heppia lutosa* (Ach.) Nyl., *Peltula radicata* Nyl., *Psora decipiens*, *Catapyrenium lachneum* (Ach.) R.Sant., and *Fulgensia fulgens*. The authors defined the habitat of these species as dried mud near the edge of a wadi, where no rain was reported to have fallen for some years.

8.9.4 Mosses

Frey and Kürschner (1988) collected and identified 40 species of bryophytes in Saudi Arabia. Their habitat and distribution are recorded in a more general treatise on bryophytes of the Arabian Peninsula. More recently, Frey et al. (1991, 1994) recorded several additional terrestrial bryophytes found in arid regions of Arabia and Jordan.

8.10 Kuwait and the Yemen

Two recent publications report on lichens of the desert areas of Kuwait. Brown (1998) studied terrestrial lichens on base-rich substrate. Common with the soil-crust species of the adjacent areas are the phycolichens *Psora decipiens*, *Squamarina lentigera*, and *Toninia sedifolia*, and the cyanolichen *Gloeoheppia turgida*, which are members of the lichen community Toninio–Psoretum (Toninion sedifoliae). Schultz et al. (2000) described 18 additional cyanolichens from Kuwait. Several of them were found on calcareous substrate, often on compacted sand in rock crevices. Species found included *Collema coccophorum*, *Gloeoheppia erosa*, *G. turgida*, *Heppia despreauxii*, *Gonohymenia algerica*, *G. iodopulchra* (Croz.) Henss., *G. mauritanica* O.L.Lange (these *Gonohymenia* species being listed as *Lichinella* spp.), *Peccania coralloides* (Massal.) Massal., *P. fontquerina* Moreno et Egea, and *Peltula radicata* Nyl. There is a striking correspondence between this flora and the cyanophilous lichen population which is reported by Marton and Galun (1981) for the Arava Valley.

According to Schultz (1998 and pers. comm.) and Belnap (pers. comm.), the terrestrial lichen vegetation of the United Arab Emirates, northern Oman, and southern Yemen resembles that of Kuwait, particularly with respect to the cyanolichens present. Under extreme desert conditions in lower altitudes, soil lichens are rare, and only cyanobacteria such as *Scytonema* seem to form soil crusts. With increasing altitude, more and more cyanolichens appear. Sand-filled clefts and rock crevices are the substrate for *Peltula*, *Heppia*, *Peccania*, *Gonohymenia*, *Collema*, and *Synalissa* species. Above 1000 m, typical soil lichen communities occur with *Psora decipiens*, *Fulgensia fulgida* (Nyl.) Szatala, and *Collema coccophorum*, and at 2000 m the ground between rocks often is totally covered by members of the Toninion alliance. The cyanolichen flora of the Yemeni island Socotra seems to be related to that reported for the Yemen (Schultz et al. 2000), whereas characteristic phycolichens of the Toninion alliance are lacking.

8.11 Conclusions

The above account indicates a high level of uniformity in the composition of the biological crusts of the deserts in the Middle Eastern countries. Several of the soil-crust lichens in the deserts of the Middle Eastern regions are also common in the Mediterranean territory, such as *Psora decipiens*, *Fulgensia fulgens*, *Collema tenax*, *Squamarina cartilaginea*, *Catapyrenium squamulosum*, and *Toninia sedifolia*. Indeed, West (1990) points out that: "if the same taxonomic rules are employed, then one usually finds that the majority of species (especially the microscopic ones) are the same when homologous environments are compared between continents" (see also Chaps. 1, 12).

Acknowledgments. This chapter was prepared under the auspices of the Humboldt-Stiftung Award to M. Galun, for which we express our gratitude.

References

Abu-Zinada AH, El Husseing TW (1975) Seasonal variations of soil microflora and their activities in the Riyadh Region, Saudi Arabia. 1. Algae. Bull Fac Sci Riyadh Univ 7:1–15

Abu-Zinada AH, Hawksworth DL, Bokhary HA (1986) The lichens of Saudi Arabia, with a key to the species reported. Arab Gulf J Sci Res Spec Publ 2:1–52

Borut S (1960) An ecological and physiological study of soil fungi of the Northern Negev (Israel). Bull Res Counc Isr Sect D 8:65–80

Brown G (1998) Notes on the lichen flora of Kuwait. Nova Hedwigia 67:267–273

Bubrick P, Galun M (1984) Cyanobiont diversity in the Lichinaceae and Heppiaceae. Lichenologist 16:279–287

Chantanachat S, Bold HC (1962) Physiological studies. II. Some algae from arid soils. The University of Texas, Publ No 6218, Austin

Danin A, Barbour MG (1982) Microsuccession of cryptogams and phanerogams in the Dead Sea area, Israel. Flora 172:173–179

Danin A, Ganor E (1991) Trapping of airborne dust by mosses in the Negev Desert, Israel. Earth Surface Processes Landforms 16:153–162

Danin A, Orshan G (eds) (1999) Vegetation of Israel. I. Desert and coastal vegetation. Bayhuys Publ, Leiden

Danin A, Bar-Or Y, Dor I, Yisraeli T (1989) The role of cyanobacteria in stabilization of sand dunes in Southern Israel. Ecol Mediterr 15:55–64

Danin A, Dor I, Sandler A, Amit R (1998) Desert crust morphology and its relations to microbiotic succession at Mt. Sedom, Israel. J Arid Environ 38:161–174

Dor I, Danin A (1996) Cyanobacterial desert crusts in the Dead Sea Valley, Israel. Algol Stud 83:197–206

El-Oqlah AA (1992) The lichen flora of Jordan. A catalogue of species and ecological notes. Dirasat 19B:109–121

El-Oqlah AA, Lahham JN (1985) Lichens from the northern part of Jordan. Nova Hedwigia 42:201–205

El-Oqlah AA, Hawksworth DL, Lahham JN (1986) Additions to the lichens of Jordan. Candollea 41:69–73

Fathi SM, El Husseiny TM, Abu-Zinada AH (1975) Seasonal variations of soil microflora and their activities in Riyadh Region. II. Fungi. Bull Fac Sci Riyadh Univ 7:17–30

Frey W, Kürschner H (1988) Bryophytes of the Arabian Peninsula and Socotra. Nova Hedwigia 46:37–120

Frey W, Kürschner H (1991) Lebensstrategien von terrestrischen Bryophyten in der Judäischen Wüste. Bot Acta 104:172–182

Frey W, Kürschner H (1992) Bryosoziologische Untersuchungen in Jordanian: terrestrische und epilitische Gesellschaften. Nova Hedwigia 54:355–378

Frey W, Kürschner H (1994) Bryosoziologische Untersuchungen in Jordanian. 2. Ergänzungen und weitere neue terrestrische und epilithische Gesellschaften aus der östlichen Wüste und Basaltregion. Nova Hedwigia 59:365–378

Frey W, Grolle R, Kürschner H (1991) Taxonomie und Verbreitung von *Asterella* P.Beauv. (Hepaticae, Marchantiidae) im Vorderen Orient. Nova Hedwigia 52:305–318

Frey W, Kürschner H, Ros RM, Guerra J, Cano MJ (1994) *Tortula mucronifera* (Pottiacaea, Musci), a new xerophyte species of the Arabian Peninsula and Jordan. Nova Hedwigia 59:345–351

Friedmann EI, Galun M (1974) Desert algae, lichens, and fungi. In: Brown GW (ed) Desert Biology. Academic Press, New York, pp 165–212

Friedmann EI, Lipkin Y, Ocampo-Paus R (1967) Desert algae of the Negev (Israel). Phycologia 6:185–196

Galun M (1963) Autecological and synecological observations on lichens of the Negev, Israel. Isr J Bot 12:179–187

Galun M (1970) The lichens of Israel. Israel Academy of Sciences and Humanities, Jerusalem

Galun M, Garty J (1972) Lichens of north and central Sinai. Isr J Bot 21:243–254

Galun M, Mukhtar A (1996) Checklist of the lichens of Israel. Bocconea 6:149–172

Galun M, Bubrick P, Garty J (1982) Structural and metabolic diversity of two desert-lichen populations. J Hattori Bot Lab 53:321–324

Hamdi YA, Yousef AN, Al-Azawi S, Al-Tai A, Al-Baquari MS (1978) Distribution of certain non-symbiotic nitrogen fixing organisms in Iraqi soils. Ecol Bull 26:110–115

Karnieli A, Tsoar H (1995) Spectral reflectance of biogenic crust developed on desert dune sand along the Israel-Egypt border. Int J Remote Sensing 16:369–374

Kidron G (1995) The impact of microbial crust upon rainfall-runoff-sediment yield relationships on longitudinal dune slopes, Nizzana, Western Negev Desert, Israel. PhD Thesis, Hebrew University, Jerusalem

Kutiel P, Lavee H, Ackermann O (1998) Spatial distribution of soil surface coverage on north- and south-facing hillslopes along a Mediterranean to extreme arid climatic gradient. Geomorphology 23:245–256

Lange OL, Kidron GJ, Büdel B, Meyer A, Kilian E, Abeliovich A (1992) Taxonomic composition and photosynthetic characteristics of the "biological soil crusts" covering sand dunes in the western Negev Desert. Funct Ecol 6:519–527

Marton K, Galun M (1981) The cyanophilous lichen population of the Arava Valley and the Judean Desert (Israel). Isr J Bot 30:125–155

Müller J (1880) Les lichens d'Egypte. Rev Mycol 10:38–44, 73–83

Rayss T, Borut S (1958) Contributions to the knowledge of soil fungi in Israel. Mycopathol Mycol Appl 10:142–174

Reichert I (1937) Eine lichenogeographische Skizze Palästinas. Verh Zool Bot Ges Wien 86/87:288–296

Reichert I (1940) A new species of *Diploschistes* from oriental steppes and its phytogeographical significance. Palest J Bot (Rehovot Ser) 3:162–182

Schubert R (1973) Notizen zur Flechtenflora des nördlichen Mesopotamien (Irak). Feddes Repert 83:585–589

Schultz M (1998) Studies on lichens from southern Yemen (Arabian Penisula). Lichenologist 30:293–297

Schultz M (1999) New localities of some *Paulia* species (Lichinaceae, lichenized Ascomycota) from Socotra (Indian Ocean). Bryologist 102:61–66

Schultz M, Brown G, Büdel B (2000) Cyanophilous lichens from Kuwait. Nova Hedwigia 70:193–216

Shachack M, Steinberger Y (1980) An algae-desert snail food chain: Energy flow and soil turn over. Oecologia 146:402–411

Werner RG (1954–1958) Notes de lichenologie libano-syrienne I–V. Bull Soc Bot Fr 101–105

West NE (1990) Structure and function of microphytic soil crusts in wild land ecosystems of arid to semi-arid regions. Adv Ecol Res 20:179–223

9 Biological Soil Crusts of Africa

I. ULLMANN and B. BÜDEL

9.1 Introduction

African climate ranges from tropical humid to hyperarid, with an aridity belt that spans about 60 % of the continent (Werger 1978). As environmental conditions are similar in a varying number of countries, this chapter is classified by biogeographical areas rather than political units. Excluded are the Atlas ranges in northwest Africa, which are part of the circum-mediterranean alpine system and treated in Chapter 6. Although climatic and edaphic conditions are favorable for the development of biological soil crusts in most African biomes (given the patchy phanerogamic vegetation), information about crust composition and distribution is still sparse. The most extensive information is for crusts of southwestern Africa (Fig. 9.1).

9.2 Deserts and Semidesert Low Shrublands

9.2.1 Namib and Karoo in Southern Africa

The Namib Desert extends as a 50–120 km wide band for approx. 1600 km along the Atlantic coast of Africa from southern Angola to the Oranje River (14°20'S to 28°30'S). The northern and southern parts are characterized by large dune fields and rocky terrain. The central Namib (21–23°S) is an almost horizontal, hummocky, gravel plain, dissected by numerous sandy washes, water runnels, and wadis (dry river beds), and containing a few isolated inselbergs and rocky ridges (Walter 1986). Mean annual rainfall is below 100 mm, with summer rainfall prevailing in the northern and central Namib, and winter rainfall prevailing in the southern Namib. Along the coastal belt, fog is common, especially from early evening till mid-morning, and fog and dew contribute substantial amounts to the available moisture. In the hyperarid coastal zone of the summer rainfall regions, precipitation

Ecological Studies, Vol. 150
J. Belnap and O.L. Lange (eds.) Biological Soil Crusts: Structure, Function, and Management

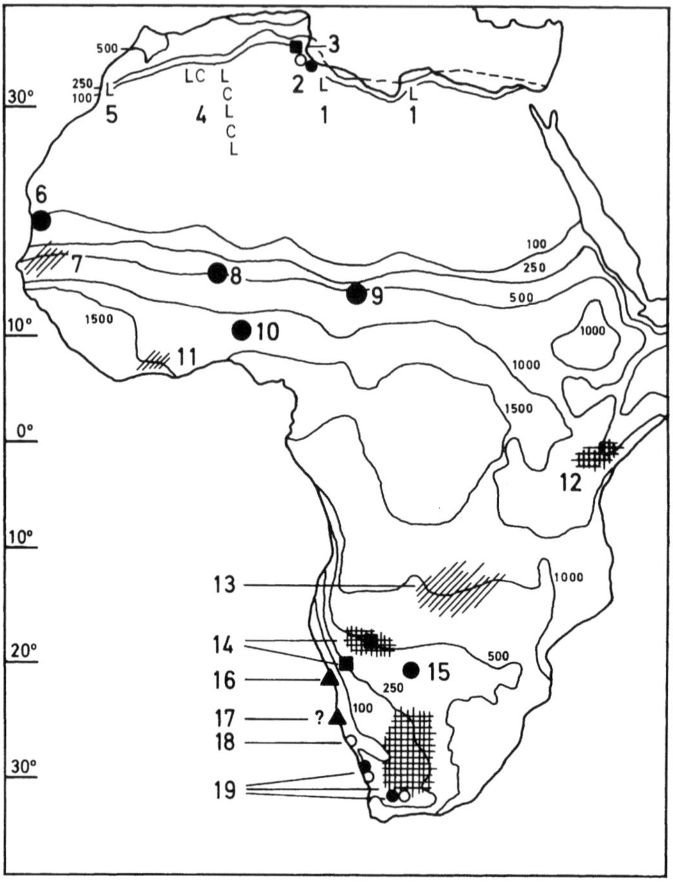

Fig. 9.1. 1–19. Recorded occurrence of biological soil crusts in Africa (as described in the text) and major zonal isohyets. *Symbols* indicate reports with geographic localization, *hatching* is used for general information. *Filled circles, hatching* Cyanobacterial crusts (no or only occasional lichens) (**2** Ullmann et al. unpubl. data; **6** Barbey and Couté 1976; **7** Roger and Reynaud 1982; **8** Hahn and Kusserow 1998, Malam Issa et al. 1999; **10** Isichei 1980; **11** Kaiser 1983, B. Büdel, unpubl. data; **13** Belnap et al. 1996; **15** Skarpe and Henriksson 1987; **19** Vogel 1955). *Open circles, cross hatching* Crusts composed of cyanobacteria and lichens (**2** Ullmann et al. unpubl. data; **12** J. Belnap et al., unpubl. data; **14** Volk 1984b; **18** Jürgens and Niebel-Lohmann 1995; **19** Vogel 1955). *Triangles* Crust composed of phycolichens only (cyanobacteria apparently only hypolithic) (**16** Schieferstein 1989; Rumrich et al. 1989; Büdel and Wessels 1991; Schieferstein and Loris 1992, Lange et al. 1994; O.L. Lange, pers. comm.; **17** I. Ullmann, pers. obs.). *Quadrats* Crusts composed of thallose liverworts (**3** Jovet-Ast and Bischler 1971; **14** Volk 1984b). *L,C* Records of crust-forming species of lichens, Cyanobacteria (**1** Crum 1993; **4** Killian and Fehér 1939; Faurel et al. 1953; **5** Egea 1989)

from fog contributes the equivalent of about 150 mm of rain, and thus is several times the actual rainfall (Werger 1986). The fog belt extends 50–60 km inland, but frequency, duration, and impact of the fog decrease with distance from the sea.

Perennial phanerogamic vegetation is extremely sparse in the fog zone of the summer-rainfall regions. On the plains of the central Namib, it is almost completely restricted to the coastline and water courses, and individual plants (xeromorphic or succulent low shrubs) are scattered. The main perennial biomass in this region is from lichens, and soil crusts are an important part of the ecosystem. The vast (tens to hundreds of km^2) patches of lichen communities in the otherwise barren landscape (clearly distinguishable on Landsat images; Wessels and van Vuuren 1986) are globally unique. The distribution and habitat-related variation in the lichen community that occurs within 30 km distance from the sea was studied by Schieferstein (1989) and Schieferstein and Loris (1992). Crust cover increases conspicuously along the coast-inland moisture gradient north of Swakopmund (22°S, mean annual rainfall 13 mm, 202–295 days $year^{-1}$ overcast, foggy, or with dewfall). Soils are sandy or gravelly, the latter being mainly gypsic. Surface cover by quartz pebbles or rubble from rocky outcrops varies, but is rarely higher than 50 %. Near the coast, a fruticose-foliose phycolichen community occurs, dominated by *Teloschistes capensis* (L.f.) Vain. and *Xanthoparmelia walteri* Knox, with crustose lichens in the spaces between these thalli and pebbles. At 3–5 km from the sea, the height and cover of *Teloschistes* decreases drastically, and the foliose *Xanthoparmelia* gains dominance. Soil surfaces, as well as pebbles and small stones, are covered with crustose lichens. Dominant terricolous lichens are *Lecidella crystallina* Vězda et Wirth (Chap. 1, Photo 11) and *Caloplaca volkii* Wirth et Vězda; the most conspicuous and abundant epilithic lichens are *Caloplaca elegantissima* (Nyl.) Zahlbr. et Stnr., *C. namibensis* Kärnef., and *Neofuscelia namaensis* (Stnr. et Zahlbr.) Essl.

Drier slopes and inland plains (further than approx. 20 km from the sea) are covered with a dense soil-crust community, interspersed with only a few small thalli of the fruticose and foliose species (Photos 51, 52). Knowledge about the composition of these crusts is limited. Apart from the species mentioned, the incomplete collections include unknown species of the genera *Acarospora, Buellia, Caloplaca, Diploschistes*, and *Lecidea* (Schieferstein 1989). Soil crusts of this type cover at least 400 km^2 (Schieferstein and Loris 1992) on the gypsum- and mica-rich gravel hummocks southeast of Swakopmund, which are separated from the coast and the humidity of the sea air by a high dunefield. In this area, the hygrochasic vagrant lichen *Xanthomaculina convoluta* (Hue) Hale (Photo 20) is well represented (Wessels and van Vuuren 1986), and windblown thalli of this species accumulate in the water channels (Walter 1986). Microclimate, water relations, and CO_2 exchange of *Xanthomaculina convoluta* and the dominant fruticose, foliose, and crustose species of the soil-crust habitats north of Swakopmund have been studied by Lange et al. (1990, 1991, 1994; see also Chaps. 18, 20).

The translucent quartz pebbles are often colonized by hypolithic cyanobacterial crusts (see Chap. 1). These are composed of *Chroococcidiopsis* spp., *Gloeocapsa* sp., *Nostoc* cf. *microscopicum* Carm. Sec. Harvey in Hook, *Pleurocapsa minor* Hansgirg, *Schizothrix* sp., *Tolypothrix* cf. *brevis* (Gardner) Geitler, and *Synechococcus elongatus* (Näg.) Näg. (Büdel and Wessels 1991). These species have well-developed mucous sheaths which may play a role in UV protection, as demonstrated for *Pleurocapsa minor*. A further component of the hypolithic crusts are diatoms. Rumrich et al. (1989) enumerated 51 species, all generalists, and confined to neither hypolithic habitats nor to arid areas. Genera reported with more than one species are *Achnanthes* (4 species), *Amphora* (2), *Cymbella* (5), *Eunotia* (2), *Fragilaria* (5), *Gomphonema* (3), *Navicula* (9), *Nitzschia* (8), and *Pinnularia* (2).

Apparently, the soil crusts of the central Namib have two unique features: namely, the absence of cyanolichens (O.L. Lange, pers. comm.), and the apparent limitation of cyanobacteria to hypolithic habitats (with a more favorable microclimate). In the southern Namib, most of the fog zone is covered by unstable dunes unsuited for soil-crust development, and stabilized parts of the dune systems were not investigated for crusts. Gravel soils have not been extensively surveyed, thus the complete species composition is also not known; however, preliminary surveys indicate that they are physiognomically very similar to those of the central Namib (e.g., at Lüderitz, I. Ullmann, pers. obs.). Jürgens and Niebel-Lohmann (1995) report the rare occurrence of the cyanolichen *Heppia* sp. in combination with *Lecidella crystallina*, *Caloplaca volkii* and *C. elegantissima*, *Lecidea* sp., and *Toninia* sp. from Alexanderbay (28°37'S).

Cyanobacteria play an important role, however, in soil crusts in the Namaqua coastal desert (the southern extension of the Namib into the Cape Province) with a mean annual winter rainfall of about 100 mm and additional fog and dew (see Photos 53, 54). A detailed study of crust organisms and their microhabitats was done by Vogel (1955) in the Knersvlakte (at 31°S, 18°30'E). Main landscape units are bare salty depressions and flat ridges, topped by calcareous sandy-silty soils and a varying density of surface quartz pebbles and small quartz rocks. These ridges support an open, succulent, dwarf-shrub vegetation. The hypolithic crusts of the pebbles and stones are rich in cyanobacteria, including *Aphanocapsa* (3 species), *Aphanothece* (2), *Chroococcus* (2), *Xenococcus kerneri* Hansg. [syn. *Xenotholos kerneri* (Hansg.) Gold-Morgan et al.], *Myxosarcina minuta* Vogel, *Stigonema minutum* (Ag.) Haas., *Tolypothrix fragilis* (Gardner) Geitler, *Scytonema ocellatum* Lyngbye, *Nostoc* "*sphaericum*" Vaucher, and 4 species of *Oscillatoria*. Co-occurring green algae were *Chlorella vulgaris* Beyerinck, *Cystococcus humicola* Näg., and *Coccomyxa hypolithica* Vogel.

Crusts dominated by cyanobacteria were found to cover hundreds of km^2 in the coastal desert and in the drier parts of the Little Karoo (Vogel 1955). These occur on the soil surface between the quartz rocks and on calcareous soils without pebbles. Vogel (1955) describes two soil-crust types in the dwarf succulent shrublands. One is more endedaphic, with a 0.5 mm thick layer of the filamentous cyanobacterium *Schizothrix* sp. (interspaced with scattered filaments of *Oscillatoria* sp. and diatoms) underneath about 1 mm surface silt and dust, followed by a layer of fungal hyphae (down to 5 mm). The other crust type is characterized by epedaphic growth of the cyanobacteria *Chroococcus westii* Boye-Petersen, *Gloeocapsa dermochroa* Näg., and the filamentous *Tolypothrix byssoidea* (Berkeley) Kirchner. In both crust types, the cyanobacterial filaments were found interwoven with fungal hyphae, apparently from both free-living fungi and the subterranean system of the lichens *Eremastrella crystallifera* (Taylor) G.Schneider, *Endocarpon* sp., *Toninia* spp., and *Lecidea* sp. (see Fig. 1.4, Photos 18, 19). From the less arid landscapes of the Little Karoo (annual rainfall up to 250 mm, Werger 1978) and from the Karoo, Vogel (1955) reports lichen soil crusts. These are associated with xerophytic and tall succulent vegetation. Dominant lichens are *Diploschistes* sp., *Heppia* sp., and *Psora decipiens* (Hedw.) Hoffm. The occurrence of cyanolichens (*Collema* sp.) also increases with rainfall towards the eastern Karoo (J. Belnap, pers. comm.).

9.2.2 The Sahara in Northern Africa

The transcontinental Saharan landscape system covers an area of more than 10 million km^2 between approx. 16–18°N and 31°N, with a northwestern boundary along the ridge of the Atlas Mountains (betweeen the coastal plains of South Morocco and South Tunisia). The climatic north and south boundaries are determined by the contemporary 100 mm isohyet and the narrow marginal belt with mean annual rainfall of 100–200 mm. About a third of this area is covered by expanded systems of unstable sand dunes, and the hyperarid continental climate is hostile towards soil crusts in further central and eastern parts of the Sahara. Environmental conditions are more favorable for crusts in both belts along the margin of the Sahara, and in the interior mountain systems (with peaks of more than 3000 m height), where rainfall and fog or dew are increased (le Houérou 1986; Monod 1986).

The only study of Saharan soil crusts is located in southern Tunisia (I. Ullmann et al., unpubl. data; see also Chap. 17). The distribution of the crusts is apparently limited by precipitation (inclusive of dew and fog) and is, therefore, restricted to the N-S-running line of the Dahar Mountains and the loess-covered lowlands of the Djeffara, which adjoin the mountains in the west and merge with the coastal plains (mean annual rainfall 100–200 mm).

Due to agricultural and pastoral land use, the occurrence of crusts is patchy, but regeneration was observed in abandoned fields. Species composition of the crusts was studied in the system of ephemeral salt lakes and dunes in the southern Djeffara (33°20'N). The bare, but Chenopodiaceae-fringed evaporitic (halitic) sebkha floors lack surface crusts. Soil crusts occur on adjacent aeolian sandy-silty deposits and on soils derived from limestone or gypsic bedrock. They are associated with a range of open, low-shrub and suffrutex plant communities and sand-accumulating taller shrubs. The authors demonstrate that the switch from cyanobacteria-dominated crusts to lichen-dominated crusts is correlated with a decrease in soil salinity. Fourteen species of cyanobacteria [e.g., *Microcoleus chtonoplastes* Thuret, *M. paludosus* (Kütz.) Gom., *Calothrix parietina* Thuret ex Born. et Flah., *Nostoc* sp., *Gloeocapsa* sp.], two green algal species, five cyanolichens [e.g., *Collema* sp., *Heppia lutosa* (Ach.) Nyl.], and nine phycolichens [e.g., *Catapyrenium lacinulatum* (Ach.) Breuss (syn. *Placidium l.*, for *Catapyrenium* taxonomy see Breuß 1996), *Diploschistes diacapsis* (Ach.) Lumbsch., *Fulgensia fulgens* (Sw.) Elenkin, and *Psora decipiens*)] were observed. The mosses *Crossidium laevipilum* Ther. et Trab., *C. crassinerve* (De Not.) Jur., *Pterygoneurum subsessile* (Brid.) Jur., and *Pseudocrossidium revolutum* (Brid.) Zander are confined to sandy soils under shrubs, where soils are enriched in organic matter and moisture from precipitated fog.

Further information can be drawn from floristic or taxonomic studies of the specific soil-crust organisms. Killian and Fehér (1939) investigated the microbiology of soils at the northern margin of the Sahara near Beni Ounif (Algeria), and along the route from Algiers to Adar (Niger). They list large numbers of bacteria, cyanobacteria, algae, and fungi (more than 40 species for each group), but soil crusts are not mentioned specifically. Although the usefulness of the data is limited (Friedmann and Galun 1974), the report of the filamentous cyanobacteria *Calothrix parietina*, *Nostoc* sp., *Phormidium* sp., and *Scytonema* sp. certainly indicates the occurrence of cyanobacterial crusts in these regions. This conclusion is supported by Focht and Martin (1979), who state that *Nostoc* and *Microcoleus* are the most important cyanobacteria in Algerian desert soils.

In an overview of the lichens of the Algerian Sahara, Faurel et al. (1953) enumerate 111 crustose species (28 % cyanolichens). The soil crust-associated genera *Dermatocarpon* (in the old circumscription; now *Catapyrenium* or *Placidium*) and *Heppia* are represented by 23 species. From the northern belt of the Sahara, the cyanolichens *Heppia lutosa*, *H. despreauxii* (Mont.) Tuck., *H. solorinoides* (Nyl.) Nyl., *Peltula patellata* (Bagl.) Swinsc. et Krog, and *P. radicata* Nyl. are reported (Egea 1989). A special type of epedaphic lichen community is found in hyperarid regions of northern Libya (Hamada al Hamra, Cyrenaica) on windswept flatlands within the range of coastal winter fogs (Crum 1993). The silty-sandy caliche soils are studded with limestone rocks, pebbles, and rocky ridges. The heavily grazed, low, very open *Anabasis articulata* shrub vegetation is associated with the potentially vagrant phycolichens *Aspicilia esculenta* (Pall.) Flag. and *A. jussufi* (Link) Mereschk. (manna lichens), which are used for sheep forage in times of severe drought. Lichen-dominated areas spread over tens of km^2, and are reported to be fairly stable with respect to extent and lichen cover, although the thalli are only loosely affixed to the soil, mainly by a chemical crust.

Crustal mosses are known from Tunisia, where they are apparently confined to coastal regions and habitats with higher soil moisture (areas with at least 100 mm mean annual rainfall). While Pottiaceae are found mainly on accumulations of sand below shrubs (see Chap. 17), the Hepaticae are associated with ephemerally ponded habitats of wadis and water reservoirs. Jovet-Ast and Bischler (1971) distinguish two communities: one is characterized by *Riccia lamellosa* Raddi, *R. atromarginata* Lev., the less frequent *R. trabutiana* Steph., and *Targonia hypophylla* L., and occurs on shallow silty-sandy soils on rocky banks and steps in wadis. The other is a community dominated by *Riccia cavernosa* Hoffm. with the less frequent *R. lamellosa*, *R. crystallina* L., and *Riellia notarisii* (Mont.) Mont. that occurs on sandy-silty wadi floors. In more arid surroundings and on weakly saline soils, *Riccia cavernosa* forms a monotypic crust. *R. cavernosa* appears to tolerate the widest range of environments; this was also noted in south Africa, where *R. cavernosa* and *R. angolensis* Steph. occur on the banks of ephemeral rivers on the outskirts of the Namib Desert (Volk 1984a). However, the drought tolerance of *R. cavernosa* is relatively low, and, experimentally, specimens did not survive more than 12 months in a desiccated state (Volk 1984a). Thus, *R. cavernosa* crusts are restricted to rivers that run for several days at least once a year.

9.3 Dry Savannas (Thorn Bush to Dry Forests)

9.3.1 Southern Africa

Thorn woodlands and dry forests cover at least 50 % of the African continent south of 10°S (Schulze and McGee 1978). Most of the area receives summer precipitation, and mean annual rainfall is >150–200 mm. To date, soil crusts in farmed regions have received less attention than those of the deserts, and information is restricted to a few specific systems.

Crusts dominated by hepatics were studied in Namibia where mean annual rainfall was 200–600 mm and the rainy season lasted 4–6 months (Volk 1984a,b). Volk describes a community with *Riccia cavernosa*, *R. angolensis*, *R. runssorensis* Steph., other *Riccia* spp., *Funaria micropyxis*, and therophytic vascular plants. This occurs on steep loamy banks and deep alluvial soils occurring in the flood plains of rivers and dams, where capillary water is available long after the rainy season. Coarsely weathered granitic and quartz sands, at the base of granite and quartz outcrops (and which obtain additional moisture from fog and dew), support a species-rich crust associated with a therophytic *Eriocaulon* community. The most important hepatics are *Riccia volkii* S.Arnell, *R. okahandjana* S.Arnell, *R. runssorensis*, *R. atropurpurea* Sim, and *Exormotheca holstii* St. (Photo 25). On shallow sandy soils over granitic layers, the drought-tolerant *Riccia okahandjana, R. atropurpurea* and *Exormotheca holstii* form a community with vascular resurrection plants. The species which occur on shallow, basic, silty-sandy soils (often on top of calcretes) in northern Namibia (e.g., *R. canescens* Steph., *R. albosquamata* S.Arnell, and *R. albolimbata* S.Arnell in the Etosha Pan) include the most drought-tolerant taxa. *R. canescens,* for example, was able to survive 7 years of drought. Distribution of hepatics (Arnell 1963; Volk 1984a) indicates that similar crusts will occur in eastern Namibia and the Kalahari. In such xerophytic crusts, the hepatics are frequently combined with lichens (*Catapyrenium* spp., *Psora* spp., *Heppia* sp., *Peltula* spp., and *Collema* spp.) and the endedaphic filaments of *Microcoleus* sp. (Volk 1984b). From Kalahari (aeolian) sands, only cyanobacterial crusts with *Nostoc* sp. and *Scytonema* sp. are reported (Skarpe and Henriksson 1987).

In dry *Acacia* savannas of southern Kenya and northern Tanzania (annual precipitation up to 400 mm, locally only 150 mm), crust development is apparently limited by edaphic factors as well as by the impact of livestock and game (J. Belnap et al., unpubl. data). On soils with no or low calcareousness, no obvious soil crusts were observed. On soils with moderate calcareousness, a cyanobacterial crust was present, except in localized areas where trampling was intense. On soils classified as high to very high in calcareousness, notable cyanobacteria and lichen cover occurred, even with repeated and intense surface disturbance by large herds of hooved mammals (e.g., on the Serengeti Plains). The most common lichens present were *Collema*, *Catapyrenium,* and *Diploschistes* spp.; hepatics were observed in places with less trampling. In addition, highly visible, surface-dwelling *Nostoc commune* colonies were present on calcareous soils where water could pool. This included very shallow soils (<50 cm) and black cotton soils with highly swelling clays (common in river valleys).

The *Brachystegia* (miombo) woodlands on the plateau of southern central Africa (altitudes above 1000 m) are less xerophytic. *Brachystegia–Julbernardia* forests with a grass undercover occur in areas where annual rainfall ranges from 600–1500 mm (Menaut 1983). In *B. spiciformis–J. globiflora* woodland on well-drained lateritic soils in Zimbabwe, cover and development of soil crusts is greatly influenced by burning frequency (Belnap et al. 1996). Total crust cover ranges approx. from 15 to 50 %, and is greatest where burning occurs every 4 years. The dominant crust organism is the filamentous cyanobacterium *Porphyrosyphon notarisii* Kütz.; lichens are absent and moss cover is low (Photos 55, 56).

9.3.2 Northern Africa (Sahel)

The dry woodland belt south of the Sahara is much narrower than that of southern Africa (Menaut 1983). Soil crusts are reported from the Sahel sensu latu (i.e., the northern transition zone to the Sahara with a mean annual rainfall <200 mm), from the thornbush belt (the True Sahel, mean annual rainfall 200–400 mm, after Monod 1986), and the southward adjoining zone (with rainfall up to 900 mm) which extends to approx. 10°N and is of the same width as the True Sahel (200–500 km). The amount of annual precipitation is irregular in the Sahel, and isohyets show wide fluctuation.

Only cyanobacterial crusts are reported from the heavily disturbed Sahel biomes. Floristic diversity and biomass of crusts apparently increase with increasing rainfall, but seem less affected by substrate (sandy versus lateritic soils). However, because of the variation in the authors' taxonomic approach, a comparison of floristic composition is restriced to the generic level. In the northern Sahel, crusts were found to contain the filamentous genera *Microcoleus, Schizothrix*, and *Symploca* (Mauritania: Barbey and Couté 1976). In crusts of the central True Sahel, *Nostoc* was the most dominant organism, occurring together with *Schizothrix*, *Scytonema*, and unspecified coccal green algae (Niger: Hahn and Kusserow 1998). In the northern part of the southern Sahel (mean annual rainfall 400–600 mm), *Scytonema*-dominated crusts with *Microcoleus* are reported from sandy-silty soils in open *Acacia senegal* savannas (Chad: Dulieu et al. 1977). These authors conjecture that crusts reduce grass cover and that crust-stabilized, elevated areas increase erosion in adjacent soil patches. Thus, they regard crust cover as an indication of soil degradation, an interpretation adopted by Hahn and Kusserow (1998). However, neither conjecture has been experimentally tested at these sites, and it is equally possible that crusts are stabilizing and protecting what little vegetated soils remain.

Crusts dominated by *Schizothrix*, with co-occurring *Aphanizomenon*, *Nostoc*, *Scytonema*, and coccal cyanobacteria and green algae, are found on lateritic soils in banded vegetation (tiger bush, Niger: Hahn and Kusserow 1998). Malam Issa et al. (1999) studied the variation of soil crusts in the bare strips between the densely vegetated bands of a tiger bush slope (Niger), and found zonation running parallel to the vegetation bands and associated with runoff/runon zones. Amongst ten filamentous species observed, *Schizothrix friesii* (Ag.) Gomont dominated in all zones. Crusts were absent from the plant litter-covered strip close to the vegetation band, and the physical crust of the next zone was only partly capped by species-poor biological crust patches. Crusts showed highest abundance and species diversity in the lower runoff deposition zone [*Microcoleus sociatus*, *Nostoc* sp., *Phormidium muscosum* Gardner, *Schizothrix penicillata* (Kütz.) Gomont, *Scytonema javanicum* Bornet, and *S. stuposum* (Kütz) Bornet], but were reduced by herbaceous plant cover (including juveniles of woody species) in the seasonally ponded zone in front of the next vegetation band. Under the shrubs, surface crusts were hardly recognizable, but endedaphic *Schizothrix friesii*, *Microcoleus sociatus*, and two *Lyngbya* species were associated with scattered mosses. Species of the *Lyngbya–Phormidium–Plectonema* group were also reported as dominants on cultivated sandy soils in Senegal (e.g., with co-occurring *Nostoc*, *Calothrix*, and *Gloeocapsa*; Roger and Reynaud 1982).

9.4 Humid Savannas

The only and initial information on soil crusts in humid tropical Africa is from regularly burnt savannas. *Scytonema myochrous* (Dillw) Ag. ex Born. et Flah. was found to dominate crusts in all Nigerian savanna types, while *Tolypothrix*, *Nostoc*, and Oscillatoriaceae occurred only occasionally. A potential nitrogen input of 3.3 to 9.2 kg ha^{-1} a^{-1} was estimated for these crusts (Isichei 1980). *Nostoc*, *Tolypothrix*, and *Calothrix* are reported from the palm savannas at Lamto (south Ivory Coast: Kaiser 1983). Dense crusts of *Scytonema* and *Schizothrix* have also been observed on developing soils in flat ephemeral rock pools on inselbergs (B. Büdel, unpubl. data).

9.5 Conclusions

Although information on the distribution of biological soil crusts is patchy to date (and also reflects the geographic preferences of researchers), soil crusts are a part of many ecosystems in Africa (not including unstable sand dune systems and rainforests). Specific crust types are associated with specific climates. In the hyperarid coastal regions of the Namib, fog moisture apparently sustains only phycolichen surface crusts (Fig. 9.1: 16, 17). In areas with more than 100 mm winter rainfall (and often additional fog precipitation), the crusts include the typical range of species found in crusts worldwide (Fig. 9.1: 2, 3, 5, 12, 14, 19). In the dry savannas of the Sahel, the observations of only cyanobacterial crusts (Fig. 9.1: 6–9) may be less climate-related, and more determined by soil surface disturbance and fire.

References

Arnell S (1963) Hepaticae of South Africa. Norstedt & Söner, Stockholm

Barbey C, Couté A (1976) Croutes à cyanophycées sur les dunes du Sahel mauritanien. Bull IFAN 38:732–736

Belnap J, Sanford RA, Lungu L (1996) Biological soil crusts: ecological roles and response to fire in Miombo woodlands of Zimbabwe. Trans Zimbabwe Sci Assoc 70: 14–20

Breuß O (1996) Ein verfeinertes Gliederungskonzept für *Catapyrenium* (lichenisierte Ascomyceten, Verrucariaceae) mit einem Schlüssel für die bisher bekannten Arten. Ann Naturhist Mus Wien 98B Suppl:35–50

Büdel B, Wessels DCJ (1991) Rock inhabiting blue-green algae/cyanobacteria from hot arid regions. Algol Stud 64:385–398

Crum H (1993) A lichenologists view of lichen manna. Contrib Univ Mich Herb 19:293–306

Dulieu D, Gaston A, Darley J (1977) Le dégradation des pâturages de la région de N'Djamena (République du Tchad) en relation avec la présence de cyanophycées psammophiles. Etude préliminaire. Rev Elev Méd Vét Pays Trop 30:181–190

Egea JM (1989) Los géneros *Heppia* y *Peltula* (Líquenes) en Europa Occidental y Norte de Africa. Bibl Lichenol 31:1–122

Faurel L, Ozenda P, Schotter G (1953) Les lichens du Sahara Algérien. Res Counc Isr Spec Publ 2:310–317

Focht DD, Martin JP (1979) Microbiological and biochemical aspects of semi-arid agricultural soils. In: Hall AE, Cannell GH, Lawton, HW (eds) Agriculture in semi-arid environment. Ecological Studies 34. Springer, Berlin Heidelberg New York, pp 119–147

Friedmann EI, Galun M (1974) Desert algae, lichens, and fungi. In: Brown GW (ed) Desert biology II. Academic Press, New York, pp 165–212

Hahn A, Kusserow H (1998) Spatial and temporal distribution of algae in soil crusts in the Sahel of W Africa: preliminary results. Wildenowia 28:227–238

Isichei AO (1980) Nitrogen fixation by blue-green algal soil crusts in Nigerian savanna. In: Rosswall T (ed) Nitrogen cycling in West African Ecosystems. Royal Swedish Academy of Science, Stockholm, pp 191–198

Jovet-Ast S, Bischler H (1971) Les hépatiques de Tunisie. Énumération, notes écologiques et biogéographiques. Rev Bryol Lichénol 38:1–125

Jürgens N, Niebel-Lohmann A (1995) Geobotanical observations on lichen fields of the southern Namib Desert. Mitt Inst Allg Bot Hamburg 25:135–156

Kaiser P (1983) The role of soil microorganisms in savanna ecosystems In: Bourlière F (ed) Tropical savannas. Ecosystems of the world 13. Elsevier, Amsterdam, pp 541–557

Killian C, Fehér D (1939) Recherches sur la phénomène microbiologie des sols désertiques. Encycl Biol 21:1–127

Lange OL, Meyer A, Zellner H, Ullmann I, Wessels DCJ (1990) Eight days in the life of a desert lichen: water relations and photosynthesis of *Teloschistes capensis* in the coastal fog zone of the Namib Desert. Madoqua 17:17–30

Lange OL, Meyer A, Ullmann I, Zellner H (1991) Mikroklima, Wassergehalt und Photosynthese von Flechten in der küstennahen Nebelzone der Namib-Wüste: Messungen während der herbstlichen Witterungsperiode. Flora 185:233-266

Lange OL, Meyer A, Zellner H, Heber U (1994) Photosynthesis and water relations of lichen soil crusts: field measurements in the coastal fog zone of the Namib Desert. Funct Ecol 8:253–264

Le Houérou HN (1986) The desert and arid zones of northern Africa. In: Evenari M, Noy-Meir I, Goodall DW (eds) Hot deserts and arid shrublands. Ecosystems of the world 12 B. Elsevier, Amsterdam, pp 101–147

Malam Issa O, Trichet J, Défarge C, Couté A, Valentin C (1999) Morphology and microstructure of microbiotic soil crusts on a tiger bush sequence (Niger, Sahel). Catena 37:175–196

Menaut J-C (1983) The vegetation of African savannas. In: Bourlière F (ed) Tropical savannas. Ecosystems of the world 13. Elsevier, Amsterdam, pp 109–149

Monod Th (1986) The Sahel zone north of the equator. In: Evenari M, Noy-Meir I, Goodall DW (eds) Hot deserts and arid shrublands. Ecosystems of the world 12 B. Elsevier, Amsterdam, pp 205–243

Roger PA, Reynaud PA (1982) Free-living blue green algae in tropical soils. In: Donmmergues YR, Diem HG (eds) Microbiology of tropical soils and plant productivity. Junk, The Hague, pp 147–168

Rumrich U, Rumrich M, Lange-Bertalot H (1989) Diatomeen als "Fensteralgen" in der Namib-Wüste und anderen ariden Gebieten von SWA/Namibia. Dinteria 20:23–29

Schieferstein B (1989) Ökologische Untersuchungen an den Flechten-Feldern der Namib-Nebelwüste. Dipl-Thesis, Univ Stuttgart-Hohenheim

Schieferstein B, Loris K (1992) Ecological investigations on lichen fields of the Central Namib. I. Distribution patterns and habitat conditions. Vegetatio 98:113–128

Schulze RE, McGee OS (1978) Climatic indices and classifications in relation to the biogreography of southern Africa. In: Werger MJA (ed) Biogeography and ecology of Southern Africa. W Junk, The Hague, pp 19–52

Skarpe C, Henriksson E (1987) Research note – nitrogen fixation by cyanobacterial crusts and associative-symbiotic bacteria in western Kalahari, Botswana. Arid Soil Res Rehab 1:55–59

Vogel S (1955) Niedere "Fensterpflanzen" in der südafrikanischen Wüste. Beitr Biol Pflanz 31:45–135

Volk OH (1984a) Beiträge zur Kenntnis der Marchantiales in Südwest-Afrika/Namibia. Nova Hedwigia 39:117–143

Volk OH (1984b) Pflanzenvergesellschaftungen mit *Riccia*-Arten in Südwestafrika (Namibia). Vegetatio 55:57–64

Walter H (1986) The Namib Desert. In: Evenari M, Noy-Meir I, Goodall DW (eds) Hot deserts and arid shrublands. Ecosystems of the world 12 B. Elsevier, Amsterdam, pp 245–282

Werger MJA (1978) The Karoo-Namib region. In: Werger MJA (ed) Biogeography and ecology of southern Africa. W Junk, The Hague, pp 231–299

Werger MJA (1986) The Karoo and southern Kalahari. In: Evenari M, Noy-Meir I, Goodall DW (eds) Hot deserts and arid shrublands. Ecosystems of the world 12 B. Elsevier, Amsterdam, pp 283–259

Wessels DCJ, van Vuuren DRJ (1986) Landsat imaginary – its possible use in mapping the distribution of major lichen communities in the Namib Desert, South West Africa. Madoqua 14:369–373

10 Biological Soil Crusts of Australia

D.J. Eldridge

10.1 Introduction

Biological soil crusts and their constituent organisms are common components of landscapes in semiarid and arid Australia (Chap. 1, Photos 57–59; Rogers 1982; Eldridge and Tozer 1997a). Crusts are generally restricted to rangelands, which occupy nearly three-quarters of the land area of Australia. The soils in this landscape are exceptionally old, and are generally shallow, infertile, and highly weathered. Rainfall has an overriding influence on the functioning of arid Australia (Stafford Smith and Morton 1990). Many landscapes react to discontinuities in rainfall by being strongly patterned into discrete soil and vegetation patches at a range of spatial scales (Ludwig et al. 1997). Plant and animal diversity is greatest within patches that receive extra resources (runon zones or fertile patches), and these patches ultimately become refugia for plants and animals during unfavorable times.

During periods of low rainfall, when many of the vascular plants have been removed from the soil surface, biological soil crusts are often the only soil organisms left to protect the soil against erosion. Crusts are important at a number of scales, ranging from a continental scale, where they may have an impact on soil carbon and nitrogen pools, to regional and local scales, where they influence soil nutrition and vascular plant germination, the flow of water through the soil, and the capacity of the soil to resist wind and water erosion.

The distribution of soil crusts and their constituent organisms results from a complex interaction of environmental and human-induced factors. At a continental and regional scale, crusts respond to broad changes in rainfall and soil chemistry (particularly pH). At a local or microsite scale (i.e., at the scale of millimeters to tens of centimeters), crust cover and composition are highly dependent on vegetation cover, soil physical and chemical properties, and local microsite variations, such as whether the site is on a small rise or depression of only a few centimeters. Some lichens and bryophytes are influenced more strongly than others by disturbance; thus grazing may lead to differences in cover and composition (see Parts VI, VII).

Ecological Studies, Vol. 150
J. Belnap and O.L. Lange (eds.) Biological Soil Crusts: Structure, Function, and Management

This chapter discusses the distribution and floristics of biological soil crusts and their constituent organisms at a range of scales in arid and semi-arid Australia, as well as their relationships with biotic and abiotic environments. Management-induced changes in crusts resulting from disturbances such as fire and grazing are treated elsewhere (see Chaps. 27–29).

10.2 Biological Soil Crusts at a Continental Scale

In Australia, researchers have tended to differentiate between crusts dominated by lichens, particularly squamulose and crustose types, and other crusts, such as those dominated by cyanobacteria, mosses, or liverworts. Crusts dominated by squamulose and crustose lichens are generally restricted to the more arid regions of Australia (Eldridge 1996a). In terms of cover and floristics, they are most strongly developed on calcareous soils (Rogers 1972a). Consequently these types of lichen crusts are most common over large areas of eastern and southern Australia, southwestern Queensland, and areas of central Australia.

The distribution of lichens is strongly influenced by climatic patterns, particularly the amount and distribution of rainfall (Rogers 1971). In some areas where summer rains are not infrequent, the wet thalli of lichens may be damaged as a result of high temperatures. Rogers (1977) suggests that this controls the distribution of the lichen *Chondropsis* sp. in eastern Australia as it occurs only in areas dominated by winter rainfall. The reduction in cover of lichens from areas subjected to increased summer rainfall, a hypothesized consequence of greenhouse and global warming, may lead to an increase in the cover and composition of cyanobacteria at the expense of most of the unstratified (homoiomerous) lichens (Rogers 1989).

The distribution of bryophytes is also related to rainfall and soil pH (Downing and Selkirk 1993). Bryophytes generally decline in dominance as rainfall declines and increase in dominance with increases in summer rainfall. Consequently, many crusts in subtropical areas are dominated by mosses and liverworts (Rogers 1994; Hodgins and Rogers 1997). However, in rangelands receiving more than about 700 mm a^{-1}, crusts generally make way for communities of mosses, liverworts, and foliose and fruticose lichens which, in an Australian sense, are generally regarded as soil mats rather than true crusts, as they are not intimately bound up with the surface soil layers and can easily be separated from the soil surface. These communities occupy open areas between trees and shrubs (Eldridge and Tozer 1997b; Eldridge et al. 2000).

Lichens generally develop more slowly than mosses and liverworts, preferring more stable areas with minimal disturbance. Like the bryophytes, lichens are thought to be more common on fine-textured soils but not on sands or self-mulching clays (Rogers and Lange 1972; Rogers 1977). Soil pH is also implicated in distribution, with a few species confined to soils with high levels of extractable calcium (Rogers 1977), i.e., highly alkaline soils. In Australia, the extent and diversity of crusts tends to increase southwards as the winter-dominant rainfall increases.

Extensive surveys in southern Australia have shown that lichens are generally restricted to loamy soils, and are absent from cracking and swelling clays, which tend to support mosses growing on the mounds associated with perennial shrubs (Rogers and Lange 1972; Eldridge and Tozer 1996; Eldridge 1998a). Apart from stable landscapes supporting mallee (*Eucalyptus* spp.) vegetation, mosses and lichens are generally absent on sandy soils. Cyanobacteria such as *Microcoleus* (Photo 1), *Nostoc*, and *Scytonema* are active on most soil types and landscapes, including unstable dunes.

10.3 Biological Soil Crusts at a Regional Scale

In Australia, most of the work on soil crusts has been conducted in winter rainfall-dominant areas of southeastern Australia dominated by calcareous soils (Eldridge and Tozer 1997a). Within southern Australia, the assessment of soil-crust communities has been largely restricted to two broad physiographic regions: (1) arid shrub-steppe and woodlands of southern to eastern Australia, and (2) the semiarid woodlands of eastern Australia.

Up until the late 1980s, the only systematic examination of lichens in soil crusts in Australia was made in the early 1970s by Rogers, who documented the distribution of terricolous lichens in arid and semiarid landscapes in eastern Australia (Rogers 1970, 1972a,b; Rogers and Lange 1971, 1972). In the early 1990s, Rogers' work was followed by a comprehensive study of both lichens and bryophytes in semiarid and arid New South Wales (Eldridge 1996a; Eldridge and Tozer 1996, 1997a,b). During the past decade, there has been renewed interest in the role of biological crusts and their associated organisms in other areas such as subtropical southern Queensland (Rogers 1994; Hodgins and Rogers 1997; Page 1997).

As well as the extensive surveys of Rogers and Eldridge, numerous regional studies have helped to improve our knowledge of the distribution and floristics of soil-crust organisms in Australia. An early account of the lichens of Central Australia was given by Müller (1893). Regional lists of lichens are found in N. Sammy (unpubl. data) for the Northern Territory, and Willis (1951) for Western Australia, as well as descriptions given in Filson and Rogers (1979). Regional lists of mosses for Western Australia are found in Stoneburner et al. (1993), for the Northern Territory in Catcheside and Stone

(1988), and New South Wales in Ramsay (1984). Regional assessments of lichens and bryophytes in soil crusts in New South Wales are given in Eldridge (1996a) and Eldridge and Tozer (1996), respectively. Regional studies of cyanobacteria and algae have been undertaken in Central Australia (Ullmann and Büdel, see Chap. 17), South Australia (Eldridge 1996c), and eastern Australia (Broady 1998).

Studies of floristics of biological soil crusts at a local scale have focused on lichens and bryophytes in New South Wales at Lake Mere (M.E. Tozer and D.J. Eldridge, unpubl.), Mungo National Park (Downing and Selkirk 1993; Eldridge 1995; Eldridge and Kinnell 1997; Eldridge et al. 1997), Cobar and Nymagee (K.C. Hynson and K.F. Wells, unpubl.), Coccopara National Park, Yathong Nature Reserve and Bunganbill State Forest (D.J. Eldridge, unpubl. data); in South Australia at Koonamore Vegetation Reserve (Rogers 1974), Maralinga (Eldridge 1998a), Middleback Station (Howarth 1983), the Mid-North Region (Hyde 1994), Maralinga (Eldridge and Ferris 1999); in Western Australia on the Nullarbor Plain (Johnson and Baird 1970); and in Victoria at Wyperfeld National Park (Scott 1982), Hattah-Kulkyne National Park (Willis 1970; Scott 1982; Ralston 1998). In Queensland, Rogers (1989) and Hodgins and Rogers (1997) described bryophyte- and cyanobacterial-dominant crusts in the semiarid subtropics. The distribution of lichen-dominant soil crusts is indicated in Fig. 10.1.

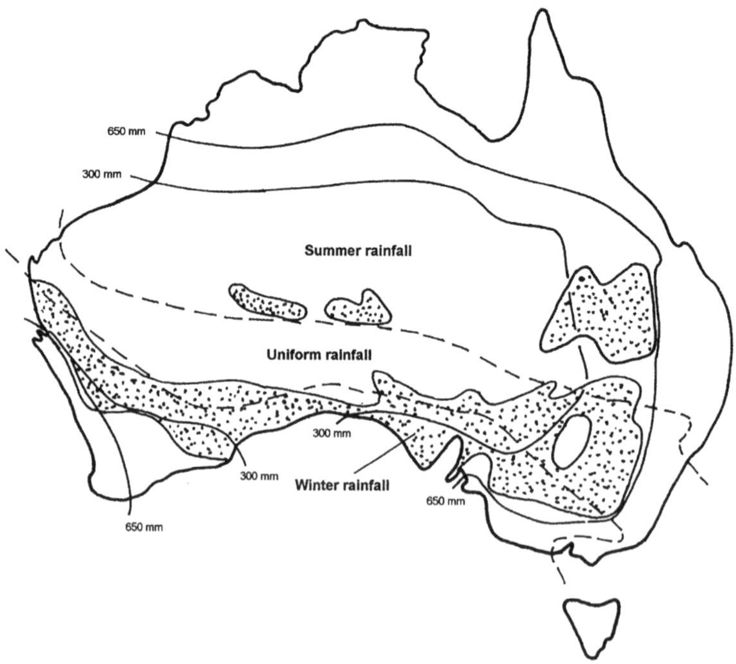

Fig. 10.1. Distribution of lichen-dominated soil crusts in Australia in relation to average annual rainfall. (After Rogers 1972a)

10.4 Biological Soil Crusts at a Landscape Scale

The distribution of organisms in arid and semiarid rangelands varies from a scale of a few kilometers to a few centimeters (see Part III, this Vol.). At the larger scale, the distribution and composition of soil crusts is affected by the regular patterning associated with groves of trees and associated grasses, aligned along the contour, and separated by essentially bare, poorly draining slopes. Water is shed from the slopes which have poor soil physical and chemical properties (runoff zones), and accumulates in the timbered groves (runon zones; Tongway and Ludwig 1994).

I am aware of only two Australian studies of soil-crust organisms in patterned landscapes. In a semiarid subtropical rangeland in Queensland, crust cover of bryophytes was generally typically higher in the timbered groves, whilst lichen cover was greater in the bare runoff zones (M.E. Tozer and D.J. Eldridge, unpubl.). In a semiarid woodland in eastern Australia, Eldridge (1999) demonstrated that crust cover was significantly greatest in the interception zones (79 %), where crust organisms such as mosses, liverworts, and cyanobacteria grew in profusion due to increased surface water. This group of organisms have been termed the opportunists (Fig. 10.2). Crust cover was least in the runon zones, which were shaded by *Callitris glaucophylla* trees, and supported only two moss species (*Fissidens vitattus* Hook.f. et Wils. and *Barbula hornschuchiana* Schultz; 7 % cover). These mosses have been termed shade tolerators. The runoff zones supported mainly the crust-

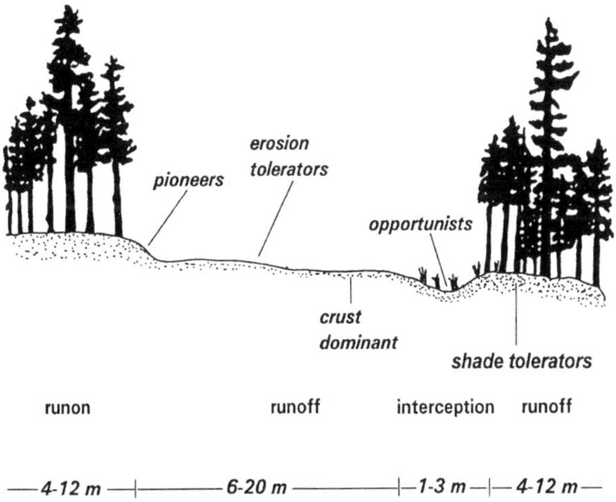

Fig. 10.2. Diagrammatic representation of the relationship between the runoff/interception/runon complex in a semiarid *Callitris glaucophylla* woodland in semiarid eastern Australia and the occurrence of broad groups of lichens and bryophytes. (After Eldridge 1999)

former group, composed of squamulose and crustose lichens, and a few liverworts or mosses.

Individual species are also strongly patterned according to landscape position. At Lake Mere in the semiarid woodlands, D.J. Eldridge and M.E. Tozer (unpubl. data) demonstrated a strong relationship between floristics and landscape position. Lichens were restricted to the runoff zones and bryophytes to the groved runon zones; landscape position was more influential than animal stocking intensity in determining the distribution of organisms. The tendency for bare runoff slopes to be dominated by lichens, and accumulating runon zones to support bryophytes, is consistent in all of these studies (Eldridge 1999; M.E. Tozer and D.J. Eldridge, unpubl. data). The Queensland data also indicated that there were major interactions between the distribution of crust organisms within the landscape and trampling-induced disturbance by grazing animals (see Chaps. 27, 29).

10.5 Biological Soil Crusts at the Microsite Scale

Soil-crust organisms are strongly influenced by changes at the micro-habitat level (i.e., small spatial scales of a few centimeters; Scott 1982). Soil factors, particularly clay content and pH, and local variations in rainfall, soil surface condition, soil nutrition, and erosion influence the distribution of both bryophytes and lichens at a microsite scale (Howarth 1983; Eldridge 1996a; Eldridge and Tozer 1997b; Eldridge et al. 1997). Anderson et al. (1982) found that abundance and species diversity increased with the clay content of the soil. In an aeolian-dominated landscape in semiarid eastern Australia, Downing and Selkirk (1993) found that the sparsest cover of bryophytes occurred on the dune sands, which were the most unstable, had the lowest pH, and were the most freely draining. They also found a remarkable similarity in the bryophyte flora in two areas which had markedly different rainfall, but where the soils had similar alkalinity (i.e., they had abundant calcium carbonate and therefore high pH).

10.5.1 Microsite Distribution of Bryophytes (Mosses and Liverworts)

Studies at a finer spatial scale (i.e., 0.1–1.0 m) reveal some of the factors strongly correlated with the distribution and abundance of bryophytes (Eldridge et al. 1995; Eldridge and Tozer 1997b; D.J. Eldridge and M.E. Tozer, unpubl. data). In a fine-scale survey over large areas of eastern Australia, Eldridge and Tozer (1997b) examined the relationship between environmental variables and the distribution of soil-crust organisms. The liverwort

Riccia nigrella DC. tended to be associated with smooth, bare, sheet-eroded areas with low infiltration and high runoff, whilst the mosses *Pottia brevicaulis* (Tayl.) C.Müll. and *Acaulon* spp. tended to be associated with sites with a capacity to trap and retain water. Different liverworts were found to have different microsite preferences (Eldridge and Tozer 1997b). Whilst *Riccia nigrella* favored moderately hard-setting, flat surfaces, *Riccia crinita* Tayl. and *Fossombronia* spp. favored cracking soils and relatively higher slopes with greater soil microrelief. These differences in preference may reflect the different strategies used by these organisms for surviving in arid environments. For example, *Riccia nigrella* has plate-like, anthocyananin-rich scales which protect it from desiccation (Scott 1982). *Fossombronia* spp. are thallose (leaf-like) liverworts lacking the hardened scales of *Riccia* species, and instead, avoid desiccation by occupying favored microsites such as drainage lines and areas below rock outcrops.

10.5.2 Microsite Distribution of Lichens

Survey results from eastern Australia indicate that the lichens *Paraporpidia glauca* (T.Taylor) Rambold, *Cladonia* sp. and *Lecidea ochroleuca* Pers. tend to be found in open areas with higher clay and organic carbon content, where vascular plant cover was sparse (Eldridge and Tozer 1997b). *Cladonia* spp., in particular, are known to occupy bare or recovering microsites where competition from vascular plants is low (Rogers 1974; Eldridge 1996a), and are known to be susceptible to shading (Brodo 1961). The foliose lichens *Heterodea* spp. are also commonly found in shaded locations.

Some lichens tend to be common in degraded, extensively eroded surfaces low in nutrients and plant cover. These include the ubiquitous pioneering lichens *Synalissa symphorea* (Ach.) Nyl., *Peltula* spp., *Collema coccophorum* Tuck., and *Heppia despreauxii* (Mont.) Tuck. (Rogers and Lange 1971; Rogers 1972b; Rogers 1974; Eldridge 1996a,b; Eldridge and Tozer 1997a,b). *Heppia despreauxii* and *Collema coccophorum* are cyanolichens capable of nitrogen fixation and therefore have a competitive advantage in areas of low nitrogen. In a semiarid patterned *Callitris glaucophylla* woodland, this pioneering group of lichens occurred midway between the lower surface of the runon zone and the runoff zones in a region of extreme surface instability (Fig. 10.2). Cyanolichens are better suited to extreme conditions due to their unique structure, which makes them more efficient at water absorption and therefore extends the period over which they can photosynthesize (Blum 1973; Galun et al. 1982). These pioneering cyanolichens are able to photosynthesize at temperatures higher than other green algal lichens (see Chap. 18), and eventually stabilize the surface, allowing the cosmopolitan mosses such as *Bryum pachytheca* C.Müll. to reinvade (Eldridge 1999).

Lichens such as *Aspicilia calcarea* (L.) Mudd., *Psora decipiens* (Hedwig) Hoffm. and *Psora crystallifera* Muell. Arg. prefer moderately alkaline soils (pH >9). Consequently, they are good indicators of soils which are high in carbonate. Likewise, *Fulgensia subbracteata* (Nyl.) Poelt is typically restricted to gypsum-rich surfaces (Eldridge 1996a; Eldridge and Tozer 1997a). Other lichens, such as *Peltula* spp. and *Psora decipiens,* are dominant on surfaces characterized by a well-developed crust, with sparse sheet erosion and extensive cracking (Eldridge and Tozer 1997b). Greater crust cover generally equates with an increased capacity to trap and retain water (and therefore moderate infiltration rates) and increased sites for vascular plant seed capture (St. Clair et al. 1984). Extensive crust cover also equates with a more stable surface in terms of water and wind erosion (Eldridge and Koen 1998; see Part V).

10.6 Floristics of the Dominant Soil-Crust Organisms

10.6.1 Lichen Floristics

Worldwide, little is known about the biodiversity of soil-crust organisms, particularly in arid and semiarid regions. Total α-diversity of the macroscopic component of the crust (i.e., lichens and bryophytes), is in the vicinity of about 250 (Table 10.1), although this number will undoubtedly increase as new species are described and more research is carried out in arid areas. Apart from some preliminary work on algae and cyanobacteria, diversity of the microscopic green algae, cyanobacteria, and bacteria is unknown.

The majority of the lichens associated with soil crusts worldwide are crustose and squamulose forms. In Australia, 43 % of soil lichens are foliose or fruticose forms. The significance of the high proportion of crustose and squamulose forms may be an adaptation to survival in dry climates (Rogers

Table 10.1. Estimates of α-diversity of lichen, moss, and liverwort families, genera, and species found in soil crusts on the Australian mainland. These figures are estimates only, and probably underestimate the number of species by 10–20 %

	Families	Genera	Species
Lichens	18	25	80
Mosses	13	38	120
Liverworts	6	7	40

1977). Eldridge (1996a) recorded 48 species of lichens from 23 genera in arid and semiarid Australia, and the resultant lichen flora shows strong similarities with other arid landscapes worldwide. Six genera (*Acarospora*, *Endocarpon*, *Catapyrenium* (*Placidium* p.p.), *Diploschistes*, *Peltula* and *Xanthoparmelia*) accounted for 57% of taxa in an extensive study of more than 500 000 km^2 of eastern Australia (Eldridge 1996a). Rogers and Lange (1972) found similar trends in lichen flora, though many more foliose lichens have since been identified due to taxonomic changes over the past 10 years (Elix 1994). Generally, the most frequently recorded soil lichens in Australia are *Psora decipiens*, *Collema coccophorum*, *Catapyrenium* spp., *Endocarpon* spp., *Peltula* spp., *Diploschistes thunbergianus* (Ach.) Lumbsch et Vezda, and *Heppia despreauxii* (Rogers and Lange 1972; Scott 1982; Eldridge 1996a, 1998b; Eldridge and Tozer 1997b; Elix and Ralston 1999). Data from Rogers (1977) for Australian lichens indicates that rainfall is a major determinant of soil-crust diversity, i.e., the total number of lichen species is equal to 0.7 times the average annual rainfall (R^2=0.54). This, however, holds true only for the southern winter rainfall-dominant region of the continent (Fig. 10.1). Within the summer-dominant subtropical areas of Australia, the flora is dominated by cyanobacteria and bryophytes, with very few soil lichens (Rogers 1989; Hodgins and Rogers 1997; M.E. Tozer and D.J. Eldridge, unpubl. data).

10.6.2 Bryophyte Floristics

In a general sense, bryophytes (mosses and liverworts) are widely distributed across the dry areas of Australia (Scott and Stone 1976; Catcheside 1980; Ramsay 1984; Catcheside and Stone 1988; Stoneburner et al. 1993; Eldridge and Tozer 1997a; Scott et al. 1997; Spence and Ramsay 2000). Mosses and liverworts are common components of the soil surface in arid and semiarid Australia (Willis 1970; Scott 1982; Howarth 1983; Downing and Selkirk 1993; Hyde 1994; Eldridge et al. 1995; Eldridge and Tozer 1996; J.H. Browne and R.F. Parsons, unpubl. data). In the higher rainfall rangelands, bryophytes generally do not form a true crust with other organisms, but occur as separate components of the ground flora (Downing 1992; Eldridge et al. 2000; Spence and Ramsay 2000).

The dominant mosses in the semiarid and arid rangelands come from the families Pottiaceae, Bryaceae, Funariaceae, Dittricaceae, and Gigaspermaceae, and the liverworts are dominated by members of the family Ricciaceae (Scott 1985). The mosses of semiarid and arid areas are typically small and cryptic, often infertile, and possess a remarkable array of strategies which enable them to survive in a hostile environment. Like the mosses, the liverworts utilize a range of strategies such as retractable scales, hairs, and pigmentation to avoid heat desiccation. A recent study of the soil-crust bryophyte flora of semiarid New South Wales indicates that it is dominated by *Bryum pachytheca, Didymodon torquatus* (Tayl.; Photo 58) Catches., *Gigaspermum repens* (Hook) Lindb. (Photo 27), *Goniomitrium enerve* Hook. et Wils., *Desmatodon convolutus* (Brid.) Grout (Photo 58), *Crossidium davidai* Catches, *Stonea oleaginosa* (Stone) Zand., *Riccia limbata* Bisch., and *R. lamellosa* Raddi (Downing and Selkirk 1993; Eldridge and Tozer 1996). In general, the cover and species number of bryophytes in soil crusts declines as the climate becomes more arid. Similarly, liverworts (and cyanobacteria) make up a greater proportion of the soil crust as the climate becomes more tropical and as the incidence of summer rainfall increases.

10.6.3 Cyanobacteria and Algae

Whilst lichens and bryophytes have been reasonably well studied, little is known of the cyanobacteria, fungi, and microscopic soil fauna. Rogers (1989) showed that soil-borne cyanobacteria are widespread in the Australian arid zone, occurring in areas receiving as little as 150 mm rainfall. Studies from arid South Australia (Eldridge 1996c) showed that five taxa (*Lyngbya, Microcoleus, Myxosarcina, Nostoc,* and *Scytonema* sp.) were common on most soil types. Studies by Ullmann and Büdel (see Chap. 17) in Central Australia showed that *Microcoleus paludosus* (Kütz.) Gomont, *Nostoc* spp., and *Scytonema* spp. were common components of crusted soils. In a study in western Queensland, the most common cyanobacteria in the soil-crust community were *Lyngbya* spp., *Microcoleus* spp., and *Porphyrosiphon* spp. The chrysophyte *Chloridella* sp. and the chlorophyte *Geminella* sp. were also common (Hodgins and Rogers (1997). Belnap's work from western New South Wales (J. Belnap, unpubl.) indicated that *Microcoleus vaginatus* (Vauch.) Gomont was one of the most widespread cyanobacteria in soil crusts.

Cyanobacteria are the least visible components of biological soil crusts and are widely distributed throughout Australia's rangelands, where they occur as symbionts of lichens. Despite their ubiquitous distribution, little is known about their floristics or distribution (Scott et al. 1997). Cyanobacteria are thought to have a preference for alkaline soils, and are more tolerant of microsites prone to regular disturbance by wind, water, and grazing animals than lichens or mosses (Metting 1991; Chap. 27). Broady (1998) confirmed the richness of soil crusts in eastern Australia, and identified up to 25 genera of cyanobacteria and green algae from only 20 soil surface samples. The dominant algae were *Microcoleus chthonoplastes* Thuret, *Leptolyngbya* sp., *Microcoleus vaginatus, Phormidium* spp., *Nostoc* spp., and *Scytonema* spp. *Chlamydocapsa* sp. and *Chlamydomonas* spp. were also common on many surfaces. Whilst little is known about soil cyanobacteria, their importance is, however, enormous. In the arid zone of South Australia, Rogers et al. (1966)

demonstrated that the cyanobacteria and lichens in crusts occupied 30 % of the area of arid Australia.

Acknowledgments. Much of this work was supported by the Department of Land and Water Conservation and the Land and Water Resources Research and Development Corporation. I thank Merrin Tozer for her comments on the manuscript. This is publication 99/25 of the Centre for Natural Resources, Department of Land and Water Conservation.

References

Anderson DC, Harper KT, Rushforth SR (1982) Recovery of cryptogamic crusts from grazing on Utah winter ranges. J Range Manage 35:355–359

Blum OB (1973) Water Relations. In: Ahmadjian V, Hale ME (eds) The lichens. Academic Press, New York, pp 381–400

Broady PA (1998) Algae in microbiotic crusts of semi-arid soils of western New South Wales. Unpubl report

Brodo IM (1961) A study of lichen ecology in Central Long Island, New York. Am Midl Nat 65:290–310

Catcheside DG (1980) Mosses of South Australia. DJ Woolman, Government Printer, Adelaide, South Australia

Catcheside DG, Stone IG (1988) The mosses of the Northern Territory, Australia. J Adelaide Bot Gard 11:1–17

Downing AJ (1992) Distribution of bryophytes on limestones in eastern Australia. Bryologist 95:5–14

Downing AJ, Selkirk PM (1993) Bryophytes on the calcareous soils of Mungo National Park, an arid area of southern central Australia. Great Basin Nat 53:13–23

Eldridge DJ (1995) Dispersal of microphytes by water erosion in an Australian semi-arid woodland. Lichenologist 28:97–100

Eldridge DJ (1996a) Distribution and floristics of lichens in soil crusts in arid and semi-arid New South Wales, Australia. Aust J Bot 44:581–599

Eldridge DJ (1996b) Cryptogamic soil crusts: fixers of the desert. Proc 1996 Workshop on rehabilitation of arid and semi-arid lands, Goldfields Land Rehabilitation Group, Boulder, WA, pp 87–92

Eldridge DJ (1996c) Biological soil crusts at Maralinga: literature review and potential for reclamation. Report to Department of Primary Industries and Energy and Australian Nuclear Science and Technology Organisation. December, 1996, Lucas Heights, NSW

Eldridge DJ (1998a) Soil crust lichens and mosses on calcrete-dominant soils at Maralinga. J Adelaide Bot Gard 18:9–24

Eldridge DJ (1998b) Trampling of microphytic crusts on calcareous soils and its impact on erosion under rain-impacted flow. Catena 33:221–239

Eldridge DJ (1999) Dynamics of moss- and lichen-dominated soil crusts in a patterned *Callitris glaucophylla* woodland in eastern Australia. Acta Oecol 20:159–170

Eldridge DJ, Ferris JM (1999) Recovery of populations of the soil lichen *Psora crenata* after disturbance in arid South Australia. Rangel J 21:194–198

Eldridge DJ, Kinnell PIA (1997) Assessment of erosion rates from microphyte-dominated calcareous soils under rain-impacted flow. Aust J Soil Res 35:475–489

Eldridge DJ, Koen TB (1998) Cover and floristics of microphytic soil crusts in relation to indices of landscape health. Plant Ecol 137:101–114

Eldridge DJ, Tozer ME (1996) Distribution and floristics of bryophytes in soil crusts in semi-arid and arid eastern Australia. Aust J Bot 44:223–247

Eldridge DJ, Tozer ME (1997a) A practical guide to soil lichens and bryophytes of Australia's dry country. Dept of Land and Water Conservation, Sydney

Eldridge DJ, Tozer ME (1997b) Environmental factors relating to the distribution of terricolous bryophytes and lichens in semi-arid eastern Australia. Bryologist 100: 28–39

Eldridge DJ, Tozer ME, Koen TB (1995) Composition of bryophytes in arid soil crusts: what can they tell us about rangeland health? Proc Vth Int Rangelands Congress, Salt Lake City, Utah

Eldridge DJ, Tozer ME, Slangen S (1997) Soil hydrology is independent of microphytic crust cover: further evidence from a wooded semiarid Australian rangeland. Arid Soil Res Rehab 11:113–126

Eldridge DJ, Semple WS, Koen TB (2000) Dynamics of cryptogamic soil crusts in a derived grassland in south-eastern Australia. Austral Ecol 25:232–240

Elix JA (1994) *Xanthoparmelia*. In: Orachard AE (ed) Flora of Australia 55, CSIRO, Canberra, pp 201–312

Elix JA, Ralston K (1999) Lichens of the Kimberley Region, Western Australia. Aust Lichenol 44:16–34

Filson RB, Rogers RW (1979) Lichens of South Australia. Government Printer, South Australia

Galun M, Bubrick P, Garty J (1982) Structural and metabolic diversity of two desert lichen populations. J Hattori Bot Lab 53:321–324

Hodgins IW, Rogers RW (1997) Correlations of stocking with the cryptogamic soil crust of a semi-arid rangeland in south-west Queensland. Aust J Ecol 22:425–431

Howarth L (1983) The ecology of perennial moss species in chenopod shrublands of Middleback Station, South Australia. BSc (Hons) Thesis, University of Adelaide, South Australia

Hyde MK (1994) A vegetation survey of disused railway corridors in the mid-north region of South Australia. July–November 1992. Nat Cons Council South Australia, Adelaide

Johnson ERL, Baird AM (1970) Notes on the flora and vegetation of the Nullarbor Plain at Forrest, Western Australia. J R Soc West Aust 53:56–61

Ludwig J, Tongway D, Freudenberger D, Noble J, Hodgkinson K (1997) Landscape ecology: function and management. CSIRO, Canberra

Metting B (1991) Biological surface features of semiarid lands and deserts. In: Skujins J (ed) Semiarid lands and deserts: soil resource and reclamation. Marcel Dekker, New York, pp 257–293

Müller J (1893) Lichenes. Trans R Soc S Aust 16:142–149

Page M (1997) Vegetation dynamics in the mulgalands: a Currawinya National Park case study. Unpubl PhD Thesis, University of Queensland, Gatton College

Ralston K (1998) Additional lichen records from Australia 33. *Xanthoparmelia eilifii* and *X. pumila* from Victoria Australasian. Lichenologist 42:9

Ramsay H (1984) Census of the New South Wales mosses. Telopea 2:455–534

Rogers RW (1970) Ecology of soil-surface lichens in arid south-eastern Australia. PhD Thesis, University of Adelaide, South Australia

Rogers RW (1971) Distribution of the lichen *Chondropsis semiviridis* in relation to its heat and drought resistance. New Phytol 70:1069–1077

Rogers RW (1972a) Soil surface lichens in arid and sub-arid south-eastern Australia. II. Phytosociology and geographic zonation. Aust J Bot 20:215–227

Rogers RW (1972b) Soil surface lichens in arid and sub-arid south-eastern Australia. III. The relationship between distribution and environment. Aust J Bot 20:301–316

Rogers RW (1974) Lichens from the T.G.B. Osborn vegetation reserve at Koonamore in arid South Australia. Trans R Soc S Aust 98:113–124

Rogers RW (1977) Lichens in hot arid and semi-arid lands. In: Seaward MRD (ed) Lichen ecology. Academy Press, London, pp 211–52

Rogers RW (1982) Lichens of arid Australia. J Hattori Bot Lab 53: 351–355

Rogers RW (1989) Blue-green algae in southern Australian rangeland soils. Aust Range J 11:67–73

Rogers RW (1994) Zonation of the liverwort *Riccia* in a temporary watercourse in subtropical semi-arid Australia. Aust J Bot 42:659–662

Rogers RW, Lange RT (1971) Lichen populations on arid soil crusts around sheep watering places in South Australia. Oikos 22:93–100

Rogers RW, Lange RT (1972) Soil surface lichens in arid and sub-arid south-eastern Australia. I. Introduction and floristics. Aust J Bot 20:197–213

Rogers RW, Lange RT, Nicholas DJD (1966) Nitrogen fixation by lichens of arid zone crusts. Nature 209: 96–97

Scott GAM (1982) Desert bryophytes. In: Smith AJE (ed) Bryophyte ecology. Chapman and Hall, London, pp 105–122

Scott GAM (1985) Southern Australian liverworts. Australian Government Publishing Service, Canberra

Scott GAM, Stone IG (1976) The mosses of Southern Australia. Australian Government Publishing Service, Canberra

Scott GAM, Entwistle TJ, May TW, Stevens GN (1997) A conservation overview of Australian non-marine lichens, bryophytes, algae and fungi. Environment Australia, Canberra

Spence JR, Ramsay HP (2000) Bryaceae. In: George AS (ed) Flora of Australia. Aust Govt Publ Serv, Canberra

Stafford Smith DM, Morton SR (1990) A framework for the ecology of arid Australia. J Arid Environ 18:255–278

St. Clair LL, Webb BL, Johansen JR, Nebeker GT (1984) Cryptogamic soil crusts: enhancement of seedling establishment in disturbed and undisturbed areas. Reclam Reveg Res 3:129–136

Stoneburner A, Wyatt R, Catcheside DG, Stone IG (1993) Census of the mosses of Western Australia. J Bryol 96:745–777

Tongway DJ, Ludwig JA (1994) Small-scale patch heterogeneity in semi-arid landscapes. Pac Conserv Biol 1:201–208

Willis JH (1951) Botany of the Russel Grimwade Expedition. Mem Nat Mus Victoria 17:75–76

Willis JH (1970) The vegetation of Hattah Lakes National Parks. The Nat Parks Auth, Victoria

11 Biological Soil Crusts of Antarctica

T.G.A. Green and P.A. Broady

11.1 Introduction

For the purposes of this chapter, Antarctica will be defined as the main continent plus islands immediately adjacent to the Antarctic Peninsula. Within this area, the terrestrial vegetation is predominantly composed of lichens, bryophytes, algae, and cyanobacteria. Only two species of phanerogams occur, *Colobanthus quitensis* and *Deschampsia antarctica*, and these are confined to the northwestern parts of the Antarctic Peninsula. Conditions seem, therefore, appropriate for the development of biological soil crusts, but the extreme desert conditions of continental Antarctica must be emphasized, and it should not be forgotten that only 0.8 % of the area is ice-free, with the remainder being permanent ice or snow.

11.2 Antarctic Climate and Vegetation

Antarctica spans around 27° latitude from the north of the Antarctic Peninsula (63°S) to the pole and has an associated large climatic range. It is the coldest, windiest, driest, and highest continent. Overall, there is a clear trend for smaller plants, lower biodiversity, and greater confinement to protected sites as latitude increases. There is general agreement that two zones can be recognized in the Antarctic which reflect both climate and vegetation. The first is the maritime Antarctic (Holdgate 1964), comprising the western side of the Antarctic Peninsula north of Alexander Island (about 69°S), and the closely adjacent islands to the west. Warmest months have mean temperatures above freezing point (0–2 °C), mean winter temperatures are rarely less than -10 °C, and precipitation, occasionally as rain, is between 350 and 500 mm (rain equivalent). The area is essentially a cold semidesert. Lichens, bryophytes, and liverworts number 300–400, 400, and 50 species, respectively. The second zone is continental Antarctica (Holdgate 1964), including the entire main Antarctic Continent, the east of the Antarctic Peninsula, and the west of the Peninsula as far north as, and including, Alexander Island. The warmest

Ecological Studies, Vol. 150
J. Belnap and O.L. Lange (eds.) Biological Soil Crusts: Structure, Function, and Management

month is below, and often substantially below, freezing. Mean winter temperatures are typically –20 °C or below, and precipitation varies from 300 mm (rain equivalent) in the coastal belt to around 120 mm at McMurdo (77°51'S) and 50 mm near the South Pole. Lichens, mosses, and liverworts number 125, 31, and 1 species, respectively.

11.3 Antarctic Soil-Crust Communities

Soil-crust vegetation in continental Antarctica is composed of bryophytes, lichens, green algae, and cyanobacteria, and is confined to ice-free areas, so-called oases (Pickard 1986), where it is very scattered and only locally abundant. The occurrence of a reliable water supply is the major determinant of vegetation growth (Kennedy 1993), but it must be in conjunction with shelter, warmth, and light. Unusual features include a variable, but regular, water supply from snow or ice melt. Peak flows, often substantial, are around midday, whereas there may be complete dryness at night with freezing temperatures (Vincent et al. 1993). The crust vegetation is species-poor, but a few of these species are common. At wetter sites, a bryophyte flush can form, whilst less rich vegetation occurs beside regularly formed or permanent snow patches and in areas with lower water flows. Algae and cyanobacteria are always present, both alone or alongside bryophytes.

11.3.1 Bryophyte Flushes

Extensive bryophyte flushes are widespread [e.g., Cape Hallett, 72°19'S, 170°16'E (Rudolph 1963), Vestfold Hills, 68°33'S, 78°15'E (Pickard 1986), and Dry Valleys, 77°37'S, 163°00'E (Schwarz et al. 1992)]. The vegetation is typically of the short moss cushion and turf sub-formation [mainly a *Bryum–Ceratodon–Hennediella* (*Pottia*) association] in a regular pattern around the main watercourses. The wettest areas are dominated by cyanobacterial crusts (Vincent et al. 1993). Slightly drier parts contain either *B. subrotundifolium* Jaeg. (previously called *Bryum argenteum* Hedw. for specimens from the Ross Sea area), *Ceratodon purpureus* (Hedw.) Brid., or *Grimmia antarctici* Card., depending on locality and environment. In Victoria Land, *B. subrotundifolium* forms particularly extensive, pure stands where there is a strong influence by birds, e.g., Beaufort Island, 76°56'S, 166°56'E (Seppelt et al. 1999) and Cape Hallett (Rudolph 1963) (see Chap. 1, Photos 60, 61). *Ceratodon* occurs in less rich areas, like Botany Bay, Granite Harbour, 77°00'S, 162°35'E. *Hennediella* (=*Pottia*) *heimii* (Hedw.) Feurnr. normally occupies the drier surrounding areas and, at the margins, can often be extensively salt-covered (sodium chloride or sulfate). *C. purpureus*, *Bryum pseudotriquetrum* (Hedw.) Gaertn., Meyer et Scherb, and *H. heimii* form extensive cushions and can grow deeply into the sub-surface. Removal of samples of *H. heimii* to a depth of 2 cm is followed by regrowth from deeper, but still viable, stems and rhizoids. In contrast, the turfs and cushions of *B. subrotundifolium* combine only with the upper soil surface and can be easily detached. Growth estimates are around 0.1 mm a^{-1} (Dry Valleys, Green et al. 1999) and 1 mm a^{-1} at Mawson Rock, 68°35'S, 154°11'E (Seppelt and Ashton 1978). Bryophytes in the drier areas are often colonized by lichens, typically *Candelariella flava* Dodge, Castello et Nimis and *Caloplaca citrina* (Hoffm.) Th. Fries.

Populations of epiphytic cyanobacteria and algae, including chlorophytes, diatoms, and xanthophyceans, are invariably associated with mosses and liverworts, both coating green tissues and extending into dead organic material below (Broady 1996). Cell numbers can be high; for instance, the unicellular chlorophyte *Pseudococcomyxa simplex* (Mainx) Fott can reach 17×10^6 cells cm^{-2} of vegetation surface at the South Orkney Islands (60°43'S, 45°38'W; Broady 1987). Cyanobacteria usually dominate the crusts (Ohtani 1986) and differences between geographical regions have been recorded. At Schirmacher Oasis (70°45'S, 11°40'E), *Stigonema minutum* (Ag.) Hass. is dominant (Pankow et al. 1991), in contrast to *Nostoc* and oscillatorialean cyanobacteria at the Vestfold Hills (68°35'S, 78°00'E; Broady 1986) and Mawson Rock (67°26'S, 62°53'E; Broady 1982). Other genera regularly recorded in crusts are *Gloeocapsa*, *Calothrix*, *Scytonema*, and *Tolypothrix*. At a local scale, dominants change in response to environmental conditions. Green crusts of chlorophytes can become prominent where salts and nutrients are elevated. *Prasiococcus calcarius* (Boye-Pet.) Vischer, *Tetracystis* sp. and other coccoid species are particularly stimulated (Broady 1983, 1986).

The factors which encourage development of epiphytic algae and cyanobacteria have not been investigated in detail. Algae and cyanobacteria may have a high tolerance to reduction in moisture supply, water logging, toxic levels of nutrients, development of salt efflorescences, and increased exposure to winddriven abrasive particles.

11.3.2 Crust Communities Dominated by Algae and Cyanobacteria

11.3.2.1 Epedaphic Communities

Epedaphic communities are common, but sporadic, throughout the ice-free regions of Antarctica, depending on the sufficient and reliable availability of water over the short summer (Kennedy 1993; Wynn-Williams et al. 1997). Communities found on saturated soils in coastal regions of continental Antarctica, such as Ross Island (Broady 1989), the Vestfold Hills (Broady 1986), and Windmill Islands region (66°20'S, 110°30'E; Ling and Seppelt 1998) are confluent black, brown, orange, blue-green, or violet mats of cyanobacteria loosely bound to the soil surface (Broady 1986, 1989). These are usually dominated by oscillatorialean cyanobacteria (species of *Oscillatoria*, *Phormidium*, *Leptolyngbya*, and *Microcoleus*) with irregularly lobed colonies of *Nostoc* cf. *commune* Vauch. as a frequent associate. In contrast, inland areas of the Vestfold Hills have granular black crusts of *Gloeocapsa* spp. in patches where soils are favorably moist. Similar crusts at the Windmill Islands are dominated by *Cyanothece* (as *Synechococcus*), *Gloeocapsa*, *Katagnymene*, and *Stigonema* (Ling and Seppelt 1998). Moist soils which do not receive obvious percolation support thin green powdery crusts. These contain diverse coccoid and filamentous chlorophytes, e.g., *Chlorella*, *Chlorococcum*, *Stichococcus*, and *Prasiococcus calcarius*. The latter seems to also prefer soils with elevated salt content.

The Dry Valleys of Southern Victoria Land, especially the northern valleys, have a more arid microclimate. In Victoria Valley, bryophytes and lichens are entirely absent, and visible black crusts, predominantly of cyanobacteria, cover no more than a few m^2 at widely scattered locations. These occur particularly on valley sides with a northern aspect where, in shallow depressions and on the lee side of boulders, drift snow collects during winter and may persist into summer. The crusts are dominated by oscillatorialeans, mainly *Microcoleus vaginatus* (Vauch.) Gom., and chroococcaleans, mainly *Gloeocapsa kuetzingiana* Naeg., with *Cyanothece aeruginosus* (Naeg.) Kom. as a frequent associate. The other main soil algal habitat is lithosols in furrows surrounding soil polygons on the valley floor that contain sparse microscopic populations of the xanthophycean unicell *Botrydiopsis* sp. (Broady 1996).

11.3.2.2 Endedaphic Crust Communities

At Larsemann Hills (69°33'S, 76°00'E), a subsurface community has been reported from lithosols of 1–2 % water content (Ellis-Evans 1997). Temperatures up to 13 °C were found at 5 mm depth and light levels were 0.005–1 % of incident light at 2.5–4 mm, the depth of peak biomass. Desiccation stress prevented development at the surface. The same stress appears to be responsible for the occurrence of hypolithic communities. These occur below translucent stones (Nienow and Friedmann 1993) and are dominated by chroococcalean and oscillatorialean cyanobacteria and the chlorophytes *Prasiococcus calcarius* and *Desmococcus* cf. *olivaceus* (Pers. ex Ach.) Laundon. They more resemble chasmoendolithic than edaphic communities, although these two communities have species in common.

11.3.3 Geothermal Sites

High-altitude, high-latitude sites in continental Antarctica occur at Mt. Rittmann (ca. 2600 m) and Mt. Melbourne (74°21'S, 164°43'E, ca. 2700 m) in Northern Victoria Land, and Mt. Erebus on Ross Island (77°32'S, 167°8'E; ca. 3500 m) where mid-summer air temperatures are ca. −10 to −20 °C (Broady 1993). Visible crusts are patchy and are primarily associated with ground around small steam vents. Warmest surfaces, at 40–60 °C, bear mats of thermophilic cyanobacteria, including the cosmopolitan thermophile *Mastigocladus laminosus* Cohn. Cooler ground is encrusted with green coccoid algae and bryophyte cushions and, at Mt. Erebus alone, moss protonema.

11.3.4 Recently Exposed Sites

New areas of lithosols have appeared following snowfield or glacier retreat over recent decades, especially at Signy Island, South Orkney Islands (Smith 1990). This ideal opportunity for research has stimulated detailed studies (Wynn-Williams 1996). Filamentous cyanobacteria, in particular, have been shown to be important in stabilizing soil surfaces by aggregating mineral fines to form small superficial "rafts", which are resistant to the physical disturbance of frost-heave events (Wynn-Williams 1993).

11.4 Conclusions

Soil-crust communities do occur in Antarctica, but are sporadic and occur only where there is a regular and reliable water supply. The most extensive form of bryophyte community, the short moss turf and cushion subformation, fits the definition of a soil crust, as do the considerable areas of algal and cyanobacterial growth both within and on the soil. There are, however, some major and consistent differences between the Antarctic communities and those at lower latitudes. First, antarctic communities are all moss, algal, or cyanobacterial-dominated, with lichens being secondary onto the drier moss cushions. Second, the crusts have a very low biodiversity. Crust diversity is strongly dependent on geographical location, since species diversity increases at lower latitudes. Third, biomass is low and there is also little development of a true organic soil. Fourth, growth rates are also low, especially in the more extreme environments (Seppelt and Ashton 1978; Green 1988; Green et al. 1999). The crusts need considerably more study, since they are one of the more common terrestrial plant communities in Antarctica.

References

Broady PA (1982) Ecology of non-marine algae at Mawson Rock, Antarctica. Nova Hedwigia 36:209–229

Broady PA (1983) Antarctic distribution and ecology of the terrestrial, chlorophytan alga *Prasiococcus calcarius* (Boye Petersen) Vischer. Polar Biol 1:211–216

Broady PA (1986) Ecology and taxonomy of the terrestrial algae at the Vestfold Hills. In: Pickard J (ed) The Vestfold Hills. Academic Press, Sydney, pp 165–202

Broady PA (1987) The morphology, distribution and ecology of *Pseudococcomyxa simplex* (Mainx) Fott (Chlorophyta, Chlorellaceae), a widespread terrestrial antarctic alga. Polar Biol 7:25–30

Broady PA (1989) Broadscale patterns in the distribution of aquatic and terrestrial vegetation at three ice-free regions on Ross Island, Antarctica. Hydrobiologia 172: 77–95

Broady PA (1993) Soils heated by volcanism. In: Friedmann EI (ed) Antarctic microbiology. Wiley-Liss, New York, pp 413–432

Broady PA (1996) Diversity, distribution and dispersal of antarctic terrestrial algae. Biodivers Conserv 5:1307–1335

Ellis-Evans JC (1997) Micro-scale distribution of photoautotrophic micro-organisms in relation to light, temperature and moisture in antarctic lithosols. In: Lyons WB, Howard-Williams C, Hawes I (eds) Ecosystem processes in antarctic ice-free landscapes. Balkema, Rotterdam, pp 89–101

Green TGA (1988) Dry Valley terrestrial plant communities – the problem of production and growth. N Z Antarct Rec 6 (Spec Suppl)

Green TGA, Schroeter B, Sancho LD (1999) Plant life in Antarctica. In: Pugnaire FI, Valladares F (eds) Handbook of functional plant ecology. Marcel Dekker, New York, pp 495–543

Holdgate MW (1964) Terrestrial ecology in the maritime Antarctic. In: Carrick R, Holdgate MW, Prévost J (eds) Biologie antarctique. Hermann, Paris, pp 181–194

Kennedy AD (1993) Water as a limiting factor in the antarctic terrestrial environment: a biogeographical synthesis. Arct Alp Res 25:308–315

Ling HU, Seppelt RD (1998) Non-marine algae and cyanobacteria of the Windmill Islands region, Antarctica, with descriptions of two new species. Algol Stud 89:49–62

Nienow JA, Friedmann EI (1993) Terrestrial lithophytic (rock) communities. In: Friedmann EI (ed) Antarctic microbiology. Wiley-Liss, New York, pp 343–412

Ohtani S (1986) Epiphytic algae on mosses in the vicinity of Syowa Station, Antarctica. Memoirs of National Institute of Polar Research special Issue No 44 (Proc 8th Symp on Polar Biology, 1985), pp 209–219

Pankow H, Haendel D, Richter W (1991) Die Algenflora der Schirmacheroase (Ostantarktika). Beih Nova Hedwigia 103:1–197

Pickard J (1986) Antarctic oases, Davis station and the Vestfold Hills. In: Pickard J (ed) Antarctic oasis; terrestrial environments and history of the Vestfold Hills. Academic Press, Sydney, pp 1–19

Rudolph ED (1963) Vegetation of the Hallett Station area, Victoria Land, Antarctica. Ecology 44:585–586

Schwarz AMJ, Green TGA, Seppelt RD (1992) Terrestrial vegetation at Canada Glacier, southern Victoria Land, Antarctica. Polar Biol 12:397–404

Seppelt RD, Ashton DH (1978) Studies on the ecology of the vegetation of Mawson Station, Antarctica. Aust J Ecol 3:373–388

Seppelt RD, Green TGA, Skotnicki M (1999) Notes on the flora, vertebrate fauna and biological significance of Beaufort Island, Ross Sea, Antarctica. Polarforschung 66: 53–59

Smith RIL (1990) Signy Island as a paradigm of biological and environmental change in antarctic terrestrial ecosystems. In: Kerry KR, Hempel G (eds) Antarctic ecosystems. Ecological change and conservation. Springer, Berlin Heidelberg New York, pp 32–50

Vincent WF, Howard-Williams C, Broady PA (1993) Microbial communities and processes in antarctic flowing waters. In: Friedmann EI (ed) Antarctic microbiology. Wiley-Liss, New York, pp 543–569

Wynn-Williams DD (1993) Microbial processes and initial stabilisation of fellfield. In: Miles J, Walton DWH (eds) Primary succession on land. Spec Publ 12, The British Ecological Society. Blackwell, Oxford, pp 17–32

Wynn-Williams DD (1996) Response of pioneer soil microalgal colonists to environmental change in Antarctica. Microb Ecol 31:177–188

Wynn-Williams DD, Russell NC, Edwards HGM (1997) Moisture and habitat structure as regulators for microalgal colonists in diverse antarctic terrestrial habitats. In: Lyons WB, Howard-Williams C, Hawes I (eds) Ecosystem processes in antarctic ice-free landscapes. Balkema, Rotterdam, pp 77–88

12 Synopsis: Comparative Biogeography of Soil-Crust Biota

B. Büdel

12.1 Introduction

Biological soil crusts most often occur under harsh conditions that include extremes in temperature and light and at least periodic lack of water. They occur in almost all arid and semiarid ecoregions worldwide, and in local arid microenvironments in other climates (e.g., temperate regions). Despite their widespread occurrence, comparing the biogeography and ecology of soil-crust biota and communities on a global basis is almost impossible. This is because most of the data were collected (1) using different methodologies, (2) using different taxonomic systems [e.g., using Drouet's (1956) keys, where he combined genera], and/or (3) were focused on a single group of organisms (i.e., lichens, algae, cyanobacteria, or mosses) to the exclusion of other groups.

In Chapters 2–11, our present knowledge about species composition and ecology of biological soil-crust biota in the different geographic regions is summarized. This information is the basis for the following synopsis.

12.2 World Distribution of Biological Soil Crusts

In Fig. 12.1, our present knowledge of the worldwide distribution of the different types of soil-crust communities is depicted. This map is based upon the information given in Chapters 2–11, upon personal experience, and that of colleagues. Single symbols indicate individual locations from which the crust communities are described; however, their distribution may be much more extensive.

Soil crusts composed of cyanobacteria, algae, lichens, mosses, or liverworts, or a combination of these organisms, occur in all biomes of the world containing an arid element that sufficiently decreases competition from phanerogamous vegetation. Only the tropical evergreen rainforest appears to

Ecological Studies, Vol. 150
J. Belnap and O.L. Lange (eds.) Biological Soil Crusts: Structure, Function, and Management

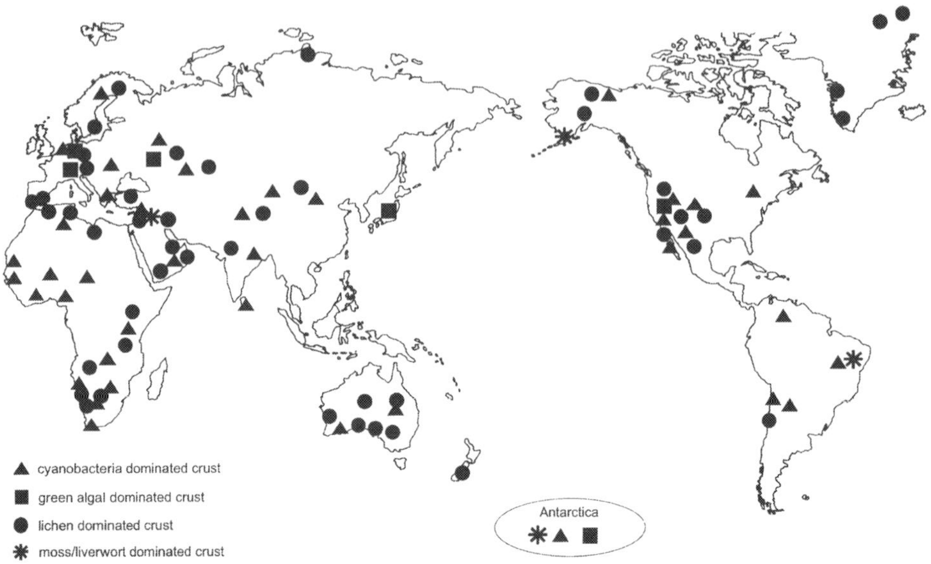

Fig. 12.1. Distribution of biological soil crusts on Earth (this map is based upon the information given in Chapters 2–11 and upon our own personal experience, as well as that of colleagues). *Single symbols* indicate individual locations, from which the crust communities are described – their distribution may be much more extended

lack biological soil crusts (Fig. 12.1). As discussed in detail in the foregoing chapters, poor sandy soils are generally dominated by cyanobacterial crusts; the proportion of lichens increases with the carbonate, gypsum, and silt contents of the substrate. If the salt influence of the soil increases, the filamentous cyanobacterium *Microcoleus chtonoplastes* becomes the main constituent of the crust (see also Chap. 17). Reliable records of moss-dominated soil crusts are very rare. Soil crusts dominated by green algae and/or other eukaryotic algal groups appear restricted to soils in temperate regions, where they appear as a successional phase, or in desert soils that are slightly acidic. However, our knowledge of this crust type is far too limited to draw any further conclusions.

12.3 Biodiversity

Many floristic studies of biological soil crusts do not distinguish between species that substantially contribute to crust formation and those that live in or on the top of crusts, but do not contribute to the formation of the crust (cf. Chap. 1). Thus, our knowledge on soil-crust biota from different geographical regions is rather dissimilar. In addition, most floristic studies conducted on biological soil crusts are restricted to one or two groups of organisms and thus are nearly always incomplete (Evans and Johansen 1999). This might be one of the reasons (besides others; see Evans and Johansen 1999) for the high species richness of different algal groups reported for soils of arid regions in North Africa, the Middle East, and Asia (Novichkova-Ivanova 1980).

In total, 35 cyanobacteria genera, 68 eukaryotic algae genera, 13 cyanolichen genera, 69 phycolichen genera, and 62 moss and liverwort genera are reported from soil crusts of the Earth (Table 12.1). The highest number of genera for algae (51, including 17 cyanobacteria), lichens (44, including 8 cyanolichens) are reported for North America (26 mosses/liverworts), followed by Europe (44 algae including 18 cyanobacteria; 34 lichens including 5 cyanolichens; 14 mosses and liverworts), Africa (30 algae including 18 cyanobacteria; 19 lichens including 4 cyanolichens; 7 mosses and liverworts), the Middle East (27 algae including 14 cyanobacteria; 16 lichens including 8 cyanolichens; 8 mosses and liverworts), Asia [26 algae including 17 cyanobacteria; 24 lichens including 7 cyanolichens; 2 (?) mosses], South America (18 algae including 15 cyanobacteria; lichens unknown; 5 mosses and liverworts), and Greenland (42 lichens including 4 cyanolichens, number of algae and mosses unknown; Table 12.1; Fig. 12.2a–c). With 42 genera, the highest number of mosses and liverworts are reported from Australia (15 algae including 10 cyanobacteria; 26 lichens including 4 cyanolichens). For Greenland, only the filamentous cyanobacterium *Nostoc* sp. has been listed (Chap. 4), probably pointing to the limited phycological investigation in that region. For Antarctica, extensive cyanobacterial crusts are reported, and even *Microcoleus vaginatus* (Chap. 1, Photo 1), a common species in soil crusts of deserts and semideserts, was found in crusts from the Dry Valleys (Chap. 11).

The most common lichens of soil crusts in arid environments of the Earth are the members of the "community of colored lichens" (e.g., *Fulgensia fulgens, Psora decipiens, Squamarina* spp., *Toninia sedifolia, Catapyrenium* spp., *Diploschistes* spp., *Endocarpon* spp., *Collema* spp., Chap. 6). This Bunte Erdflechtengesellschaft occurs worldwide in comparable habitats like the Central European steppe formations, the Mediterranean region, and the North American, African, Asian, and Australian deserts and semideserts.

Cyanobacterial genera, in many cases also species, are generally more widely distributed than eukaryotic algae (Table 12.1). The two most common genera are *Microcoleus* (*M. vaginatus, M. paludosus, M. chtonoplastes, M. sociatus*) and *Nostoc* (*N. commune*). Other common genera are *Calothrix, Lyngbya, Oscillatoria, Phormidium, Scytonema,* and *Tolypothrix*. Common green algae in soil crusts are *Chlorella, Chlorococcum, Coccomyxa,* and *Klebsormidium*. Typical mosses of soil crusts are *Bryum* species, *Crossidium* spp., and *Tortula* spp., and among the liverworts, the genus *Riccia* is the most widespread, followed by *Fossombronia*.

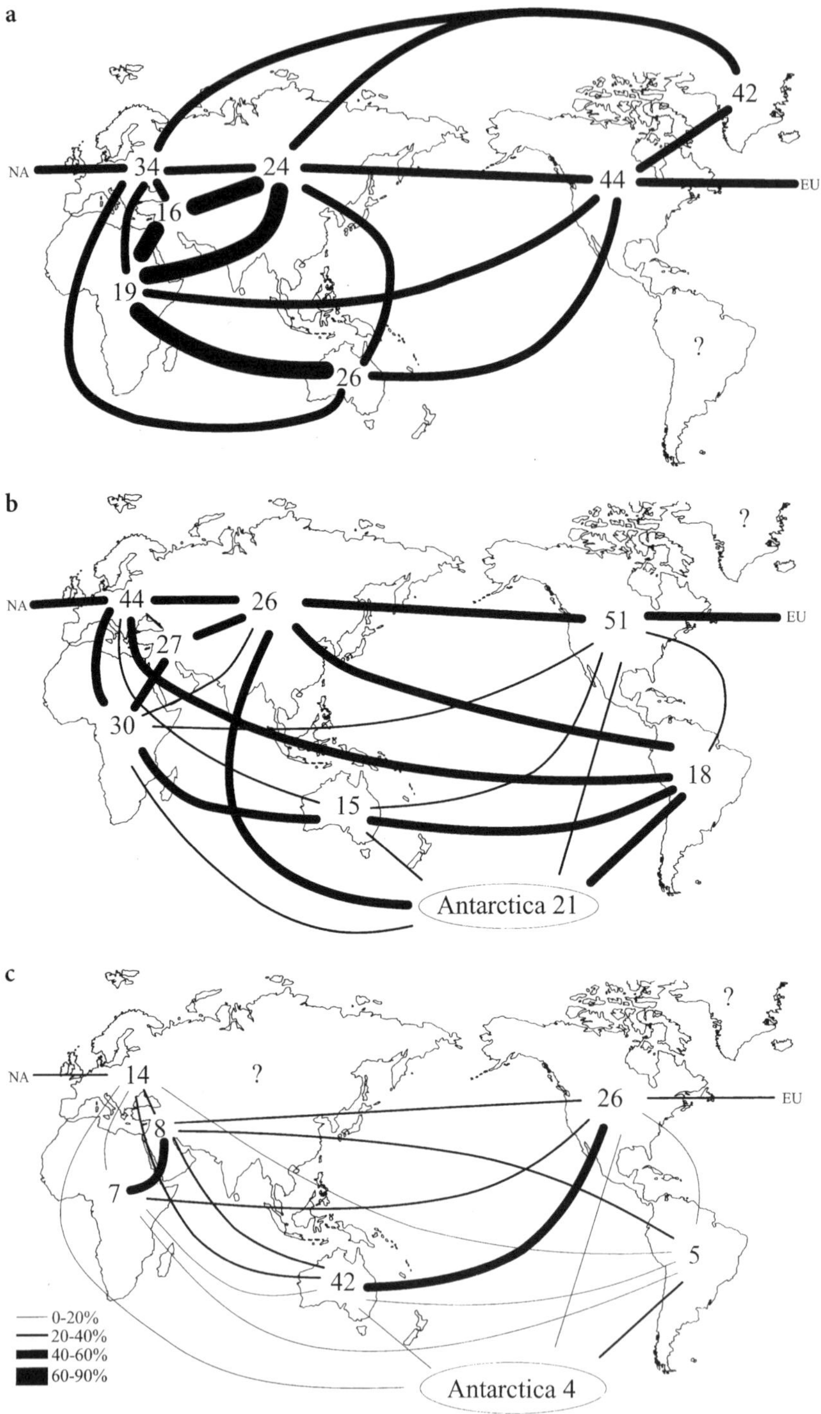

Fig. 12.2a–c. Floristic similarity on genus basis according to the Sörensen coefficient. *Numbers* indicate the absolute number of genera. *NA* North America; *EU* Europe. **a** Lichens. **b** Algae and cyanobacteria. **c** Mosses and liverworts

Table 12.1 List of lichen, cyanobacteria, algal, moss and liverwort genera reported from soil crusts of different geographical regions according to our present knowledge (generated from the information given in Chapters 2–11)

	North America	Europe	Africa	Asia	Middle East	Australia	South America	Arctic Greenland	Antarctica
Phycolichens									
Acarospora	+		+	+	+	+		+	
Aspicilia	+		+	+		+			
Baeomyces	+	+		+				+	
Buellia	+	+	+	+	+	+		+	
Caloplaca	+	+	+			+		+	+
Candelariella	+							+	+
Catapyrenium	+	+	+	+	+	+		+	
Cetraria	+	+							
Cladonia	+	+				+		+	
Dermatocarpon	+								
Desmazieria	+								
Diploschistes	+	+	+	+	+	+		+	
Endocarpon	+	+	+	+		+		+	
Fulgensia	+	+	+	+	+	+		+	
Gypsoplaca	+							+	
Heteroplacidium	+								
Lecanora	+							+	
Lecidoma	+	+						+	
Lepraria	+							+	
Leprocaulon	+								
Leptochidium	+								
Massalongia	+								
Megaspora	+							+	
Multiclavula	+								
Ochrolechia	+							+	
Pannaria	+								
Phaeophyscia	+							+	
Physconia	+							+	
Placynthiella	+	+							
Psora	+	+	+	+	+	+		+	
Siphula	+					+			
Solorina	+	+		+				+	
Squamarina	+	+		+	+				
Texosporium	+								
Trapeliopsis	+	+							
Xanthoparmelia	+		+	+		+			
Arthrorhaphis		+						+	
Bacidia				+					
Catolechia		+							
Chondropsis						+			
Cladia						+			

Table 12.1. (*continued*)

	North America	Europe	Africa	Asia	Middle East	Australia	South America	Arctic Greenland	Antarctica
Coelocaulon		+							
Dacampia								+	
Dactylina								+	
Dibaeis		+						+	
Eremastrella			+			+			
Fuscopannaria		+							
Gyalecta		+						+	
Heterodea						+			
Involucropyrenium		+							
Lecidea		+	+	+		+		+	
Lecidella			+					+	
Leproloma								+	
Mycobilimbia		+						+	
Neofuscelia			+			+			
Paraporpidia						+			
Parmelia				+					
Pertusaria								+	
Phaeorrhiza								+	
Polyblastia		+							
Protoblastenia		+						+	
Psoroma		+							
Rinodina		+		+				+	
Stereocaulon								+	
Teloschistes								+	
Toninia		+	+	+	+	+		+	
Trapelia						+			
Xanthoria								+	
Cyanolichens									
Collema	+	+	+	+	+	+		+	
Gloeoheppia	+				+				
Heppia	+	+	+	+	+	+			
Leptogium	+	+		+				+	
Peltigera	+	+						+	
Peltula	+	+	+	+	+	+			
Pseudopeltula	+								
Psorotichia	+			+	+				
Synalissa			+		+	+			
Gonohymenia					+				
Leciophysma								+	
Lichinella				+					
Peccania				+	+				

Table 12.1. (*continued*)

	North America	Europe	Africa	Asia	Middle East	Australia	South America	Arctic Greenland	Antarctica
Cyanobacteria									
Anabaena	+	+		+			+		
Calothrix	+	+	+	+	+	+	+		+
Chroococcus	+	+	+						
Coccochloris	+								
Cyanothece	+								+
Gloeothece	+	+							
Lyngbya	+	+	+	+	+	+	+		
Microcoleus	+	+	+	+	+	+	+		+
Nostoc	+	+	+	+	+	+	+	+	+
Oscillatoria	+	+	+	+	+		+		+
Phormidium	+	+		+	+	+	+		+
Plectonema	+			+			+		
Porphyrosiphon	+		+			+	+		
Schizothrix	+	+	+	+	+		+		
Scytonema	+	+	+	+	+	+	+		+
Synechococcus	+	+	+						
Tolypothrix	+	+	+	+	+		+		+
Tychonema		+							
Aphanocapsa			+	+	+				
Aphanothece			+	+					
Chlorogloea					+				
Chroococcidiopsis			+	+	+				
Crinalium		+							
Cylindrospermum		+		+					
Gloeocapsa		+	+		+	+			+
Gloeothece									
Katagnymene									+
Leptolyngbya						+			+
Mastigocladus									+
Myxosarcina			+		+	+			
Nodularia		+		+			+		
Pleurocapsa			+						
Stigonema							+		+
Symploca			+	+			+		
Algae									
Astasia	+								
Bracteacoccus	+			+	+				
Chlorella	+	+	+	+					+
Chlorococcum	+	+		+		+			+
Coccomyxa	+	+	+	+					
Deasonia	+								
Desmococcus	+			+					+

Table 12.1. (*continued*)

	North America	Europe	Africa	Asia	Middle East	Australia	South America	Arctic Greenland	Antarctica
Dictyococcus	+								
Ellipsoidion	+								
Fottea	+								
Friedmannia	+				+				
Fritschiella	+								
Hantzschia	+	+							
Klebsormidium	+	+		+	+				
Leptosira	+	+							
Lobococcus	+								
Macrochloris	+				+				
Myrmecia	+			+					
Navicula	+	+	+						
Nitzschia	+		+						
Oedogonium	+								
Palmellopsis	+								
Palmogloea	+								
Pinnularia	+		+						
Pleurastrum	+								
Pleurochloris	+								
Protococcus	+				+				
Protosiphon	+				+				
Rhizoclonium	+								
Stauroneis	+								
Stichococcus	+	+			+				+
Trochiscia	+				+				
Ulothrix	+				+				
Xanthonema	+	+							
Zygogonium		+					+		
Achnanthes			+						
Amphora			+						
Botrydiopsis		+							+
Botrydium					+				
Caloneis		+							
Chlamydocapsa						+			
Chlamydomonas		+		+		+			
Chloridella		+				+			
Chlorosarcinopsis					+				
Cosmarium		+					+		
Cylindrocystis							+		
Cymbella		+	+						
Cystococcus			+						
Diploneis		+							
Elliptochloris		+							
Euastrum		+							

Table 12.1. (*continued*)

	North America	Europe	Africa	Asia	Middle East	Australia	South America	Arctic Greenland	Antarctica
Eunotia		+	+						
Fragilaria			+						
Geminella						+			
Gloeobotrys		+							
Gloeocystis		+							
Gomphonema			+						
Heterococcus		+							
Interfilum						+			
Monocilia				+					
Penium		+							
Prasiococcus									+
Pseudococcomyxa									+
Radiosphaera					+				
Tetracystis									+
Trebouxia					+				
Tribonema		+							
Vaucheria		+							
Mosses and liverworts									
Astomum	+								
Athalamia	+								
Barbula	+	+				+			
Brachythecium	+	+							
Bryum	+			+	+	+	+		+
Ceratodon	+	+				+			+
Crossidium	+		+		+	+			
Didymodon	+					+			
Encalypta	+								
Ephemerum	+					+			
Funaria	+		+			+			
Grimmia	+				+				+
Homalothecium	+								
Microbryum	+								
Phascum	+					+			
Plagiochasma	+					+			
Pleurozium	+	+							
Polytrichum	+	+				+			
Pseudocrossidium	+		+		+				
Pterygoneurum	+		+			+			
Reboulia	+								
Riccia	+	+	+		+	+	+		
Steerecleus	+								
Syntrichia	+								
Tortula	+	+		+	+	+			

Table 12.1. (*continued*)

	North America	Europe	Africa	Asia	Middle East	Australia	South America	Arctic Greenland	Antarctica
Weissia	+					+			
Acaulon						+			
Aloina						+			
Anthelia								+	
Archidium						+			
Asterella						+			
Blasia		+							
Bruetelia						+			
Bryobartramia						+			
Campylopus						+			
Cephaloziella		+							
Dawsonia						+			
Desmatodon						+			
Dicranum						+			
Dicremoloma						+			
Ditrichum						+			
Eccremidium						+			
Exormotheca			+						
Fissidens						+	+		
Fossombronia		+			+	+			
Gigaspermum		+			+	+			
Goniomitrium						+			
Hedwigia						+			
Henediella									+
Hyophiladelphus							+		
Hyophyla							+		
Hypnum		+				+			
Leptobryum						+			
Letochloa						+			
Lophocolea						+			
Pleuridium						+			
Pottia						+			+
Rhacomitrium		+							
Stonea						+			
Targonia			+						
Thuidium						+			
Tortella		+							
Trichostomum						+			
Triquetrella						+			

12.4 Floristic Affinities

Because of the use of different species concepts, different protocols for isolation and identification, and most importantly, the rather selective focusing on certain geographical regions in soil-crust research, we are restricting similarity comparisons between geographic regions to the genus level. Even with this precaution, conclusions reached in the similarity analysis shown in Fig. 12.2 should be interpreted conservatively, as they may change rapidly with increasing information on any of the groups of organisms included in the analysis.

For the similarity analysis presented in Fig. 12.2, we used the Sörensen coefficient (Qs) that weights matches in taxon composition between two samples. It is calculated using the formula:

$$Qs(\%) = \frac{2 \cdot C}{A + B} \cdot 100,$$

where C = number of taxa matching; A = number of taxa in region A; B = number of taxa in region B.

The strongest relation of lichen genera was found between Asia, the Middle East, and Africa, showing 60-90 % similarity. The lichen genera of Australia and Africa showed a similarly close relationship. Less related were the lichens of Australia with those from Europe and North America (Fig. 12.2a). The strong relationship between Asia, the Middle East, and Africa may be due to the high proportion of arid lands, the presence of a land connection, and relatively short dispersal distances. The similarity between Africa and Australia may be a result of their previous land connection (as the supercontinent Gondwana connections; for a review see Galloway 1991).

The relatively evenly distributed similarity of algal genera of different regions of the Earth (40–60 %), as is suggested from Fig. 12.2b, might be due to (1) the phylogenetical age of the groups involved, enabling distribution over the landmass of Pangea, and/or (2) airborne dispersal (Hoffmann 1996; Kristiansen 1996). Airborne algae, found up to 3300 m above ground, have included many soil crust-forming genera of algae and cyanobacteria (e.g., Schlichting 1969; Kristiansen 1996). However, there are more algal genera in common from Australia with Africa and South America than in any other part of the world, thus pointing to the old Gondwana relationship.

Although mosses and liverworts can form an important part of biological soil crusts, the information available is scarce for most regions. Sufficient data for analysis are only available for North America, Europe, the Middle East, and Australia. Mosses and liverworts of the temperate regions of Europe and North America have 35 % similarity. Much stronger levels of similarity were found between the Middle East and Africa (40 %) and between North America and Australia (41 %; Fig. 12.2c). The close relationship of North America and Australia, however, cannot be explained yet. It is probably due to the high degree of investigation of crust mosses and liverworts in both regions and reveals the lack of knowledge in other parts of the world. Only further investigations will answer such questions.

12.5 Conclusions

Although biological soil crusts occur in many biomes of the world and are a widespread phenomenon (as can be seen in Fig. 12.1), our knowledge of their ecological function, their role in ecosystems, as well as their floristics and phytogeography is still rather incomplete or even unknown in many regions. For future floristic and phytogeographical research in biological soil crusts, the application of a standardized and commonly used protocol would contribute greatly towards a reliable international basis of comparison. In addition, scientists need to use modern floras for determination, and need to distinguish between organisms that contribute significantly to soil-crust formation and species that are regularly or irregularly accompanied with the first group. This would lead to a much better understanding of the large-scale distribution patterns, the ecological role, and the functioning of soil crusts and soil-crust organisms.

References

Drouet FA (1956) Revision of the coccoid Myxophyceae. Butler Univ Bot Stud 12:1–218

Evans RD, Johanson JR (1999) Microbiotic crusts and ecosystem processes. Crit Rev Plant Sci 18:183–225

Galloway DJ (1991) Phytogeography of Southern Hemisphere lichens. In: Nimis PL, Crovello TJ (eds) Quantitative approaches to phytogeography. Kluwer Academic Publishers, Dordrecht, pp 233-262

Hoffmann L (1996) Geographic distribution of freshwater blue-green algae. Hydrobiologia 336:33–40

Kristiansen J (1996) Dispersal of freshwater algae – a review. Hydrobiologia 336:151–157

Novichkova-Ivanova LN (1980) Soil algae of the Sahara–Gobi Desert region. Nauka Publishing House, Leningrad (in Russian)

Schlichting HE (1969) The importance of airborne algae and protozoa. J Air Pollut Control Assoc 19:946–951

Part II

Heterotrophic Components of Biological Soil Crusts

13 Soil Fungi as Components of Biological Soil Crusts

J.S. STATES, M. CHRISTENSEN, and C.L. KINTER

13.1 Introduction

In early studies of biological crusts, free-living fungi of hot-desert soils were identified as conspicuous and important components of the crust structure (Fletcher and Martin 1948). Subsequent investigations revealed extensive and persistent growth of both hyphae and algal filaments that strongly influenced the aggregation of soil particles and the crust-forming properties of soils collected from a wide range of arid and semiarid environments (Thornton et al. 1956; Bond and Harris 1964; Cameron 1971). Surprisingly, recent investigations rarely or only incidentally mention fungi in characterizations of soil-crust structure and function. In this chapter, we examine the extant literature regarding the nature of fungal soil-crust relationships and address the possible functional roles attributable to this group of abundant and highly diverse microorganisms. Our report includes the results of an examination of the vertical profile of some typical biological crusts, and the soils beneath them, in search of a fungal community composition indicative of an ecological specificity for the crust environment.

13.2 Classification and Descriptive Terminology of Soil-Inhabiting Fungi

The fungi most commonly encountered as soil inhabitants include yeasts, molds, endophytes, facultative plant pathogens, and mycorrhizal fungi. Soils and affiliated soil crusts serve as a vast repository for the assimilative and reproductive units of these fungi. Many of these fungal propagules may have resided in the soil for decades, dating back to the time of soil genesis. The species adapted to the immediate soil environment and its interactive biological processes are principal species in the soil community, in contrast to a vast number of species that are present but infrequent (Christensen 1989).

Ecological Studies, Vol. 150
J. Belnap and O.L. Lange (eds.) Biological Soil Crusts: Structure, Function, and Management

Most of the principal soil fungi are the primary decomposers of dead plants and plant products (saprotrophs), whereas others form symbiotic, often host-specific associations with living plants either as parasites (pathogens) or as mutualists, as in mycorrhizae and lichens.

Because of the strong metabolic affinities for particular carbohydrate-based organic compounds, the presence of different species and strains of soil fungi has been observed strongly correlated with the predominant cover vegetation (Christensen 1981; States 1981). It is logical to postulate a similar correlation in soil crusts where the primary carbon and energy source is contributed by free-living algae, the photobionts of lichens, and bryophytes. More will be said of this relationship later.

Fungi are ubiquitous, albeit cryptic, components of biological soil crusts and are highly diverse. They have been classified on the basis of biotrophic interactions with algae and plants in combination with their ecological niche (Hawksworth 1988; Hawksworth et al. 1995). We have applied such a categorization to crust-inhabiting microfungi discussed in this chapter. They are defined as follows: terricolous/humicolous, saprotrophic on organic matter intermingled with inorganic soil particles; bryophyllous/herbicolous, saprotrophic on relatively intact mosses/herbaceous plant tissue; fungicolous/lichenicolous, growing on other fungi/lichens as parasites, parasymbionts, or saprobes, and mycorrhiza-forming, mutualistic association with vascular plant roots. As in most biological classifications, it is not possible to place all observed taxa in discrete categories. Many biotrophic associations are highly variable, i.e., facultative or obligate. Furthermore, there is today little more than a cursory understanding of the life histories of fungi inhabiting soil crusts.

13.3 Fungi/Crust Associations

Biological soil crusts are widespread in arid and semiarid regions of the world and are relatively unexplored ecological systems which contain a wide diversity of microorganisms (see Chap. 1). The nature of biotic crust composition has been studied and described primarily in terms of the predominant photo-autotrophic components (cyanobacteria, algae, lichens, mosses), with little consideration of the heterotrophic components (bacteria, actinomycetes, fungi, microfauna). In a search of the literature we have identified and described five soil-crust types for which there is information on resident decomposer fungi. Because there was little available data specifically treating crust-inhabiting fungi, we conducted a survey to determine the species composition of several representative biotic crusts in arid grasslands of Wyoming and Utah. The results of the survey have been incorporated in Tables 13.1 and

13.2 and in the discussion of fungal crust associations below (States and Christensen 2001).

1. *Physical Crusts (Rain or Desiccation Crusts).* Physical crusts are typically formed by deposition and compaction of inorganic particulate and are initially devoid of life (see Chap. 15). The deposition and incorporation of organic particulate from nearby vegetated soils provide an energy source for microbial decomposers. Favorable moisture conditions promote spore germination and growth of bacteria, actinomycetes, and fungi, a situation which enables them to precede cyanobacteria and algae in the formation of biotic crusts (Fletcher and Martin 1948; Bond and Harris 1964). Microcolonies of fungi have been observed to develop on rock surfaces in the absence of algae (Staley et al. 1982) and free-living algae and fungi have been found to colonize rocks and rock fragments in stabilized surface soils (Garty 1992). In Antarctica, crusts from soils left behind the receding melt line of glaciers contained bacteria and fungi which, in some instances, were observed to have preceded algae and cyanobacteria in crust formation (Cameron 1971).

2. *Cyanobacterial/Algal Crusts.* Aggregations of bacteria and actinomycetes have been reported in association with mucilaginous sheaths formed by cyanobacteria (Durrell and Shields 1961; Cameron 1971). We have also seen extensive actinomycete growth on abandoned algal sheaths buried beneath the active crust veneer in Wyoming and Utah soils (J.S. States, unpubl.). It is likely that bacteria and actinomycetes serve as principal reducers of algal biomass (Frankland 1974). Cyanobacteria, especially *Microcoleus* and *Scytonema*, appear to have a major role in crust development (Johansen 1993), but actinomycetes and fungi may be important as well.

3. *Lichen Crusts.* A variety of saprotrophic, fungicolous, and lichenicolous fungi have been reported to be associated with lichens and their organic residues (Hawksworth 1982; Alstrup and Hawksworth 1990; Triebel et al. 1991; Moller and Gams 1993; Wasser et al. 1995; Zhurbenko 1996). A list of fungal isolates from cool-desert lichen dominated crusts in Wyoming and Utah is provided in Table 13.1 (States and Christensen 2001). These fungi were observed on lichens from intact crusts examined immediately following collection and following a 2-week incubation period in illuminated, humid chambers.

4. *Bryophyte Crusts.* The fungal composition of bryophyte-dominated crusts is essentially unstudied. We observed an interesting and relatively common crust community in which lichens were entirely absent. Like desiccation

Table 13.1. Fungi associated with lichen/fungi and moss/herb components of some biotic soil crusts[a] from cool-desert grass-shrub communities in Utah (U) and Wyoming (W)

Fungal categories	Species		
Lichenicolous/ fungicolous	*Ascochyta* sp. U *Mycosphaerella* sp. W *Sclerococcum* sp. ? W *Trichoderma* spp. U, W	*Cladosporium* spp U, W *Myrothecium* sp. W *Stachybotrys* sp. W	*Monodictys* sp. U, W *Phaeospora* W *Taeniollela* sp. U, W
Bryophylous/ herbicolous	*Aureobasidium pullulans* U,W *Bipolaris spp.* U,W *Cladosporium* spp.U, W *Fusarium* spp. U, W *Leptosphaeria* spp. U, W *Mycosphaerella* W *Phoma* spp. U, W *Trichoderma* spp. W	*Acremonium* spp. U, W *Cyphellostereum* sp. W *Embellisia* spp U,W *Graphyllium permundum* U *Melanospora* sp. W *Myrothecium* sp. U, W *Preussia* sp. W *Ulocladium* spp. U, W	*Alternaria* spp. U,W *Chaetomium* spp. U,W *Epicoccum sp.* U,W *Heteroconium* ? W *Monodictys* sp. U,W *Papulaspora* sp. U,W *Stachybotrys* sp. W

[a] Lichen dominated crusts containing a mixture of cyanobacteria, green algae, mosses, and vascular plant litter.

Table 13.2. Principal taxa in biotic crusts from Wyoming and Utah (States and Christensen 2000). An asterisk (*) signifies occurrence at six or more subsites (frequency 60–100 %)

Species[a]	Frequency/density[b]		Occurrence in noncrust soils		
	Wyoming	Utah	Worldwide surveys[c]	Wyoming[d]	Utah-Arizona[e]
Zygomycetes					
St 94. *Mortierella* sp.	20/4	–	–	–	–
Ascomycetes					
R 10 *Chaetomium perlucidum*	10/1	–	–	+	–
C 18 *Macroventuria wentii*	–	30/7	–	–	–
St 20 *Pleospora richtophensis*	10/4	–	–	–	–
St 44 *Graphyllium permundum*	30/5	–	–	–	–
C 94 *Kalmusia utahensis*	–	50/5	–	–	–
Pale Hyphomycetes					
St 11 *Aspergillus leporis*	10/5	–	–	+	+
C 13 *Aspergillus ustus*	–	10/4	+	–	+
*St 74 *Pseudozyma sp.*	40/6	70/19	–	–	–
C 101 *Chrysosporium pannorum*	10/1	10/4	+	+	+
St 1 *Fusarium flocciferum*	50/17	–	–	–	–
St 2 *Fusarium sp.*	10/16	–	–	–	–
C 8 Yeast II	–	20/5	–	–	–
Coelomycetes and dark Hyphomycetes					
C 16 *Alternaria tenuissima*	–	30/10	+	–	+
St 32 *Bipolaris ravenelii*	50/14	–	–	–	+
St 43 *Cladosporium herbarum*	20/4	–	+	+	+
C 30 *Cladosporium macrocarpum*	–	20/8	–	+	–

Table 13.2. (*continued*)

Species[a]	Frequency/density[b]		Occurrence in noncrust soils		
	Wyoming	Utah	Worldwide surveys[c]	Wyoming[d]	Utah-Arizona[e]
Coelomycetes and dark Hyphomycetes					
C 12 *Embellisia clamydospora*	–	20/13	–	+	–
*St 27 *Embellisia telluster*	80/32	90/26	–	–	–
St 6 *Epicoccum purpurascens*	30/10	–	–	–	+
C 23 *Phoma fimeti*	–	50/27	?	+	–
C 22 *Phoma leveillei*	–	10/2	–	–	–
*St 80 *Phoma sp.*	70/22	–	–	–	–
*St 68 *Phoma anserina*	90/20	–	–	–	–
C 25 *Phoma nebulosa*	–	10/2	–	–	–
C 2 *Ulocladium chartarum*	–	10/5	?	–	+
St 34 *Ulocladium multiforme*	10/2	20/4	–	?	+
Mycelia sterilia, dark					
St 21	30/7	–	Mycelia sterilia not listed	Mycelia sterilia 22 % of all taxa	Mycelia sterilia not listed
St 38	30/4	–			
*St 47	70/17	–			
St 49	30/3	–			
*St 51, C 32	100/30	30/8			
St 78	50/7	–			
C 26	–	30/5			
*C34	–	60/11			
C 57	–	40/5			
*C 60	–	60/9			
C 92	–	20/4			

Species[a]	Frequency/density[b]		Occurrence in noncrust soils		
	Wyoming	Utah	Worldwide surveys[c]	Wyoming[d]	Utah-Arizona[e]
Mycelia sterilia, light					
St 65	10/2	–			
C 3	–	30/14			
C 56	–	10/4			
C 91	–	10/2			

[a] Culture numbers. St = Stewart crusts WY. C = Canyonlands crusts UT.
[b] Frequency % = sites of occurrence/total sites sampled × 100. Density = number of isolates.
[c] Christensen (1981). Fungal survey of 33 soils worldwide.
[d] Christensen, unpubl. Survey of soil fungi in sagebrush grasslands in southwest Wyoming
[e] States (1978)

crusts, the crust structure is typically thin, and often curled on drying. The crusts frequently develop in hoof-print depressions of grazed, arid grasslands. A clearly visible mat of organic debris is bound together by a network of *Microcoleus* sp. and interwoven fungus mycelium. They apparently form as a result of an incremental collection of organic matter deposited by wind and precipitation events. Mosses are present on stabilized crusts that have developed a dense mineralized organic component.

5. *Mixed Biotic Crusts.* Over time, persistent, undisturbed soil crusts develop a diverse and complex community structure, comprised of a mixture of crust-forming bryophytes, lichens, algae, cyanobacteria, and associated decomposer microorganisms (see Chap. 1). Fungal diversity is greatly enhanced by the addition of organic matter and mineral nutrients made available by the unique and characteristic producer organic components. The addition of bryophytes and vascular plants to algae/lichen dominated crusts brings a greater diversity of fungi, especially humicolous, lignicolous and fungicolous saprobes (Table 13.1). Also to varying degrees, the non-reproductive mycelia of mycorrhizae and plant pathogens and other symbiotic fungi are represented in the soil matrix (Went 1973; Friese and Allen 1991). In our crust/noncrust survey, these terricolous fungi appear among the dark pigmented (dematiaceous) and light pigmented (moniliaceous) sterile (nonsporulating) isolates recorded in Table 13.2.

13.4 Fungal Communities of Biological Crusts in Utah and Wyoming

Results from a recent, detailed study of the soil fungi in Wyoming and Utah crusts are shown in Table 13.2. Ten crust samples were collected at each of two sites (Stewart site, in west central Wyoming; Canyonlands site, in south-central Utah). Following isolation by suspension plating and particle plating and a careful sorting by macromorphology, the quantitatively prevalent species among 637 isolates were identified. Of the 92 and 106 apparent taxa identified from among 323 and 314 isolates, respectively, 43 taxa in Table 13.2 were the prevalent and characteristic fungi, collectively accounting for 441 or 69 % of all isolates.

In support of our contention of uniqueness in the case of the aggregate species occurring within and beneath the two crusts, we note that only 9 (36 %) of the 25 identified species in the Wyoming-Utah study (Table 13.2) are among the 450 species treated in the *Compendium of Soil Fungi* (Domsch et al. 1980) as “species most likely to be encountered in soil fungal analyses”. In addition, several of our most prevalent species may be undescribed. The relatively high numbers of ascomycete genera (*Kalmusia, Macroventuria, Pleospora*) and black chlamydosporic forms are an especially noteworthy finding. None of these appeared among 294 species recorded as common in 33 desert grassland and forest communities of soil fungi analyzed by Christensen (1981) and they were not prevalent in surveys of noncrust cool-desert soils in Wyoming and Utah–Arizona (M. Christensen unpubl.; States 1978). Five of the 11 genera of the dark-pigmented coelomycetes and hyphomycetes isolated in this study (Table 13.2, *Alternaria, Bipolaris, Cladosporium, Embellisia, Phoma*) reportedly have teleomorphs in the Loculascomycetes (Hawksworth et al. 1995). Approximately 37 % of the principal fungi in the crust soils were sterile forms (*mycelia sterilia*, Table 13.2) and approximately two thirds of those were both black-pigmented and chlamydosporic.

13.5 The Role of Fungi in Crust Structure and Function

Soil fungi have a propensity for forming intertwined strand-like aggregations of hyphae and for developing specialized hyphae for attachment and penetration of substrates, both of which promote crust development. For example, stabilization of sand dunes and colonization of the soil surface by lichens was reported to be facilitated by the action of fungal hyphae in accretion of sand dune particles (James et al. 1977). As is the case for cyanobacteria (Belnap and Gardner 1993), adhesive, extracellular polysaccharides on fungal hyphae serve to bind soil particles together (Bond and Harris 1964; Schulten 1985). Arid soil aggregation and stabilization functions have been reported for a broad range of fungi including: lichen-forming fungi (Ahmadjian and Jackobs 1981), endomycorrhizal fungi (Friese and Allen 1991); ectomycorrhizal fungi (Thornton et al. 1956), and humicolous and lignicolous saprotrophic fungi (Fletcher and Martin 1948; Went 1973).

Fungi are major contributors to carbon flow in soil ecosystems and, as such, are important agents in ecosystem processes and community succession (Christensen 1989). This certainly must be the case for soil crusts, but there is a need for detailed studies. We found actinomycetes and bacteria to be present in considerable abundance on cyanobacteria and green algae (J.S. States and M. Christensen, unpubl.), but few fungi were observed to colonize these organisms. Several fungi colonized lichens and an even greater number grew on mosses and vascular plant litter (Table 13.1). Because of consistent decomposer presence in various stages of crust development, it is highly unlikely that the entire soil carbon contribution to crusts can be attributed to only producers as has been suggested by Beymer and Klopatek (1991). Also, a variety of soil microfauna, particularly nematodes, collembola, and dipteran larval stages, are mycophagous on lichens and mycelial mats (see Chap. 14), thus indicating the existence of a well-developed food web in biotic crusts.

13.6 Conclusions

Principal fungi isolated from biotic crusts can be categorized as fungicolous (growing on fungi or lichens, some possibly as lichen mycobionts), bryophylous/herbicolous (growing on mosses and herbs), and terricolous saprotrophs (colonists of organic matter incorporated in soil). The sexual and sterile forms of saprotrophic fungi were found to be most characteristic in soil-crust communities. Thus, pleosporaceous genera *Embellisia*, *Phoma, Bipolaris,* and four sterile dark-pigmented species (*mycelia sterilia*) with abundant chlamydospores were the most abundant representative fungi in crust soils of Utah and Wyoming. In addition to their decomposer role, fungi are important participants in aggregation of soil particles, and contribute to soil stabilization and crust development in the absence of cyanobacteria and lichens. Fungal biomass serves as a source of food for crust dwelling microfauna. Fungi therefore constitute an important component of the crust food web.

Much remains to be learned of the composition and functional roles of the decomposer community in biotic crusts. Areas of research recommended for future emphasis include: (1) improvements in isolation protocols and identification of the microbiota coupled with an assessment of functional roles, (2) standardization of approaches in obtaining measures of fungal species distribution, abundance, and diversity within a standardized classification of biological soil crusts.

References

Ahmadjian V, Jackobs JB (1981) Relationship between fungus and algae in the lichen *Cladonia cristatella* Tuck. Nature 389:169–172

Alstrup V, Hawksworth DL (1990) The lichenicolous fungi of Greenland. Meddelelser om Grønland. Bioscience 31:1–90

Belnap J, Gardner JS (1993) Soil microstructure in soils of the Colorado Plateau: the role of the cyanobacterium *Microcoleus vaginatus*. Great Basin Nat 53:40–47

Beymer RJ, Klopatek JM (1991) Potential contribution of carbon by microphytic crusts in pinyon-juniper woodlands. Arid Soil Res Rehabil 5:187–198

Bond RD, Harris RD (1964) The influence of the microflora on physical properties of soils. I. Effects associated with filamentous algae and fungi. Aust J Soil Res 2: 111–122

Cameron RE (1971) Antarctic soil microbial and ecological investigations. In: Quam LO, Porter HD (eds) Research in the Antarctic. American Association for the Advancement of Science, Washington DC, pp 137–189

Christensen M (1981) Species diversity and dominance in fungal communities. In: Wicklow DT, Carroll GC (eds) The fungal community: its role and organization in the ecosystem. Marcel Dekker, New York, pp 185–199

Christensen M (1989) A view of fungal ecology. Mycologia 81:1–19

Domsch KH, Gams W, Anderson PH (1980) Compendium of soil fungi, vols 1,2. Academic Press, London

Durrell LW, Shields LM (1961) Characteristics of soil algae relating to crust formation. Trans Am Microsc Soc 80:73–79

Fletcher JE, Martin WP (1948) Some effects of algae and molds in the rain crust of desert soils. Ecology 29:95–100

Frankland JC (1974) Decomposition of lower plants. In: Dickinson CH, Pugh GJF (eds) Biology of plant litter decomposition, vol 1. Academic Press, London, pp 3–36

Friese CF, Allen MF (1991) The spread of VA mycorrhizal fungal hyphae in the soil: inoculum types and external hyphal architecture. Mycologia 38:409–418

Garty J (1992) The postfire recovery of rock-inhabiting algae, microfungi, and lichens. Can J Bot 70:301–312

Hawksworth DL (1982) Secondary fungi in lichen symbiosis: parasites, saprophytes and parasymbionts. J Hattori Bot Lab 52:357–366

Hawksworth DL (1988) The variety of fungal-algal symbioses, their evolutionary significance, and the nature of lichens. Bot J Linn Soc 96:3–20

Hawksworth DL, Kirk PM, Sutton BC, Pegler DN (1995) Ainsworth and Bisby's dictionary of the fungi. 8th edn. CAB International, Wallingford

James PW, Hawksworth DG, Rose F (1977) Lichen communities in the British Isles: a preliminary conspectus. In: Seaward MRP (ed) Lichen ecology. Academic Press, New York, pp 295–413

Johansen JF (1993) Cryptogamic crusts of semiarid and arid lands of North America. J Phycol 29:140–147

Moller C, Gams W (1993) Two new hyphomycetes isolated from Antarctic lichens. Mycotaxon 48:441–450

Schulten JA (1985) Soil aggregation by cryptogams of a sand prairie. Am J Bot 72: 1657–1661

Staley JT, Palmer F, Adams JB (1982) Microcolonial fungi: common inhabitants on desert rocks? Science 215:1093–1095

States JS (1978) Soil fungi of cool-desert plant communities in northern Arizona and southern Utah. J Ariz Acad Sci 13:13–17

States JS (1981) Useful criteria in the description of fungal communities. In: Wicklow DT, Carroll GC (eds) The fungal community: its role and organization in the ecosystem. Marcel Dekker, New York, pp 185–199

States JS, Christensen M (2001) Fungi associated with biological crusts in desert grasslands of Utah and Wyoming. Mycologia 93(3):432-439

Thornton RH, Cowie JD, McDonald DC (1956) Mycelial aggregation of sand under *Pinus radiata*. Nature 177:231–232

Triebel D, Rambold G, Nash TH III (1991) On lichenicolous fungi from continental North America. Mycotaxon 42:263–296

Wasser SP, Nevo E, Vinogradova ON, Navrotskaya IL, Ellanskaya IA, Voltz PA, Virchenko VM, Tsarenko PM, Kondratyuk SY (1995) Diversity of cryptogamic plants and fungi in Evolution Canyon, Nahal Oren, Mount Carmel Natural Preserve, Israel. Isr J Plant Sci 43:367–383

Went FW (1973) Rhizomorphs in soil not connected with fungal fruiting bodies. Am J Bot 60:103–110

Zhurbenko M (1996) Lichens and lichenicolous fungi of the northern Krasnoyarsk Territory, central Siberia. Mycotaxon 58:185–232

14 Microbes and Microfauna Associated with Biological Soil Crusts

J. BELNAP

14.1 Introduction

Microbial populations play a critical role in the regulation of nutrient cycling and energy flow in ecosystems, as they mediate decomposition and subsequent mineralization rates which, in turn, regulate nutrient availability and primary production (Zak and Freckman 1991). Even in regions where water is considered limiting, nutrient availability can restrict plant growth (Romney et al. 1978). Extreme temperatures and/or low soil moisture restrict soil food web development. As soil favorability increases, edaphic taxa generally appear in the following order: pigmented bacteria > actinomycetes > algae and cyanobacteria > fungi, protozoa, other bacteria > lichens > mosses and microarthropods (Cameron et al. 1970).

The trophic structure of soil food webs is very important in soil nutrient cycles. Soil primary producers are lichens, mosses, green algae, and cyanobacteria. These organisms, along with plant material, are both grazed directly and decomposed by soil biota. During decomposition, early-colonizing yeast and bacteria are grazed by nematodes and protozoans, while mites control nematode numbers. Later stages of decomposition are dominated by fungi, which are grazed by nematodes, collembola, and mites. Thus, trophic relationships among soil biota become major regulators of decomposition and mineralization in soils (Ingham et al. 1985).

There is also a synergy between bacteria, fungi, protozoa, and cyanobacteria. Grazing by protozoa stimulate cyanobacterial nitrogen fixation (Ghabbour et al. 1980). Addition of heterotrophic bacteria and fungi to soils significantly increases cyanobacterial biomass (Schiefer and Caldwell 1982). Bacteria and fungi release nutrients and scavenge cyanobacterial "wastes", including O_2, enhancing cyanobacterial N fixation (see Chap. 19). Nitrogen provided by the cyanobacteria, in turn, increases microbial decomposition activity (Lynch and Harper 1983). The objective of this chapter is to discuss the relationship between biological soil crusts and other heterotrophic components of the soil food web, such as bacteria, fungi, protista, nematodes, and soil invertebrates.

Ecological Studies, Vol. 150
J. Belnap and O.L. Lange (eds.) Biological Soil Crusts: Structure, Function, and Management

14.2 Biological Soil Crusts as a Food Source for Other Soil Food Web Components

Bacteria and Fungi. Bacteria are generally numerically dominant in desert soil food webs. Bacteria that can form desiccation-resistant structures are the most common genera in deserts (Kieft 1991). Bacteria are the major decomposer/consumers of soil-crust cyanobacteria (J.S. States, unpubl.). Fungi occur in smaller number than bacteria, but often have similar biomass in desert soils (see Chap. 13). Fungi can form desiccation-resistant spores and grow in air-filled soil pores (Kieft 1991). Yeasts and molds are less prominent in hotter deserts than cold deserts (Cameron and Blank 1967). Fungi specialize on more recalcitrant material than bacteria, and thus are more common in shrub communities and on litter with higher C:N ratios (Kieft 1991). Actinomycetes, especially *Streptomyces*, use the cyanobacterium *Microcoleus* as a food source, and in preference to *Scytonema*. While common soil fungi are found in all crust types, unique saprotrophic fungi are involved in the decomposition of lichens and bryophytes (J.S. States, unpubl.).

Protista. Protists, including amoebae, ciliates, and flagellates are also important in soil food webs, and eat soil cyanobacteria and algae (Tchan and Whitehouse 1953; Whitford 1996). *Nasula picta* and *Thecamoeba* have been observed eating *Microcoleus* in surface soils collected from under *Spinifex* plants in a hot Australian desert (S.S. Bamforth, pers. comm.).

Nematodes. Nematodes are important consumers in soil food webs. Functional groups include bacterivores, omnivore-predators, plant-feeders, and fungivores. Numbers and types of nematodes present are dependent on the amount and type of food sources (Yeates 1979; Zak and Freckman 1991). Nematodes are known to feed on algae and cyanobacteria (Table 14.1; Tchan and Whitehouse 1953).

Arthropods. Arthropods are the most abundant herbivores and detritivores in many terrestrial ecosystems, and are important in regulating rates of decomposition and mobilization of nutrients (Parkinson 1983). Mites are the major group found in deserts, and can constitute up to 90 % of the microfauna (Wallwork 1972). Microarthropods feed on algae, fungi, cyanobacteria (specifically *Microcoleus*), bacteria, other invertebrates, and plant detritus, and thus are important regulators of soil microflora (Table 14.1). Microarthropods are also important in the dispersal of soil microflora. They carry bacteria and fungal cells and spores, lichen fragments, and soredia on their bodies (Bailey 1976; Parkinson 1983; Steinberger 1991).

Table 14.1. Soil arthropod taxa that feed (+= degree of feeding) on cyanobacteria, algae, and fungi from the Chihuahuan Desert and shortgrass steppe. (After Walter 1988 and J. States, unpubl.)

Taxon	Degree of feeding	Habitat	
		Desert	Steppe
Coleoptera			
Anotylus sp.	+++		×
Collembola			
Hypogastura scotti	+++		×
Folsomia elongata	+++		×
Oribatida			
Pilogalumna sp.	+++		×
Haplozetes sp.	+++		×
Ceratozetes sp.	+++		×
Pergalumna sp.	+++	×	
Peloribates sp.	+++	×	
Oppia sp.	+++	×	
Astigmata			
Tyrophagus zachvakini	+++		×
Tyrophagus similis	+++		×
Endeostigmata			
Alicorhagia fragilis	++		×
Alicorhagia usitata	++		×
Prostigmata			
Eupodes spp.	+++	×	×
Linopodes sp.	+++	×	
Paratydeus sp.	+	×	
Tydeidae	+++	×	×
Tardigrades	++	×	×

Macroarthropods are also important in desert ecosystems. However, few are known to be dependent on biological soil crusts directly for habitat or food. Exceptions include a species of isopod, two snails, Australian termites, and a mole cricket, each of which feeds on biological soil crusts. Fecal material from these species can be an important source of nitrogen for the ecosystems in which they occur (Steinberger 1991).

14.3 Distribution of Food-Web Organisms in Desert Soils Relative to Biological Soil Crusts

On a regional level, the distribution, size, and composition of soil food-web components are controlled by regional rainfall and temperature patterns. On a local scale, the occurrence of different taxa is controlled by the distribution of soil moisture, temperature, and organic matter, all of which are influenced by the presence of biological soil crusts (Binet 1981; Whitford 1996).

Distribution of soil microorganisms is correlated with distribution of soil organic matter. In deserts, where soil organic matter is very low (generally <1 %), occurrence of microorganisms is expected to be closely tied to that of biological soil-crust development, as crusts are a major source of soil C and N in these ecosystems. This is especially true in plant interspaces, where input of plant litter is low (see Chaps. 18, 19). Indeed, studies worldwide have generally found numbers of bacteria, fungi, protozoa, nematodes, and microarthropods greatest at the soil surface, except in extreme environments and in soils next to plant roots (Antarctica, hyperarid deserts; Table 14.2). This difference is especially pronounced in plant interspaces

Studies directly addressing the influence of biological soil crusts on soil microflora have shown similar results (Table 14.2). In gypsum soils on the Colorado Plateau, USA, soil surfaces (0–2 cm) in lichen-crusted plant interspaces were found to have larger bacterial and fungal populations than underlying soils (2–5 cm). Adjacent alluvial soils with no lichens showed no significant differences between surface soils and subsurface soils (K.B. McKnight et al., unpubl.). In general, crusts on sandy soils had greater active bacteria and fungi than underlying soils (4–6 cm) in interspaces and under plants (J. Belnap, unpubl.). In Tunisia, Skujins (1984) reported that microbial biomass is greater, but species diversity less, in surface than in subsurface soils.

In Great Basin, US, alluvial soils, indicators of microbial activity (dehydrogenase activity, respiration, and proteolysis) were five to ten times higher in the upper 3 cm of soil that contained soil-crust organisms than the rest of soil profile (Rychert and Skujins 1973) although plate counts measured fewer organisms at the surface (Kovda et al. 1979). This was also seen by Bolton et al. (1993). No significant differences were seen in heterotroph, actinomycete, or fungal numbers between surface and subsurface soils. However, indicators of microbial activity were much higher in the surface soils (microbial C, N, ATP, respiration, dehydrogenase with glucose, and phosphatase).

Soils experiencing extreme cold and hot temperatures are exceptions to this pattern, as they tend to have higher microbial numbers below, rather than at, the soil surface. Bacteria, fungi, and protozoa numbers were higher at 2–15 cm than 0–2 cm in Antarctic dry valleys (Cameron et al. 1970). Higher

Table 14.2. Abundance and activity of soil organisms on the soil surface (0–2 cm) relative to underlying soil (generally >4 cm)

Habitat	Hot deserts													Cool deserts							
	Colorado Desert			Mojave							Sonoran	Tunisia		Colorado Plateau					Great Basin		
Soil food web components	*Sand*			*Sand*							*Sand*	*Silt*	*Clay*	*Sand*					*Silt*		
	I^1	I^1	I^1	I^1	I^1	I^1	I^1	I^1	IP^3	I^3	I^4	IP^5	IP^5	IP^6	IP^6	I^7	GP^7	P^7	I^8	P^8	IP^9
Aerobic bacteria	9	6	3	3	1																
Anaerobic bacteria	5580	10	<1	2	<1														1		
Bacteria									1			4	51	3	2	3	<1	3	1		
Fungi	16	<1	7	10	3				1		4	2	136	3	8	2	0	1	4		
Nematodes				3	1	3	2	1	4	2				2	3						
Microarthropods				2																	
Flagellates											28										
Amoebae											40										
Ciliates											3										
Microbial activity																				5–10	2–3

Numbers are the ratio of surface to subsurface values measured.
I = soil from plant interspaces; *P* = soil from under the plant canopy; *IP* = soil from under plant canopies and in the interspaces combined; *G* = gypsum.
From [1]Cameron and Blank 1967 (multiple sites); [2]Freckman et al. 1974; [3]Rundel and Gibson 1996, J. Belnap unpubl.; [4]Bamforth 1984; [5]Skujins 1984; [6]J. Belnap unpubl.; [7]McKnight unpubl.; [8]Rychert and Skujins 1973; [9]Bolton et al. 1993.
Abundance and activity is generally greater on the surface in most deserts, with the exception of Antarctica and hyperarid regions (not shown).

microbial numbers are found at depth in some hot Middle East deserts (Binet 1981). Cameron and Blank (1967) found fungal populations greater at depth at two of six sites in the Mojave Desert.

Biological soil crusts provide habitat for larger organisms as well. Microbial-feeding nematodes are highest in surface soils on the Colorado Plateau, Great Basin, Sonoran, and Mojave Deserts in the US (Zak and Freckman 1991; J. Belnap, unpubl.). Yeates (1979) reported a high correlation of nematode abundance and moss cover ($r^2=0.86$; $P<0.001$). In both polar regions, microarthropods occur commonly in mosses and lichens at the surface (Crawford 1981). In warm deserts, mosses provide habitat for cryptostigmatid, prostigmatid, and mesostigmatid mites, as well as collembolans, nematodes, rotifers, tardigrades, and some protozoans (Yeates 1979). Microarthropod numbers are found to be greatest at soil surfaces in Australian, Great Basin, Mojave, Colorado Plateau, and Chihuahuan Deserts (Wood 1971; Whitford et al. 1981; Steinberger 1991; Rundel and Gibson 1996; J. Belnap, unpubl.). Franco et al. (1979) found 80 % of collected microarthropods in the top 10 cm of soils during most of the year in the Mojave. Progstigmata mites, tardigrades, isopods, snails, mole crickets, tenebrionid beetles, protozoans, termites, and ants have all been recorded feeding on cyanobacteria, lichens, and mosses (Bailey 1976; Ghabbour et al. 1980; Loria and Herrnstadt 1980; Rogers et al. 1988; Steinberger 1991).

Limited data indicate that the level of crust development influences microorganism numbers. In Colorado, USA, lichen-moss crusts supported three times as many bacterial-feeding nematodes and seven times as many fungal-feeding nematodes as adjacent cyanobacteria-only crusts (Anderson et al. 1984). In the Arctic, crusts with liverworts had more bacteria and fungi than crusts with only cyanobacteria (Smith and Griggs 1932).

14.4 Conclusion

As can be seen from the above discussion, there is strong evidence that biological soil crusts are important in the maintenance of desert soil food webs. Early successional crusts (those dominated by cyanobacteria) support a less complex decomposition biota than crusts with lichen and bryophyte species. The role of the biological soil crusts appears especially important in the large interspaces between plants, as these areas have limited input of organic matter. In addition, plant roots are continuously extracting nutrients from interspace soils, reducing their fertility over time. Biological soil crusts help maintain the fertility of these interspaces by contributing both carbon and nitrogen to the interspace areas, thus partially counteracting the tendency of nutrients and microorganisms to concentrate under shrubs.

Acknowledgments. The help of Michelle Schmid, Sue Phillips, Elaine Ingham, and Andy Moldenke is gratefully acknowledged.

References

Anderson RV, Ingham RE, Trofymow JW, Coleman DC (1984) Soil mesofaunal distribution in relation to habitat types in a shortgrass prairie. Pedobiologia 26: 257–261

Bailey RH (1976) Ecological aspects of dispersal and establishment in lichens. In: Brown DH, Hawksworth DL, Bailey RH (eds) Lichenology: progress and problems. Academic Press, London, pp 215–247

Bamforth SS (1984) Microbial distributions in Arizona deserts and woodlands. Soil Biol Biochem 16:133–137

Binet P (1981) Short-term dynamics of minerals in arid ecosystems. In: Goodall DW, Perry RA (eds) Arid land ecosystems, vol 2. Cambridge University Press, Cambridge, pp 325–356

Bolton JH, Smith JL, Link SO (1993) Soil microbial biomass and activity of a disturbed and undisturbed shrub-steppe ecosystem. Soil Biol Biochem 25:545–552

Cameron RE, Blank GB (1967) Soil studies – microflora of desert regions. VIII. Distribution and abundance of desert microflora. Jet Prop Lab Space Progr Summary, No 37–44, 4:193–201

Cameron RE, King J, David CN (1970) Soil microbial ecology of Wheeler Valley, Antarctica. Soil Sci 109:110–120

Crawford CS (1981) Biology of desert invertebrates. Springer, Berlin Heidelberg New York

Franco PJ, Edney EB, McBrayer JF (1979) The distribution and abundance of soil arthropods in the northern Mojave Desert. J Arid Environ 2:137–149

Freckman DW, Sher SA, Mankau R (1974) Biology of nematodes in desert ecosystems. US Int Biol Progr, Desert Biome Res Memo 74–35, Utah State University, Logan, UT

Ghabbour SI, El-Ayouty EY, Khadr MS, El-Tonsi AMS (1980) Grazing by microfauna and productivity of heterocystous nitrogen-fixing blue-green algae in desert soils. Oikos 34:209–218

Ingham RE, Trofymow JA, Ingham ER, Coleman DC (1985) Interactions of bacteria, fungi, and their nematode grazers: effects on nutrient cycling, and plant growth. Ecol Monogr 55:119–140

Kieft TL (1991) Soil microbiology in reclamation of arid and semi-arid lands. In: Skujins J (ed) Semiarid lands and deserts: soil resource and reclamation. Marcel Dekker, New York, pp 209–256

Kovda VA, Samoilova EM, Charley JL, Skujins J (1979) Soil processes in arid lands. In: Goodall DW, Perry RA, Howes KMW (eds) Arid-land ecosystems: structure, functioning and management. Cambridge University Press, Cambridge, pp 439–470

Loria M, Herrnstadt J (1980) Moss capsules as food of the harvest ant. Bryologist 83:524–525

Lynch JM, Harper SHT (1983) Straw as a substrate for cooperative nitrogen fixation. J Gen Microbiol 129:251–253

Parkinson D (1983) Functional relationships between soil organisms. In: Lebrun P, Andre HM, De Medts A, Gregoire-Wibo C, Wauthy G (eds) Proc VIII Int Coll Soil Zoology. Dieu-Brichart, Louvain-la-Neuve, Belgium, pp 153–165

Rogers LE, Fitzner RE, Cadwell LL, Vaughn BE (1988) Terrestrial animal habitats and population responses. In: Rickard WH, Rogers LE, Vaughn BE, Liebetrau SF (eds) Shrub-steppe: balance and change in a semi-arid terrestrial ecosystem. Elsevier, Oxford, pp 181–256

Romney EM, Wallace A, Hunter RB (1978) Plant response to nitrogen fertilization in the northern Mojave Desert and its relationship to water manipulations. In: West NW, Skujins J (eds) Nitrogen in desert ecosystems. Dowden, Hutchinson & Ross, Stroudsburg, PA, pp 232–243

Rundel PW, Gibson AC (1996) Ecological communities and processes in a Mojave Desert ecosystem, Rock Valley, Nevada. Cambridge University Press, Cambridge

Rychert RC, Skujins J (1973) Microbial activity in acid soils. Utah Sci 34:96–98

Schiefer GE, Caldwell DE (1982) Synergistic interaction between *Anabaena* and *Zooloea* spp. in carbon dioxide-limited continuous cultures. Appl Environ Microbiol 44:84–87

Skujins J (1984) Microbial ecology of desert soils. In: Marshall CC (ed) Advances in microbial biology. Plenum Press, New York, pp 49–91

Smith NR, Griggs RF (1932) The microflora of the ash of Katmai Volcano with especial reference to nitrogen fixing bacteria. Soil Sci 34:365–373

Steinberger Y (1991) Biology of arid region soils – faunal components. In: Skujins J (ed) Semiarid lands and deserts: soil resource and reclamation. Marcel Dekker, New York, pp 173–192

Tchan YT, Whitehouse JA (1953) Study of soil algae. II. The variation of the algal populations in sandy soil. Proc Linn Soc NSW 78:160–170

Wallwork JA (1972) Distribution patterns and population dynamics of the microarthropods of a desert soil in southern California. J Anim Ecol 41:291–310

Walter DE (1988) Nematophagy by soil arthropods from the shortgrass steppe, Chihuahuan Desert and Rocky Mountains of the central United States. Agric Eco-Syst Environ 24:307–316

Whitford WG (1996) The importance of the biodiversity of soil biota in arid ecosystems. Biodiv Conserv 5:185–195

Whitford WG, Freckman DW, Elkins NZ, Parker LW, Parmalee R, Phillips J, Tucker S (1981) Diurnal migration and responses to simulated rainfall in desert soil microarthropods and nematodes. Soil Biol Biochem 13:417–425

Wood TG (1971) The distribution and abundance of *Folsomides deserticola* (Collembola: Isotomidae) and other micro-arthropods in arid and semi-arid soils in southern Australia, with a note on nematode populations. Pedobiologia 11:446–468

Yeates GW (1979) Soil nematodes in terrestrial ecosystems. J Nematol 11:213–229

Zak JC, Freckman DW (1991) Soil communities in deserts: microarthropods and nematodes. In: Polis G (ed) Ecology of desert communities. University of Arizona Press, Tucson, AZ, pp 55–88

Part III

Structure of Biological Soil Crusts: Microscale to Landscape

15 Comparative Structure of Physical and Biological Soil Crusts

J. Belnap

15.1 Introduction

The presence of physical and/or biological soil crusts alters many characteristics of the soil surface, and thus can play a defining role in many ecosystem functions (Greene and Ringrose-Voase 1994; Issa et al. 1999). The presence of a physical crust can seal and smooth surfaces, thus decreasing rainfall infiltration and increasing the volume and velocity of water runoff (Sumner and Stewart 1992). Physical crusts often inhibit vascular plant seedling establishment. Smooth biological crusts, like physical crusts, can also control local hydrology by smoothing and partially sealing soil surfaces (Kidron and Yair 1997). In contrast, soil surfaces roughened by biological soil crusts can increase rainfall infiltration, decrease water runoff volume and velocity, and retain seeds and organic matter (Loope and Gifford 1972; J. Belnap, unpubl.). Thus, understanding the factors that control the form of physical or biological soil crusts is essential in interpreting how the presence of these crusts may influence ecosystem functions (Mücher et al. 1988).

15.2 Physical (Inorganic) Soil Crusts

Nonbiotic soil-surface crusts, or physical crusts, are a major structural feature in many arid regions. The properties of physical crusts, and their manner of formation, have been studied for years, primarily because of the detrimental effect these crusts have on agriculture crops (Sumner and Stewart 1992). These crusts are transient soil-surface layers which range in thickness from less than 1 mm to a few cm, and which are structurally different from the material immediately beneath them (Fig. 15.1).

There are four main causes of physical crusts: the impact of raindrops, compressional forces such as animal trampling or vehicular traffic, evaporative processes (forming chemical crusts), and trapped gas bubbles (forming

Ecological Studies, Vol. 150
J. Belnap and O.L. Lange (eds.) Biological Soil Crusts: Structure, Function, and Management

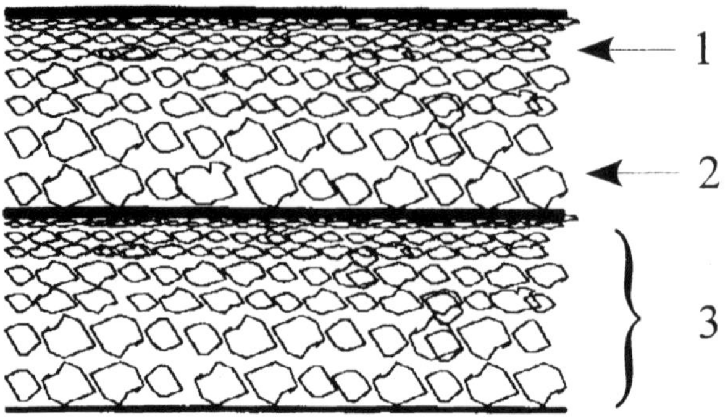

Fig. 15.1. Physical crusts are often layered as successive events break down larger soil aggregates into small particles. Water-ponding allows the small particles to rise to the surface and they then bind together to form a crust. This figure shows how breakdown of larger aggregates (*arrow 2*) by raindrops creates finer particles (*arrow 1*) that float to the soil surface, creating a physical crust. This can happen multiple times (*arrow 3*). Layers are generally a few mm thick. (After Sumner and Stewart 1992)

vesicular crusts). Physical crusts can also be created by any combination of these factors. Most physical crusts are formed by raindrop impact hitting unprotected soil surfaces. Raindrops break up soil aggregates and create smaller particles. These smaller particles can then wash into spaces between the remaining larger particles, clogging soil pores and reducing infiltration rates by as much as 90 %. This can occur within the first few minutes of a rainstorm. As drying takes place, surface tension pulls the components together, forming a dense, strong layer (Sumner and Stewart 1992). Thackett and Pearson (1965) showed physical crusts that formed under simulated rainfall had a dense surface layer 1–3 mm thick that was coated with a thin layer of well-oriented clay. The physical crust was underlain by a more porous structure and the water permeability of the underlying material was about five times that of the surface 0–5 mm. Rain-formed crusts are thicker when raindrops are larger, as bigger drops have more energy and can "blast" deep into the soil, thus destroying the original structure to a greater depth. Most rain-formed crusts are less than 5 mm thick. Because large pores are absent in a physical crust, they usually have low saturated hydraulic conductivity and limited infiltration, thus increasing the velocity and amount of water runoff. This generally also increases soil loss.

Compressional forces, such as those generated by machinery or hooves of grazing animals, can also destroy soil aggregate structure and cause a physical soil crust (Lemos and Lutz 1957). Trampling compacts the soil aggregates into a comparatively impermeable surface layer, especially when soils are moist. These compacted surfaces have reduced infiltration rates and increased surface runoff. In this sense, they function hydrologically in a manner similar to raindrop-induced crusts. Chemical crusts are very common in more arid regions (Miller 1971). These crusts are often harder than those caused by raindrop or compressional impacts. As water evaporates, compounds such as salts, lime, and silica are deposited at the soil surface. Chemical crusts often form on top of raindrop or compressional crusts, which both decrease infiltration and cause water to pool on the surface. As this water evaporates, chemical crusts are formed. Vesicular crusts are common in arid soils that contain a high percentage of silt. They are characterized by numerous air pockets formed just under the soil surface, and can significantly impede water infiltration.

Physical crusts may form on soils of almost any texture except coarse sandy soils that contain very low silt and clay (Lemos and Lutz 1957). Soils especially susceptible to physical crusting are those with low structural and aggregate stability, including those with low organic matter, high silt, and/or high salt (e.g., sodium or calcium carbonates). Because organic matter increases aggregate stability (through cementation of particles and moderation of forces that reduce aggregate stability) and incorporated plant residues form planes of weakness within the soil, soils with high organic matter do not often form physical crusts. Sand particles do not form bonds between grains; many clays shrink and swell, breaking soil bonds. In contrast, silts form strong bonds on drying, and have low swelling and shrinking properties; thus, silty soils readily form physical crusts. High sodium or calcium carbonate contents can result in surface chemical crusts on any soil type.

Because physical crusts can reduce water infiltration and plant establishment, they are generally regarded as undesirable in a management situation (Sumner and Stewart 1992). However, once formed, they are very difficult to remove. Land managers are often told to use intensive grazing to break up physical crusts (Savory 1988). While livestock trampling does break the crust for a short time, the physical crust will reform after the first minutes of the next intense rainstorm. In addition, intense trampling further breaks and compacts the soil aggregates, leading to even greater long-term physical crusting. To effectively address a physical soil crusting problem, livestock grazing systems must promote greater soil aggregate stability (Thurow 1991). Therefore, management systems that increase soil surface protection and soil organic matter content through increased vascular plant and biological soil-crust cover are the only lasting solution to physical crust reduction (Blackburn 1983).

In contrast, there are situations where physical crusting can play an essential role in structuring ecosystems. In hyperarid regions, plant interspaces are often covered by a combination of physical and biological soil crusts that reduce water infiltration. Runoff from these crusted plant interspaces can increase water to nearby vascular plants, thus providing water essential for their survival. Nearby areas that lack physical and biological crusting often also lack well-developed vegetation (Tongway and Ludwig 1990; Cornet et al. 1992; Zaady and Shachak 1994). Ancient peoples in the Middle East such as the Nabateans utilized these principles to support crops in arid regions. Studying remainders of ancient farm systems in the field, Evenari et al. (1982) reconstructed and reintroduced "runoff farming" in the Central Negev in Israel.

15.3 External Morphology and Internal Structure of Biological Soil Crusts

Both the external morphology and internal structure of biological soil crusts are very different from physical crusts. Thus, biological soil crusts generally have a very different influence on ecosystem processes. The following sections discuss the factors that influence the structure of biological soil crusts, including, climate, soil texture, soil chemistry, aeolian deposition, and disturbance history.

15.3.1 External Morphology: Major Types of Biological Soil Crusts

Biological soil crusts in different habitats can differ greatly in appearance, species composition and ecological function. Crust classification systems have been proposed for crusts in the western US (Johansen 1993) and Australia (Eldridge and Greene 1994). The classification presented below allows for worldwide comparisons and is based on the general literature and personal observations by J. Belnap and O.L. Lange. This classification is based on external crust morphology, as this aspect is easily distinguished visually and also greatly affects many ecosystem functions. This classification also represents a potential evapotranspiration (PET) gradient. Four crust groups are recognized: smooth, rugose, rolling, and pinnacled (Table 15.1). Smooth crusts occur in hyperarid and arid hot deserts with the highest PET (Chap. 1, Photo 49); rugose crusts occur in hot, arid deserts with slightly lower PET (Photos 51, 53). Both crust types occur where soil freezing does not occur. In contrast, pinnacled crusts occur in arid to semiarid cool deserts (Photo 31), where soils freeze during cold winters and lower PET supports a higher biomass of mosses and lichens than the hot deserts. Rolling crusts occur in even colder semiarid cool and cold deserts (Photo 29), where soils freeze in winter and even lower PET supports a larger biomass of lichens, mosses, and vascular plants than found in less moist deserts. Exceptions to this distribution pattern are the smooth and rugose crusts which occur in temperate regions (with very low PET, soil freezing) as a successional phase after

Table 15.1. Definitions of crust types

Topography	Surface Roughness	Organisms Involved	Region of Occurrence
Smooth no frost heaving	0–1 cm	Cyanobacteria, algae, etc. inside the soil. No lichens or mosses	Hyperarid habitats without soil freezing, e.g., hyperarid Australia, dunes in the Negev (partially fixed); or as a successional stage in temperate regions; physical and chemical encrustations may be involved (Photos 49, 50)
Rugose no frost heaving	1–3 cm	Scattered patches of lichens and mosses (in addition organisms inside the soil)	Arid habitats without soil freezing, e.g., Australia, Central Negev, Coastal Fog Zone of the Namib, Mediterranean area (Photos 48, 51, 52, 53) or as a successional stage in temperate regions (Photos 9, 55)
Rolling frost heaving	3–5 cm	Fairly continuous cover of lichens and mosses (in addition to organisms inside the soil)	Habitats with winter freezing, often in clay soil or with high moss-lichen cover, e.g. northern Great Basin, USA (Photos 29, 30)
Pinnacled frost heaving	5–15 cm	Cyanobacteria, algae, etc. inside the soil with or without lichens and mosses	Habitats with winter frost, e.g. Colorado Plateau, central Great Basin, USA, areas with low moss-lichen cover (Photos 31, 32, 33)

disturbance. This classification system is highly generalized, and the four different types represent a continuum; thus, intermediate crust types will be found between these four groups. These classifications are based on late successional stages of crusts; in frequently disturbed areas, smooth or rugose crusts can be seen in any geographic region.

Smooth Crusts. Smooth crusts consist mostly of endedaphic cyanobacteria, algae, and fungi. Soil surfaces are mostly mineral particles and extremely flat, as the cyanobacteria bind the sand grains together, creating an even smoother soil surface than where the cyanobacteria are absent (Photo 50). Smooth crusts occur in hyperarid and arid regions, where precipitation is very low, temperatures are very high, and soils never freeze (e.g., northwestern

Australia, the Arava Valley and Nizzana dunes in Israel, the Atacama Desert in Chile, and the central Sahara Desert). This type of crust can also be found in areas that are frequently disturbed in any region.

Rugose Crusts. Rugose crusts occur in arid and semiarid regions with lower PET than areas with smooth crust; however, both occur in regions that lack soil freezing. Like smooth crusts, rugose crusts are dominated by cyanobacteria, green algae and fungi, but, in addition, contain sparse patches of lichens and mosses growing on the more or less even soil surface (Photos 48, 52, 54). This type of crust occurs in areas such as the Sonoran and Mojave Deserts in the US, and the Central Negev in Israel. Rugose crusts also occur as a successional stage in temperate climates, where they can be dominated by mosses or by an epedaphic layer of filamentous algae (Photos 9, 55). Such crusts are found in mediterranean vegetation types, open woodlands (e.g., pine barrens of US, miombo woodlands of Africa), and steppe formations of Europe.

Pinnacled Crusts. Pinnacled crusts are dominated by endedaphic cyanobacteria (such as *Microcoleus* sp.), but can locally support up to 40% lichen and moss cover. These crusts are characterized by strikingly pedicelled mounds that are formed as the soils that are uplifted by frost-heaving are differentially eroded by downward-cutting water (Photos 32, 33). These castle-like mounds can be up to 15 cm high, and have delicate tips that are less than 4 mm across (J. Belnap, unpubl.). Lichens, mosses, small rocks, or concentrations of cyanobacteria often act as a cap for these tips, offering greater resistance to erosion than adjacent soil (Fig. 15.2a). Cyanobacteria occurring in the center of these tips still receive sufficient light for photosynthesis; thus, these mounds have a greatly increased surface area for colonization by crustal organisms. Pinnacled crusts occur in mid-latitude, cool deserts such as the Colorado Plateau and mid-latitude China. This crust type is the most vulnerable to soil surface disturbance, as the frost-heaved surface is easily broken and churned, often burying crustal organisms (see Chap. 27).

Rolling Crusts. Rolling crusts occur in colder regions, such as the northern Great Basin in the US, the northern Mongolian steppes, and the Arctic, where low PET supports biological soil crusts that are heavily dominated by lichens, mosses, and/or thick mats of cyanobacteria. The upward frost-heaving of the soil is counteracted by the cohesive, thickly encrusted mats of lichens and mosses and the many surface roots of vascular plants; thus, rather than pinnacled surfaces, this combination creates a rough, rolling surface (R. Rosentreter, unpubl.; Photo 30). These types of crusts are sometimes easily detached from the soil surface, making them vulnerable to soil surface disturbance.

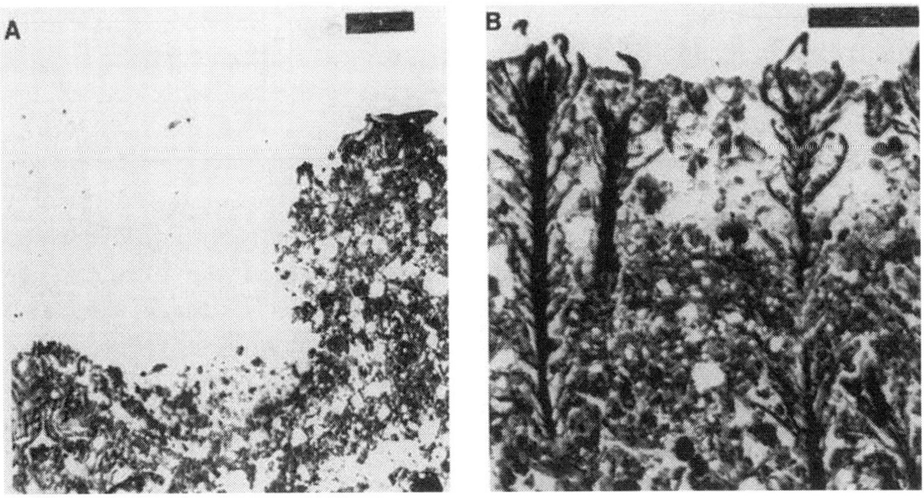

Fig. 15.2A,B Thin-sections of soil crusts from Australia. **A** Lichens can create pinnacled surfaces by protecting underlying soil from erosion. **B** Surface roughness created by lichens and mosses traps sediment. Unlike lichens, mosses can grow up through deposited sediment. *Bar* 1 mm. (Photos courtesy D. Eldridge)

15.3.2 Other Factors Influencing the External Morphology of Biological Soil Crusts

Soil Texture and Chemistry. Soil texture and chemical characteristics can partially or almost completely override the effects of frost-heaving or biological components in defining the external morphology of a soil crust. Crusts with similar species and climate on fine-textured soils are flatter than crusts on adjacent coarser soils, and on finer soils, cyanobacteria are concentrated closer to the surface, as light is quickly limited at depth (Garcia-Pichel and Belnap 1996; see Chap. 16). In soils with a weak crystalline structures (e.g., soils derived from calcite and gypsum), polysaccharide materials can combine with the dissolved minerals when soils are wet and, upon drying, sheaths are formed that are part organic and inorganic (Belnap and Gardner 1993; see Sect. 15.3.3). The internal strength of these soils resists winter frost-heaving and does not form the highly dissected surfaces that may be found in adjacent sandy soils. In addition, soils with a high carbonate content almost always support a higher lichen and moss cover (see Part I), and thus a rougher soil surface, than adjacent soils with lower carbonate levels (J. Belnap et al., unpubl.).

Soils with a high content of shrink-swell clays (e.g., bentonite) often create a very unstable soil surface. In deserts, biological crusts on these soils generally consist only of large filamentous cyanobacteria (e.g., *Microcoleus*), except in moist microhabitats where a limited cover of mosses or lichens can be found (J. Belnap, unpubl.). On these soils, surface morphology is controlled more by physical and chemical characteristics of the soil, with little contribution from the biological components. As with gypsiferous soils, soils with high clay content do not form highly pinnacled surfaces.

Physical crusts are often colonized by biological crusts. In soils with heavy physical crusting, the surface morphology of crusts is mostly controlled by soil physical and chemical characteristics, with the biological components having only a limited effect on external appearance of the crust (Mücher et al. 1988). If only cyanobacteria are present, the resultant crusts are smooth. If some lichens colonize as well, the slight roughening of the soil surface creates a rugose crust. Pinnacles can form in soils that have low to moderate physical crusting.

Disturbance History. Intensity, type, and time since soil surface disturbance can control the external morphology of biological soil crusts. Crustal components are brittle and easily crushed, especially when dry (Belnap and Gardner 1993). Damage to buried sheath material is nonrepairable, as living cyanobacteria are no longer present at depth to regenerate sheath materials (Harper and Marble 1988). When adjacent areas are disturbed, sediment can be blown or washed in on top of soil-crust organisms. The roughened soil surface often traps this material (Fig. 15.2b). While mosses can often grow up through this sediment (Fig. 15.2b), and the larger cyanobacteria can glide up to the surface of the new deposits, lichens and the smaller cyanobacteria are less mobile. Thus, if deposition rates are high, these latter organisms will die.

Intense disturbance results in bare soil. When colonized by large filamentous cyanobacteria, smooth crusts are formed (see Chap. 27). In regions with frost-heaving, some surface microtopography appears after the first winter, creating a rugose crust (Belnap 1995). As time passes, a combination of soil deposition, frost-heaving, differential erosion, and the colonization of lichens and mosses results in thicker crusts with greater surface microtopography (Belnap 1998; Photo 32). However, if disturbance continues, crusts will stay in early successional stages (i.e., a thin veneer of cyanobacteria only; see Chap 27).

15.3.3 Internal Structure of Biological Soil Crusts

In most desert soils, large filamentous cyanobacteria are often the dominant species, providing the matrix in which other components of the biological crust are embedded. The scanning electron micrographs below present a detailed look at the internal structure of soil crusts and the role the dominant cyanobacterium *Microcoleus vaginatus* plays in soils from the Colorado Plateau (Belnap and Gardner 1993).

In Fig. 15.3a, the sheaths of *Microcoleus* can be seen winding among the sand particles, connecting the individual grains together, even though sheaths are dry. These sheaths act much like fibers in fiberglass, conferring great tensile strength to the soil. This high tensile strength can be seen in both the photo and the photo b, where a single cyanobacterial strand can hold a sand grain aloft (Fig. 15.3a, b). With moisture, the cyanobacterial sheath swells and becomes turgid, appearing like a fishing net on the soil surface (Fig 15.3 c). This "net" holds soil particles firmly in place, reducing the possibility of soil loss by water or wind erosion.

Large filamentous cyanobacteria such as *Microcoleus* occur either as single filaments, or as a group of filaments, surrounded by a large polysaccharide sheath (Fig. 15.3d,e). The individual cells of the filaments can be clearly seen in Fig. 15.3e. When wetted, the polysaccharide sheaths swell and the filaments within are mechanically extruded from the sheath. As the soils dry, filaments retract into the sheath. The exposed portions of the filaments then secrete additional polysaccharide material. When dry, the cyanobacterial filaments are completely encased in the polysaccharide sheaths. Migrations to the surface, whether in response to rain or additions of sediment, result in sheath buildup underneath the soil surface. Visual examination of undisturbed soil crusts on sandy soils of the Colorado Plateau shows cyanobacterial sheath material to occur as deep as 10 cm below the surface of the soil (Belnap and Gardner 1993). Since most living biomass is in the top 4 mm of soil (Tchan and Whitehouse 1953; Garcia-Pichel and Belnap 1996), sheath material below that depth must represent remnants of cyanobacterial crusts once found near or at the soil surface and later buried by sediments. Though no longer associated with living filaments, such sheath material is still capable of binding soil particles together and still increases nutrient and moisture retention of associated soil.

Fig. 15.3A–E. (*see next page*) *Microcoleus vaginatus* in sandy soils: **A** Sheaths adhere firmly to sand grain surfaces, binding the grains together (note the intimate connection between the sheath and grain surfaces. *Bar* 100 μm. **B** Inset photo shows the strength of the sheath even when dry, as a single sheath holds a sand grain aloft. *Bar* 500 μm. **C** When wetted (using freeze substitution), sheaths and filaments swell and cover the soil surface. *Bar* 100 μm. **D** Note the large extracellular sheath surrounding the living filaments. *Bar* 10 μm. **E** Closeup of *M. vaginatus* sheath and filaments (note the cell walls). *Bar* 10 μm. (After Belnap and Gardner 1993)

Fig. 15.4. (*see next page*) **A** Multiple sheaths wrap around a sand grain, holding it firmly in place. *Bar* 50 μm. **B** Angular fine particles may be incorporated into sheath material. *Bar* 10 μm. **C** In fine-textured soils such as this limestone-derived substrate, many small particles adhere to the sheath material. *Bar* 10 μm. **D** In gypsiferous soils, sheaths can be made of both organic and inorganic material, and the sheaths are often coated with gypsum crystals. *Bar* 10 μm. (After Belnap and Gardner 1993)

Fig. 15.3A–E

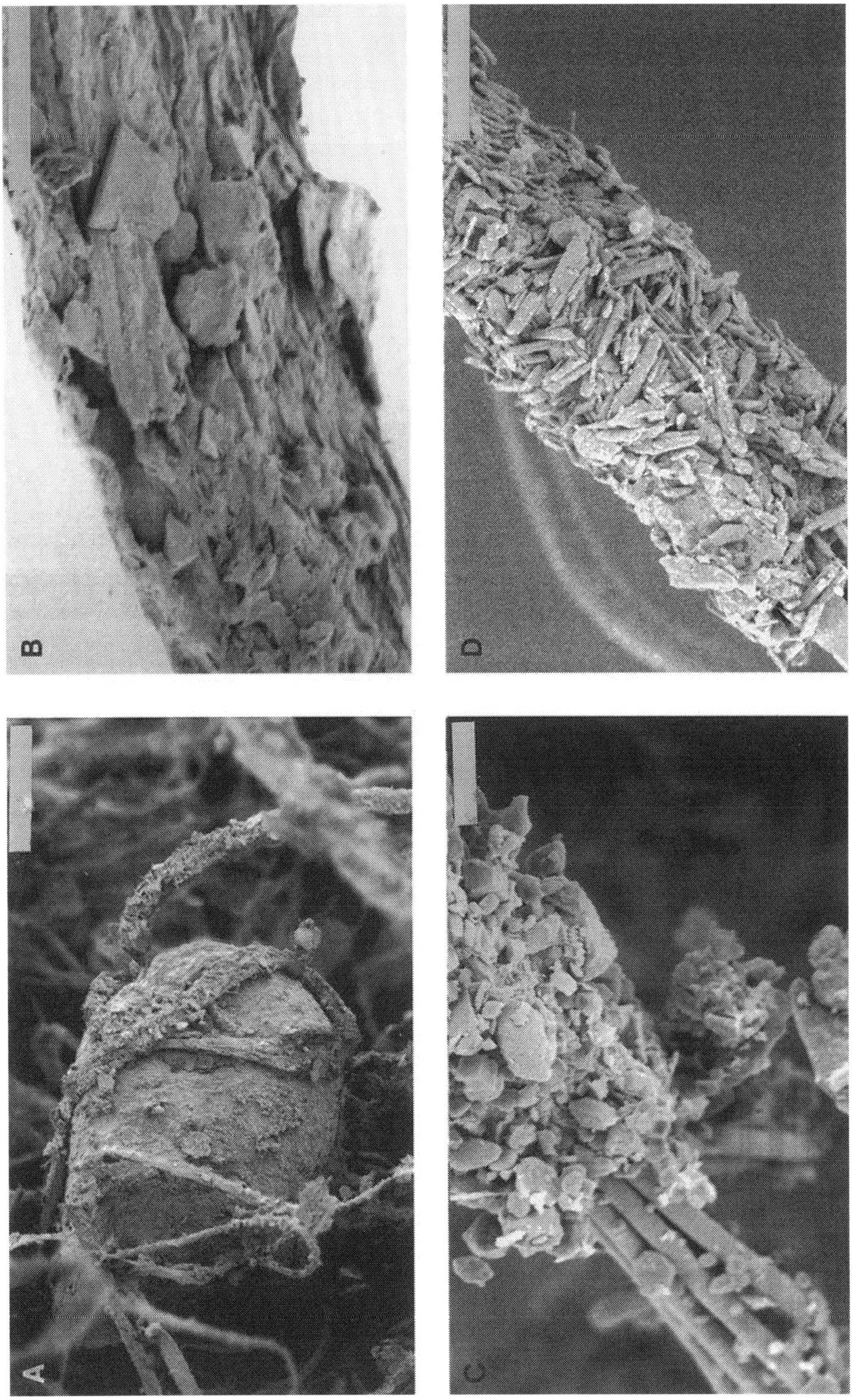

Fig. 15.4

In Fig. 15.4a, the extensive connections between sand grains by the sheath material can be seen, and multiple sheaths can often be seen attached to the same sand grain. Clay particles are also bound to, and incorporated into, the polysaccharide sheath material (Fig 15.4b). This may be a mechanism by which cyanobacteria increase mineral availability to themselves and vascular plants. Positively charged macronutrients are bound to both negatively charged clay particles and the sheath materials; thus, they are held in the upper soil horizons in a form readily available to vascular plants, instead of being leached away by water or bound in a chemically unavailable form (Belnap and Gardner 1993). In limestone (Fig. 15.4c), many fine particles are bound to the outside of the sheath. In gypsum (Fig. 15.4d), gypsum crystals often reprecipitate on the outside of the sheath. In many cases, the gypsum crystals combine with polysaccharide material, and take on a distinctive shape as a result.

Once soils are stabilized by the large, filamentous cyanobacteria, the number of smaller cyanobacteria and green algae increases (Dor and Danin 1996). Cyanobacteria, green algae, and fungi all secrete polysaccharide material that sticks to soil particles. Once lichens and mosses colonize, their rhizines and rhizoids also contribute internal structure to the crust and increase soil stability (Fig 15.2). Unlike lichens, cyanobacteria and mosses can grow upwards through deposited sediment. However, this ability is limited, especially in areas with low precipitation, and the crustal organisms can easily be buried and die (Chap. 27).

15.3.4 Other Factors Influencing the Internal Structure of Biological Soil Crusts

Cyanobacterial Layering. Well-developed crusts show a vertical layering of cyanobacterial species (Davey and Clarke 1992; J. Belnap, unpubl.; see Chap. 3). Smaller, less mobile organisms (e.g., *Nostoc, Scytonema, Chroococcidiopsis*) are most often found at the soil surface, and contain specialized pigments for UV-protection. The larger cyanobacteria (e.g., *Microcoleus, Phormidium*) are found below the soil surface when soils are dry. When soils are wetted, these large cyanobacteria partially leave their sheaths and glide up to the soil surface. As soils dry, they retreat back into their sheaths below the soil surface (Tchan and Whitehouse 1953; Dor and Danin 1996). These species generally lack UV-protective pigments. A similar layering of species is found in intertidal cyanobacterial mats (Cohen and Rosenberg 1989).

Aeolian Deposition. Aeolian deposition is often the dominant source of fine soil particles in desert soils (Danin and Yaalon 1982; Reynolds et al. 1998). Many authors report that airborne silt and clay are trapped by sticky

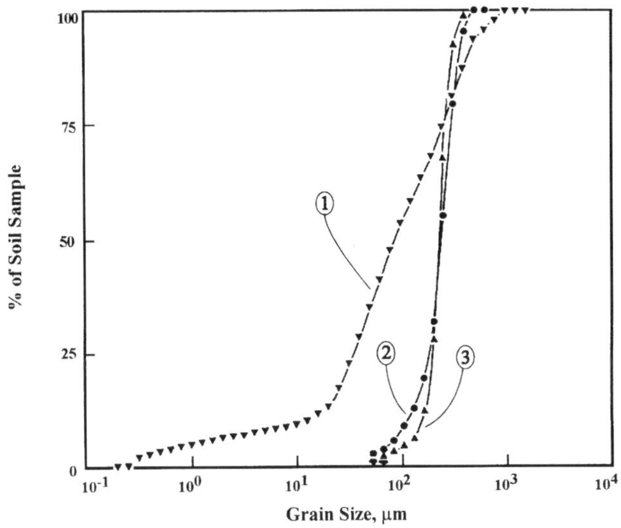

Fig. 15.5. Cumulative size distribution curves for soil particles in three dune samples. Note the greater amount of silt and clay particles in the crusted soil (*curve 1*) relative to the sand directly underneath the crust (*curve 2*) or from a nearby active dune (*curve 3*). (After Verrecchia et al. 1995)

cyanobacterial sheaths, by frost-heaved surfaces, and by the protruding moss stems and lichen thalli. This results in a thin layer of silt and clay on the crust surface (Figs. 15.2 and 15.5). Silt particles increase soil fertility and water-holding capacity (Nieboer and Richardson 1981; Puckett 1985; Danin and Ganor 1991; Davey and Clarke 1992; Verrecchia et al. 1995; although see Littman 1997).

15.4 Conclusions

The external morphology of biological soil crusts is determined by multiple factors. Of these, climate and disturbance history are probably the most important, followed by soil texture and chemistry. The morphology of the soil crust determines how susceptible the crust is to soil-surface disturbance, as frost-heaved soils are easily compacted and churned, thus burying crust organisms. Morphology also determines how soil crusts influence ecosystem functions such as hydrology, vascular plant relations, and water and wind erosion.

Much of the internal structure of crusts in most deserts is provided by large filamentous cyanobacteria. They aggregate soils and provide great tensile strength. Other contributors to internal structure are microfungi, and anchoring structures of mosses and lichens. Silt is often increased on crusted surfaces, increasing site fertility.

Acknowledgments. Many thanks to Sasha Reed, Michelle Schmid, and Sue Phillips for helpful comments and graphics.

References

Belnap J (1995) Surface disturbances: their role in accelerating desertification. Environ Mon Assess 37:39–57

Belnap J (1998) Choosing indicators of natural resource condition: a case study in Arches National Park, Utah, USA. Environ Manage 22:635–642

Belnap J, Gardner JS (1993) Soil microstructure in soils of the Colorado Plateau: the role of the cyanobacterium *Microcoleus vaginatus*. Great Basin Nat 53:40–47

Blackburn WH (1983) Influence of brush control on hydrologic characteristics of range watersheds. In: MacDaniel KC (ed) Proceedings, Brush management symposium, Texas Tech Press, Lubbock USA, pp 73–88

Cohen Y, Rosenberg E (1989) Microbial mats: physiological ecology of benthic microbial communities. American Soc Microbiol, Washington DC

Cornet AF, Montana C, Delhoume JP, Lopez-Portillo J (1992) Water flows and the dynamics of desert vegetation stripes. In: Hansen AJ, di Castri F (eds) Landscape boundaries: consequences for biotic diversity and ecological flows. Springer, Berlin Heidelberg New York, pp 327–345

Danin A, Ganor E (1991) Trapping of airborne dust by mosses in the Negev Desert, Israel. Earth Surf Process Landforms 16:153–162

Danin A, Yaalon DH (1982) Silt plus clay sedimentation and decalcification during plant succession in sands of the Mediterranean coastal plain of Israel. Isr J Earth-Sci 31:101–109

Davey MC, Clarke KJ (1992) Fine structure of a terrestrial cyanobacterial mat from Antarctica. J Phycol 28:199–202

Dor I, Danin A (1996) Cyanobacterial desert crusts in the Dead Sea Valley, Israel. Algol Stud 83:197–206

Eldridge DJ, Green RSB (1994) Microbiotic soil crusts: a review of their roles in soil and ecological processes in the rangelands of Australia. Aust J Soil Res 32:389–415

Evenari M, Shanan L, Tadmor N (1982) The Negev – the challenge of a desert. Harvard University Press, Cambridge

Garcia-Pichel F, Belnap J (1996) Microenvironments and microscale productivity of cyanobacterial desert crusts. J Phycol 32:774–782

Greene RSB, Ringrose-Voase AJ (1994) Micromorphological and hydraulic properties of surface crusts formed on a red earth soil in the semiarid rangelands of eastern Australia. In: Ringrose-Voase AJ, Humphreys GS (eds) Soil micromorphology: studies in management and genesis. Elsevier, New York, pp 763–776

Harper KT, Marble JR (1988) A role for nonvascular plants in management of arid and semiarid rangelands. In: Tueller PT (ed) Vegetation science applications for rangeland analysis and management. Kluwer Academic Publishers, Boston, pp 135–169

Issa OM, Trichet J, Defarge C, Coute A, Valentin C (1999) Morphology and microstructure of microbiotic soil crusts on a tiger bush sequence (Niger, Sahel). Catena 37:175–196

Johansen JR (1993) Cryptogamic crusts of semiarid and arid lands of North America. J Phycol 29:140–147

Kidron GJ, Yair A (1997) Rainfall-runoff relationship over encrusted dune surfaces, Nizzana, western Negev, Israel. Earth Surf Process Landforms 22:1169–1184

Lemos P, Lutz JF (1957) Soil crusting and some factors affecting it. Soil Sci Soc Am Proc 21:485–491

Littman T (1997) Atmospheric input of dust and nitrogen into the Nizzana sand dune ecosystem, north-western Negev, Israel. J Arid Environ 33:433–457

Loope WL, Gifford GF (1972) Influence of a soil microfloral crust on select properties of soils under pinyon-juniper in southeastern Utah. J Soil Water Conserv 27:164–167

Miller DE (1971) Formation of vesicular structure in soil. Soil Sci Soc Am Proc 35:635–637

Mücher HJ, Chartres CJ, Tongway DJ, Greene RSB (1988) Micromorphology and significance of the surface crusts of soils in rangelands near Cobar, Australia. Geoderma 42:227–244

Nieboer E, Richardson DHS (1981) Lichens as monitors of atmospheric deposition. In: Eisenreich SJ (ed) Atmospheric pollutants in natural waters. Ann Arbor Science Publishers, Ann Arbor, pp 339–388

Puckett KJ (1985) Temporal variation in lichen element levels. In: Brown DH (ed) Lichen physiology and cell biology. Plenum Press, New York, pp 211–225

Reynolds R, Belnap J, Reheis M, Mazza N (1998) Eolian dust on the Colorado Plateau – magnetic and geochemical evidence from sediment in potholes and biologic soil crust. Proc Conf on Dust Aerosols, Loess, Soils, and Global Change. Seattle, WA, Oct 11–14, 1998

Savory A (1988) Holistic resource management. Island Press, Covelo, California

Sumner ME, Stewart BA (1992) Soil crusting: chemical and physical processes. Lewis Publishers, Boca Raton, FL

Tchan YT, Whitehouse JA (1953) Study of soil algae. II. The variation of the algal populations in sandy soil. Proc Linn Soc NSW 78:160–170

Thackett JL, Pearson RW (1965) Some characteristics of soil crusts formed by simulated rainfall. Soil Sci Soc Am Proc 99:407–413

Thurow TL (1991) Hydrology and Erosion In: Heitschmidt RK, Stuth JW (eds.) Grazing management: an ecological perspective. Timber Press, Portland, OR, pp 151–152

Tongway DJ, Ludwig JA (1990) Vegetation and soil patterning in semiarid mulga lands of eastern Australia. Aust J Ecol 15:23–34

Verrecchia E, Yair A, Kidron GJ, Verrecchia K (1995) Physical properties of the psammophile cryptogamic crust and their consequences to the water regime of sandy soils, north-western Negev Desert, Israel. J Arid Environ 29:427–437

Zaady E, Shachak M (1994) Microphytic soil crust and ecosystem leakage in the Negev Desert. Am J Bot 81:109

16 Small-Scale Environments and Distribution of Biological Soil Crusts

F. Garcia-Pichel and J. Belnap

16.1 Introduction

On a large geographic scale, variability in climate and habitats results in differences in crust-community composition (see Part I). Within a single region or locality, additional heterogeneity may result from differences in physical parameters such as elevation, soil characteristics, or microclimate gradients (Kutiel et al. 1998). Even at a very small scale of 1 m or 1 cm, patchiness in the distribution of organisms can be considerable (Grondrin and Johansen 1993). This chapter is devoted to those variations in environmental variables occurring at spatial scales of millimeters to meters.

16.2 Distribution and Environments on a Millimeter Scale

Some microenvironments are inherent to the nature of the physical substrate, resulting from the interaction of physicochemical parameters with the abiotic or biotic constituents of the crusts. For example, vertical gradients of solar radiation within the soil are caused by absorption and scattering of incident light. Other microenvironments result from metabolic activities of microorganisms, acting under conditions in which mass transport phenomena are dominated by diffusion; this involves localized gradients of chemical species. The presence of inherent microenvironments is likely to determine the microscale distribution of organisms, which, in turn, will determine the nature and distribution of chemical microenvironments. Functionally stable, steady-state microbial communities of biological crusts will be those that take optimal advantage of the diversity of microniches offered under a given set of environmental parameters.

Microsensors for light, O_2 and H_2S, pH, NH_4^+, and NO_3^-, offering spatial resolution of several micrometers, have been essential for measuring these microenvironments (see reviews by Revsbech and Jørgensen 1986; Kühl and

Ecological Studies, Vol. 150
J. Belnap and O.L. Lange (eds.) Biological Soil Crusts: Structure, Function, and Management

Revsbech 2001; Klimant et al. 1997). While most microsensor studies have examined permanently submerged microbial communities, such as marine microbial mats, biofilms, and sediments, there have been some studies with soil crusts. Here, we review what is known about microgradients in soil crusts, identify features that are particular to these topsoil communities, and discuss the implications of these features.

16.2.1 Optical Microenvironments

Biological soil crusts are ecosystems dominated by phototrophic organisms, including cyanobacteria, algae, lichens, or mosses. For this reason, the optical characteristics of the medium are of upmost importance for the organisms living inside the soil. Optical microenvironments have two aspects: incident irradiance and light gradients within the crusts. Whereas incident irradiance is likely to vary significantly in the horizontal scale, vertical light gradients with depth are more predictable and lend themselves more easily to interpretations and generalizations.

Incident Irradiance. While incident irradiance is constant on a flat surface, soils roughened by biological soil crusts may have considerable surface microtopographies due to mounds, fissures, depressions, and overhangs.

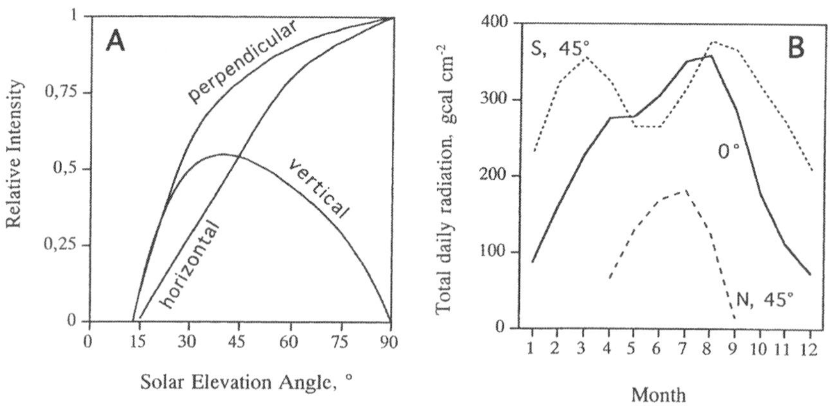

Fig 16.1A,B. Typical temporal and spatial variations in intensity of the incident irradiance, potentially conducive to the formation of optical microenvironments at the surface of crusts with marked microtopography. **A** Variations in incident direct radiation with angle of solar elevation for surfaces oriented vertically, horizontally, and perpendicular to the sun. Data are from Koller (1965), calculated for monochromatic radiation at 350 nm, and may be strongly wavelength-dependent. **B** Seasonal variation of the daily insolation falling on either a horizontal surface (0°), or on a north- or south-oriented surface (45° inclination; Jaag 1945)

Thus, incident irradiance at the soil surface may deviate significantly from that of the large-scale slope orientation, or that which would occur on a flat soil surface. These variations in exposure create microniches and affect daily and yearly photoperiods, soil temperature, soil moisture, and maximal irradiance (Fig. 16.1). In addition, the ratio of direct to diffuse light varies with soil surface angle with respect to solar elevation. This has been shown to increase the shorter wavelength components of incoming light (Castenholz and Garcia-Pichel 2000).

Light Gradients Within Crusts. Light reaching phototrophic organisms within the soil environment is subject to intense attenuation due to the high density of mineral and biogenic particles. Thus, photosynthetically active radiation levels are a function of both absorption and multiple scattering by mineral particles and the biota. In habitats optically analogous to soil crusts, such as sediments and microbial mats, photic zones (where the scalar irradiance, E_0, is attenuated to 1 % of that incident) vary between 3.1 mm for quartz sand and 0.45 mm for silty muds (Garcia-Pichel and Bebout 1996). In the ultraviolet (UV) at 310 nm, the corresponding depths are only 1.25 and 0.23 mm, as shorter wavelengths penetrate less deeply than longer wavelengths. As light penetrates into the soil or sediment and successive scatter events randomize direction of the photons, light fields become increasingly diffuse and maxima in scalar irradiance (larger than that incident) build up close to the surface. Below this surficial zone, E_0 attenuates quasi-exponentially. This is governed mostly by grain-size distribution of the mineral particles, with smaller grains resulting in shallower light penetration. In the case of quartz sand, scalar irradiance in the subsurface zone increases linearly with wavelength (Garcia-Pichel and Bebout 1996). In the same way in which diaphanous quartz pebbles may locally increase the photic zones to several cm (Cameron 1966), a clear sand grain may locally increase light penetration. Light is more scattered and penetrates less in dry than in wet soils. In general, there are two distinct optical zones: a surface zone, where light-trapping occurs, and where the maximum scalar irradiance, intensity, and downward directionality of the light is reached, and a second, subsurface zone where intensities decay exponentially, light fields redden, and light is close to diffuse.

16.2.2 Anoxic and Hyperoxic Microenvironments

Water logged soil crusts can be temporary functional analogues of benthic microalgal communities or microbial mats. The respiratory and photosynthetic activities of microorganisms in cyanobacterial crusts from sand dunes can be sufficiently high to overwhelm diffusive transport of oxygen and thus create spatially localized environments where oxygen might be depleted or

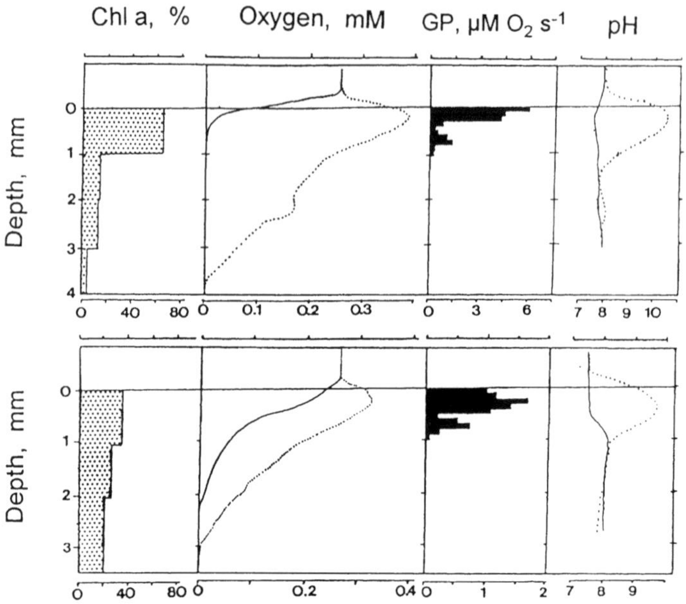

Fig 16.2. Microenvironments associated with respiratory and photosynthetic metabolism in two cyanobacterial crusts (Utah). *Upper panel: Microcoleus vaginatus* crust (subsurface populations); *lower panel* crust containing subsurface *M. vaginatus* and surface *Nostoc* and *Scytonema*; *left* relative distribution of chlorophyll *a* content (related to total content in the profile); *middle left* O_2 concentration profiles in dark (solid line) and in the light (dotted line); *middle right* depth distribution of gross photosynthesis and *right* pH in the light (*dotted line*) and in the dark (*solid line*). (After Garcia-Pichel and Belnap 1996)

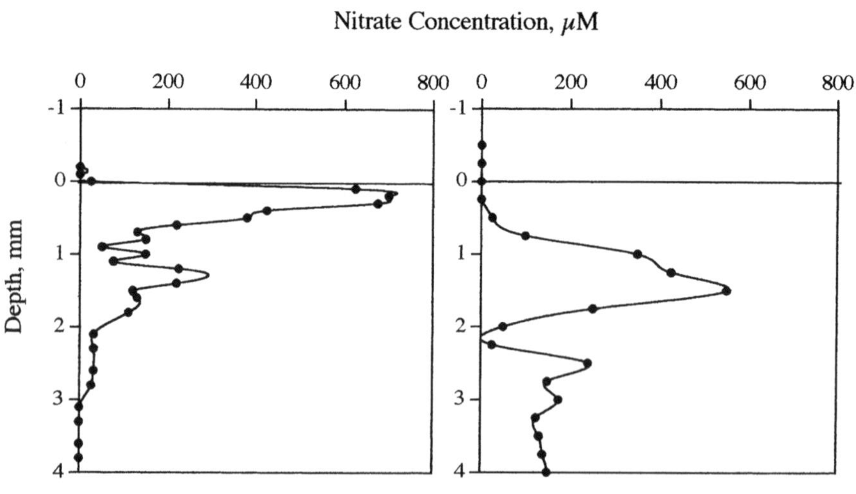

Fig. 16.3. Profiles of nitrate concentration in two crusts (same types as Fig. 16.2), incubated in the light. Such complex profiles indicate the presence of spatially separated nitrate sources and sinks within the crusts

supersaturated (deWinder 1990; deWinder et al. 1989). In wet, illuminated cyanobacterial crusts on sandy arid soils, the photosynthetically active zones (ca. the top 1 mm) reach steady states in which molecular oxygen is supersaturated. By contrast, the photosynthetically inactive, deeper zones (from 1 to 3–4 mm deep) are dominated by respiratory processes, and molecular oxygen is partially or completely depleted (Fig. 16.2; Garcia-Pichel and Belnap 1996). In the absence of photosynthesis (wet crusts in the dark) anoxic zones can be much shallower (0.7–2 mm). There is some horizontal small-scale variability in supersaturated zones and the depths at which anoxia occurs. Due to the vertical forcing of photosynthesis by inherent light gradients or to the unidirectional supply of oxygen from the atmosphere, the formation of oxygen microenvironments with such typical vertical distribution is likely to be a general characteristic of soil crusts. Microsensor profiling of O_2 allows the measurement of net and gross rates of respiration within the crusts, and is discussed in Garcia-Pichel and Belnap (1996). The extent and intensity of gradients mostly reflect total microbial activity. Other influencing factors include internal gas bubbles acting as oxygen reservoirs, cracks linking oxic and anoxic areas, and large mineral particles that impede diffusive equilibria.

16.2.3 Microenvironments of Chemical Species Other Than Oxygen

As a result of CO_2 uptake in photosynthesis, and its release in respiration, microscale changes in soil pH are likely to occur whenever oxygen microgradients are formed if the natural buffering capacity of the soils is low. Alkalinization of the top 1 mm of soil crusts by as much as 2–3 pH units in the light, and acidification by 0.5–1 pH units in the dark, have been directly measured (Fig. 16.2) in sandy soils, but data are lacking for other soil types. Marked vertical microgradients in nitrate concentration have also been measured (Fig 16.3).

16.2.4 Biological and Biogeochemical Consequences of Microenvironments

Vertical gradients in light intensity play an important role in defining the vertical distribution of phototrophs. Some cyanobacteria display phenotypic traits typical of shade-adapted organisms, including high levels of photosynthetic pigments and an absence of sunscreen compounds. These species colonize the reddened, low-intensity optical refugia found 0.5–1 mm below the soil surface (e.g., *Microcoleus vaginatus*; Garcia-Pichel and Belnap 1996). During periods of low incident irradiance (i.e., cloudy conditions) these cyanobacteria move upwards, using finely tuned photoresponses (Garcia-

Pichel and Castenholz 2001). In contrast, some cyanobacteria synthesize large amounts of sunscreens (e.g., *Nostoc, Scytonema, Chlorogloeopsis, Calothrix;* Scherer et al. 1988; Garcia-Pichel and Castenholz 1991, 1993) and can survive on the soil surface. However, even high levels of sunscreens do not always adequately protect these organisms from intense irradiance, as mortality can be 30–40 % during summer (see Chap. 30). Soil lichens and mosses must also cope with this harsh surface microenvironment, and there is evidence that microtopographical features create more favorable habitats necessary for the establishment of certain lichens (Danin et al. 1998; see Sect. 16.3 and 16.4 below). The presence of heterogeneities, boundaries, and gradients within an ecosystem is thought to contribute significantly to the potential organismal diversity that an ecosystem will contain (Nübel et al. 1999), and, similarly, the large diversity of microorganisms in soil crusts may reflect the variety of microniches present (reviewed by Evans and Johansen 1999).

The presence of internal anoxic zones has obvious implications for the mineralization of organic matter (both allochtonous and, probably primarily, autochtonous) in biological soil crusts. Fermentation or anaerobic respiratory metabolisms may play a significant role in the mineralization of organic matter. The availability of final electron acceptors other than oxygen in the crusts is likely to determine which processes may dominate (Garcia-Pichel and Belnap 1996). In prospective research, B. Thamdrup and F. Garcia-Pichel (unpubl.) followed the concentrations of several possible electron acceptors and products of anaerobic metabolism in sandy soil crusts (similar to those discussed above), incubated under dark anoxia. Within several hours of incubation there was significant depletion of the oxidized nitrogenous compounds NO_3^- + NO_2^- (on the order of 100 to 300 μmol m^{-2} h^{-1}), with contemporary increases in NH_4^+. Concomitant emissions of NO and N_2O during the incubation occurred at rates of 2–3 mol m^{-2} h^{-1}, which could be increased ten-fold by the addition of acetylene. Thus, denitrification may take place in the soil crusts, largely powered by the formation of anoxic microzones. However, the high soil temperatures required for significant denitrification rates preclude nitrogen fixation by crust organisms. Thus, denitrifiers would depend on ammonia available through other processes. As levels of free ammonia in desert soils are generally low, overall N loss by denitrification may be limited under field conditions (Chap. 19). Even though iron was abundant in the crusts (4 nmol m^{-2} as amorphous iron oxides and almost 50 mmol m^{-2} as crystalline oxides), no iron reduction activity could be measured. Similarly, no sulfate or manganese reduction could be detected.

Photosynthesis by crustal organisms removes CO_2 from soil pore water. Such depletion of protons in (poorly buffered) wet soils may affect the solubilization and mobilization of silica, the establishment and germination of higher plant seeds (Garcia-Pichel and Belnap 1996) and/or the volatilization of ammonia (Evans and Johansen 1999).

16.3 Distribution and Environments on a Centimeter Scale

In cool and cold deserts, frost-heaving of crusted surfaces results in greatly roughened soil surfaces on a centimeter scale, creating a wide range of microhabitats. Mound size can vary greatly, with mounds in well-developed pinnacled crusts up to 15 cm high, 10 cm long, and 5 cm wide. Surveys in southeastern Utah show that crust components are not evenly distributed on these mound surfaces. Cyanobacterial biomass, total lichen cover, and cover of the lichens *Collema tenax*, *C. coccophorum*, *Catapyrenium squamulosum*, *Toninia sedifolia*, and *Squamarina lentigera* are up to 30 % higher on the ENE aspects of mounds when compared to other aspects. Distribution of the lichen *Psora decipiens* shows no pattern. Cover of the mosses *Syntrichia caninervis* and *Pterogoneurum* spp. is greatest on NNW aspects, while cover of *Bryum* spp. is highest on both the NNW and ENE aspects. Colonization of lichens is also highest on ENE aspects (George et al. 2000; M.A. Bowker et al., unpubl. data).

Crust distribution patterns are probably explained by the different microhabitats on the various mound aspects. When light, soil temperature, and soil moisture are compared between different mound aspects throughout the year, the ENE aspect is the most favorable for net carbon gain and nitrogen fixation. Data suggests that micronutrients may also vary between mound aspects (J. Belnap et al., unpubl.).

16.4 Distribution on a Meter (or Greater) Scale

Little work has been done on meter-scale distribution patterns of soil-crust organisms in hot deserts. It has long been noted that mosses and lichens prefer polar-facing slopes where water availability is highest (Alexander and Calvo 1990). However, soil and plant community characteristics can influence lichen and moss cover and distribution on the meter scale. In sandy and sandy-loam soils of southeastern Utah (M.A. Bowker et al., unpubl.), lichen and moss richness and cover was greatest on flat benches with extremely shallow (<20 cm), sandy-loam soils with wide vascular plant spacing (*Pinus-Juniperus*-mixed shrub, 1–2 m apart) and on deep (>100 cm), sandy-loam soils with closely spaced grasses (*Hilaria jamesii*) when compared to directly adjacent sandy soils with wider plant spacing (>1 m) or sandy soils >50 cm, such as *Stipa* bunch grasslands or *Coleogyne* shrublands. Analysis of these data indicates that total lichen and moss cover is probably not determined as much by plant spacing or soil depth as by other factors such as soil texture and perhaps soil carbonate content (M.A. Bowker et al., unpubl.).

The location of vascular plants did influence the distribution of lichens and mosses. Sites with sparse vegetation showed a slight increase in lichen cover under plant canopies. However, where plants were closely spaced and lichen cover was high, cover of all lichen species was greatly reduced under plant canopies (relative to plant interspaces), perhaps due to reduced light via litterfall or shading. Moss cover was consistently higher on the NNE sides of plant canopies compared to SSW sides and plant interspaces. SSW sides of plant canopies did not support significantly more moss cover than plant interspaces, suggesting that increased nutrients found under plant canopies did not influence moss distribution as much as increased moisture availability (due to decreased radiation and air/soil temperatures).

16.5 Prospects and Outlook

Examination of crustal microenvironments can greatly assist in the understanding of distribution, abundance, and functioning of soil crusts. However, appropriate techniques for microscale investigations are not widely available and the number of published reports is low. Much remains to be learned, such as small-scale dynamics of nitrogen cycling and water availability. On a larger scale, little is specifically known about how variation in small-scale habitat differences affects distribution and abundance of crust species in most deserts. As such information is crucial for effective management and restoration of biological soil crusts, this area needs further scientific attention.

Acknowledgments. Many thanks to Michelle Schmid and Sue Phillips for editorial assistance.

References

Alexander RW, Calvo A (1990) The influence of lichens on slope processes in some Spanish badlands. In: Thornes JB (ed) Vegetation and erosion. John Wiley, Chichester, UK, pp 385–398

Cameron RE (1966) Desert algae: soil crusts and diaphanous substrata as algal habitats. Jet Propuls Lab Techn Rep 32-971:1–41

Castenholz RW, Garcia-Pichel F (2000) Cyanobacterial responses to UV-radiation. In: Whitton BA, Potts M (eds) The ecology of cyanobacteria: their diversity in time and space. Kluwer Academic Publishers, Dordrecht, pp 591–611

Danin A, Dor I, Sandler A, Amit R (1998) Desert crust morphology and its relations to microbiotic succession at Mt. Sedom, Israel. J Arid Environ 38:161–174 de Winder B (1990) Ecophysiological strategies of drought-tolerant phototrophic microorganisms in dune soils. PhD Thesis, University of Amsterdam, Amsterdam

de Winder B, Pluis J, DeReus L, Mur L (1989) Characterization of a cyanobacterial algal dune crust in the coastal dunes in the Netherlands. In: Cohen Y, Rosenberg E (eds) Microbial mats: physiological ecology of benthic microbial communities. Am Soc Microbiology, Washington, DC, pp 77–83

Evans RD, Johansen JR (1999) Microbiotic crusts and ecosystem processes. Crit Rev Plant Sci 18:183–225

Garcia-Pichel F, Bebout BM (1996) The penetration of ultraviolet radiation into shallow water sediments: high exposure for photosynthetic communities. Mar Ecol Prog Ser 131:257–262

Garcia-Pichel F, Belnap J (1996) Microenvironments and microscale productivity of cyanobacterial desert crusts. J Phycol 32:774–782

Garcia-Pichel F, Castenholz RW (1991) Characterization and biological implications of scytonemin, a cyanobacterial sheath pigment. J Phycol 27:395–409

Garcia-Pichel F, Castenholz RW (1993) Occurrence of UV-absorbing, mycosporine-like compounds among cyanobacterial isolates and an estimate of their screening capacity. Appl Environ Microbiol 59:163–169

Garcia-Pichel F, Castenholz RW (2001) Photomovements of microorganisms in benthic and soil microenvironments. In: Häder DP et al. (eds) Photomovements. Comprehensive Series in Photosciences, vol 1. Elsevier, Amsterdam, pp 401–418

George DB, Davidson DW, Schliep KC, Patrell-Kim J (2000) Microtopography of microbiotic crusts on the Colorado Plateau, and the distribution of component organisms. Western North American Nat 60:349–354

Grondrin AE, Johansen JR (1993) Microbial spatial heterogeneity in microbiotic crusts in Colorado National Monument. I. Algae. Great Basin Nat 53:24–30

Jaag O (1945) Untersuchungen über die Vegetation und Biologie der Algen des nackten Gesteins in den Alpen, im Jura und im schweizerischen Mittelland. Beitr Kryptogamenflora Schweiz 9:1–558

Klimant I, Kühl M, Glud RN, Holst G (1997) Optical measurement of oxygen and temperature in microscale: strategies and biological applications. Sensors Actuators B 38–39:29–37

Koller LR (1965) Ultraviolet radiation. Wiley, New York

Kühl M, Revsbech NP (2001) Biogeochemical microsensors for boundary layer studies. In: Boudreau BP, Jørgensen BB (eds) The benthic boundary layer. Oxford University Press, Oxford, UK pp 180–210

Kutiel P, Lavee H, Ackermann O (1998) Spatial distribution of soil surface coverage on north and south facing hillslopes along a mediterranean to extreme arid climatic gradient. Geomorphology 23:245–256

Nübel U, Garcia-Pichel F, Kühl M, Muyzer G (1999) Spatial scale and the diversity of benthic cyanobacteria and diatoms in a salina. Hydrobiologia 401:199–206

Revsbach NP, Jørgensen BB (1986) Microelectrodes: their use in microbial ecology. Adv Microb Ecol 9:293–352

Scherer S, Chen TW, Bilger P (1988) A new UV-A/B protecting pigment in the terrestrial cyanobacterium *Nostoc commune*. Plant Physiol 88:1055–1057

17 Ecological Determinants of Species Composition of Biological Soil Crusts on a Landscape Scale

I. ULLMANN and B. BÜDEL

17.1 Introduction

Structure and composition of biological soil crusts varies with substrate, climate, and successional processes (see Chaps. 1–11), and specific factors controlling the distribution and abundance of crustal organisms may vary across different scales (see Chap. 16). On a landscape scale (i.e., from meters to kilometers), variation and limitation of soil crusts are mainly related to precipitation, soil chemistry and texture, topography, cover of phanerogamic vegetation, and disturbance. However, because of the problems with taxonomic identifications (see Chap. 1), most relevant studies omit analyses of floristic variation or treat only with a fraction of the flora present.

The aim of this chapter is to demonstrate the variation of crusts along ecological gradients, and to present relationships between environmental factors and floristic composition (based on determination of uncultured field material) and distribution patterns of crusts dominated by lichens or cyanobacteria, respectively. The focus is on nonsuccessional studies in landscapes with distinct spatial patterns occurring frequently in semiarid and arid environments, especially ephemeral salt lakes (=playas, Arab.: sebkhas) and adjoining dune systems. Examples are presented from northern Africa and southern and central Australia.

17.2 Ecological Determinants on a Local Scale

Soil chemistry and physical properties of soil, runon/runoff rain water displacement, vegetation cover, and topographic microsite variation are the most important nonanthropogenic environmental factors that influence floristic variation in soil crusts on a local scale (e.g., Danin and Barbour 1982; Harper and Marble 1988; Eldridge 1999; Malam Issa et al. 1999). Studies in semiarid areas in both the Northern and Southern Hemisphere (I. Ullmann et

Ecological Studies, Vol. 150
J. Belnap and O.L. Lange (eds.) Biological Soil Crusts: Structure, Function, and Management

al., unpubl. data) show that at playa margins, there is a similar pattern in short-distance zonation of both soil crusts and phanerogamic vegetation. Species richness, biomass (expressed as chlorophyll *a* content), and dominant organisms of the biological crusts are correlated with substrate salinity, carboniferous or gypsic soil components, and texture of surface soil. Comparative studies in late dry season were done in north Africa (Tunisia) and south Australia in winter-rainfall areas of similar latitude (33°), similar annual rainfall (approx. 200 mm), and similar grazing and trampling pressure (moderate to high, sheep and goat versus kangaroos and rabbits, respectively).

Soil salinity and soil surface texture apparently limit the occurrence of coherent soil crusts in the sebkhas of south Tunisia. An impressive example of zonation along a salinity gradient in calcareous substrates was obvious within the first 20 m of a transect of 1 m width at the southwestern margin of the small Sebkha Mhabeul (Table 17.1). This transect went from the inner border of vegetation at the edge of the evaporitic sebkha floor into the *Artemisia herba-alba – Helianthemum kahiricum* formation on the low terrace plateau approx. 2.5 m above the floor. Four zones were easily distinguished. In zone 1 (transect 0–4 m), crust organisms were absent from the hypersaline compacted deposits in the belt of succulent Chenopodiaceae at the edge of the sebkha floor. In zone 2 (transect 4–7 m), the porous evaporitic soils of high salinity lacked higher plants, but contained scattered cells or filaments of halotolerant cyanobacteria (especially *Chroococcidiopsis* sp. and *Microcoleus chtonoplastes*, see also Ehrlich and Dor 1985). In zone 3 (transect 7–12 m), coherent cyanobacterial crusts were present in soils with a high silt component and where soil salinity was less than 50 % of the former zone, but the vegetation was still restricted to halotolerant species. Halotolerant coccoid and filamentous cyanobacteria were codominant in these crusts, and the first tiny lichenized (often imperfectly) organisms occurred here. In zone 4 (transect 12–22 m), the switch to nonsaline sandy-silty soils was indicated by the replacement of halophytes by the open low shrub community with *Artemisia herba-alba*. Here, crusts covered up to 50 % of the soil surface and are characterized by lower diversity of cyanobacteria than in zone 3 and a high proportion of both cyanolichens and phycolichens.

A consistent response of soil-crust composition to soil chemistry is shown in Fig 17.1. Samples were taken in the system of sebkhas north of Zarzis within a radius of 5 km around the Sebkha Mhabeul. All substrates were calcareous, with one hypergypsic site. The independence of salinity (Na: $r^2 = 0.77$) and carbonate or gypsum contents of soils (Ca, S, C: $p < 0.05$) is indicated by the rotational positions of the chemical variables in the ordination diagram. The most common crust-forming species were the cyanobacteria *Microcoleus paludosus*, *Calothrix parietina*, *Nostoc* sp., *Gloeocapsa* sp., and the cyanolichen *Collema* sp. Cyanobacterial crusts with *Microcoleus chtonoplastes* were confined to saline (evaporite-affected) silty-sandy or sandy substrate,

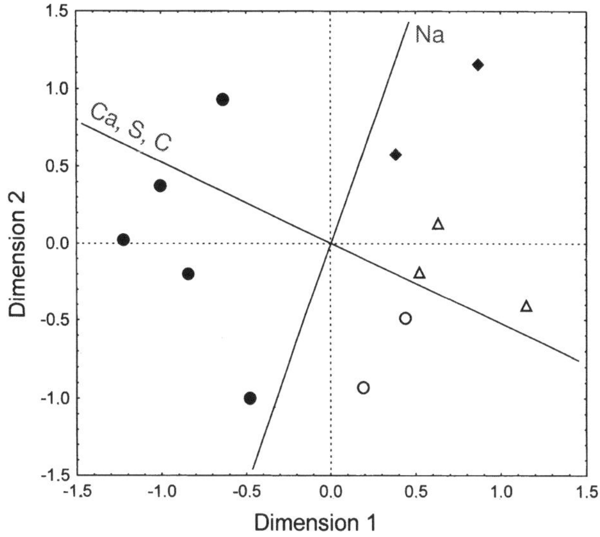

Fig. 17.1. Two-dimensional MDS ordination of soil crusts from 12 different habitats in the sebkha system near Zarzis (south Tunisia) and rotational positions of peak correlation for the four soil variables sodium (Na), calcium (Ca), sulfur (S), and carbonate-carbon (C). Species lists were combined from presence-absence data of five samples for each habitat. *Diamonds* Crust dominated by cyanobacteria, with/dominated by *Microcoleus chtonoplastes; triangles* crusts dominated by cyanobacteria, with abundant *Microcoleus paludosus; open circles* proportion of phycolichens in number of crust species 25–30 %; *filled circles* proportion of phycolichens in number of crust species 35–50 %. STATISTICA for Windows 5.5 (Statsoft Inc., 1999, Tulsa, OK) was used for nonmetric Multi-Dimensional Scaling (a robust technique for indirect gradient analysis; e.g., Feoli and Orlóci 1991). Rotational correlation is a linear form of trend surface analysis which locates the best fit between a habitat variable and the coordinates of the two-dimensional ordination (Dargie 1984). Intersite resemblance was based on the coefficient of community difference derived from the Sørensen index (Dargie 1984)

whereas those dominated by *M. paludosus* occurred on incompletely stabilized sand dunes and on substrates with lower salinity. The proportion of lichens, especially of phycolichens, increased with carbonate, gypsum and silt content of the substrate. The most frequent species were *Catapyrenium lacinulatum* (syn. *Placidium l.*) and the cosmopolitan calciphilous species listed in Table 17.1. Mosses, predominantly *Crossidium laevipilum*, were restricted to wind-deposited sand dunes and nebkas (pyramidal aeolian deposits around shrubs, covering the lower parts of the branches) where organic matter concentrates. However, the clearly observed relationship between crust composition and phanerogamic vegetative community type seen at the m scale (see Table 17.1) was not found at the km scale. Only the crusts dominated by *Microcoleus chtonoplastes* occurred exclusively within a plant community dominated by *Frankenia thymifolia*.

Table 17.1. Composition of soil crusts, phanerogamic layer, and substrate characteristics along a transect at the southern margin of the Sebkha Mhabeul de Zarzis (south Tunisia).

Crust species												
Cyanobacteria, coccoid												
Chroococcidiopsis sp.			×	×								
Gloeocapsa sp.												
Xenococcus sp.			×		×	×						
Staniera cf. cyanosphaera					×	×	×					
Cyanobacteria, filamentous												
Microcoleus chtonoplastes			×	×								
Microcoleus paludosus			×	×								
Calothrix parietina												
Nostoc sp.				×								
Tychonema sp.						×	×					
Lichens												
Lichinella sp. (C)						×	×					
Catapyrenium lacinulatum												
Heppia lutosa (C)												
Squamarina lentigera												
Collema sp. (C)												
Psora decipiens												
Diploschistes diacapsis												
Lecidea sp.												
Fulgensia fulgens												
Number of crust species	0	0	4	4	8	10	9	7	10	10	10	10
Chlorophyll *a* (mg m^{-2})	0	0	0.1	0.1	31	30	25	45	44	36	40	40

Substrate characteristics

Slope (°)	2	3	7	7	10	15	12	5	5	2	2	1
Chemical factors of substrate * [g (kg dw)$^{-1}$]												
Sodium	15	19	8,2	4.2	1.9	1.6	1.2	0.2	0.2	0.3	0.2	0.2
Sulfur	42	64	93	60	33	53	0.2	0.2	0.2	0.2	0.2	0.2
Calcium	68	93	126	82	62	79	38	23	30	33	32	28
Soil surface	Smooth, caked with salt crystals			Porous with gypsum and salt crystals			Irregular porous, rugged, smooth silty / sandy				Smooth silty / sandy	
Parent material	Evaporitic-detritic deposits			Evaporitic deposit on gypsum clays			Gypsum clays / calc. silts				Calcareous silts aeolian deposits (sand and loess)	

Vascular plants

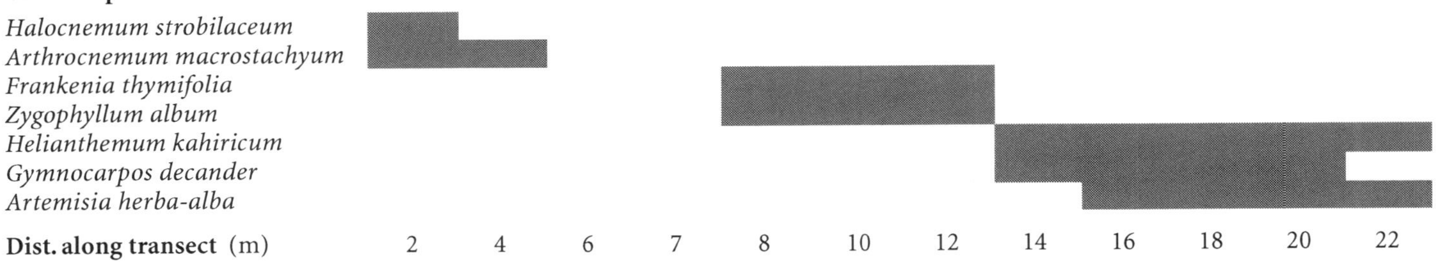

Species abundance is shown semiquantitatively: scattered individuals (×), regularly occurring to frequent (light bars), dominant species (dark bars). (C) cyanolichen.

* samples from a 2-cm-deep layer below the analyzed crusts.

Soil sodium is used to express salinity, the switch from gypsic calcareous to nongypsic calcareous substrate is indicated by the drop in sulfur and calcium.

Table 17.2. Composition of soil crusts in different habitat types within a 50-m transect on the shore of a small playa, northwest Lake Gilles, South Australia.

Crust species				
Algae, coccoid				
Chlorococcum sp.	×			
Cyanobacteria, coccoid				
Gloeocapsa alpina		■	▒	▒
Cyanobacteria, filamentous				
Microcoleus paludosus	■	■	▒	▒
Calothrix parietina	■	▒	▒	▒
Nostoc sp.	▒	▒	▒	▒
Lichens				
Collema coccophorum (C)		▒	▒	▒
Catapyrenium lacinulatum		▒	▒	▒
Toninia sedifolia			▒	▒
Diploschistes diacapsis			▒	■
Psora decipiens			▒	■
Fulgensia fulgens				■
Aspicilia sp.				▒
Endocarpon simplicatum		×		
Synalissa sp. (C)				×
Mosses				
Desmatodon convolutus	▒	▒	▒	▒
Didymodon torquatus		▒	▒	▒
Gigaspermum repens			▒	▒
Crossidium geheebii				×
Crossidium davidai				×
Phascum robustum				×
Number of crust species*	4 + 1	7 + 2	9 + 3	12 + 6
Vascular plants				
Halosarcia pergranulata ssp. *pergr.*	■			
Halosarcia halocnemoides	■			
Aristida contorta		■		
Acacia papyrocarpa			■	
Callitris cf. preissii			■	
Zygophyllum aurantiacum				■
Cover (%)	60	25	clumps	40
Chemical factors of substrate** [g (kg dw)$^{-1}$]				
Sodium	0.7	0.05	0.05	0.1
Carbon (carbonate-C)		2.2	2.7	8.7
Calcium	0.3	0.4	0.8	13.5
Habitat type	1	2	3	4

Table 17.2 (*continued*)

Species abundance is shown semiquantitatively: scattered individuals (x), regularly occurring to frequent (light bars), dominant species (dark bars).
(C) cyanolichen.
* Second number refers to mosses; ** samples from a 2-cm-deep layer below the analyzed crust.
The low values for soil Na seen at habitat type 1 were probably caused by leaching in the course of unseasonal heavy rainfalls previous to sampling

Composition and local differentiation of soil crusts in south Australia (in comparable settings) were strikingly similar to those in south Tunisia, as shown in Table 17.2. Predominant substrates at the site at northeastern Lake Gilles (sw Port Augusta) are carboniferous red aeolian sands (habitat types 1–3) and calcareous outcrops (habitat type 4). Both substrates are patchily enriched by organic matter (litter). The crust matrix was composed of the same filamentous cyanobacteria that are important in similar substrates in south Tunisia (*Microcoleus paludosus*, *Calothrix parietina*, *Nostoc* sp.), and, as in south Tunisia, *Collema* and *Catapyrenium lacinulatum* were the most frequent lichens. Lichens were absent from the crusts on the nebkas (up to 80 cm high) under the halophytic shrubs (habitat type 1), but litter accumulation promoted the occurrence of mosses in all four habitat types. The crust-forming moss species found are typical for low rainfall areas of Australia (see Chap. 10). Increase in diversity and cover of both lichens and mosses was clearly related to increase in soil calcium and soil carbonate (Table 17.2).

17.3 Ecological Determinants on a Regional Scale

On regional scales, the distribution, biodiversity, and potential maximal biomass of biological soil crusts are determined by interacting geological and geomorphological features and climatic factors. The most prominent determinants are elevation, soil parent material and related soil pH, rock cover (e.g., desert pavement), temperature regimes and continentality, and amount and patterns of precipitation, inclusive of dew and fog (Johansen 1993; Lange et al. 1994; Evans and Johansen 1999; J. Belnap, unpubl. data). A recent study (I. Ullmann and B. Büdel, unpubl. data) along an aridity gradient showed that rainfall and continentality influenced crustal composition on a regional scale without obliterating the influence of substrate in any region.

The study was located in southern and central Australia, where the isohyets run more or less parallel to the south coast. Transects were run from the coast into the interior in west and south Australia (Esperance-Laverton, 121–122° longitude; Port Lincoln-Lake Eyre, 134–138° longitude) and additional samples were taken along their central Australian (road) connection. Sampling sites were spread between 25° and 35° southern latitude, with the northernmost sites located in the Amadeus Basin. Mean annual rainfall in the investigated area (mainly winter rainfall, but irregular in central Australia) ranges between 125 mm and >400 m (western coast), and dewfall occurs even in the arid parts (Beadle 1981). At most sites, crusts were collected from different habitat types (e.g., Table 17.2). Samples were taken within plots of 10 or 25 m^2 (according to relief and vegetation), with each sample consisting of five subsamples of 10 cm^2. The survey was not designed to test variation within highly developed crusts, and thus mosses were not included in the analysis.

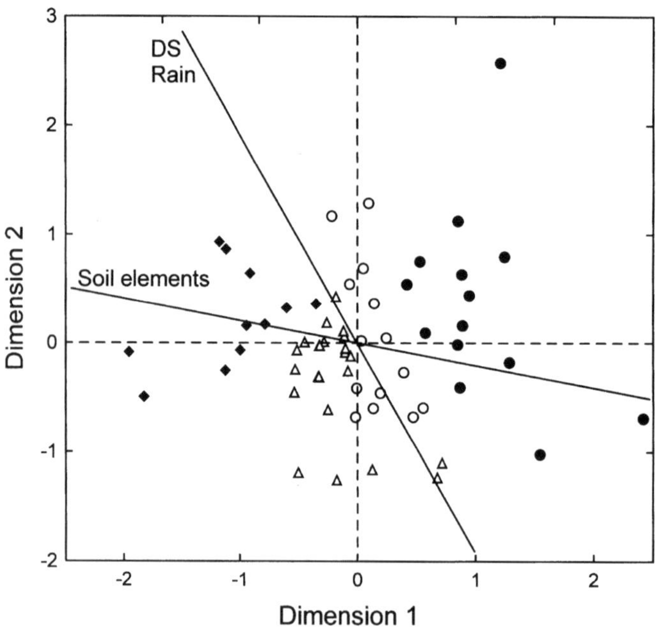

Fig. 17.2. Two-dimensional MDS ordination of 66 soil-crust samples, based on absence-presence data of cyanobacteria, algae, and lichens. Samples were collected along coast-inland transects from southern and southwestern to central Australia. Species lists were combined from presence-absence data of the five subsamples per sample. Rotational positions of peak correlation are given for four soil variables (Na, Ca, S, carbonate-C), mean annual rainfall (Rain), and distance from coast (*DS*) as an indicator for continentality. *Diamonds* Crusts dominated by cyanobacteria, containing *Microcoleus chtonoplastes* or lack *Microcoleus* (type A); *triangles* crusts dominated by cyanobacteria, with/dominated by *Microcoleus paludosus* (type B); *open circles* proportion of phycolichens in number of crust species <30 % (type C); *filled circles* proportion of phycolichens in number of crust species >35 % (type D)

Table 17.3. Composition of the soil crust types shown in Fig. 17.2

Crust type	A	B	C	D
Nostoc sp.	black bar	black bar	black bar	black bar
Microcoleus paludosus	×	black bar	black bar	black bar
Calothrix parietina	black bar	gray bar	gray bar	gray bar
Scytonema ocellatum	gray bar	gray bar	gray bar	gray bar
Collema sp. (C)	×	×	gray bar	gray bar
Scytonema myocrous	×	×	gray bar	gray bar
Gloeocapsa alpina	gray bar	×	×	gray bar
Collema coccophorum (C)	×	×	gray bar	gray bar
Chlorococcum sp.	×	×		×
Gloeodinium sp.	gray bar	×		gray bar
Gloeocapsa kützingiana	×	×		×
Schizothrix telephoroides	×		×	×
Catapyrenium lachneum (P)	×		gray bar	gray bar
Heppia lutosa (C)		×	gray bar	gray bar
Synalissa sp. (P)		×	×	×
Peltula radicata (C)		×	×	×
Lyngbya sp.	gray bar	×		
Chroococcidiopsis sp.		gray bar	×	
Phormidium sp.		×	×	
Tychonema sp.		×	×	
Stigonema ocellatum		×		×
Catapyrenium lacinulatum (P)			gray bar	black bar
Psora decipiens (P)			×	black bar
Endocarpon simplex (P)			×	gray bar
Gloeocapsa sanguinea			×	×
Peltula patellata (C)			×	×
Heppia despreauxii (C)			×	×
Xenococcus sp.			×	×
Cladia aggregata (P)			×	×
Porocyphus sp. (C)			×	×
Microcoleus chtonoplastes	black bar			
Gloeocapsa muralis	gray bar			
Gloeocapsa punctata	×			
Fulgensia fulgens (P)				black bar
Toninia sedifolia (P)				gray bar
Diploschistes diacapsis (P)				gray bar
Eremastrella crystallifera (P)				gray bar

Species occurring only once in all 66 samples. A: *Gloeocapsa fusco-lutea*, *Microcoleus sociatus minor*, *Oscillatoria* sp., *Schizothrix* sp., *Lichina* sp. (C), *Catapyrenium squamulosum* (P), B: *Gloeocapsa* sp., *Collema flaccidum* (C), C: *Trebouxia arboricola*, *Scytonema hofmanni*, *Zygogonium ericetorum*, *Synalissa symphorea* (C), *Peccania coralloides* (C), D: *Stigonema minutum*, *Peltula obscurans* (C), *Aspicilia* sp. (P), *Diploschistes muscorum* (P), *Siphula* sp. (P), *Squamarina* sp. (P).

Species are classified with respect to their relative frequency within the respective crust type: <20 % (X), 20–50 % (gray bars), >50 % (black bars).

(C): cyanolichen; (P): phycolichen.

The change from the dominance of cyanobacteria to dominance of lichens in crust composition that was seen for individual samples is also obvious in the species diversity of the four crust types. Numbers of species occurring in the crust types are 23, 21, 31, and 36 in crust types A–D, respectively. The relationship ($p<0.05$) between the composition of individual soil-crust samples and the tested habitat variables is shown in Fig. 17.2. The ratio of cyanobacteria to lichens is 3:1, 2:1, 1:1, and 1:2 from Type A to D, respectively (Table 17.3). This ratio was predominantly determined by substrate variables (similar to the local scale), whereas distribution of the cyanobacteria-dominated crusts was more affected by climatic factors (as indicated by the rotational axes in Fig. 17.2).

Basic crust species were *Microcoleus paludosus, Nostoc* sp. (relative frequency 100 % in B and C, 93 % in D) and *Scytonema ocellatum* (relative frequency was between 40 and 50 % for all types). However, *M. paludosus* was uncommon in crust type A, which was confined to coastal and near-coast sites. Crusts with *Microcoleus chtonoplastes* were restricted to clayey fringes of small playas, but were absent from beach dunes or sandy deposits on coastal cliffs. *Calothrix parietina* (the most frequent species of type A) was less frequent at inland sites than coastal sites (24–28 % versus 91 %), even in the cyanobacteria-dominated crusts of type B, which occurred on calcareous and nonbasic sandy substrates. Numbers of species ranged from two to five, but were higher (to nine) in plots which received runoff from calcareous or gypsic surroundings. Most of the sites of these crusts were evaporite-affected and/or subject to occasional ponding, and 50 % of the samples were associated with halophytic plant communities dominated by *Halosarcia* spp. (most frequent was *H. halocnemoides, H. pergranulata,* and *H. indica*) or *Frankenia* spp. This was the only crust type found in the arid lands around Lake Eyre. Lichen-rich crusts (types C and D), which were associated with a wide range of vascular plant communities, occurred on soils which were less evaporite-affected. Type D crusts were confined to nonsaline calcareous or gypsic substrate, including stabilized dunes.

17.4 Conclusions

On a local scale, species composition of biological soil crusts, in particular the transition from cyanobacteria-dominated to lichen-dominated crusts, showed distinct spatial patterns that are highly correlated with substrate characteristics, even along an aridity gradient. Striking floristic similarities were found in similar environments in the Northern Hemisphere (Tunisia) and in the Southern Hemisphere (Australia). Reports of distribution of major crust components encountered in this study from other hot arid areas (e.g., Israel) indicate similar relationships with respect to soil components and salinity dependencies (Danin and Barbour 1982; Danin et al. 1989). This strengthens the view that individual soil-crust components are highly adapted to a specific range of habitat conditions. On the other hand, many of the crust organisms have a worldwide distribution in arid lands; consequently, similar patterns of crust types and community structure recur in similar landscapes, even in quite different biogeographic areas (see Chap. 12).

Acknowledgments. We thank W. Frey and D. Eldridge for the identification of mosses, and R. J. Chinnoc for valuable help with the identification of Australian Chenopodiaceae and Aizoaceae. The field surveys (I.U.) were substantially supported by the Dept. of Geography, University of Würzburg (Tunisia), and the Dept. of Botany, University of Adelaide (Australia).

References

Beadle NCW (1981) The vegetation of Australia. G Fischer, Stuttgart

Danin A, Barbour MG (1982) Microsuccession of cryptogams and phanerogams in the Dead Sea area, Israel. Flora 172:173–179

Danin A, Bar-Or Y, Dor I, Ysraeli T (1989) The role of cyanobacteria in stabilization of sand dunes in Southern Israel. Ecol Mediterr 15:55–64

Dargie TCD (1984) On the integrated interpretation of indirect site ordinations: a case study using semi-arid vegetation in southeastern Spain. Vegetatio 55:37-55

Eldridge DJ (1999) Dynamics of moss- and lichen-dominated soil crusts in a patterned *Callitris glaucophylla* woodland in eastern Australia. Acta Oecol 20:159–170

Evans RD, Johansen JR (1999) Microbiotic crusts and ecosystem processes. Crit Rev Plant Sci 18:183–225

Ehrlich A, Dor I (1985) Photosynthetic microorganisms of the Gavish Sabkha. In: Friedman GM, Krumbein WE (eds) Hypersaline ecosystems. The Gavish Sabkha. Ecological Studies 53. Springer, Berlin Heidelberg New York, pp 296–321

Feoli E, Orlóci L (eds) (1991) Computer assisted vegetation analysis. Kluwer, Dordrecht

Harper KT, Marble JR (1988) A role for nonvascular plants in management of arid and semiarid rangelands. In: Tueller PT (ed) Vegetation science applications for rangeland analysis and management. Kluwer, Dordrecht, pp 135–169

Johansen JR (1993) Cryptogamic crusts of semiarid and arid lands of North America. J Phycol 29:140–147

Lange OL, Meyer A, Zellner H, Heber U (1994) Photosynthesis and water relations of lichen soil crusts: field measurements in the coastal fog zone of the Namib Desert. Funct Ecol 8:253–264

Malam Issa O, Trichet J, Défarge C, Couté A, Valentin C (1999) Morphology and microstructure of microbiotic soil crusts on a tiger bush sequence (Niger, Sahel). Catena 37:175–196

Part IV

Biological Soil Crusts as an Ecosystem Component: Carbon and Nitrogen Acquisition and Interaction with Vascular Plants

18 Photosynthesis of Soil-Crust Biota as Dependent on Environmental Factors

O.L. Lange

18.1 Introduction

Limited productivity of vascular plant communities in semiarid and arid regions generally results in low soil organic carbon content. In these situations, crust biota can be the most important autotrophic contributors of fixed carbon, delivering this carbon to the soil ecosystem by leaching and decaying processes. The photosynthetic carbon assimilation of crustal organisms thus plays an important role by contributing to the humus reservoir of the soil, supporting heterotrophic soil life, and supplying nutrients for the phanerogamous vegetation (see Beymer and Klopatek 1991 and Chaps. 19–21).

The microenvironment in which the soil-crust biota are living – the uppermost millimeters of the topsoil of desert and steppe formations – is one of the most extreme habitats where autotrophic organisms are found. They are exposed to high and low temperatures, long-lasting and extensive desiccation, and excessive sun radiation. Soil-crust organisms must have high resistance to extreme temperatures, large water loss, and extensive light stress (see Chap. 1). Their photosynthetic and respiratory activity is restricted to short periods of hydration. Therefore, another requirement for their existence is the capability to begin metabolism as soon as moisture becomes available, and to make use of even very small amounts of water for photosynthetic carbon gain. Only a small subset of species of cyanobacteria, algae, lichens, and bryophytes have developed the necessary adaptations for existence in the soil-crust environment. Evidently, all of them are poikilohydrous in nature.

Ecophysiological work which analyzes the special functional features of soil-crust biota existence and their photosynthetic production is rare. We are far from a complete understanding of their characteristics, and the existing literature covers only a minimal number of the different soil-crust types, organisms, geographical regions, and climatic situations. Most of the published work reports on laboratory measurements. Photosynthetic performance of total soil-crust samples, containing a variety of different organisms, has been studied with material from the US (Beymer and Klopatek 1991;

Ecological Studies, Vol. 150
J. Belnap and O.L. Lange (eds.) Biological Soil Crusts: Structure, Function, and Management

Jeffries et al. 1993a,b; Garcia-Pichel and Belnap 1996), from savannas in Venezuela (San José and Bravo 1991), from sand dunes in The Netherlands (De Winder 1990), and from Israel (Lange et al. 1992). Ecophysiology of different lichen species growing in the soil-crust environment was investigated with material from Israel (Lange et al. 1993), central Europe (Lange et al. 1995), and from the US (Lange et al. 1997a, 1998). Field measurements of CO_2 exchange in soil-crust organisms under natural conditions are known from Israel (Lange et al. 1970), Namibia (Lange et al. 1994b), from central European habitats (Lange et al. 1997c; Lange 2000a,b), and from the United States (Phillips and Belnap 1998).

The majority of these studies concern lichens. Therefore, the following report necessarily concentrates on the performance of lichens, with a few studies on soil cyanobacteria and mosses. Performance of soil-crust cyanobacteria and green algae in culture will be included only in those cases where they are ecologically relevant.

Abbreviations. DR: dark respiration; MCP: moisture compensation point of CO_2 exchange; NP: net photosynthesis; PPFD: incident photosynthetically active photon flux density; WC: water content.

18.2 Soil-Crust Chlorophyll Content

Due to the large variability and differences in soil-crust types, biomass density, and organisms involved, area-related crust chlorophyll content varies greatly. Table 18.1 illustrates the range for a variety of crust types and organisms reported from different parts of the world. Crusts dominated by *Microcoleus,* and containing other cyanobacteria and chlorophyta within the soil, but without lichens and mosses above the surface (see Chap. 1 for definitions), generally show maximal chlorophyll contents below 100 mg m^{-2}. This chlorophyll is mainly concentrated in the uppermost layer of the soil. Garcia-Pichel and Belnap (1996; see Chap. 15) found more than ca. 75 % of total chlorophyll within the first 2 mm of soil. The presence of mosses and lichens or of mat-forming algae increases area-related chlorophyll content of the soil-crust system greatly, with maxima higher than 900 mg m^{-2} chlorophyll reported. This is more than the chlorophyll content given for an average leaf of a C3 plant (Lawlor 1990), and according to Björkman (1981), maximum quantum yield of photosynthetic CO_2 uptake for C3 leaves is reached at 500–700 mg chl m^{-2}. Therefore, with respect to chlorophyll content, the ground in areas with limited vascular plant cover and large amounts of soil crusts (e.g., in the coastal fog zone of the Namib Desert) is covered by a continuous sheet of photosynthetic machinery equivalent to a layer of phanerogamous leaves.

Table 18.1. Area-based chlorophyll content ($a + b$ or a) for different crust types (data with high values selected): total crusts with multiple components (mixed crust, with or without lichens/mosses), crusts with one dominant species, or single soil lichens

Material	Chlorophyll content mg m^{-2} $a + b$	a	Remarks
Mixed crust with dense population of chlorophyta (*Klebsormidium* spp.)		986–498	B. Büdel and A. Lukešová (unpubl.) Open pine forest, Czech Republic
Diploschistes muscorum (phycolichen)	981–416 526 (a)		O.L. Lange (unpubl.) Local steppe formation, Würzburg, Germany
Crust dominated by *Lecidella crystallina* (phycolichen)	508		Lange et al. (1994b) Namib Desert, Namibia
Mixed crust with cyanobacteria		319 (a)–32 (a)	De Winder (1990) Sand dunes, The Netherlands
Psora cerebriformis (phycolichen)	381–133 255 (a)		Lange et al. (1997a) Colorado Plateau, Utah, USA
Mixed crust with cyanobacteria, lichens, mosses		Maximal 180 102 (a), 107 (a)	I. Ullmann and B. Büdel, (unpubl.) North Africa, Tunisia
Crust dominated by *Collema tenax* (cyanolichen)		169–130	O.L. Lange (unpubl.) Local steppe formation, Würzburg
Crust dominated by *Collema tenax* (cyanolichen)		155–58.9 90.8 (a)	Lange et al. (1998) Colorado Plateau, Utah, USA
Mixed crust with cyanobacteria, dominated by *Microcoleus paludosus*		Maximal 82 60 (a)–50 (a)	I. Ullmann and B. Büdel (unpubl.) Central Australia
Mixed crust with chlorophyta, cyanobacteria and lichens		64 (a)	Beymer and Klopatek (1991) Juniper-pinyon woodland, Arizona, USA
Mixed crust dominated by *Microcoleus vaginatus*		53 (a)	Belnap et al. (1994) Central Utah
Mixed crust dominated by cyanobacteria		51(a)–17(a)	Kidron (1995) Sand dunes, Negev, Nizzana, Israel
Mixed crust with *Microcoleus vaginatus, Nostoc,* etc.		31.6 (a), 6.3 (a)	Garcia-Pichel and Belnap (1996) Southeastern Utah
Leaf of typical C3 plant, average	500		Lawlor (1990)

Single measurements, range of population or crust type within one area (limits separated by dash) and/or averages (a) are indicated. Textbook number for higher plants is included for comparison.

18.3 CO_2 Exchange as Dependent on Environmental Factors

18.3.1 Response of Photosynthesis and Respiration to Mode and Degree of Hydration

Desiccated poikilohydrous organisms are dormant and, in this state, are highly resistant to extreme environmental conditions such as extreme temperatures and excessive light. Active metabolism is possible only after moistening, which quickly restores respiratory and photosynthetic capacity. Hydration of soil-crust organisms can take place either through liquid water (in the form of rain, fog, and dew) or by water vapor uptake from highly humidified air.

It is known that – as far as studied – cyanobacteria and cyanolichens on the one hand, and aerophillic chlorophyta and phycolichens on the other, respond quite differently to different sources of water. The latter are able to achieve net photosynthetic carbon gain through water vapor uptake alone (i.e., they require only very low degrees of hydration), whereas cyanobionts need liquid water for positive net photosynthesis (Lange et al. 1986). A similar difference seems to exist between phycolichens and xerophilic mosses (see Rundel and Lange 1980). Figure 18.1 shows a rehydration experiment with green-algal and cyanobacterial soil-crust lichens. At time zero, dry thalli were exposed to a stream of almost-water-saturated air. After less than 2 h, CO_2 uptake of green-algal *Caloplaca tominii* and *Fulgensia bracteata* began, reaching almost maximal rates after 50 h. In contrast, under identical conditions, there was absolutely no NP with the two cyanobacterial samples (*Gonohymenia* spp. and *Heppia lutosa*) any time during exposure to high humidity. Seventy-two h after the start of humidification, all lichens were sprayed with water (arrows in Fig. 18.1), and only then was NP of the cyanobacterial lichens activated. At the same time, dark respiration of all the lichens was increased by the application of liquid water. The same phenomenon was found for total soil crusts dominated by the cyanobacterium *Microcoleus sociatus* (Lange et al. 1992), which also needed liquid water for carbon gain. (However, it seems that cultivated *Microcoleus* performs differently; see Lange et al. 1994a; Mazor et al. 1996.) The ability to use water vapor for photosynthetic productivity favors the existence of aerophillic green algae and phycolichens over cyanobacterial crust organisms in coastal areas with low rain precipitation but frequently elevated air humidity.

Independent of the mode of hydration, a low MCP is advantageous for making use of even very small hydration events. Indeed, all of the green-algal soil-crust lichens tested have an extremely low MCP, especially after water vapor uptake (Lange et al. 1997a). *Diploschistes muscorum,* as an extreme example, is able to begin photosynthetic carbon gain at a WC as low as

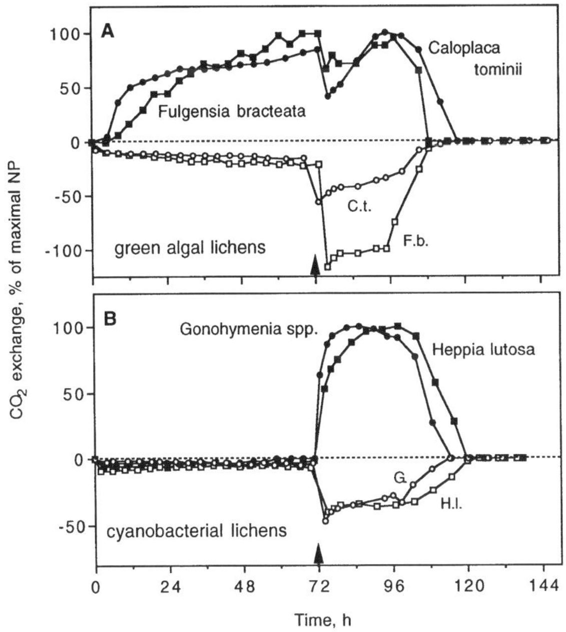

Fig. 18.1A,B. Time courses of CO_2 exchange, related to maximal net photosynthesis, in the light (*solid symbols*) and in the dark (*open symbols*) during water vapor uptake and subsequent spraying with liquid water (*arrows*). **A** Green-algal *Caloplaca tominii* (Dzungurian Basin, China, leg. I. Ullmann), *Fulgensia bracteata* (Cyprus). **B** Cyanobacterial *Heppia lutosa, Gonohymenia collemoides*, and *G. algerica* (Judean Desert). (**B** after Lange et al. 1993)

0.04 mm (precipitation equivalent: water content on an area basis, $l\,m^{-2}$), in equilibrium with a relative air humidity of ca. 85 %, i. e., at a water potential of about −200 bar. In surprising contrast, free-living cyanobacteria and cyanolichens need much more water to reach their MCP. Contrary to expectation, Brock (1975) reports that NP of *Microcoleus* from soil crusts is not especially adapted to low water potentials, since its photosynthesis was virtually inhibited at a water potential of −28 bar and still markedly reduced at −18 bar. Cyanolichens such as *Collema tenax* have an MCP of 0.22 mm WC, which is much higher than that of phycolichens (Lange et al. 1998). There are several studies reporting moisture requirements for the start of photosynthesis in cyanobacterial soil crusts. However, comparison is difficult due to the differences in expressing soil WC. Cyanobacterial-dominated soil crusts from the Negev (Lange et al. 1992) had their MCP at a relatively high WC of 0.1 mm. Nevertheless, this is low enough to allow photosynthetic carbon gain and growth after most dewfall evants, which the average provide 0.16 mm precipitation per single dew event in that region.

After reactivation has taken place, actual rates of NP of soil-crust biota are controlled by the degree of hydration. Figure 18.2 portrays the dependence of CO_2 exchange of *Squamarina lentigera* on thallus WC. NP begins at a low WC of 0.08–0.09 mm. Increasing WC causes a steep rise in NP until optimal CO_2 fixation takes place at around 0.5 mm WC. However, not only lack of water reduces CO_2 fixation; at natural external CO_2 partial pressure, excessive, suprasaturating water content also hinders photosynthetic carbon gain. NP is low only when the lichen is fully saturated with water, simulating conditions after a rain shower. The thallus has to lose two thirds of its maximum WC before a strong increase in NP takes place. Thus, maximal NP is possible only within a small window of thallus hydration. This depression of photosynthesis at suprasaturation is a frequent characteristic of many lichens. Potential quantum yield of photosynthetic CO_2 fixation remains high when actual NP is reduced by high WC (Lange et al. 1996), and the depression can be removed experimentally by high external CO_2 concentration (Fig. 18.2). This leads to the conclusion that the observed suprasaturation depression is caused by increased thallus diffusion resistances because of pathway blockage by water, thus reducing CO_2 supply for the carboxylating photobionts. A similar mechanism often causes a severe depression of NP at high WC for soil crusts with mixed populations. Figure 18.3 shows WC dependence of NP for three crust types from savanna habitats in Venezuela (San José and Bravo 1991). CO_2 fixation is reduced at WC of more than 85 %, with optimal NP taking place

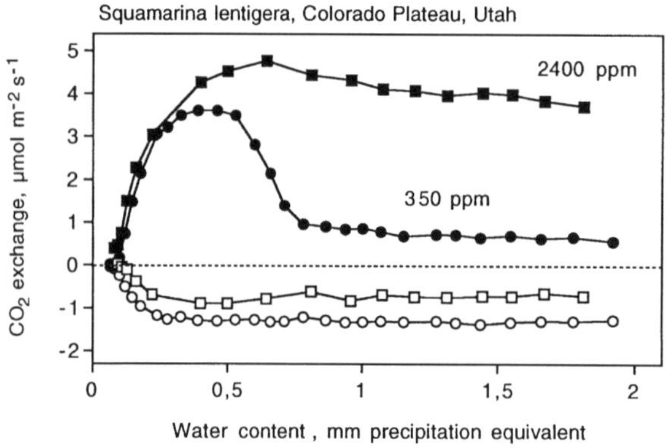

Fig. 18.2. Dependence of net photosynthesis (*closed symbols*) and dark respiration (*open symbols*) at different levels of external CO_2 concentration (as indicated) on thallus water content (mm precipitation equivalent) for *Squamarina lentigera* (Colorado Plateau, Utah). The highest WC indicated equals the maximal water-holding capacity of the thallus; 17 °C, 500 µmol $m^{-2} s^{-1}$ PPFD. (After Lange et al. 1997a)

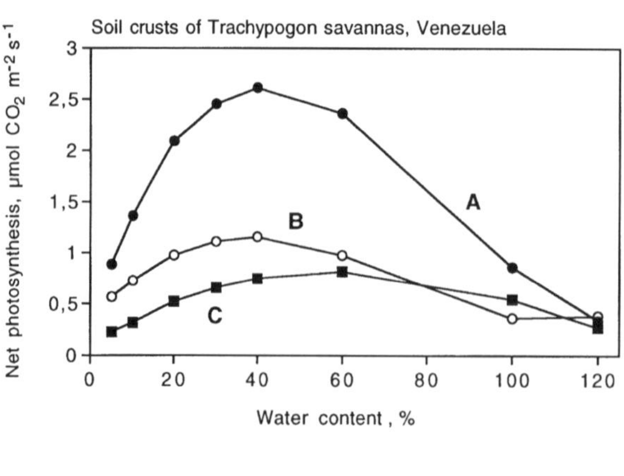

Fig. 18.3. Dependence of net photosynthesis on thallus water content (% of oven dry weight) for three soil-crust types (*A, B, C* Different coverage of several cyanobacteria and one green alga) of the *Trachypogon* savannas of the Orinoco Llanos, Venezuela. (After San José and Bravo 1991)

only after loss of more than 50 % of maximal WC. This fact is of great ecological significance, as it reduces carbon gain of soil crusts at water saturation, for instance after a rain shower, and high rates of photosynthesis become possible only after substantial loss of water.

Many free-living and lichenized cyanobacteria and microalgae possess photosynthetic CO_2-concentrating mechanisms (CCM, see Badger and Price 1992; Badger et al. 1993). This has also been found in cyanobacterial soil crusts in savannas of Venezuela (Ziegler and Lüttge 1998) and with soil-crust lichens (Lange et al. 1999a). The mechanism increases the internal CO_2 partial pressure around the carboxylating enzymes, improving photosynthetic efficiency (as in C4 plants). At the same time, CCMs decrease the CO_2 concentration around the photosynthesizing cells. This increases diffusion gradients for CO_2 transport and helps overcome the adverse impacts of suprasaturation. Therefore, there is a high diversity in response of different soil-crust organisms to high WC. In contrast to lichens with virtually no NP at high WC, there are other species, especially in the genus *Diploschistes,* whose NPs are not affected at even the highest thallus WC. We are not yet able to explain these differences in terms of anatomical or physiological features of the different lichens (Lange et al. 1999a).

Photosynthetic production of soil-crust lichens depends not only on WC-dependent NP, but also on the water-holding capacity of the lichen thalli. It ultimately determines how long the lichen remains wet and, therefore, the duration of metabolic activity. Figure 18.4 illustrates this interaction for three different photosynthetic strategies found in soil-crust lichens. The green-algal *Fulgensia fulgens,* with a very low MCP, is capable of using the slightest hydration by dew, fog, or even high air humidity alone. However, its water-

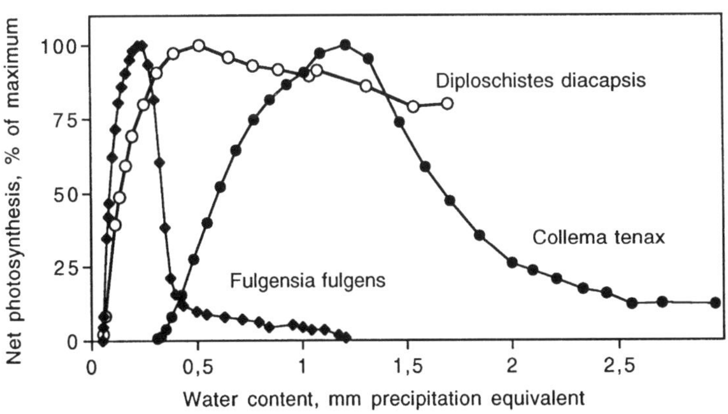

Fig. 18.4. Dependence of net photosynthesis (% of maximum) on water content for different types of soil-crust lichens: *Fulgensia fulgens* (local steppe formation, Würzburg, Germany), *Diploschistes diacapsis,* and *Collema tenax* (both Colorado Plateau, Utah)

holding capacity is small and NP is heavily depressed at suprasaturation. One can predict that such lichen types are best adapted to regions where slight moistening is a frequent event. In contrast, both moisture requirements and water-holding capacity of the gelatinous cyanobacterial *Collema tenax* are much higher. This species begins photosynthetic carbon gain at a WC when optimal hydration for *Fulgensia* is already surpassed. However, with its high water storage, *Collema* can make better use of heavy moistening, giving it long-lasting periods of activity. Such cyanobacterial species are frequently found in deserts and semideserts, where rain is the predominant source of moisture, even if it is sparse. *Diploschistes* species are highly favored in all habitats through lack of suprasaturation depression and through substantial water-holding capacity. Soil-crust species of this genus are very widely distributed, ranging from the temperate and mediterranean regions, across different types of deserts, and into the cold steppe formations in Asia and the US (see Chap.12).

18.3.2 Initial CO_2 Exchange Responses of Soil Crusts upon Rewetting

The life of soil-crust biota is characterized by a frequent shift between long periods of inactivation during desiccation and short periods of hydration. Resaturation of poikilohydric organisms often results in an initial, transient phase of elevated CO_2 release before steady-state gas exchange is attained. With lichens, the first response is a short "rewetting burst" of CO_2 output, most probably a purely physical process (Farrar and Smith 1976). Subsequently, there follows a phase of elevated CO_2 release, the "resaturation respiration" (Smith and Molesworth 1973), which can last for a few minutes to several hours, and which might be caused by physiological repair processes of the cells.

Soil lichens show this response upon rewetting, but CO_2 loss usually is low and does not impact their carbon gain in nature (Sect. 18.4). Resaturation of soil crusts containing a variety of species in the soil matrix is a much more complex process. Oxygen-exchange measurements have shown that for *Microcoleus, Nostoc*, and *Scytonema* species both NP and DR start within minutes after rewetting (Garcia-Pichel and Belnap 1996). This agrees with fluorescence monitoring, which revealed that quantum yield of photosystem II of initially dry chlorophyta and cyanobacteria becomes reactivated within seconds after rehydration, attaining final steady-state level several to 20 min later, respectively (Lange et al. 1989). However, when a dry crust system involving a mixture of autotrophic and heterotrophic organisms is rewetted, there is often a steep burst of CO_2 release, even in the light. When soil-crust samples from the Negev were kept dry for 4 days, it took between 4 and 5 h after rewetting until the compensation point was surpassed (Fig. 18.5; see

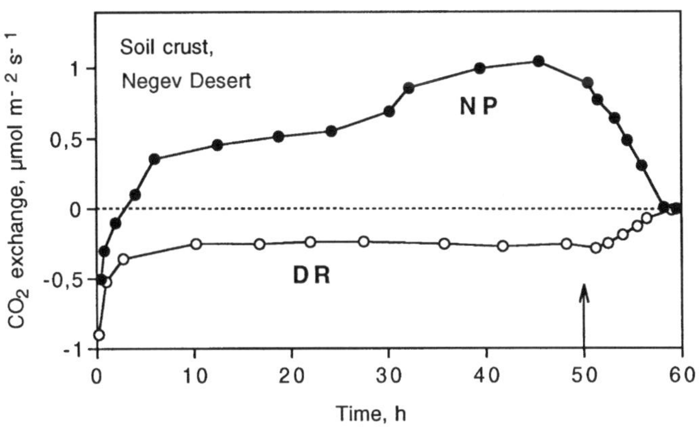

Fig. 18.5. Time courses of CO_2 exchange in the light (*NP, solid symbols*) and in the dark (*DR, open symbols*) of soil crusts (mean of several samples) from sand dunes of the Negev Desert (Israel). Crusts were saturated with water at time zero and became dry after ca. 50 h (*arrow*). (Data partly from Lange et al. 1992)

Lange et al. 1992). Subsequently, there was a second period of about 30 h during which NP stayed almost constant at a low level or increased only slightly. During a third phase, CO_2 uptake increased and, after almost 2 days, NP was almost doubled. During this last phase, a large amount of growth took place. Cells of *Macrochloris multinucleata* (originally reported as *Chlorococcum* sp.) lost most of their hematochrome, trichomes of *Microcoleus* were moving out of their sheaths to the crust surface, and young colonies of *Nostoc* established. Similar processes were observed by Garcia-Pichel and Belnap (1996; crusts from southern Utah). Such greening of the crusts occurs also in the field, after rain. Growth of photosynthesizing biomass and the improved light exposure on the soil surface is certainly responsible for the dramatic increase in NP in the third phase after rehydration, while the initial increase is due to physiological processes, probably in combination with physical and chemical reactions in the soil matrix.

Jeffries et al. (1993a,b) rehydrated *Microcoleus*- and *Scytonema*-dominated soil crusts (from Utah) after dryness of ca. 3 months, let them dry again for ca. 6 days, and then subjected them to a second rewetting cycle. Upon the first rewetting, there was a steep initial outburst of CO_2. Crusts took longer than 1 day to reach the compensation point and 2 days to attain full activity. After the second rehydration, they still took 1 day to reach full activity. This clearly shows that recovery from desiccation depends on the length of the foregoing dry period, and that rewetting processes can greatly impact photosynthetic production of soil crusts. However, the limited data available do not yet allow any generalizations on this process.

18.3.3 Light Response of Net Photosynthesis

Lichen-dominated crusts show saturation-type light responses; according to light requirements, they exhibit "sun-plant" characteristics. They are well-protected against damage from excessive light. No indication of persistent photoinhibition has been reported. Under natural conditions, light stress usually is connected with strong water loss, and desiccation is an effective photoprotection (Demmig-Adams et al. 1990a). However, hydrated thalli also seem to be well-protected. An extensive study was performed with the foliose *Peltigera rufescens*, a cyanolichen that colonizes soil in xerothermic formations of central Europe, including the Mediterranean region (Leisner et al. 1995, 1996). This species was able to acclimate to seasonal changes in illumination and to withstand exposure to full insolation while moist without irreversible damage to photosystem II. With phycolichens, rapid formation of the carotenoid zeaxanthin is clearly involved in photoprotection (Demmig-Adams et al. 1990b), and it is hypothesized that accumulation of canthaxanthin is responsible for avoiding photoinhibitory damage in cyanolichens (Lange et al. 1999b). The influence of UV radiation on soil cyanobacteria is discussed in Chapters 16 and 30.

In agreement with the WC dependence of CO_2 exchange, the highest rates of NP achieved by soil-crust lichens were at various PPFDs, depending on their hydration level. Figure 18.6 shows light-response curves at different WC for the gelatinous soil-crust species *Collema tenax*, one of the most common terricolous lichens in the Intermountain Area of the US. At maximal water-holding capacity, NP was strongly suppressed (curve 1, Fig. 18.6). Supra-saturation of the thallus impacted NP at both saturating and limiting PPFD, even near the light compensation point. Decreasing WC increased the level of NP (curves 2, 3) and maximal rates were achieved at WC of ca. 1.2 mm (curve 4). Subsequent decrease of NP as WC declined was obvious (curves 5–7), until the desiccated thalli lost total photosynthetic capacity. Light saturation was also strongly dependent on thallus hydration (Fig. 18.7). At low and high WC, photosynthetic CO_2 uptake of the lichen was already light-saturated at low PPFD levels. At a mid-range of WC, the highest levels of PPFD were necessary for maximal NP. In some cases, light saturation of the optimally hydrated lichen was not yet fully attained at 2000 $\mu mol\ m^{-2} s^{-1}$ PPFD.

In contrast to maximal NP and light saturation, the light compensation point of CO_2 exchange has proved to be rather insensitive to the degree of hydration for soil-crust lichens. For a series of species studied (Lange et al. 1994b, 1997a, 1998), it has ranged between 27 (*Lecidella crystallina*, Namib) and 100 $\mu mol\ m^{-2} s^{-1}$ PPFD (*Diploschistes diacapsis*, Utah), and was ca. 65 $\mu mol\ m^{-2} s^{-1}$ PPFD for soil crusts from the Negev (Lange et al. 1992). Such compensation points are extremely high and corroborate sun-plant photosynthetic features.

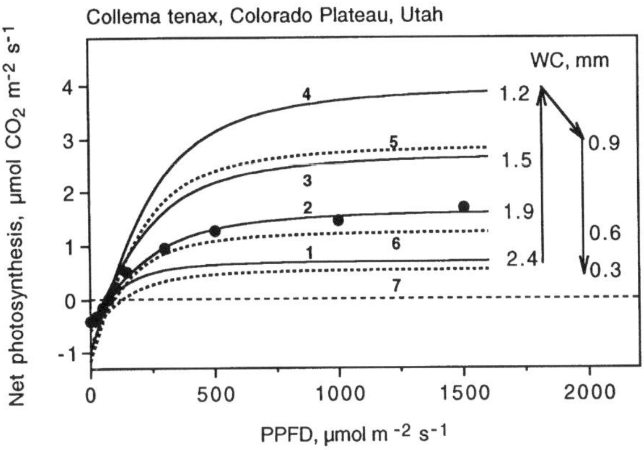

Fig. 18.6. Response of net photosynthesis to incident photon flux density (PPFD) of *Collema tenax* (Colorado Plateau, Utah). *Curves* are fitted by the Smith function; actual data points are given for one example only. *Sequence of curve numbers and arrows* indicate decreasing steps of water content (WC); *curve number 1* means maximum water-holding capacity of lichen. (After Lange et al. 1998)

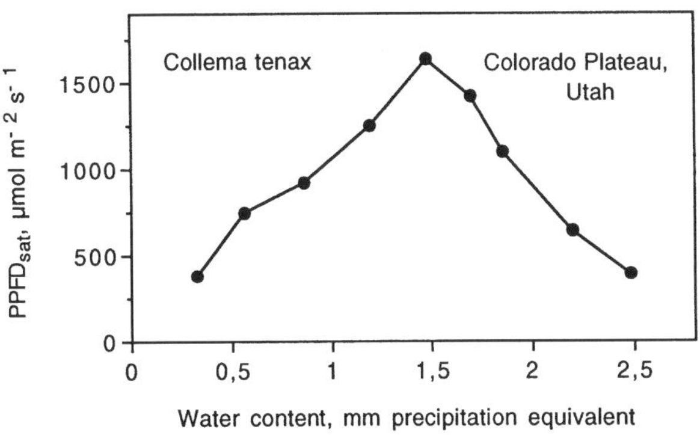

Fig. 18.7. Dependence of light saturation for maximal net photosynthesis ($PPFD_{sat}$) on thallus water content in *Collema tenax* (Colorado Plateau, Utah). (After Lange et al. 1998)

However, the above characterization of light response of soil biota applies only for the sun-exposed colonizer of the crust habitat. There is a steep vertical gradient in light intensity within the uppermost layers of the soil crusts, and at 1 mm from the surface, less than 1 % of PPFD remaines (Garcia-Pichel and Belnap 1996). Thus, many autotrophic organisms within the crust are living in shade and must be adequately adapted (see Chaps. 15, 16).

18.3.4 Temperature Response of Photosynthesis and Respiration

Soil-crust biota occur within a large gradient of habitats and temperatures, ranging from hot deserts to cold steppe and boreal environments. Studies of temperature responses of metabolic activity are rare, so no generalizations are possible. However, measurements with soil lichens from the Colorado Plateau show that their potential photosynthetic productivity is adapted to a broad range of temperatures, from near freezing to temperatures higher than 40 °C (Fig. 18.8; see Chap. 1). Green-algal *Psora cerebriformis* and *Diploschistes diacapsis* can maintain high NP down to 2 °C, while the optimal range is bet-

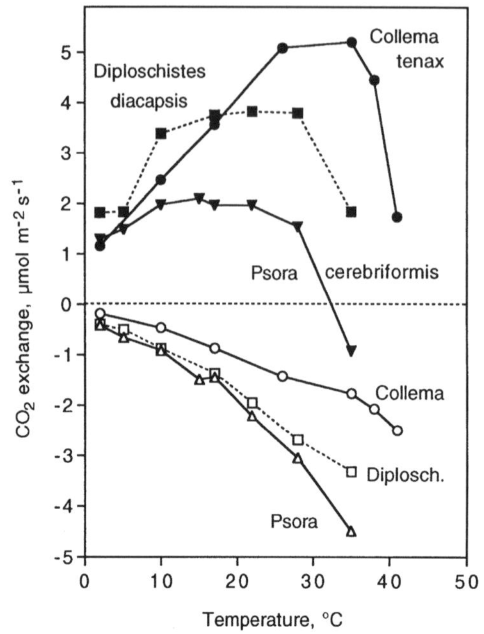

Fig. 18.8. Dependence of net photosynthesis (*filled symbols*) and dark respiration (*open symbols*) on temperature at optimal water content for *Collema tenax, Diploschistes diacapsis,* and *Psora cerebriformis* (Colorado Plateau, Utah). (After Lange et al. 1998)

ween 10 and 28 °C. Such temperatures might appear rather low for desert organisms, living in a hot environment. However, high temperatures usually occur when the lichen thalli are dry. Temperatures of photosynthesizing, moist lichens are much lower due to evaporative cooling and low irradiance during and briefly after rainfall or after early morning dew. In contrast, the cyanobacterial *Collema tenax* seems to be adapted to higher temperatures. At 41 °C, its NP is still at 30 % of optimal rates. This is an exceptional performance, as lichen photosynthetic metabolism usually tends to work most efficiently at much lower temperatures (Kappen 1988). This is apparently a special feature of gelatinous, slow-drying lichens, which may often be exposed to high sun radiation before final desiccation. Adaptation to high temperatures is also reported for savanna soil crusts from Venezuela (San José and Bravo 1991). There, maximal NP took place at 45 °C, and CO_2 was still fixed at 50 °C.

18.4 Patterns of CO_2 Exchange Under Natural Conditions

Only a few studies have measured CO_2 exchange of soil-crust organisms under natural conditions. Diel CO_2 exchange courses of *Diploschistes steppicus* were investigated in the central Negev by using conditioned gas exchange chambers (Lange et al. 1970). Performance of lichen-covered soil crusts in the coastal fog zone of the Namib was studied by using CO_2 porometers with short-time enclosure for 1 to 2 min (Lange et al. 1994b). Phillips and Belnap (1998) measured CO_2 exchange of soil crusts composed of cyanobacteria, algae, lichens, and mosses in the Canyonlands National Park, Utah, USA by using a canopy assimilation chamber. An automatic klapp-cuvette was used for continuous monitoring of soil lichens at a quasinatural, local steppe formation in the Würzburg Botanic Garden (Germany; Lange et al. 1997c; see also Chap. 20). By means of fluorescence measurements, in a similar habitat, metabolic activity of the terrestrial *Peltigera rufescens* was monitored for a whole year (Leisner et al. 1996; Lange et al. 1999b).

Photosynthesis and respiration of soil-crust biota in the field represent a complicated response to the complex environment. However, the degree of hydration dominates all the other factors. Thus, typification of the various response patterns is best based on the fashion of moistening. Some of the more important diel response types are discussed below.

For many arid and semiarid soil-crust habitats, dew and fog are the main sources of hydration. This is especially the case in the coastal fog zone of the Namib, where rain precipitation is negligible, but where soil lichens cover large areas of the landscape. The thalli are dry during most of the day, and moistening from fog and/or dew allows only a short period of photosynthetic activity. Figure 18.9 shows a diel time course of CO_2 exchange for the crustose *Caloplaca volkii* on a favorable summer day. At midnight, no CO_2 exchange of the desiccated lichen was detectable. Then fog began to drift over the coastal plain and dew condensation took place. The lichen crust absorbed water and

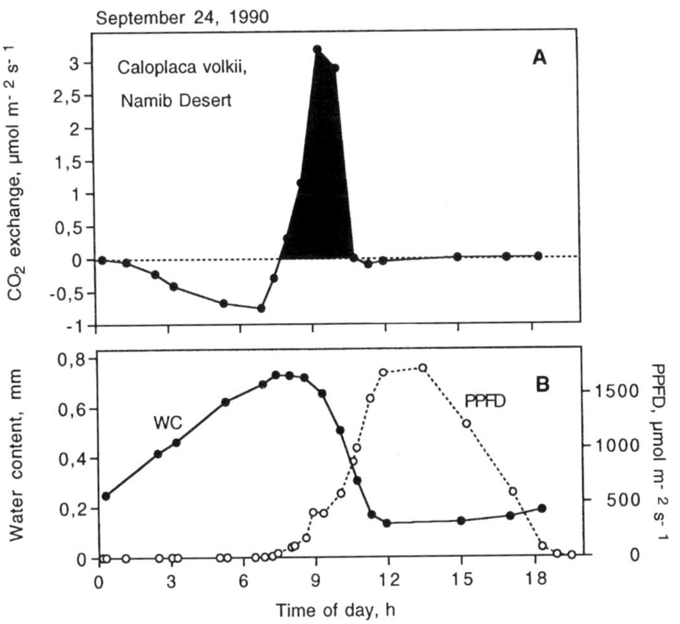

Fig. 18.9A,B. Diel time courses of **A** CO_2 exchange and **B** water content (*WC*) and incident photon flux density (*PPFD*) for soil crust covered by *Caloplaca volkii* (Namib Desert, Namibia) on a day with heavy dew and fog. (Data from Lange et al. 1994b)

DR was activated. As WC of the lichen rose, respiration increased to a maximum at sunrise. The well-hydrated lichen then started photosynthesis. Light compensation was reached at 07.40 h, and CO_2 uptake rose steeply to a maximum at about 10.00 h. High radiation, increasing temperatures, and decreasing air humidity rapidly dried the lichen, and NP declined quickly. The moisture compensation point was reached about 45 min after the CO_2 uptake maximum, and subsequently the soil crust became so dry that no CO_2 exchange could be detected for the remainder of the day. Nevertheless, nocturnal carbon loss was more than compensated by diurnal photosynthesis so that the daily net carbon balance was positive.

Photosynthetic activity of the Namib soil crusts differed from day to day according to the specific hydration conditions. Figure 18.10 shows the performance of crusts covered with *Lecidella crystallina* over 12 days. Dew alone, or in combination with fog of varying intensity, resulted in morning peaks of photosynthesis with different amounts of carbon gain. Four days were so dry that no NP was measurable. On one day (Sept. 11), lichen WC increased later in the afternoon due to water vapor uptake at higher air humidity, and this allowed another small peak of NP before sunset.

A short morning peak of NP after nocturnal hydration through dew and fog is a regular response pattern of lichens not only in arid and semiarid areas, but in many other regions (Kappen 1988). It is the most frequent mode of metabolic activation of soil lichens during the drier seasons in the Mediterranean area and in central Europe. Figure 18.11 depicts such a morning peak for *Cladonia convoluta,* a terrestrial green-algal species, on an early spring day in Würzburg after a cold night with much dew. High rates of NP were attained for a short period of time. However, hydration through dew usually is only sufficient for activation of phycobionts; cyanobacterial lichens need more water. CO_2 exchange of the cyanobacterial *Collema cristatum,* growing within the same ecosystem as *Cladonia,* was monitored under identical conditions (Fig. 18.11). No NP was detected. The disadvantage for *Collema* is compensated for when ample hydration has taken place and the gelatinous species can profit from its high water-holding capacity. When a nocturnal summer rain was followed by a bright, warm day (Fig. 18.12), *Cladonia* photosynthesized for 5 h after sunrise until its thallus was dry again. Under identical conditions, the time period of carbon gain for *Collema* was longer than 8 h and its primary production amounted to more than twice that of *Cladonia.* These differences in water economy illustrate the functional reasons for differences in habitat preference by representatives of both of these soil-lichen types (see Lange 2000b).

As discussed, high thallus WC may depress NP of soil-crust lichens. This can result in pronounced reduction of carbon gain. After heavy rain during a night in July (Fig. 18.13), *Fulgensia fulgens* was thoroughly wetted and then additionally hydrated by showers until noon. After sunrise, NP of the

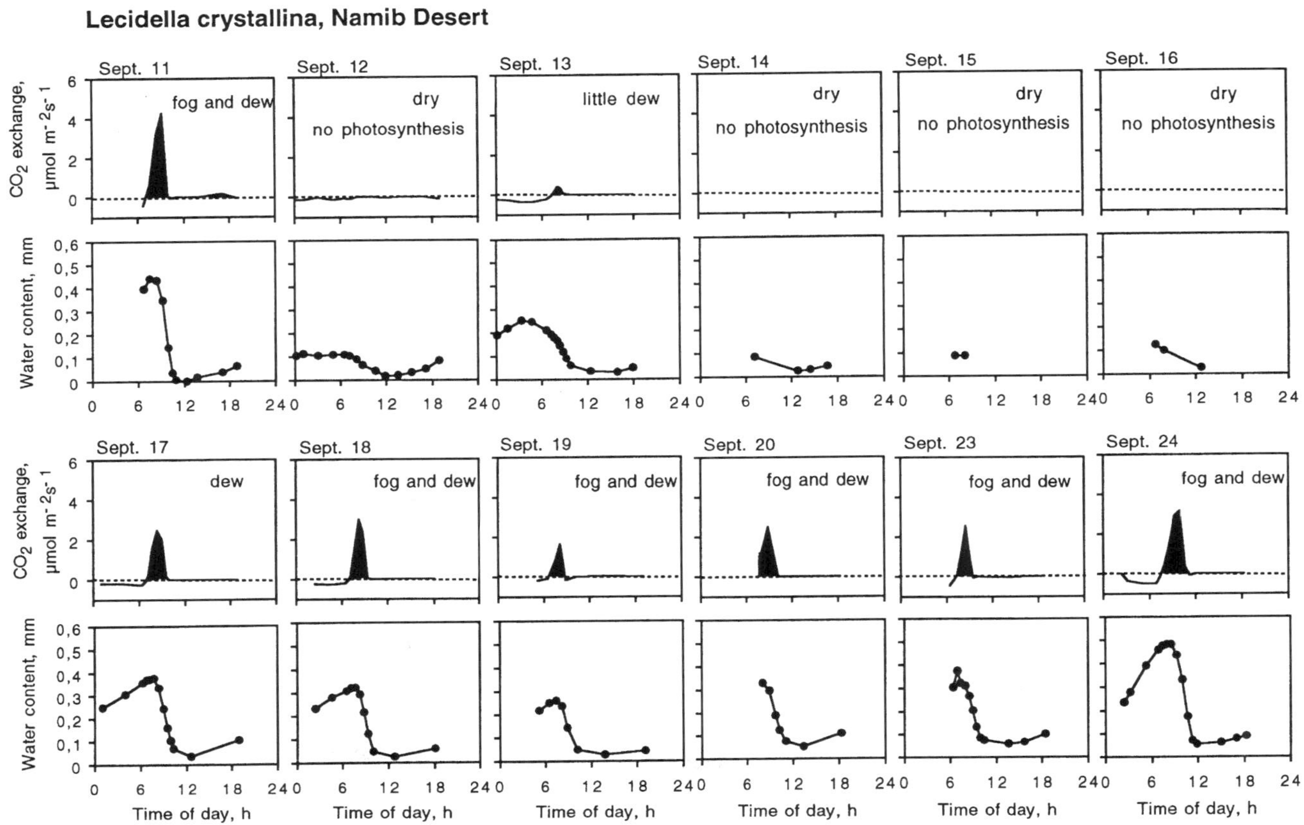

Fig. 18.10. Diel time courses of CO_2 exchange and water content during the period from 11 to 24 September 1990 for soil crusts covered by *Lecidella crystallina* (Namib Desert, Namibia). (After Lange et al. 1994b)

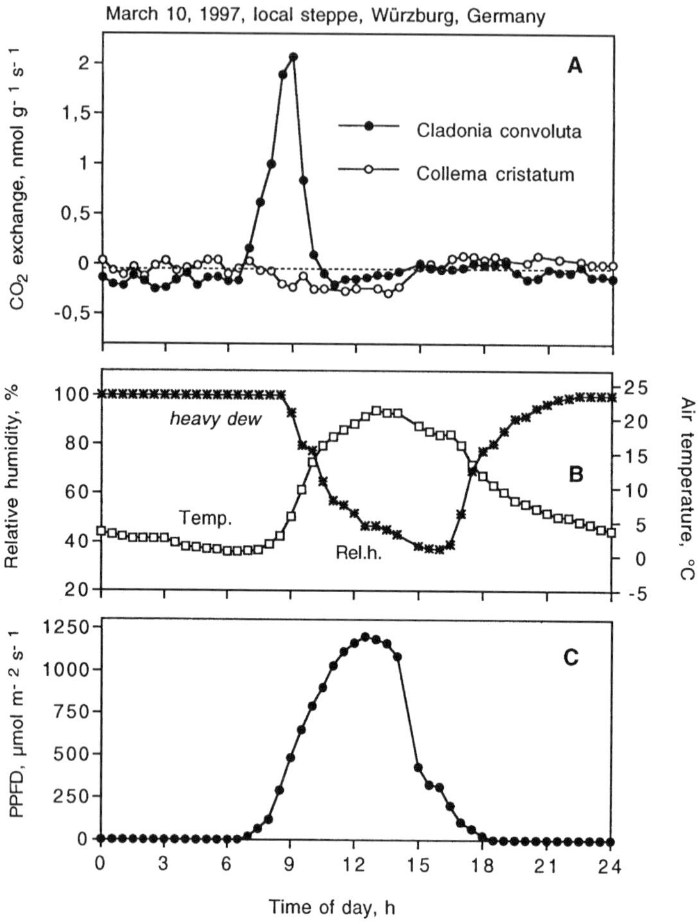

Fig. 18.11A–C. Diel time courses of **A** CO_2 exchange, **B** air temperature and relative humidity, and **C** incident photon flux density (PPFD) for *Cladonia convoluta* and *Collema cristatum* (quasinatural local steppe formation, Würzburg, Germany) on a day with heavy dew

suprasaturated lichen rose, but only until 08.00 h when it had reached around 0.5 µmol CO_2 $m^{-2}s^{-1}$ at ca. 100 µmol $m^{-2}s^{-1}$ PPFD. Over the next 5 h, NP remained almost completely insensitive to incident PPFD, as the lichen was light-saturated. The thallus, meanwhile, was drying and, at 15.00 h, NP suddenly increased. At that point, WC had reached the optimal value so that peak photosynthesis was possible until all metabolic activity ceased due to total desiccation. This phenomenon occurs frequently, and high WC severely limits photosynthetic carbon gain of lichens like *Fulgensia*. In contrast, CO_2 exchange of *Diploschistes muscorum* is independent of hydration-induced diffusion resistances. This species is capable of using even the highest degrees of water saturation for favorable production.

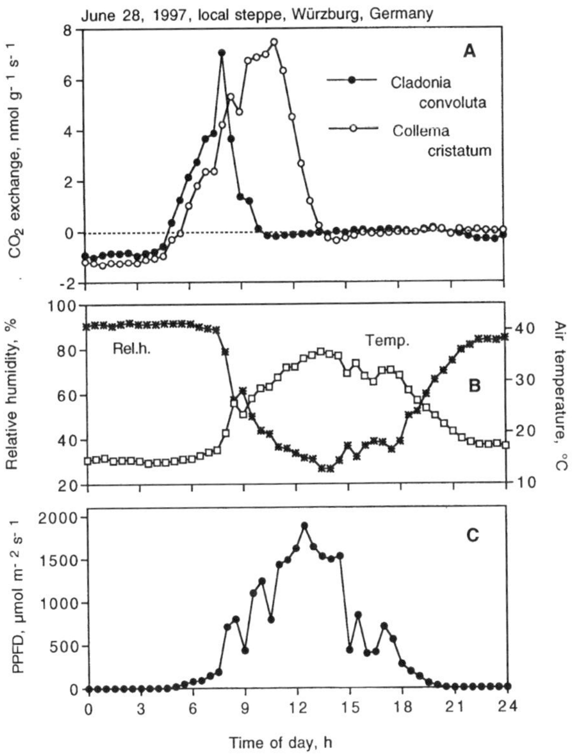

Fig. 18.12A–C. As Fig. 18.11, but on a dry day with rain in the foregoing night

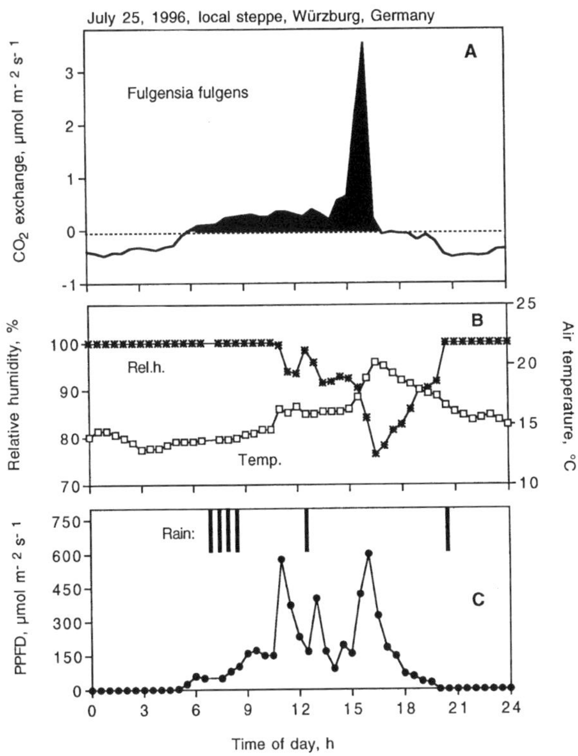

Fig. 18.13A–C. As Fig. 18.11, but for *Fulgensia fulgens* on a rainy day with drying of the lichen after noon. Occurrence of rain is indicated in **C**.

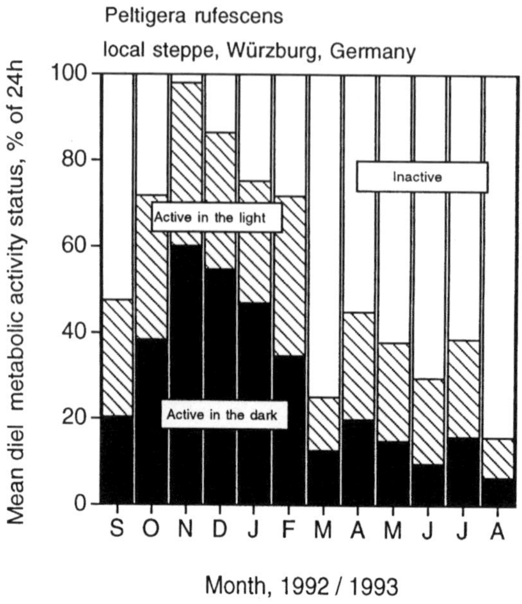

Fig. 18.14. Monthly means of daily activity status (as determined by continuous monitoring of chlorophyll fluorescence) from September 1992 to August 1993 for *Peltigera rufescens* (quasinatural local steppe formation, Würzburg, Germany). (After Lange et al. 1999b)

Annual distribution of hydration-dependent metabolic activity for lichens in the Würzburg steppe formation is determined by the seasonal weather rhythm, as shown for *Peltigera rufescens* in Fig. 18.14. The lichen was inactivated through desiccation (or freezing) for 46.5 % of the total time of the year. For the remaining 53.5 % of the time, it was active in the light for 25.6 % and in the dark for 27.9 % of time. This temperate soil-crust habitat represents an intermediate situation between the extremes of arid desert regions, where metabolic activity might be reduced to less than 10 % of total time, and the boreal and arctic regions, where soil-crust organisms will experience longer periods of annual hydration. However, no data are available for a detailed comparison of long-term environmental conditions for soil crusts in different geographical regions. Such information would substantially add to our understanding of crust performance.

18.5 Maximal Rates of Soil-Crust Net Photosynthesis

Reported maximal rates of net photosynthesis (NP_{max}) under optimal conditions vary greatly and span the range of 2 orders of magnitude between 0.111 and 11.5 μmol CO_2 $m^{-2}s^{-1}$ (Table 18.2). Differences are due to different crust structure, organisms involved, and density of crust colonization. Lowest NP_{max} occurs in crust types where the bulk of the autotrophic biomass are cyanobacteria and chlorophyta that are predominantly positioned within the

Table 18.2. Rates of area-based maximal net photosynthesis at natural ambient CO_2 concentration, optimal (favorable) temperature, light, and water content for total crusts with multiple components (mixed crust), with one dominant lichen species, or for single soil lichens

Material	Maximal rates of area-based net photosynthesis $\mu mol\ CO_2\ m^{-2}\ s^{-1}$	Remarks
Mixed crust with dense population of chlorophyta (*Klebsormidium* spp.)	11.5	Lab: B. Büdel and A. Lukešová (unpubl.) Open pine forest, Czech Republic
Collema tenax (cyanolichen)	7.0	Lab: Lange et al. (1998) Colorado Plateau, Utah, USA
Crust dominated by *Lecidella crystallina* (phycolichen)	5.9	Field experiment: Lange et al. (1994b) Namib Desert, Namibia
Diploschistes muscorum (phycolichen)	5.9	Lab: Lange et al. (1997b) Local steppe formation, Würzburg, Germany
Squamarina lentigera (phycolichen)	5.2	Lab: Lange et al. (1997a) Colorado Plateau, Utah
Fulgensia fulgens (phycolichen)	5.2	Lab: Lange et al. (1995) Local steppe formation, Würzburg
Diploschistes muscorum (phycolichen)	5.1	Field: O.L. Lange (unpubl.) Local steppe formation, Würzburg
Squamarina lentigera (phycolichen)	4.8	Lab: Lange et al. (1995) Local steppe formation, Würzburg
Mixed crust, cyanobacteria and chlorophyta	4.70, 2.32	Lab: San José and Bravo (1991) Orinoco Llanos, Venezuela
Squamarina lentigera (phycolichen)	4.2	Field: O.L. Lange (unpubl.) Local steppe formation, Würzburg
Fulgensia fulgens (phycolichen)	3.5	Field: O.L. Lange (unpubl.) Local steppe formation, Würzburg
Crust dominated by *Gloeoheppia turgida* (cyanolichen)	3.0	Lab: Lange et al. (1993) Judean Desert, Dead Sea, Israel

Table 18.2. (*continued*)

Material	Maximal rates of area-based net photosynthesis µmol CO_2 m^{-2} s^{-1}	Remarks
Mixed crust with cyanobacteria, algae, lichens, mosses	1.5	Field experiment: Phillips and Belnap (1998) Canyonlands National Park, Utah
Mixed crust with *Microcoleus vaginatus, Nostoc*, etc.	1.17, 0.41	Lab: Garcia-Pichel and Belnap (1996) Southeastern Utah
Mixed crust with cyanobacteria and chlorophyta	1.12–0.72	Lab: Lange et al. (1992) Negev, sand dunes, Israel
Mixed crust, *Microcoleus*- and *Scytonema*-dominated	0.111–0.187	Lab: Jeffries et al. (1993a) Kaiparowits Basin, Utah
Average maximum of leaves of C3 crop plants	20 – 40	Larcher (1995)
Average maximum of needles of evergreen coniferous trees	4 – 8 (12)	Larcher (1995)

Measurements (source of material is indicated) were done under natural (field) or under controlled conditions (lab). Dashes separate range of data of the same crust type; commas separate maxima for different crust types. Textbook data for higher plants are included for comparison soil. Sparse concentration of organisms and limited light penetration do not allow NP much higher than 1 μmol CO_2 $m^{-2}s^{-1}$. Exposure of free-living or lichenized cyanobacteria and/or algae on top of the soil surface greatly increases CO_2 fixation. Among the lichens, *Collema tenax* shows the highest area-related NP_{max}, which is certainly correlated with its nitrogen-fixing ability (Chap. 19). Phycolichens show a range of photosynthetic capacity between 3 and 6 μmol CO_2 $m^{-2}s^{-1}$. It is worthwhile to note that NP_{max} for the same species is similar or only slightly lower when field data are compared with measurements under controlled conditions (see *Diploschistes muscorum, Squamarina lentigera,* and *Fulgensia fulgens*). This shows that, under the varying natural conditions, optimal combination of weather factors do occur – even if only for short periods of time.

Cryptogams are often thought of as organisms with extremely low photosynthetic capacity (e.g., Larcher 1995). This is certainly not true for soil-crust organisms or crust communities. With 11.5 μmol CO_2 $m^{-2}s^{-1}$, NP_{max} of the dense population of green algae colonizing the topsoil in a temperate pine forest is about 50% of the average NP_{max} of leaves of C3 crop plants (Table 18.2), and substantially larger than area-related photosynthetic capacity of needles of conifers. NP_{max} of soil-crust lichens is often well in the upper range or even higher than maximal area-related leaf NP rates of the phanerogamous vegetation characterizing the same site (Lange et al. 1998). Even crusts from the Negev, with autotrophs mainly within the topsoil layer, fix more than 20% of the daily morning NP maximum of the coexisting desert shrub *Zygophyllum dumosum.* Thus, photosynthetic capacity of poikilohydric soil crusts is very high. However, in contrast to homoiohydric plants, periods of metabolic activity of the soil microphytes are always restricted to the short periods when they are sufficiently hydrated. Nevertheless these highly specialized poikilohydric crust organisms contribute substantially to the carbon budget of their arid and semiarid ecosystems where vascular plant production is low (see Chap. 20).

Acknowledgments. This work was supported by the Deutsche Forschungsgemeinschaft within projects of the Sonderforschungsbereich 251 der Universität Würzburg.

References

Badger MR, Price GD (1992) The CO_2 concentration mechanism in cyanobacteria and microalgae. Physiol Plant 84:606–615

Badger MR, Pfanz H, Büdel B, Heber U, Lange OL (1993) Evidence for the functioning of photosynthetic CO_2-concentrating mechanisms in lichens containing green algal and cyanobacterial photobionts. Planta 191:57–70

Belnap J, Harper KT, Warren SD (1994) Surface disturbance of cryptobiotic soil crusts: nitrogenase activity, chlorophyll content, and chlorophyll degradation. Arid Soil Res Rehabil 8:1–8

Beymer RJ, Klopatek JM (1991) Potential contribution of carbon by microphytic crusts in pinyon-juniper woodlands. Arid Soil Res Rehabil 5:187–198

Björkman O (1981) Responses to different quantum flux densities. In: Lange OL, Nobel PS, Osmond CB, Ziegler H (eds) Physiological plant ecology, I. Encyclopedia of plant physiology. Springer, Berlin Heidelberg New York, pp 57–107

Brock TD (1975) Effect of water potential on a *Microcoleus* (Cyanophyceae) from a desert crust. J Phycol 11:316–320

Demmig-Adams B, Adams WW III, Czygan F-C, Schreiber U, Lange OL (1990a) Differences in the capacity for radiationless energy dissipation in the photochemical apparatus of green and blue-green algal lichens associated with differences in carotenoid composition. Planta 180:582–589

Demmig-Adams B, Máguas C, Adams WW III, Meyer A, Kilian E, Lange OL (1990b) Effect of high light on the efficiency of photochemical energy conversion in a variety of lichen species with green and blue-green phycobionts. Planta 180:400–409

De Winder B (1990) Ecophysiological strategies of drought-tolerant phototrophic micro-organisms in dune soils. Academisch Proefschrift, University of Amsterdam, Amsterdam

Farrar JF, Smith DC (1976) Ecological physiology of the lichen *Hypogymnia physodes*. III. The importance of the rewetting phase. New Phytol 77:115–125

Garcia-Pichel F, Belnap J (1996) Microenvironment and microscale productivity of cyanobacterial desert crusts. J Phycol 32:774–782

Jeffries DL, Link SO, Klopatek JM (1993a) CO_2 fluxes of cryptogamic crusts. I. Response to resaturation. New Phytol 125:163–173

Jeffries DL, Link SO, Klopatek JM (1993b) CO_2 fluxes of cryptogamic crusts. II. Response to rehydration. New Phytol 125:391–396

Kappen L (1988) Ecophysiological relationships in different climatic regions. In: Galun M (ed) Handbook of lichenology, II. CRC Press, Boca Raton, pp 37–100

Kidron G (1995) The impact of microbial crust upon rainfall-runoff-sediment yield relationships on longitudinal dune slopes, Nizzana, western Negev desert, Israel. PhD Thesis, The Hebrew University of Jerusalem, Jerusalem (English summary)

Lange OL (2000a) Photosynthetic performance of a gelatinous lichen under temperate habitat conditions: long-term monitoring of CO_2 exchange of *Collema cristatum*. Bibl Lichenol 75:307–332

Lange OL (2000b) Die Lebensbedingungen von Bodenkrusten-Organismen: Tagesverlauf der Photosynthese einheimischer Erdflechten. (Conditions of life for soil-crust organisms: diel courses of photosynthesis of central European terrestrial lichens) Hoppea: 61:423–443

Lange OL, Schulze E-D, Koch W (1970) Experimentell-ökologische Untersuchungen an Flechten der Negev-Wüste. III. CO_2-Gaswechsel und Wasserhaushalt von Krusten- und Blattflechten am natürlichen Standort während der sommerlichen Trockenperiode. Flora 159:525–538

Lange OL, Kilian E, Ziegler H (1986) Water vapor uptake and photosynthesis of lichens: performance differences in species with green and blue-green algae as phycobionts. Oecologia 71:104–110

Lange OL, Bilger W, Rimke S, Schreiber U (1989) Chlorophyll fluorescence of lichens containing green and blue-green algae during hydration by water vapor uptake and by addition of liquid water. Bot Acta 102:306–313

Lange OL, Kidron GJ, Büdel B, Meyer A, Kilian E, Abeliovich A (1992) Taxonomic composition and photosynthetic characteristic of the 'biological soil crusts' covering sand dunes in the western Negev Desert. Funct Ecol 6:519–527

Lange OL, Büdel B, Meyer A, Kilian E (1993) Further evidence that activation of net photosynthesis by dry cyanobacterial lichens requires liquid water. Lichenologist 25:175–189

Lange OL, Meyer A, Büdel B (1994a) Net-photosynthesis activation of a desiccated cyanobacterium without liquid water in high air humidity alone. Experiments with *Microcoleus sociatus* isolated from a desert soil crust. Funct Ecol 8:52–57

Lange OL, Meyer A, Zellner H, Heber U (1994b) Photosynthesis and water relations of lichen soil crusts: field measurements in the coastal fog zone of the Namib Desert. Funct Ecol 8:253–264

Lange OL, Reichenberger H, Meyer A (1995) High thallus water content and photosynthetic CO_2 exchange of lichens. Laboratory experiments with soil-crust species from local xerothermic steppe formations in Franconia, Germany. In: Daniels F, Schulz M, Peine J (eds) Flechten Follmann. University of Cologne, Cologne, pp 139–153

Lange OL, Green TGA, Reichenberger H, Meyer A (1996) Photosynthetic depression at high thallus water content in lichens: concurrent use of gas exchange and fluorescence techniques with a cyanobacterial and a green algal *Peltigera* species. Bot Acta 109:43–50

Lange OL, Belnap J, Reichenberger H, Meyer A (1997a) Photosynthesis of green algal soil crust lichens from arid lands in southern Utah, USA: role of water content on light and temperature responses of CO_2 exchange. Flora 192:1–15

Lange OL, Green TGA, Reichenberger H, Proksch P (1997b) Do secondary substances in the thallus of a lichen promote CO_2 diffusion and prevent depression of net photosynthesis at high water content? Oecologia 112:1–3

Lange OL, Reichenberger H, Walz H (1997 c) Continuous monitoring of CO_2 exchange of lichens in the field: short-term enclosure with an automatically operating cuvette. Lichenologist 29:259–274

Lange OL, Belnap J, Reichenberger H (1998) Photosynthesis of the cyanobacterial soil crust lichen *Collema tenax* for arid lands in southern Utah, USA: role of water content on light and temperature response of CO_2 exchange. Funct Ecol 12:195–202

Lange OL, Green TGA, Reichenberger H (1999a) The response of lichen photosynthesis to external CO_2 concentration and its interaction with thallus water status. J Plant Physiol 154:157–166

Lange OL, Leisner MRL, Bilger B (1999b) Chlorophyll fluorescence characteristics of the cyanobacterial lichen *Peltigera rufescens* under field conditions. II. Diel and annual distribution of metabolic activity and possible mechanisms to avoid photoinhibition. Flora 194:413–430

Larcher W (1995) Physiological plant ecology. Springer, Berlin Heidelberg New York

Lawlor DW (1990) Photosynthese. Thieme, Stuttgart

Leisner JMR, Bilger W, Lange OL (1995) Does photoinhibition occur under natural field conditions in a cyanobacterial lichen? In: Mathis P (ed) Photosynthesis: from light to biosphere, IV. Kluwer, Dordrecht, pp 259–262

Leisner JMR, Bilger W, Lange OL (1996) Chlorophyll fluorescence characteristics of the cyanobacterial lichen *Peltigera rufescens* under field conditions. I. Seasonal patterns of photochemical activity and the occurrence of photosystem II inhibition. Flora 191: 261–273

Mazor G, Kidron GJ, Vonshak A, Abeliovich A (1996) The role of cyanobacterial exopolysaccharides in structuring desert microbial crusts. FEMS Microbiol Ecol 21:121–130

Phillips S, Belnap J (1998) Shifting carbon dynamics due to the effect of *Bromus tectorum* invasion on biological soil crust. Bull ESA, 205. 1998 Meeting Abstracts

Rundel PW, Lange OL (1980) Water relations and photosynthetic response of a desert moss. Flora 169:329–335

San José JJ, Bravo CR (1991) CO_2 exchange in soil algal crusts occurring in the Trachypogon savannas of the Orinoco Llanos, Venezuela. Plant Soil 135:233–244

Smith DC, Molesworth S (1973) Lichen physiology. XIII. Effects of rewetting dry lichens. New Phytol 72:525–533

Ziegler H, Lüttge U (1998) Carbon isotope discrimination in cyanobacteria of rocks of inselbergs and soils of savannas in the Neotropics. Bot Acta 111:212–215

19 Factors Influencing Nitrogen Fixation and Nitrogen Release in Biological Soil Crusts

J. Belnap

19.1 Introduction

Nitrogen (N) occurs in the atmosphere as N_2, a form that is not useable by vascular plants. N_2 must first be "fixed", or reduced, to ammonia (NH_4^+) by prokaryotic organisms such as eubacteria and cyanobacteria. Thus, an important feature of the cyanobacteria and cyanolichens in soil crusts is their ability to fix atmospheric N. As this fixation is an anaerobic process, most cyanobacterial fixation takes place in heterocysts, which are specialized, thick-walled cells with enhanced respiration and no oxygen-producing photosystem II (Paerl 1990). Heterocystic genera commonly occurring in soil crusts include *Anabaena, Calothrix, Cylindrospermum, Dicothrix, Hapalosiphon, Nodularia, Nostoc, Plectonema, Schizothrix*, and *Scytonema* (Harper and Marble 1988). Nitrogen fixation has also been demonstrated in non-heterocystous soil genera such as *Lyngbya, Microcoleus, Oscillatoria, Phormidium,* and *Tolypothrix* (Rogers and Gallon 1988; Belnap 1996), although this may be a result of associated bacteria (Steppe et al. 1996). Nonheterocystic species can exclude oxygen in several ways: (1) behaviorally by clumping; (2) spatially or chemically within a cell; (3) temporally, by fixing at night when no oxygen is being evolved by photosynthesis; or (4) through a combination of these (Paerl 1978; Rogers and Gallon 1988; Paerl 1990). Bacteria associated with cyanobacteria may also contribute to N inputs by scavenging oxygen (thus creating anaerobic microzones for the cyanobacteria) or by fixing N themselves. This has been demonstrated for *Microcoleus vaginatus* isolated from soil crusts (Steppe et al. 1996). Soil lichens with cyanobacterial photobionts also fix N. Common N-fixing soil lichens include *Nostoc*-containing *Collema* spp. and *Peltigera* spp. and *Scytonema*-containing *Heppia* spp. Cyanobacteria also live as epiphytes on soil mosses and phycolichens; thus, this consortium of organisms can show fixation activity (Peters et al. 1986).

N fixation activity is generally measured as $^{15}N_2$ incorporation or by the acetylene (C_2H_2) reduction assay (ARA). As $^{15}N_2$ uptake is a direct measure of N fixation, it is the most reliable method. However, it is costly and requires

Ecological Studies, Vol. 150
J. Belnap and O.L. Lange (eds.) Biological Soil Crusts: Structure, Function, and Management

highly controlled conditions; thus, most studies use ARA. In the presence of C_2H_2, the nitrogenase enzyme produces C_2H_4 (ethylene); thus, C_2H_4 production is a measure of nitrogenase enzyme activity (NA). Because NA is so widely used as a surrogate for nitrogen fixation, these terms will be used synonymously in this chapter.

There are some problems associated with ARA. C_2H_2 can affect physiological functioning of the organism, especially when exposure lasts longer than 6 h (David and Fay 1977). The biggest issue is that conversion of ARA data to absolute amounts of N_2 fixed requires calibration by $^{15}N_2$, which is seldom done. This is essential, as reported conversion rates in the literature for the C_2H_4:N_2 ratio vary from 2 to 56, while conversion ratios for naturally-occurring soil cyanobacteria generally range from 1.9 to 6.1 (Potts 1984). However, two recent studies indicate that conversion ratios can be much lower and also can be seasonally variable. S. Phillips and J. Belnap (unpubl. data) found that desert *Nostoc commune* sheets had an average conversion ratio of 0.31. Liengen (1999) reported conversion ratios varied with environmental conditions for both *Nostoc* sheets (0.11–0.4) and free-living soil cyanobacteria (0.02–0.07) in the Arctic. He also analyzed *Anabaena* in culture and obtained a ratio close to 4. These studies may explain why reported rates are so variable, as many studies have used cultured specimens or aquatic species, while actual rates in soils may be much lower. In addition, environmental conditions may lead to large variations in conversion rates. If these more recent results apply to other ecosystems or soil cyanobacterial species, N fixation rates may be currently underestimated by an order of magnitude. This matter clearly needs additional investigation.

19.2 Environmental Controls of Nitrogen Fixation

19.2.1 Importance of Preexisting Conditions

Many studies have addressed the effects of different environmental factors on N fixation in soil cyanobacteria and cyanolichens. Because past conditions influence current fixation activity through the amount of ATP, reductants, nitrogenase enzyme, and N compounds present in the cyanobacteria, it is essential that past environmental conditions be taken into account when measuring current activity. In light of what we now know about the effects of preexperimental conditions, the seeming contradiction between many data sets can be resolved. Unfortunately, this also makes many data sets unuseable, as the preexperimental condition of the organisms was not taken into account by researchers.

The importance of pre-experimental environments on N fixation is well-illustrated by a recent study in southeastern Utah. NA was measured bi-weekly over a 2-year period in *Microcoleus* crust, *Nostoc-Scytonema-Microcoleus* crust, and *Collema* crust (Fig. 19.1). Under standard conditions, specimens in all three crust types showed no or low NA if no rain had fallen within 3 days previous to collection. If rain had fallen within 3 days of collection, NA values were strongly related (r^2=0.93) with the average air temperature of the previous 3 days when air temperatures were between 1 and 26 °C (see temperature Sect. 19.2.4). There were some exceptions to this pattern, as discussed in the figure legend.

19.2.2 Photosynthesis, Carbon Storage, and Dark Fixation

Photosynthesis/Carbon Storage. N fixation depends on photosynthesis to provide ATP for energy and carbon compounds as electron donors (Paul and Clark 1996). Consequently, both the duration and rate of N fixation depend on past and current conditions that influence cyanobacterial carbon balance, such as moisture, temperature, light intensity, and supply of assimilates. In the laboratory, exposure to higher light intensity and higher carbon fixation rates before NA assays correlates with higher light and dark N fixation rates in soil crusts (Jones 1977b). Microprobes show incorporation of ^{15}N is correlated with gross photosynthetic rates in *Nostoc* (Dodds 1989).

Dark Fixation. All soil-crust cyanobacteria capable of light N fixation show dark-fixation activity as well. This includes mixed cyanobacterial crusts and *Nostoc commune* from soils in Hawaii, South Africa (Jones 1977a,b,d), Germany (Potts et al. 1987), Antarctica (Davey and Marchant 1983), the Arctic (O.C. Bliss, pers. comm), China (Scherer et al. 1984), Canada (Coxson and Kershaw 1983a,b), and mixed crusts from Utah (J. Belnap, unpubl.). Rates and duration of dark fixation is limited by endogenous carbon reserves, as polyglucose reserves are mobilized to produce ATP and reductant for N fixation (Millbank 1978). Carbon limitation generally results in dark N fixation being at lower rates than light N fixation in soil crusts from all studied regions (Jones 1977a,b; Rai et al. 1981).

19.2.3 Moisture and Rehydration

Moisture. Cyanobacteria are physiologically active only when wet; consequently, all N-fixing activities in soil cyanobacteria and cyanolichens are ultimately controlled by moisture (Kershaw 1985; Nash 1996). Moisture can be a result of rainfall, snow melt, dew, or fog. Liquid water is necessary for

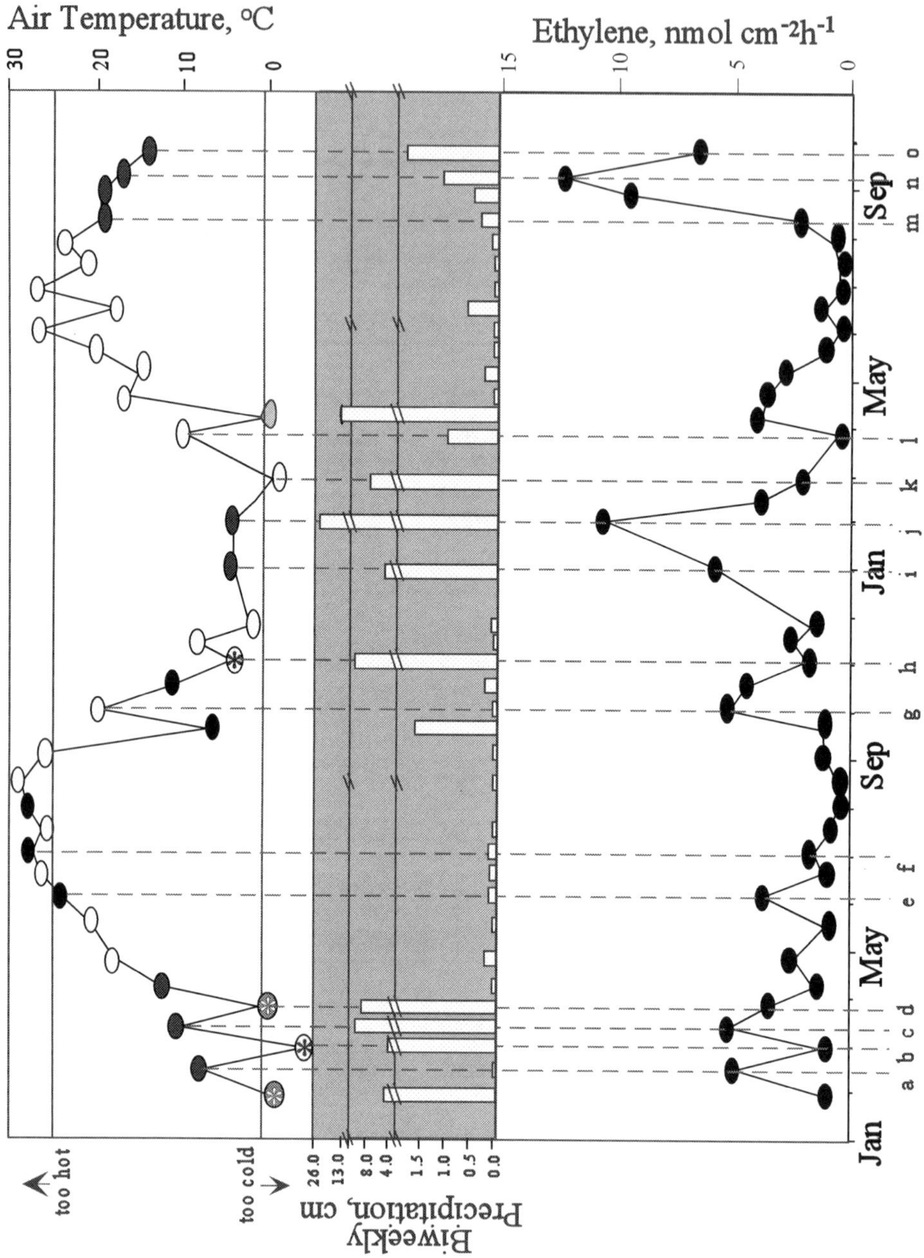
Air Temperature, °C
Ethylene, nmol cm-2h-1
30
20
10
0
15
10
5
0
Biweekly Precipitation, cm
26.0
13.0
8.0
4.0
1.5
1.0
0.5
0.0
too hot
too cold
Jan
May
Sep
Jan
May
Sep
a b c d e f g h i j k l m n o

Fig. 19.1

cyanobacterial carbon fixation (see Chap. 18). Because N fixation requires the products of photosynthesis, availability of liquid water ultimately determines the amount of N fixation.

Moisture levels needed to initiate and optimize N fixation vary widely with species, habitats, and pre-collection conditions. Soil-crust cyanobacteria show NA at values ranging from a water content of 6 % dry weight to total water immersion (Jones 1977d; Kershaw 1985; Belnap et al. 1999). Most soil cyanolichens, including *Collema* species, require a water content of at least 80 % dry weight for initiation of net carbon fixation activity (Lange et al. 1998). NA in *Microcoleus-Collema* soil crusts drops rapidly below soil water potentials of –33 kPa, with a 50 % reduction by –100 kPa (Rychert et al. 1978). NA rates at a given moisture content vary within and between species (DuBois and Kapustka 1983; Kershaw 1985). Jones (1977d) found *Nostoc commune* in subtropical soils had maximal NA rates at 22–126 % water content by weight, while Coxson and Kershaw (1983b) showed maximal NA rates at 500 % water content in *Nostoc* from a northern temperate grassland soil. *Nostoc commune* from Mongolian soils had the highest NA rates when totally saturated (Belnap et al. 1999).

As noted above, pre-collection conditions strongly influence NA levels at a given moisture content, presumably by affecting ATP and carbon stores (Kershaw 1985). Constant hydration can also affect nitrogen-fixation rates,

Fig. 19.1. Biweekly measurement of acetylene reduction assay (ARA) for 2 consecutive years in field-collected *Collema* from southeastern Utah, measured in the lab under standard conditions (fully moistened, light, 25 °C). *Top panel* Average air temperatures for the sampling date. Samples were collected in early morning. Sample dates with rainfall within 3 days of sample collection are indicated by circle color: *dark* rain; *gray* snow; *clear* none; * snowmelt that added water to soils; *middle panel* Amount of precipitation. Precipitation was measured 10 miles from the sample site; thus amounts recorded only estimate those at the study site; *bottom panel* Nitrogenase activity (NA). Samples were collected, moistened, and incubated under light at 25 °C for 4 h. In general, if soils had been moist within 3 days of collection, NA levels were highly correlated with daily average temperature (r^2=0.93) unless temperatures were below 1 °C or above 26 °C. The following letters refer to the *vertical lines labeled at the bottom of the figure: a–d* when temperatures are above 1 °C and soils are moist, NA is observed. Even if soils are moist, low air temperatures preclude ARA activity; however, once temperatures increase, ARA activity resumes. Levels are positively correlated with temperature; *e* temperatures are at the maximum for NA. Soils are moist, but this moisture follows a long dry period, and thus NA levels are moderate; *f* air temperatures exceed the maximum, and although soils are moist, no NA is detected; *g* this sample point was anomalous, as no soil moisture was recorded, but moderate NA were still observed; *h* low temperatures and soil moisture result in low NA levels; *i–k* snow melts, and NA levels soar until air temperatures get too low; *l* in spite of optimal air temperature, lack of moisture precludes NA; *m–o* rain after a long dry period initiates NA. As soils continue to receive moisture, NA increases, although air temperatures are similar. As temperatures decrease, so does NA with NA rates lower in soil *Nostoc commune* kept continuously hydrated (Jones 1989). Massive glucose efflux was shown in lichens kept continuously hydrated, which could lead to depletion of energy reserves necessary for N fixation (Kershaw 1985). Thus, long periods of high soil moisture may reduce overall N fixation of soil crusts, such as in subtropical or arctic regions. Crusts may also fix less N in ecosystems where dense plant litter keeps organisms wet for extended periods of time. Alternation between wet and dry soils, as occurs in most deserts, may lead to greater N fixation per unit "moist time" than occurs in ecosystems with constantly hydrated soils. However, no study has tested this hypothesis.

Rehydration. Respiration and photosynthesis in cyanobacteria begin almost immediately upon wetting (Chap. 18). However, there is generally a lag time between wetting and the initiation of N fixation, with a further lag time before maximal fixation rates are reached. Lag time length is species-dependent, and is also influenced by past and current environmental conditions. There are few data on lag times for initiation of N fixation in crustal species. Free-living and lichenized *Nostoc*, desiccated for a few days, takes minutes to 1 h to initiate NA activity (Kershaw 1985; Dodds et al. 1995). NA in *Scytonema* soil crust begins within 10 min of wetting (Rychert and Skujins 1974; Potts 1980; Jeffries et al. 1992). However, initial NA rates are generally extremely low.

Time to reach maximal fixation rates after rehydration varies widely. Comparison between reported studies must be done with caution, as many authors do not report pre-collection conditions. At 25 °C, a *Nostoc-Anabaena* soil crust showed maximum NA 36 h after being wetted, while Englund (1978) showed maximal NA in a *Nostoc-Scytonema-Microcoleus* crust was reached 25 h after wetting. *Nostoc* has been reported to reach maximum NA at 1–5 h (Jones 1977d; Fritz-Sheridan 1988; Dodds et al. 1995). *Tolypothrix* crusts showed maximal NA rates 6 h after wetting (Fritz-Sheridan 1988). However, none of these studies accounted for potential increases in cyanobacterial biomass.

The longer crustal organisms are desiccated, the more time is required for N fixation to both begin and to reach maximal rates after rewetting. Dodds et al. (1995) reported *Nostoc* samples desiccated for a few days reached maximal NA rates 3 h after hydration, while *Nostoc* desiccated for several months took 8–12 h to reach maximal rates (Nash 1996). Jeffries et al. (1992) found that *Microcoleus* and *Scytonema* crusts required 4–6 days of light and hydration after 6 months' desiccation for maximal NA rates. Scherer et al. (1984) showed that after 2 years of desiccation, *Nostoc flagelliforme* and *N. commune* from northwest China reached maximal NA rates after 120–150 h, while material dried for 2 days required 30–50 h. Again, however, these studies did not measure possible increases in biomass during incubation.

Coxson and Kershaw (1983b) found recovery of maximal NA rates in *Nostoc* after desiccation was also temperature dependent. At 14 and 21 °C, 14 h were needed to recover previously measured maximal rates. Maximum rates were never recovered at 7 °C. Previous exposure to high temperatures also results in a longer time to reach maximal rates. *Scytonema-Microcoleus* crusts exposed to air temperatures of 35 °C for 4 days required wet, light, and cooler conditions (<26 °C) for up to 3 days before maximal NA rates were reached (J. Belnap, unpubl.).

The lag time between wetting and fixation activity probably reflects levels of carbon stores, nitrogenase enzymes, heterocysts, and photodamage. If there are sufficient carbon stores in the heterocysts, there may be no lag time for initiation of NA. However, if carbon stores are depleted (through prolonged desiccation or having been wet in the dark), photosynthates may first be used to build carbohydrate or energy reserves, or repair photodamage, and then secondly to service or rebuild the nitrogenase enzyme (Scherer et al. 1984, 1986; Fritz-Sheridan 1988; Dodds et al. 1995; Nash 1996). Consequently, during dry times of the year, or during very dry years, soil crusts may need to be wet longer before N fixation begins, or before reaching optimal N fixation rates, than during wet times of the year.

19.2.4 Temperature

Given sufficient moisture and light, N fixation rates are generally controlled by temperature. Most soil cyanobacteria and cyanolichens are capable of N fixation between –5 and 30 °C. Optimum temperatures reported generally range from 20–30 °C for crusts in the Arctic, Antarctica, Scotland, Canada, South Africa, and subalpine regions (Jones 1977b; Stewart et al. 1977; Coxson and Kershaw 1983a; Davey and Marchant 1983; DuBois and Kapustka 1983; Fritz-Sheridan 1988; Lennihan et al. 1994). However, studies of temperature optimum must be interpreted carefully. Constant hydration levels are difficult to maintain in material at higher temperatures and the relationship between tissue moisture and NA is seldom linear; thus, changes in thallus moisture need consideration (Kershaw 1985). Air, and not thallus, temperatures are often reported; however, thallus temperatures can be 15–25 °C warmer than air temperatures on sunny days, while close to the same temperature as air on cloudy days (Belnap 1995). Thus, results can be difficult to interpret if weather conditions are not reported. In addition, Coxson and Kershaw (1983b) warn that, although soil *Nostoc* showed an optimum at 35 °C air temperature for several hours, the long-term optimum for this species was 28 °C.

Minimum air temperatures for NA varies. Field measurements of NA in free-living and lichenized *Nostoc* (in *Leptogium*) have been recorded at –7.6 and 0 °C, respectively (Horne 1972; Davey and Marchant 1983). *Stigonema* and

Scytonema crusts showed no NA at 0 °C (Isichei 1980). Freezing can substantially affect NA rates. After freezing, recovery of NA is dependent on conditions before and after the freezing event. Freezing has the ability to damage nitrogenase in *Nostoc* (DuBois and Kapustka 1983; Scherer et al. 1984). Low temperatures can reduce photosynthetic rates and thus reduce available ATP and reductant pools, creating a lag time after freezing before N fixation is initiated (Kershaw 1985). When *Nostoc* was frozen for 3 days and then thawed at different temperatures, recovery of NA to prefreeze levels was much faster at 25 and 30 °C when compared to lower temperatures. Light exposure before freezing reduced recovery time, as well as protein synthesis inhibition, indicating the need for de novo nitrogenase production (DuBois and Kapustka 1983; Scherer et al. 1984).

Above a minimum temperature that is somewhat species-dependent, N fixation rates show a strong, positive response to increasing air temperature until an upper limit is reached, after which rates quickly decline. As seen in Fig. 19.1, *Collema* crusts in southeastern Utah showed potential NA values that were strongly related with average air temperatures of the previous 3 days (r^2=0.93) if crusts had received precipitation within 3 days prior to collection. With one exception, NA was observed only when air temperatures were between 0 and 26 °C. NA in *Scytonema-Microcoleus* and *Microcoleus* crusts from the same area was related to values obtained for *Collema* crusts (r^2=0.86), indicating that similar environmental conditions control NA for different crust types in this region. Similar upper temperature limits are reported for other species from other environments. *Nostoc-Microcoleus-Schizothrix* crust showed higher NA at 20 °C than at 39 °C (Englund 1978). *Stigonema* from Brazil and *Scytonema* from Nigeria showed no NA at 0 °C, maximum NA at 25–30 °C, and a sharp decline above 30 °C. *Scytonema* crusts from Nigeria showed maximal NA at 40 °C, declining to 0 at 45 °C (Isichei 1980). *Nostoc commune-Cylindrospermum* crust from Scotland showed very low NA at 0 °C, reaching maximum NA at 15-25 °C. A slight decline was seen at 35 °C. Subalpine *Tolypothrix* crusts showed NA at 5 °C, with maximum NA at 20–30 °C (Stewart et al. 1977; Fritz-Sheridan 1988). Loftis and Kurtz (1980) reported that NA ceased in mixed cyanobacterial crusts when soil-surface temperatures reached 45–50 °C. The thallus temperature of *Nostoc commune* was always below 38 °C when NA was observed (Coxson and Kershaw 1983b). Because higher temperatures inhibit N fixation, fixation in desert crusts is generally higher during cooler fall, winter, and spring temperatures. Conversely, low temperatures in the polar regions restrict N fixation in soil crusts to the summer season (Horne 1972; Davey and Marchant 1983).

19.2.5 Light

Many studies have examined the effects of light on N fixation of cyanobacteria and lichens. Unfortunately, many of these data are not directly comparable, as different light measurement units were used and/or moisture status, temperature, and pre-experimental conditions of organisms were not controlled. For example, pre-experimental conditions can influence distribution and concentrations of photosynthetic and UV-screening pigments, which, in turn, can greatly alter light attenuation and photosynthetic rates in cyanobacterial colonies and lichen tissue (Dodds 1989; Dodds et al. 1995; Garcia-Pichel and Belnap 1996). Light optima can also be expected to vary with soil characteristics and with cyanobacterial distribution within the substrate, colony, or thallus. In spite of these complications, we can draw some conclusions useful for understanding the effects of light on N fixation.

Soil cyanobacteria and cyanolichens generally reach light saturation for maximum NA at relatively low light levels (Fig. 19.2). Below this level, there is

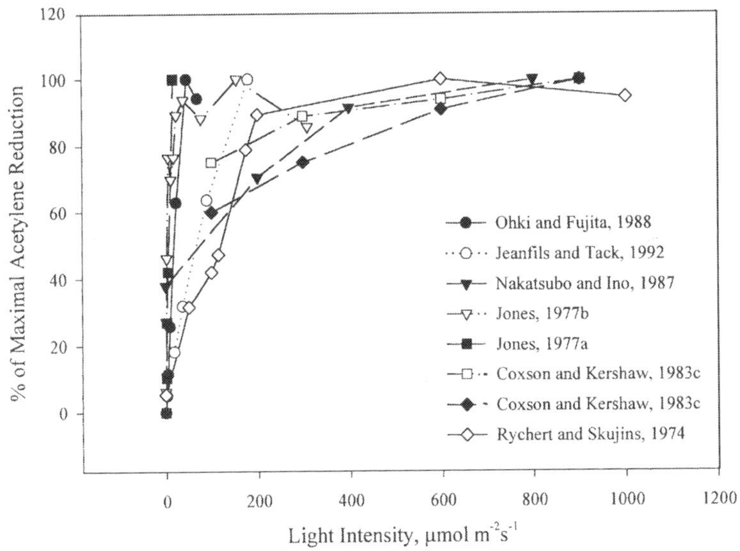

Fig. 19.2. The influence of light (PPFD) on nitrogen fixation activity in cyanobacterial soil crusts from the Arctic, Antarctic, tropics, Canada, and the Great Basin (USA). Different light intensity data were converted to photosynthetically active photon flux density using conversion factors for various light sources. However, differences in lamps and sensors make data conversion and comparison of experiments conducted with different light sources imprecise, and errors of +20 % are not uncommon (McCree 1981). As can be seen, optimal nitrogenase activity is obtained at low PPFD, and then maintained throughout the levels measured (Rychert and Skujins 1974; Jones 1977a,b; Coxson and Kershaw 1983c; Nakatsubo and Ino 1987; Ohki and Fujita 1988; Jeanfils and Tack 1992)

a general, but not always linear, correlation between light levels and nitrogen fixation rates, presumably due to the reduced availability of photosynthetic products to support N fixation (Kershaw 1985). Free-living and lichenized *Nostoc* can grow in very low light (see Chap. 16; Dodds et al. 1995; Garcia-Pichel and Belnap 1996). In many studies, *Nostoc* showed an increase in NA up to 100–150 μmol m^{-2} s^{-1} PPFD, after which maximal NA rates are maintained (Jones 1977b; DuBois and Kapustka 1983; Fritz-Sheridan 1988; Lennihan et al. 1994). Other species, including *Calothrix, Stigonema*, and *Scytonema*, also reach maximal NA rates at lower PPFD levels, which are maintained at higher light as well (Griffiths and Gallon 1987; Fritz-Sheridan 1987). Rychert et al. (1978) found that a *Microcoleus-Scytonema-Nostoc-Collema* crust reached optimal NA rates at 200 μmol m^{-2} s^{-1}; these rates were sustained at higher light as well. In contrast, two studies have found higher fixation at relatively higher light levels: *Nostoc* showed maximal NA at 400–800 μmol m^{-2} s^{-1} (Nakatsubo and Ino 1987), while Isichei (1980) found no saturation in *Scytonema* crusts from Nigeria, as NA increased up to 34 000 lx (ca. one third maximum sunlight), which was the highest light level tested.

Depression of N fixation by high light has been reported. However, most of these reports are from cyanobacteria in culture, aquatic habitats, marine mats, or cyanobacteria found on leaf surfaces (Smith 1984; Fernandes and Leganes 1990). Some studies in soil systems are often cited as evidence that high light can limit N fixation (for example, see Reddy and Giddens 1975; Jones 1977c; Rychert et al. 1978; Coxson and Kershaw 1983a). However, high air temperatures (>27 °C), low moisture content of the organisms, and/or failure to account for differential crustal biomass could also explain these results. Any suppression of NA at higher light levels has yet to be conclusively demonstrated.

19.2.6 Interaction of Temperature, Moisture, and Light

Based on the above discussion, environmental controls on N fixation are clearly hierarchical. As with photosynthesis (see Chap. 18), the ultimate control on the duration and rate of fixation is the moisture content of the organism. With sufficient hydration, the presence and accessibility of photosynthetic products and nitrogenase enzymes controls fixation activity. This requires access to either intracellular carbon stores (or adequate light to replenish carbon stores). With sufficient hydration and access to carbon stores, temperature then controls fixation rates until carbon stores are exhausted.

This hierarchy is shown clearly by laboratory studies done by Belnap (Fig 19.1), as well as field work done by Coxson and Kershaw (1983c; Fig 19.3). Both studies simultaneously recorded field temperatures, rainfall, thallus moisture, and NA in the soil crusts. Both data sets clearly show NA drops as

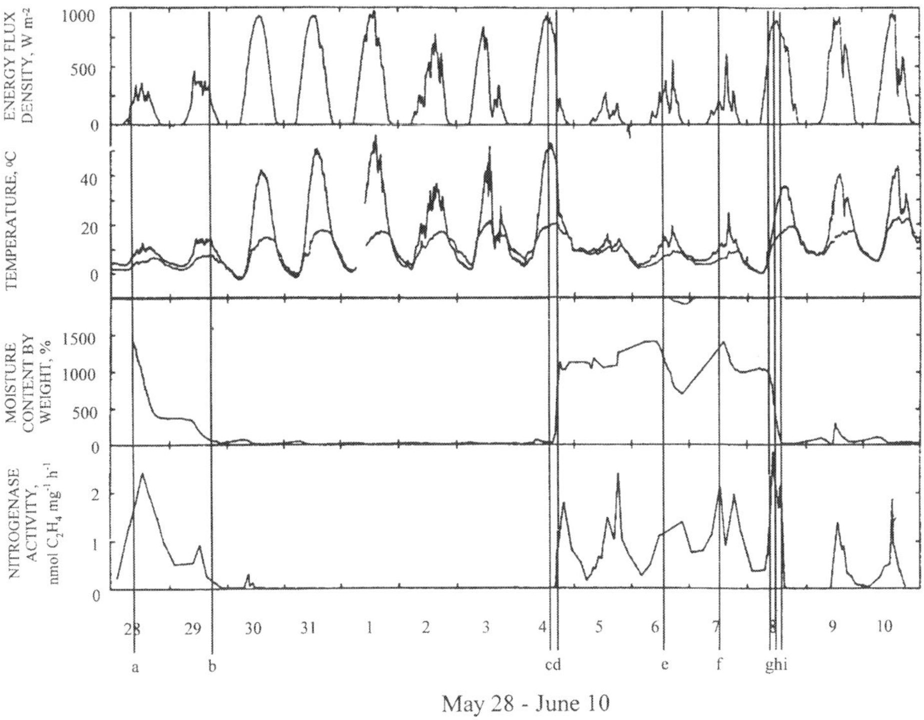

Fig. 19.3. Environmental controls on nitrogen fixation measured under field conditions in the Arctic. (After Coxson and Kershaw 1983c). The following interpretation demonstrates how ARA levels are determined in a hierarchical fashion. Dates are May 28 (5/28) through June 10 (6/10). *Top panel* Surface energy flux; *second panel* thalli (*top line*) and air (*bottom line*) temperature; *third panel* moisture content by weight of the soil crust (note smaller-scale *inset* at top); *bottom panel* measured NA. The following letters refer to the *vertical lines in the figure: a* as moisture contents drop (5/28–29), so does ARA, in spite of favorable temperatures and light; *b* from 5/29–6/4, the soil crusts are not wet and so no ARA is observed, although light and temperatures are favorable; *c* on 6/4, crusts are moist. However, temperatures are well above 30 °C, and little ARA is observed; *d* however, as soon as the temperature drops below 30 °C, ARA soars; *e–g* from 6/4–6/8, crusts are moist. Increases in temperatures and light increase ARA values, while drops in temperature and light are followed by drops in ARA, with some lag time noted; *h* on 6/8, crusts experience the most optimal conditions during this experiment (moisture 1000 %, air temperature 25 °C, light 800 W m^{-2}); this is reflected in the highest ARA recorded during this time; *i* shortly thereafter, the crusts dry out and ARA stops, in spite of favorable light and temperature

moisture drops, in spite of favorable temperatures and light. With sufficient hydration, rising temperatures initially increase NA values, while drops in temperature and light are followed by drops in NA. At air temperatures above 26–30 °C, little NA is observed. As soon as the temperature drops, NA is resumed. Although this overall pattern is clear in these two data sets, there are still some anomalous data points that resist explanation, indicating that refinement of these relationships is still needed.

19.2.7 Other Factors

Snow. Little or no NA has been reported from cyanobacteria or cyanolichens under snow (Englund and Meyerson 1974; Alexander and Kallio 1976; Crittenden and Kershaw 1979), despite the ability of nitrogenase to function unimpaired after thalli had been exposed to temperatures of –40 °C (Kershaw 1985). Laboratory experiments with material collected from under snow show normal NA rates when incubated with light and warmer temperatures, while material kept dark and warm showed no recovery of NA (Huss-Danell 1977; MacFarlane and Kershaw 1980). This would suggest that low light and temperature, resulting in low energy and reductant pools, are responsible for lack of NA under snow cover, not inactivation of the nitrogenase enzyme (Coxson and Kershaw 1983c; Kershaw 1985). Recovery of N fixation after freezing is discussed in Sect. 19.2.4 above.

Seasonality. Seasonality in NA rates of temperate soil crusts has been reported by multiple authors, with reported rates being lower during hot and cold periods (summer and winter), and higher in late spring and early fall (Stewart 1967; Johnson 1982; Huss-Danell 1978; Coxson and Kershaw 1983b; DuBois and Kapustka 1983; Kershaw 1985). Summer in temperate regions is generally a time of low moisture and high temperatures, factors that both limit metabolic activity time and lead to inactivation of nitrogen fixation. Low winter temperatures limit both carbon and N fixation. Early fall and late spring, on the other hand, are the times when soil moisture and moderate temperatures most often occur together, and therefore would be when both carbon and N fixation rates are the highest (Coxson and Kershaw 1983b; Kershaw 1985) or greatest cyanobacterial biomass (Jones 1981). Cyanobacterial biomass is greatest after warm (but not hot) wet periods, such as are found in late spring–early summer or early–late fall (Lynn and Cameron 1972; Johansen and Rushforth 1985; Johansen et al. 1993). These studies suggest that there is no endogenous component to nitrogen fixation, but that rates are dictated mostly by different combinations of moisture and temperature (Kershaw 1985).

Salinity and pH. There are few data available on the effect of salinity on N fixation in soil cyanobacteria. A study comparing species in saline soils showed that 15 N-fixing species belonging to six genera preferred soils with high electrical conductivity (70 mmhos cm^{-1}). Exceptions were *Microcoleus, Phormidium, Nostoc, Tolypothrix, Gloeotrichia,* and *Hapalosiphon,* which were found in soils with low conductivity (6.5 mmhos cm^{-1}; Ali and Sandhu 1972). Crust development was found to be greater on soils with higher electrical conductivity when compared to nearby soils with lower conductivity (62 vs. 39 mmhos cm^{-1} soil; Anderson et al. 1982).

Growth and N fixation in soil cyanobacteria is greatest at pH 7 or above (Brock 1973; Davey and Marchant 1983; DuBois and Kapustka 1983), although some depression of NA has been seen at pH 8–10 (Granhall 1970). Stewart et al. (1977) found NA in soil *Stigonema* increased from low rates at pH 4 to maximal rates at pH 8; a slight decrease was noted at pH 10. Isichei (1980) reported maximal NA in *Scytonema* crusts at pH 6–9, with declines at pH 4, 5, and 10. *Anabaena* and *Nostoc* can fix N down to pH 4, with an optimum at pH 7 or above (Granhall 1981). NA depression at low pH is likely due to reduced photosynthetic capacity (Shapiro 1973) or changes in the nitrogenase enzyme (Stewart et al. 1977). Soil crusts occur in soils with a wide range in pH. Based on the above information, crusts on alkaline soils (dominant in most deserts) are expected to fix more N per unit surface area than crusts on more acidic soils.

Soil Chemistry. Soil chemistry has been shown to affect N fixation in cyanobacteria. Most studies have focused on N compounds in the medium regulating NA. Elevated ammonium depresses N fixation, while mixed results have been obtained with elevated nitrate. However, similar results were obtained when NaCl was added, suggesting that a nonspecific salt effect may be partially responsible (Delwiche and Wijler 1956; Reddy and Giddens 1975; Klubek and Skujins 1980).

Lime additions have also been found to enhance NA on pH 6 soils (Reddy and Giddens 1975). However, it is not known whether the stimulation was due to increases in pH or Ca. Phosphorus additions have been shown to stimulate NA in cultures, in pot trials, and in field experiments, perhaps through a stimulation of ATP synthesis (Granhall 1981; Metting 1981; Fritz-Sheridan 1988; Dodds et al. 1995). N fixation in soil *Nostoc* (both free-living and lichenized by *Collema*) is stimulated by low concentrations of As, Ni, Pb, Pd, and Zn. However, longer-term exposure to Cd, Pb, and Zn inhibits fixation (Henriksson and DaSilva 1978). Alexander et al. (1978) report additions of Mb and Co stimulate NA. Boron, Mg, Mo, Va, and Fe are also important for N fixation. However, high levels of Fe, Al, Mn, Cu, Zi, and Hg adversely affect NA (Granhall 1981; Dodds et al. 1995).

Bacterial Associations. Bacterial associations may be essential for growth and N fixation in some species of cyanobacteria. Bacteria may scavenge oxygen and thus create anaerobic microzones, replenish CO_2, contribute bacterially fixed N, and secrete growth factors and vitamins required by the cyanobacteria (Paerl 1982). Bacteria inside the sheath of *Microcoleus vaginatus* isolated from Utah soil crusts appear to contribute fixed N (Steppe et al. 1996).

Associations with Mosses and Vascular Plants. Cyanobacteria can grow epiphytically on soil mosses and lichens. These cyanobacteria may be better protected from desiccation and high temperatures, both of which can limit N fixation activity. Moss tissue may also protect cyanobacteria from contact with low pH substrates. Mosses both secrete alkaline substances and use the high cation-exchange-capacity of their leaves to buffer the pH of leaf surfaces (Englund 1978; Davey and Marchant 1983; Smith 1984). Light, however, may be limiting, due to shading by the moss tissue (Smith 1984).

Interactions with vascular plants may inhibit or stimulate N fixation in cyanobacteria and cyanolichens. While overhanging plant cover can moderate desiccation and high temperatures and prolong fixation time, leachate from some shrubs may inhibit N fixation (Skujins and Klubek 1978). In addition, as vegetation limits light and soil space, crust cover, and therefore N inputs to the soils, declines (Kapustka and Rice 1978).

19.3 Nitrogen Release to the Surrounding Substrate

Extracellular release of nitrogenous compounds occurs in both cyanobacteria and lichens (Stewart 1970; Millbank 1982). It is especially pronounced in N-fixing organisms, which release 5–70 % of N fixed. In cultures of *Nostoc, Scytonema,* and *Anabaena*, the proportion of fixed N that is released is greater in young cultures than in older cultures, indicating that release is not due to cell turnover. Release is greater when pH, temperature, light, CO_2, or ionic conditions are not optimal, or during wetting after desication (Henriksson 1957; Jones and Stewart 1969a,b; Millbank 1982; Jeanfils and Tack 1992). Some of the liberated N is reassimilated by the cyanobacteria or lichen, but *Nostoc* has a limited ability to do so (Meeks et al. 1985).

Nitrogen release to the surrounding substrate has been documented in many diverse soil-crust organisms, including *Scytonema* from Nigeria, *Nostoc commune* from New Mexico (Dodds et al. 1995; J. Belnap and F. Garcia-Pichel, unpubl.), *Microcoleus-Nostoc-Scytonema-Collema* crusts from southeastern Utah (F. Garcia-Pichel and J. Belnap, unpubl.) and from northern Utah (Klubek et al. 1978), *Nostoc* soil crusts (Stewart 1967), *Collema tenax* from Europe (Henriksson 1957) and southern Utah (J. Belnap and F. Garcia-Pichel, unpubl.), and *Peltigera canina* (Millbank 1978). N compounds released include small amounts of amide, peptides, and free amino acids, with most as nitrate and ammonium. Released N is readily taken up by surrounding organisms, including vascular plants, fungi, actinomycetes, and bacteria (Stewart 1967; Jones and Stewart 1969a,b; Rogers and Burns 1994).

Many hypotheses have been advanced to explain extracellular release of N. Dodds et al. (1995) suggest that loss by outward diffusion may be unavoidable in low N environments. Millbank (1982) and Potts (1984) consider N release as an unavoidable result of membrane deformation with desiccation and rewetting, as is seen with ^{14}C, sucrose, fructose, mannitol, glucose, erythritol, glycerol, ribitol, and polyols in cyanobacteria, mosses, and lichens (Farrar 1976; Coxson et al. 1992). Others hypothesize that since N compounds can be toxic, release may be necessary if assimilation, conversion or storage capability in cells is limited (Stewart and Rogers 1977; Bergmann et al. 1992; Silvester et al. 1996). N release may also attract beneficial organisms.

19.4 Nitrogen Budgets

As discussed above, ample data demonstrate that biological soil crusts can contribute significant amounts of fixed nitrogen to the ecosystems in which they occur. However, very little is known about the relationship between gaseous N losses and the presence of biological soil crusts. Some studies report that denitrification losses may be significant in arid ecosystems (Peterjohn and Schlesinger 1991) and volatilization rates minimal relative to soil N pools. Klubek et al. (1978) reported volatilization rates were 0.4 % of fixed $^{15}N_2$ after weeks of incubation. Evans et al. report low rates in undisturbed *Scytonema-Nostoc-Collema* crusts (Chap. 20). Klubek et al. (1978) found 75–80 % of fixed $^{15}N_2$ was lost from dead, decomposing crusts (kept dark and wet for 10 weeks). Because volatilization was extremely low, they concluded the losses must have been from denitrification. N_2O emissions from three sites in the Sonoran Desert (Guilbault and Matthias 1998) showed fluxes ranging from −1.6 to 7.3 ng N_2O-N $m^{-2}\,s^{-1}$, with a mean flux of 2.4 N_2O-N $m^{-2}\,s^{-1}$. Rates of 0.02 ng N_2O-N $m^{-2}\,s^{-1}$ were observed in Great Basin soils (Mummey et al. 1994). However, it is not known if soil crusts were present in these latter two studies. Field measurement of gaseous losses in wetted, living soil crusts of southeastern Utah (K.T. Bradley et al., unpubl.) during fall and spring (when soil temperatures are relatively cool) show quite low denitrification and volatilization rates. N_2O emissions from either *Microcoleus* or *Scytonema-Nostoc-Collema* crusts peaked at 7.1 to 13.9 ng N_2O-N $m^{-2}\,s^{-1}$, while NO_x fluxes averaged 0.07 ng NO_x-N $m^{-2}\,s^{-1}$. However, summer measurements (when soil temperatures are relatively high, but dry rapidly) showed a relatively high NO_x flux, while N_2O fluxes were very low. As discussed previously in this chapter, most N inputs from crusts in this ecosystem occur in fall, winter, and spring, with only minimal fixation occurring during summer. Thus, in this ecosystem, N inputs and N losses appear decoupled in time. Nitrogen fixed and released appears readily available to plants and microbes from fall until summer (that is, most of the growing season), at which point much of the available N appears lost through volatilization. This would explain why plants growing in crusted soils have higher N than those growing in uncrusted soils, but, at the same time, why N in crusted soils does not accumulate over time.

A similar scenario might be expected in other deserts that receive most of their rain during cool periods. However, regions with rainfall dominant during warm times, or hot deserts where soils are warm most of the year, may be quite different. In these deserts, most N fixed and released by crusts is likely to occur at the same time as N losses are greatest. Thus, in these areas, higher N losses relative to N inputs, and less transfer of newly fixed N to plants and microbial biomass is a likely result, as volatilization and dentrification processes would be competing simultaneously with plants and microbes for the newly released N.

19.5 Conclusions

Arid and semiarid soils are generally N-limited and biological soil crusts can often be important sources of N input. For this reason, it is important to understand factors that influence N fixation and release in soil crusts. Given adequate soil nutrients and alkaline pH, the major factors controlling N-fixation rates are temperature, light, and moisture. As crust organisms are only active when wet, moisture is the most limiting factor. When sufficiently moist, access to intracellular carbon stores then controls fixation rates. Because of this reliance on carbon stores, N-fixation rates are highly dependent on conditions experienced prior to experimental measurements. With sufficient hydration and access to carbon stores, temperature controls fixation rates.

Other factors also influence N-fixation rates. High soil salinity, high soil N, and low pH can depress fixation rates. Phosphorus additions can stimulate fixation rates. Micronutrients have been shown to both increase and decrease fixation rates. Association with other species has also been shown to influence N-fixation rates. Bacterial symbionts may increase rates. Associations with mosses, lichens, and higher plants can both increase or decrease rates, depending on conditions.

Much of the N fixed by crustal species is released soon after fixation. Released N has been shown to be utilized by surrounding organisms, including vascular plants, fungi, actinomycetes, and bacteria. Little is known about denitrification or volatilization losses from biological soil crusts, but amounts appear to be very low, and decoupled in time from N inputs.

Acknowledgments. Sue Phillips, Sasha Reed, Michelle Schmid, and Otto L. Lange provided helpful comments and graphics.

References

Alexander V, Kallio S (1976) Nitrogenase activity in *Peltigera aphthosa* and *Stereocaulon paschale* in early spring. Rep Kevo Subarct Res Stn 13:12–15

Alexander V, Billington M, Schell DM (1978) Nitrogen fixation in arctic and alpine tundra. In: Tieszen LL (ed) Vegetation and production ecology of an Alaskan arctic tundra. Springer, Berlin Heidelberg New York, pp 539–558

Ali S, Sandhu GR (1972) Blue-green algae of the saline soils of the Punjab. Oikos 23:268–272

Anderson DC, Harper KT, Holmgren RC (1982) Factors influencing development of cryptogamic soil crusts in Utah deserts. J Range Manage 35:180–185

Belnap J (1995) Surface disturbances: their role in accelerating desertification. Environ Monit Assess 37:39–57

Belnap J (1996) Soil surface disturbances in cold deserts: effects on nitrogenase activity in cyanobacterial-lichen soil crusts. Biol Fertil Soils 23:362–367

Belnap J, Ojima D, Phillips S, Barger N (1999) Biological soil crusts of inner and outer Mongolia: impacts of grazing and precipitation on nitrogen cycles. In: Eldridge DJ, Tongway D, King G (eds) Proc 6th Int Rangeland Congress, Townsville, Australia

Bergmann B, Johansson C, Söderbäck E (1992) The *Nostoc-Gunnera* symbiosis. New Phytol 122:379–400

Brock TD (1973) Lower pH limits for existence of blue-green algae: evolutionary and ecological implications. Science 179:480–483

Coxson DS, Kershaw KA (1983a) The pattern of in situ summer nitrogenase activity in terrestrial *Nostoc commune* from *Stipa-Bouteloua* grassland, southern Alberta. Can J Bot 61:2686–2693

Coxson DS, Kershaw KA (1983b) Rehydration response of nitrogenase activity and carbon fixation in terrestrial *Nostoc commune* from *Stipa-Bouteloua* grassland. Can J Bot 61:2658–2668

Coxson DS, Kershaw KA (1983c) Nitrogenase activity during chinook snow-melt sequences by *Nostoc commune* in *Stipa-Bouteloua* grassland. Can J Bot 29:938–944

Coxson DS, McIntyre DD, Vogel HJ (1992) Pulse release of sugars and polyols from canopy bryophytes in tropical montane rain forest (Guadeloupe, French West Indies). Biotropica 24:121–133

Crittenden PD, Kershaw KA (1979) Studies on lichen-dominated systems. XXII. The environmental control of nitrogenase activity in *Stereocaulon paschale* in spruce-lichen woodland. Can J Bot 57:236–254

Davey A, Marchant HJ (1983) Seasonal variation in nitrogen fixation by *Nostoc commune* Vaucher at the Vestfold Hills, Antarctica. Phycologia 22:377–385

David KAV, Fay P (1977) Effects of long-term treatment with acetylene on nitrogen-fixing microorganisms. Appl Environ Microbiol 34:640–646

Delwiche CC, Wijler J (1956) Non-symbiotic nitrogen fixation in soil. Plant Soil 7:113–129

Dodds WK (1989) Microscale vertical profiles of N_2 fixation, photosynthesis, O_2, chlorophyll *a* and light in a cyanobacterial assemblage. Appl Environ Microbiol 55:882–886

Dodds WK, Gudder DA, Mollenhauer D (1995) The ecology of *Nostoc*. J Phycol 31:2–18

DuBois JD, Kapustka LA (1983) Biological nitrogen influx in an Ohio relict prairie. Am J Bot 70:8–16

Englund B (1978) Effects of environmental factors on acetylene reduction by intact thallus and excised cephalodia of *Peltigera aphthosa* Willd. Ecol Bull 26:234–246

Englund B, Meyerson H (1974) In situ measurements of nitrogen fixation at low temperatures. Oikos 25:283

Farrar JF (1976) The lichen as an ecosystem: observation and experiment. In: Brown DH, Hawksworth DL, Bailey RH (eds) Lichenology: progress and problems. Academic Press, New York, pp 385–406

Fernandes VE, Leganes F (1990) Regulatory effects of pH and incident irradiance on the level of nitrogenase activity in the cyanobacterium *Nostoc*. J Plant Physiol 135:623–627

Fritz-Sheridan RP (1987) Nitrogen fixation on a tropical volcano, La Soufriere. I. Nitrogen fixation by *Scytonema* sp. and *Stereocaulon virgatum* Ach. during colonization of phreatic material. Biotropica 19:297–300

Fritz-Sheridan RP (1988) Physiological ecology of nitrogen fixing blue-green algal crusts in the upper-subalpine life zone. J Phycol 24:302–309

Garcia-Pichel F, Belnap J (1996) Microenvironments and microscale productivity of cyanobacterial desert crusts. J Phycol 32:774–782

Granhall O (1970) Acetylene reduction by blue-green algae isolated from Swedish soils. Oikos 21:330–334

Granhall U (1981) Biological nitrogen fixation in relation to environmental factors and functioning of natural ecosystems. In: Clark FE, Rosswall T (eds) Terrestrial nitrogen cycles. Ecol Bull 33, Swedish Nat Res Council, Stockholm, pp 131–144

Griffiths MSH, Gallon JR (1987) The diurnal pattern of dinitrogen fixation by cyanobacteria in situ. New Phytol 107:649–657

Guilbault M R, Matthias A D (1998) Emissions of N_2O from Sonoran desert and effluent-irrigated grass ecosystems. J Arid Environ 38:87–98

Harper KT, Marble JR (1988) A role for nonvascular plants in management of arid and semiarid rangelands. In: Tueller PT (ed) Vegetation science applications for rangeland analysis and management. Kluwer Academic Publishers, Dordrecht, pp 135–169

Henriksson E (1957) Studies in the physiology of the lichen *Collema*. I. The production of extracellular nitrogenous substances by the algal partner under various conditions. Physiol Plant 10:943–948

Henriksson E, DaSilva EJ (1978) Effects of some inorganic elements on nitrogen fixation in blue-green algae and some ecological aspects of pollution. Z Allg Mikrobiol 18: 487–494

Horne AJ (1972) The ecology of nitrogen fixation on Signy Island, South Orkney Islands. Bull Br Antarct Surv 27:1–18

Huss-Danell K (1977) Nitrogen fixation by *Stereocaulon paschale* under field conditions. Can J Bot 55:585–592

Huss-Danell K (1978) Seasonal variation in the capacity for nitrogenase activity in the lichen *Stereocaulon paschale*. New Phytol 81:89–98

Isichei AO (1980) Nitrogen fixation by blue-green algae soil crusts in Nigerian savanna. In: Rosswall T (ed) Nitrogen cycling in West African ecosystems. NFR, Stockholm, pp 191–199

Jeanfils J, Tack JP (1992) Identification and study of growth and nitrogenase activity of nitrogen-fixing cyanobacteria from tropical soil. Vegetatio 103:59–66

Jeffries DL, Klopatek JM, Link SO, Bolton H Jr (1992) Acetylene reduction by cryptogamic crusts from a blackbrush community as related to resaturation and dehydration. Soil Biol Biochem 24:1101–1105

Johansen JR, Rushforth SR (1985) Cryptogamic soil crusts: seasonal variation in algal populations in the Tintic Mountains, Juab County, Utah, USA. Great Basin Nat 45:14–21

Johansen JR, Ashley J, Rayburn WR (1993) Effects of rangefire on soil algal crusts in semiarid shrub-steppe of the lower Columbia Basin and their subsequent recovery. Great Basin Nat 53:73–88

Johnson GV (1982) Nitrogen fixation by soil crusts in a Chihuahuan desert ecosystem. N M J Sci 22:23

Jones K (1977a) Acetylene reduction by blue-green algae in sub-tropical grassland. New Phytol 78:421–426

Jones K (1977b) Acetylene reduction in the dark by mats of blue-green algae in sub-tropical grassland. Ann Bot 41:807–812

Jones K (1977 c) The effects of light intensity on acetylene reduction by blue-green algal mats in sub-tropical grassland. New Phytol 78:427–431

Jones K (1977d) The effects of moisture on acetylene reduction by mats of blue-green algae in sub-tropical grassland. Ann Bot 41:801–806

Jones K (1981) Diurnal acetylene reduction by mats of blue-green algae in sub-tropical grassland: use of short-term and long-term in situ assays. New Phytol 88:73–78

Jones K (1989) Interactions between desiccation and dark nitrogen fixation in tropical *Nostoc commune*. New Phytol 113:1–5

Jones K, Stewart WDP (1969a) Nitrogen turnover in marine and brackish habitats. III. Production of extracellular N by *Calothrix*. J Mar Biol Assoc UK 49:475–488

Jones K, Stewart WDP (1969b) Nitrogen turnover in marine and brackish habitats. IV. Uptake of the extracellular products of the nitrogen fixing algae *Calothrix*. J Mar Biol Assoc UK 49:701–716

Kapustka LA, Rice EL (1978) Symbiotic and asymbiotic N_2-fixation in a tall grass prairie. Soil Biol Biochem 10:553–554

Kershaw KA (1985) Physiological ecology of lichens. Cambridge University Press, London

Klubek B, Skujins J (1980) Heterotrophic nitrogen fixation in arid soil crusts. Soil Biol Biochem 12:229–236

Klubek B, Eberhardt PJ, Skujins J (1978) Ammonia volatilization from Great Basin Desert soils. In: West NE, Skujins JJ (eds) Nitrogen in desert ecosystems. Dowden, Hutchinson and Ross, Stroudsburg, pp 107–129

Lange OL, Belnap J, Reichenberger H (1998) Photosynthesis of the cyanobacterial soil-crust lichen *Collema tenax* from arid lands in southern Utah, USA: role of water content on light and temperature responses of CO_2 exchange. Funct Ecol 12:195–202

Lennihan R, Chapin DM, Dickson LG (1994) Nitrogen fixation and photosynthesis in high arctic forms of *Nostoc commune*. Can J Bot 72:940–945

Liengen L (1999) Conversion factor between acetylene reduction and nitrogen fixation in free-living cyanobacteria from high arctic habitats. Can J Microbiol 45:223–229

Loftis SG, Kurtz EB (1980) Field studies of inorganic nitrogen added to semiarid soils by rainfall and blue-green algae. Soil Sci 129:150–155

Lynn RI, Cameron RE (1972) The role of algae in crust formation and nitrogen cycling in desert soils. Research Memorandum RM 73–40, vol 3, International Biome Programme, Utah State University 2.3.4.6:1–10, Logan, UT

MacFarlane JD, Kershaw KA (1980) Physiological-environmental interactions in lichens. IX. Thermal stress and lichen ecology. New Phytol 84:669–685

McCree KJ (1981) Photosynthetically active radiation. In: Lange OL, Nobel PS, Osmond

CB, Ziegler H (eds) Physiological plant ecology I: responses to the physical environment. Springer, Berlin Heidelberg New York, pp 41–55

Meeks JC, Enderlin CS, Joseph CM, Chapman JS, Lollar MWL (1985) Fixation of $[^{15}N]N_2$ and transfer of fixed nitrogen in the *Anthoceros-Nostoc* symbiotic association. Planta 164:406–414

Metting FB (1981) The systematics and ecology of soil algae. Bot Rev 47:195–312

Millbank JW (1978) The contribution of nitrogen fixing lichens to the nitrogen status of their environment. Environmental role of nitrogen-fixing blue-green algae and asymbiotic bacteria. Ecol Bull 26, Swedish Nat Res Council, Stockholm, pp 260–265

Millbank JW (1982) The assessment of nitrogen fixation and throughput by lichens. III. Losses of nitrogenous compounds by *Peltigera membranacea*, *P. polydactyla* and *Lobaria pulmonaria* in simulated rainfall episodes. New Phytol 97:229–234

Mummey DL, Smith JL, Bolton H (1994) Nitrous oxide flux from a shrub-steppe ecosystem: sources and regulation. Soil Biol Biochem 26:279–286

Nakatsubo T, Ino Y (1987) Nitrogen cycling in an Antarctic ecosystem. 2. Estimation of the amount of nitrogen fixation in a moss community on East Ongul Island. Ecol Res 2:31–40

Nash TH III (1996) Lichen biology. Cambridge University Press, Cambridge

Ohki K, Fujita Y (1988) Aerobic nitrogenase activity measured as acetylene reduction in the marine non-heterocystous cyanobacterium *Trichodesmium* grown under artificial conditions. Mar Biol 98:111–114

Paerl HW (1978) Light-mediated recovery of N_2-fixation in the blue-green algae *Anabaena* spp. in O_2-supersaturated waters. Oecologia 32:135–139

Paerl HW (1982) Interactions with bacteria. In: Carr NG, Whitton BA (eds) The biology of cyanobacteria. University of California Press, Berkeley, pp 441–461

Paerl HW (1990) Physiological ecology and regulation of N_2 fixation in natural waters. Adv Microb Ecol 11:305–343

Paul EA, Clark FE (1996) Soil microbiology and biochemistry. Academic Press, London

Peterjohn WT, Schlesinger WH (1991) Factors controlling denitrification in a Chihuahuan desert ecosystem. Soil Soc Am J 55:1694–1701

Peters GA, Toia RE Jr, Calvert HE, Marsh BH (1986) Lichens to *Gunnera* – with emphasis on *Azolla*. Plant Soil 90:17–34

Potts M (1980) Blue-green algae (Cyanophyta) in marine coastal environments of the Sinai Peninsula: distribution, zonation, stratification and taxonomic diversity. Phycologia 19:60–73

Potts M (1984) Nitrogen fixation in mangrove forests. In: Dov Por F, Dor I (eds) Hydrobiology of the Mangal: the ecosystems of the mangrove forests. Dr W Junk, The Hague, pp 155–162

Potts M, Olie JJ, Nickels JS, Parsons J, White DC (1987) Variation in phospholipid ester-linked fatty acids and carotenoids of desiccated *Nostoc commune* (cyanobacteria) from different geographic locations. Appl Environ Microbiol 53:4–9

Rai AN, Rowell P, Stewart WDP (1981) Nitrogenase activity and dark CO_2 fixation in the lichen *Peltigera aphthosa* Willd. Planta 151:256–264

Reddy GB, Giddens J (1975) Nitrogen fixation by algae in fescuegrass soil crusts. Soil Sci Soc Am Proc 39:654–656

Rogers L, Gallon J (1988) Biochemistry of the algae and cyanobacteria. Clarendon Press, Oxford

Rogers SL, Burns RG (1994) Changes in aggregate stability, nutrient status, indigenous microbial populations, and seedling emergence following inoculation of soil with *Nostoc muscorum*. Biol Fertil Soils 18:209–215

Rychert RC, Skujins J (1974) Nitrogen fixation by blue-green algae-lichen crusts in the Great Basin desert. Soil Sci Soc Am Proc 38:768–771

Rychert RC, Skujins J, Sorensen D, Porcella D (1978) Nitrogen fixation by lichens and free-living microorganisms in deserts. In: West NW, Skujins J (eds) Nitrogen in desert ecosystems. Dowden, Hutchinson, Ross, Stroudsburg, PA, pp 20–30

Scherer S, Ernst A, Chen TW, Böger P (1984) Rewetting of drought-resistant blue-green algae: time course of water uptake and reappearance of respiration, photosynthesis and nitrogen fixation. Oecologia 62:418–423

Scherer S, Chen TW, Böger P (1986) Recovery of adenine-nucleotide pools in terrestrial blue-green algae after prolonged drought periods. Oecologia 68:585–588

Shapiro J (1973) Blue-green algae: why they become dominant. Science 179:382–384

Silvester W, Parsons R, Watt P (1996) Direct measurement of release and assimilation of ammonia in the *Gunnera-Nostoc* symbiosis. New Phytol 132:617–625

Skujins J, Klubek B (1978) Nitrogen fixation and cycling by blue-green algae-lichen-crusts in arid rangeland soils. In: Granhall U (ed) Environmental role of nitrogen-fixing blue-green algae and asymbiotic bacteria. Ecol Bull 26, Swedish Nat Res Council, Stockholm, pp 164–171

Smith VR (1984) Effects of abiotic factors on acetylene reduction by cyanobacteria epiphytic on moss. Appl Environm Microbiol 48:594–600

Steppe TF, Olson JB, Paerl HW, Litaker RW, Belnap J (1996) Consortial N_2 fixation: a strategy for meeting nitrogen requirements of marine and terrestrial cyanobacterial mats. FEMS Microbiol Ecol 21:149–156

Stewart WDP (1967) Transfer of biologically fixed nitrogen in a sand dune slack region. Nature 214:603–604

Stewart WDP (1970) Algal fixation of atmospheric nitrogen. Plant Soil 32:555–588

Stewart WDP, Rogers GA (1977) The cyanophyte-hepatic symbiosis. II. Nitrogen fixation and the interchange of nitrogen and carbon. New Phytol 78:459–471

Stewart WDP, Sampaio MJ, Isichei AO, Sylvester-Bradley R (1977) Nitrogen fixation by soil algae of temperate and tropical soils. In: Döbereiner J, Burris RH, Hollaender A, Franco AA, Neyra CA, Scott DB (eds) Limitations and potentials for biological nitrogen fixation in the tropics. Plenum Press, New York, pp 41–63

20 Biological Soil Crusts and Ecosystem Nitrogen and Carbon Dynamics

R.D. Evans and O.L. Lange

20.1 Introduction

Biological soil crusts are found in diverse ecosystems. They reach their best development in deserts and polar/alpine regions where harsh environments limit growth by higher plants allowing the ground surface to be exposed to direct sunlight (see Chap. 1). Coverage of soil crusts in these ecosystems can approach 100% in plant interspaces and total coverage may exceed that of higher plants (Kleiner and Harper 1972; Harper and Marble 1988). The capability of the dominant species in soil crusts to conduct photosynthesis and nitrogen (N) fixation, coupled with their high coverage, has led many researchers to conclude that they are dominant components of ecosystem carbon (C) and N cycles (Harper and Marble 1988; West 1990; Evans and Johansen 1999). However, relatively few studies have assessed the role of biological soil crusts in ecosystem C and N dynamics, and thus it is difficult to make broad generalizations regarding their role in these dynamics, due to variation in crust species composition and environments both within and between ecosystems (West 1990).

20.2 Nitrogen Dynamics

20.2.1 Nitrogen Input

The composition of biological soil crusts is diverse, and only those crusts with cyanobacteria or cyanolichens are capable of N fixation (see Chap. 19). Soil crusts composed of green algae, phycolichens, and mosses are common in many regions of the world, but these organisms are not capable of N fixation. In addition, ways of measuring N fixation vary (see discussion below). It is therefore not surprising that estimates of N input into ecosystems from biological soil crusts exhibit considerable variation (Evans and Johansen 1999).

Ecological Studies, Vol. 150
J. Belnap and O.L. Lange (eds.) Biological Soil Crusts: Structure, Function, and Management

Rychert et al. (1978) reviewed N input for several ecosystems. Estimates for the Sonoran Desert were 7 to 11 kgN ha^{-1} a^{-1} (Mayland et al. 1966) and 13 to 18 kgN ha^{-1} a^{-1} (MacGregor and Johnson 1971). In contrast, values for the Great Basin range from 10 to 100 kgN ha^{-1} a^{-1} (Rychert and Skujins 1974; Rychert et al. 1978; West and Skujins 1977). Estimates of Jeffries et al. (1992), ranged from 0.7 to 3.6 kgN ha^{-1} a^{-1} for the Colorado Plateau. Few studies have been conducted outside of North America. Rychert et al. (1978) estimate that annual input for a crust in Australia is 1.3 kgN ha^{-1} a^{-1}, while Isichei (1980) measured rates of N fixation of 3.3 to 9.2 kg ha^{-1} a^{-1} in Nigerian savannas. Liengen (1999), using field-applied ^{15}N, estimates arctic crusts can fix up to 7–10 kgN ha^{-1} day^{-1}.

One variable often not accounted for is the spatial variation in N fixation due to different species composition. In the Mojave Desert, nitrogenase activity for mixed-lichen crusts was over 400 % greater than *Microcoleus*- or moss-dominated crusts, although the different types were found within meters of each other (Fig. 20.1). Evans and Belnap (1999) observed similar differences among different crust communities on the Colorado Plateau; nitrogenase activity for *Collema*-dominated crusts had rates of 50.5±9.2, compared to 26.7±6.2, and 12.7±2.6 nmol C_2H_2 m^{-2} h^{-1} for crusts dominated by *Microcoleus* and moss, respectively (see Chap. 19). Jeffries et al. (1992) showed maximum rates of N fixation for *Microcoleus*-dominated crusts were 6.3 nmol N m^{-2} s^{-1}, while rates for *Scytonema* crusts were 9.4 nmol N m^{-2} s^{-1}. These authors also accounted for seasonal variation in climate by using extremes of 50- and 180-day growing seasons. Such variable results emphasize the importance of accounting for variation in crust composition, density and seasonal variation when estimating total rates of N input.

Methods of measuring N fixation have changed over time. Early experiments were done with ^{15}N, giving absolute rates of N fixation. However, due to cost and difficulty, most researchers now use the acetylene reduction technique. This technique is useful for relative rates of nitrogenase activity, but many authors caution that it is difficult to extrapolate these values to total

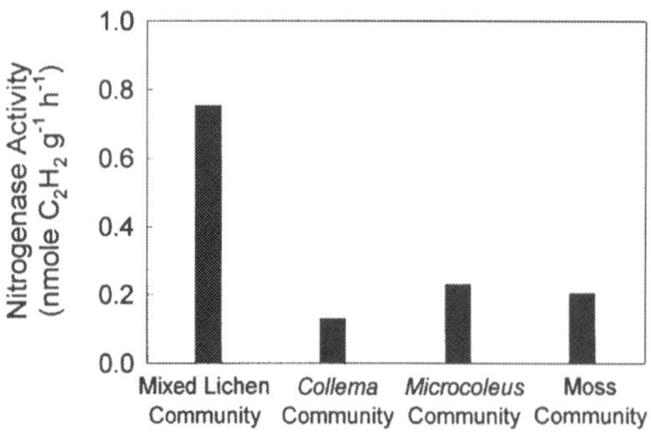

Fig. 20.1. Nitrogenase activity of four different biological soil-crust communities at the Nevada Desert FACE Facility (NDFF). Each community is named for characteristic components of the crust. Each *bar* represents the mean of five observations

ecosystem N inputs (West 1990). Multiple studies have shown the conversion ratio of ethylene to N is highly variable (0.1 to 6 for terrestrial species), depending on species and environmental conditions (see Chap. 19). Liengen (1999) recently showed that while cyanobacteria (*Anabaena*) in culture had a conversion ratio close to the theoretical value of 3, *Nostoc commune* living in soils and on the soil surface had conversion ratios of 0.02 to 0.07 and 0.11 to 0.44, respectively, depending on environmental conditions. He hypothesized that the ratio varied according to the amount of polysaccharide material present that affected diffusion of acetylene into, and ethylene out of, the fixation site. Thus, without calibration by ^{15}N studies, absolute rates of N inputs cannot be calculated with confidence.

The relative contribution of N fixation to the total N in biological soil crusts can be estimated by measuring natural variation in N isotope composition (Shearer and Kohl 1986). Organisms that assimilate 100 % of their N from N fixation will have $\delta^{15}N$ values between −2 and 0 ‰ (Shearer and Kohl 1986). In contrast, organisms that rely on inorganic N in the soil often have $\delta^{15}N$ values greater than 2 ‰. Evans and Belnap (1999) found a strong negative correlation between soil crust $\delta^{15}N$ and nitrogenase activity for three soil-crust communities in Canyonlands National Park. Crusts with the greatest nitrogenase activity had $\delta^{15}N$ values closest to 0 ‰, indicating relatively greater input from N fixation (Fig. 20.2A). A survey of crust $\delta^{15}N$ values found large variation in the importance of N fixation (Fig. 20.2B). Crusts samples were collected from throughout North America, and at sites in Africa and Australia (Fig. 20.3). Twenty-four of the samples had $\delta^{15}N$ values from −2 to 2 ‰, indicating significant input from N fixation for these crust communities (Fig.

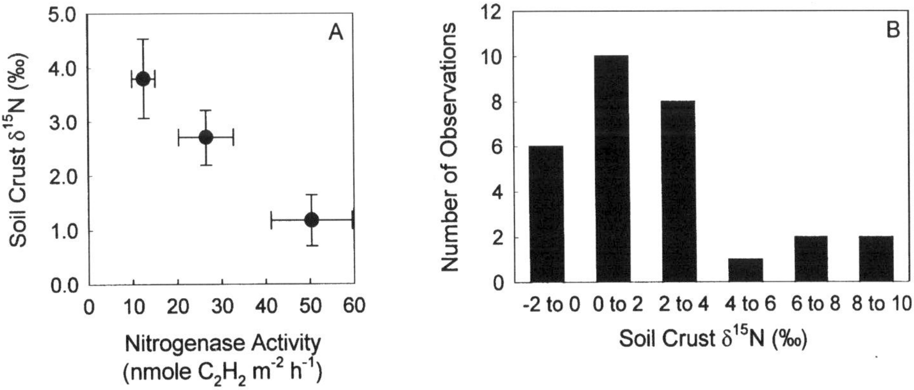

Fig. 20.2. A Correlation between biological soil-crust nitrogen isotope composition ($\delta^{15}N$) and nitrogenase activity. Each *point* represents the mean (±SE) of five observations. (Based on data presented in Evans and Belnap 1999). **B** The $\delta^{15}N$ of biological soil crusts from ecosystems presented in Fig. 20.3

Fig. 20.3. Location of collection sites for biological soil-crust $\delta^{15}N$ values presented in Fig. 20.2. The locations are: *1* Idaho National Engineering Lab; *2* Dugway Proving Ground; *3* Mono Lake; *4* Nevada Desert FACE Facility; *5* Zion National Park; *6* Bryce Canyon National Park; *7* Coral Pink Sand Dunes State Park; *8* Canyonlands National Park; *9* Mesa Verde National Park; *10* Joshua Tree National Monument; *11* Yuma Proving Ground; *12* Organ Pipe National Monument; *13* White Sands; *14* Fort Bliss; *15* Ozark Highlands; *16* Big Bend National Park

20.2B). In contrast, N fixation is not likely to be an important source of N input for the five communities with $\delta^{15}N$ values greater than 4‰. There was no pattern of $\delta^{15}N$ values among sites, again emphasizing the importance of identifying the species composition of the crust communities.

Evans and Ehleringer (1993) quantified the relative contributions of atmospheric deposition and N fixation by biological soil crusts to the total ecosystem N budget in a semiarid woodland by measuring soil N content and $\delta^{15}N$. Soil N content decreases during decomposition because of mineralization and subsequent gaseous loss by volatilization, nitrification, and denitrification. The loss of N is accompanied by an increase in soil $\delta^{15}N$, because each of these processes discriminates against ^{15}N causing an enrichment of ^{15}N in the remaining soil organic matter over time. As soil organic matter decomposes, it migrates down through the soil profile causing an overall decrease in soil N content, and an increase in soil $\delta^{15}N$, with depth

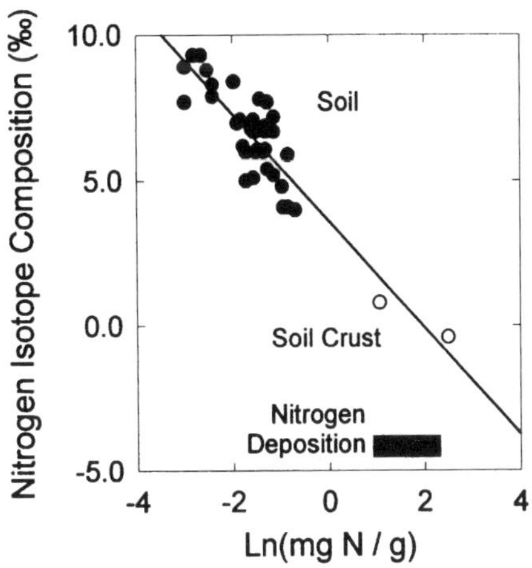

Fig. 20.4. The nitrogen isotope composition versus ln (nitrogen content) for soil, biological soil crusts, and nitrogen deposition located in a pinyon-juniper community in southern Utah. *Points* for biological soil crusts represent the mean of five observations. Values for atmospheric deposition were determined by measuring the nitrogen isotope composition of mosses growing on rocks whose only source of nitrogen is deposition. (Based on data in Evans and Ehleringer 1993 and R.D. Evans, unpublished data)

(Nadelhoffer and Fry 1988). The corresponding changes in soil N content and $\delta^{15}N$ have been shown to follow Rayleigh distillation kinetics in soils without horizon development. The dominant N source (N fixation or atmospheric deposition) will be the one that is found closest to the predicted relationship because it represents the origin for soil organic N. The $\delta^{15}N$ of N assimilated during N fixation will retain its original isotopic composition of 0 ‰ (Shearer and Kohl 1986), while the $\delta^{15}N$ of atmospheric deposition is often lower than −3 ‰ (Heaton 1986; Evans and Belnap 1999). Evans and Ehleringer (1993) plotted soil values from the woodland and found the predicted linear relationship (Fig. 20.4). Values for atmospheric deposition were estimated by measuring the N content and isotope composition of nonfixing species growing on rocks, so their only possible source of N was from deposition. Based on these results, N fixation by biological soil crusts represents the dominant source of N input into this ecosystem; in contrast to atmospheric deposition, values for soil crusts were found directly along the linear relationship established by soil $\delta^{15}N$ and nitrogen concentration. Their conclusion was that N fixation by the biological soil crust was the dominant form of N input into this ecosystem.

20.2.2 Nitrogen Transformations

Nitrogen input from biological soil crusts can influence ecosystem N cycles through leakage of N from organisms comprising the crust, or through the decomposition of crust material. Direct leakage of N assimilated during N fixation can be significant. Mayland et al. (1966) observed that 1 to 2 % of the total N in soil crusts can be located in the extracellular N pool, and Kershaw (1985) reports that 19 to 28 % of the N supplied as a ^{15}N label was subsequently leaked into the surrounding medium. Klubek et al. (1978) also exposed biological soil crusts to a ^{15}N label and observed that as much as 20 % of the label leaked to the extracellular NH_4^+ pool. While the experiments of Klubek et al. (1978) were conducted under conditions not conducive to crust activity, they nevertheless point out the potential for significant N leakage from these organisms.

Laboratory experiments have demonstrated that N transfer can occur directly from biological soil crusts to vascular plants, fungi, actinomycetes, and bacteria (Mayland et al. 1966; Mayland and McIntosh 1966; Stewart 1967; Jones and Stewart 1969a,b; Rogers and Burns 1994). Field experiments are necessary to test if these results are applicable to natural ecosystems where ammonium and nitrate can be lost by volatilization, nitrification, or denitrification (Evans and Johansen 1999). Plants growing in association with biological soil crusts can have higher concentrations of N than plants in uncrusted soils (Harper and Pendleton 1993; Belnap 1995; Belnap and Harper 1995). Experiments are needed to test if this is the direct result of N transfer from biological soil crusts to higher plants, or if biological soil crusts enhance root growth or N transformations by modifying the soil microenvironment (Harper and Pendleton 1993).

Soil N content is generally greater on soils with biological soil crusts (Fletcher and Martin 1948; Cameron and Fuller 1960; Callison and Brotherson 1985; Evans and Belnap 1999). Some of these comparisons have been made between disturbed and undisturbed sites, and observed differences may be due to greater N loss due to disturbance (Kleiner and Harper 1972; Evans and Ehleringer 1993, 1994; Evans and Belnap 1999) rather than greater nitrogen input on sites with soil crusts. The greatest rates of N transformation often occur under vascular plants (with or without soil crusts) rather than in crusted interspaces, as nutrients from the decomposition of crust and plant material are often redistributed by wind, runoff, and plant uptake to areas underneath the plants (Jackson and Caldwell 1993; Schlesinger et al. 1996; Bolton et al. 1990; Zaady et al. 1996). Nitrogen mineralization, C and N concentrations, and microbial biomass are also generally greater under higher plants than adjacent crusted interspaces (Bolton et al. 1990; Evans and Ehleringer 1994). The litter found in these islands of fertility may actually conserve N that was assimilated by biological soil crusts because immobilization prevents gaseous N loss (Zaady et al. 1996).

20.2.3 Gaseous Nitrogen Loss

Gaseous loss of nitrogen from biological soil crusts may be significant. West and Skujins (1977) summarized the nitrogen cycle for a cold-desert ecosystem and estimate that 80 % of the nitrogen assimilated by soil crusts is subsequently lost to the atmosphere during nitrification, denitrification, and volatilization. However, this number most probably is an overestimate (see below). Biological soil crusts provide ideal environments for denitrification following rain events. Water content, and carbon and NO_3^- availability (Peterjohn and Schlesinger 1991) determine rates of denitrification in desert ecosystems. Biological soil crusts can exhibit high rates of leakage of inorganic and amino acid nitrogen into extracellular spaces, and the gelatinous sheaths of cyanobacteria would provide a carbon source for denitrifying organisms. Microbial activity during the dark also creates anoxic conditions that would promote denitrification (Garcia-Pichel and Belnap 1996).

In spite of its potential importance, few studies have quantified denitrification from biological soil crusts (Evans and Johansen 1999). West and Skujins' (1977) estimate of gaseous N loss was based on work by Klubek et al. (1978), who applied a ^{15}N label to soil crusts and found that 75 to 80 % was lost over a 10-week period. Label in inorganic nitrogen pools and NH_3 from volatilization did not account for the loss, and their conclusion was that the loss must have occurred during denitrification. Denitrification was not measured in this experiment; rather, the goal was to understand decomposition of biological soil crusts. Algal-lichen crusts were allowed to decompose in total darkness over a 10-week period, but crust organisms are usually perennial and do not decompose suddenly under anaerobic conditions, so this may not represent conditions that occur in the field. It is clear that direct measurements of denitrification from intact biological soil crusts under field conditions are needed to address this important question.

There have also been very few measurements of volatilization from biological soil crusts (Evans and Johansen 1999). Volatilization of NH_3 is favored when temperature, NH_4^+ concentration, and pH are high (Haynes and Sherlock 1986; Schlesinger and Peterjohn 1991). It is possible that volatilization is an important process in biological soil crusts because NH_4^+ can leak into the extracellular pool, and pH can increase over two units in the top 1 mm of the crust during the day (Garcia-Pichel and Belnap 1996). Klubek et al. (1978) found that volatilization was not significant in laboratory experiments, but NH_4^+ was probably not available because the crusts were grown in the dark thereby preventing photosynthesis and ultimately nitrogen fixation. In contrast, considerable nitrogen volatilization can be observed under field conditions (Fig. 20.5). Removal of biological soil crusts resulted in a 68 % decrease in NH_3 loss. The greater rates of volatilization for soils with crusts were most likely due to leakage of NH_4^+ from the crusts. Both NH_4^+ and NO_3^- concentrations were significantly greater for the soil + crust plots, while no differences were observed for soil moisture, soil temperature, or pH.

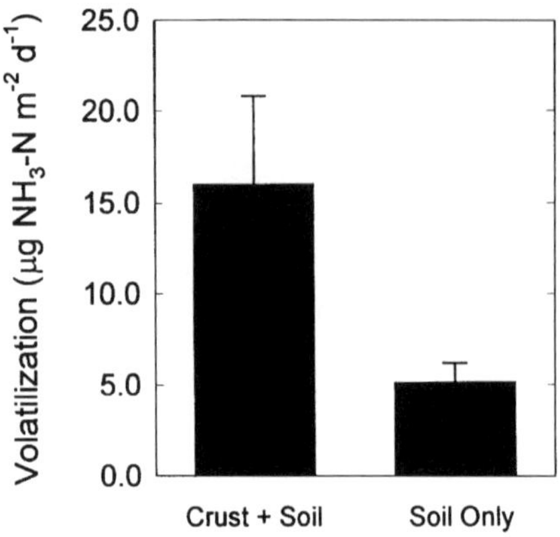

Fig 20.5. Nitrogen volatilization from plots located on the Colorado Plateau. *Crust + Soil* signifies intact plots, while biological soil crusts were removed from *Soil Only* plots. Each *bar* represents the mean (±SE) of ten observations

20.3 Carbon Dynamics

20.3.1 Seasonal Courses of Environmental Conditions and Carbon Exchange of Soil-Crust Lichens: a Case Study

There is only one long-term study available in the literature in which microclimatic conditions and CO_2 exchange of soil-crust organisms were monitored under at least semi-natural conditions (Lange and Green 1996; Lange 2000a,b). This study was restricted to soil lichens and was conducted under temperate, Central-European conditions, where communities of biological soil crusts colonize open patches between the phanerogamous vegetation in local steppe formations (see Chap. 6).

The experimental species *Cladonia convoluta, Diploschistes muscorum, Fulgensia fulgens,* and *Squamarina lentigera* are common European and Mediterranean soil phycolichens. The cyanobacterial *Collema cristatum* grows on calcareous boulders, small stones and, together with *Co. tenax*, on soils of the same community. The epilithic, green algal *Lecanora muralis* colonizes sandstone slabs and was included in the monitoring program. All but *Cl. convoluta* also occur in North American soil-crust communities (see Chap. 2). Microclimate and CO_2 exchange of these lichens were continuously recorded in a field habitat (Botanical Garden Würzburg, northern Bavaria) using two automatic klapp-cuvettes (Lange et al. 1997) which both allowed one reading every 30 min. More than 1100 diurnal courses for the time period from June 1995 to August 1997 were recorded.

Thallus water content controlled the metabolic activity of the lichens. Water came from different sources (rain, dew, frost, and snow), and these poikilohydric organisms were constantly alternating between activation and inactivation due to drying and wetting. In a given year, the thalli were dry and metabolically inactivate 18% of all days (counted from 0 to 24h and averaged over all species and all days of the measuring period). Dry days were most common in summer; however, with the exception of November, totally dry days also occurred during winter (Fig. 20.6). The lichens were hydrated on 82% of all days. Rain regularly hydrated the lichens during daylight hours

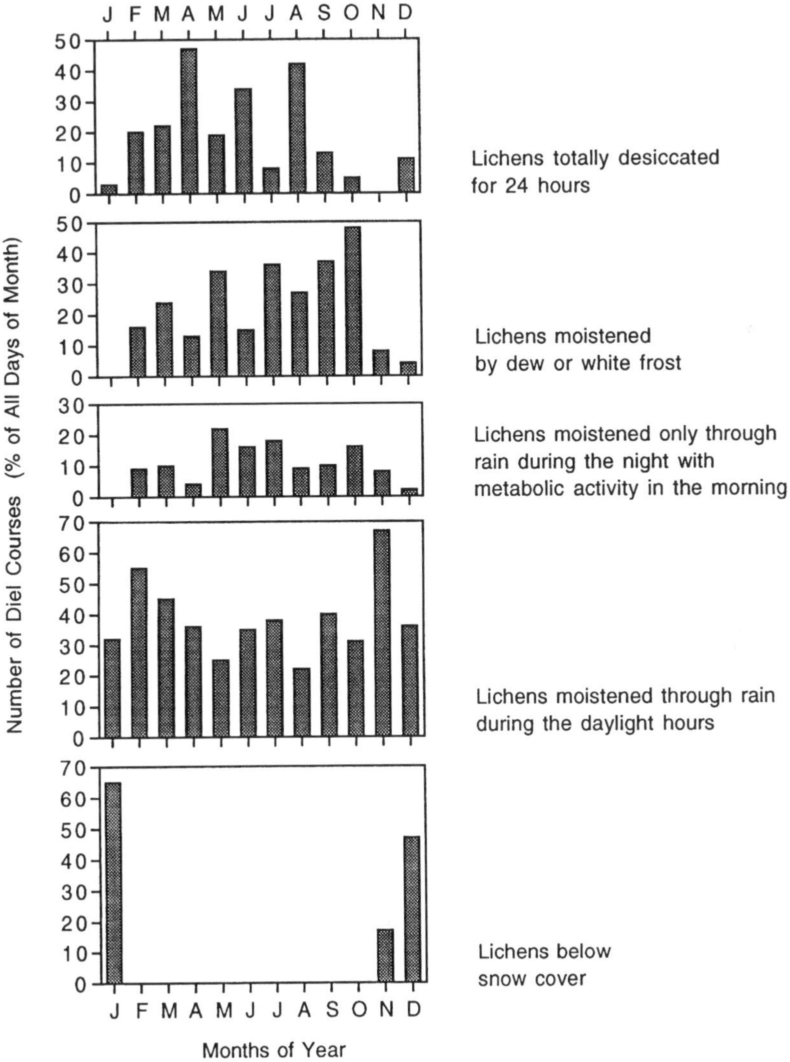

Fig. 20.6. Weather condition types indicating hydration of soil-crust lichens in their quasinatural habitat in the Botanic Garden Würzburg, Germany, northern Bavaria (ca. 750 diel courses, spread over the time period from March 1995 to August 1997). Indicated are the numbers of days (counted from 0 to 24h) in percent of all days of the months. (After Lange 2000b)

throughout the year (39 % of all days). Dew and white frost were sources of water on 22 % of all days, which resulted in short but efficient activation of phycolichen photosynthesis with maximum frequency in late summer and early fall (Fig. 20.6). Nocturnal rain, which allowed photosynthetic activity during the early morning after sunrise, occurred on 10 % of all days, and for a similar number of days the lichens were covered by snow, often frozen under hydrated conditions. Lichen response patterns of CO_2 exchange for these different climatic situations are discussed in Chapter 18.

Under identical habitat conditions, time periods of photosynthesis and dark respiration varied greatly between the different lichen species, mainly due to their specific morphology (Table 20.1). The foliose-fruticose *Cl. convoluta* had the longest period of activity: it photosynthesized or respired for 65.6 % of the year, being dry only for 34.4 % (time resolution of 30 min). This is not unexpected, since, once hydrated, its compact cushion retained moisture and prevented rapid water loss. The other species had flat, crustose, or indistinctly lobate thalli. Among them, the gelatinous cyanolichen *Co. cristatum* had active CO_2 exchange (and thus, potential N fixation) 57.3 % of the time, more than 1.6 times greater than *L. muralis* or *F. fulgens*. Respiration time of *Collema cristatum* exceeded net photosynthetic time by 1.5 times, mainly because the high water-holding capacity of its gelatinous thallus kept the lichen hydrated at night. In contrast, *L. muralis*, with its thin and quickly drying thallus, respired only 1.13 times longer than it photosynthesized. *Co. cristatum* also profited from its high water-holding capacity: after heavy nocturnal precipitation, it was often able to photosynthesize the entire following day without rain, while the faster-drying phycolichens experienced hardly any activity at all. This was especially true in the winter. Thus, the C balance of individual lichen species was determined by pronounced species-specific differences in duration and patterns of hydration, together with

Table 20.1. Time span (in % of total time period of recording) during which the lichens were metabolically active, either with photosynthetic carbon gain or with respiratory carbon loss, the remainder of the time being inactive. Numbers of diel time courses with continuous recording of CO_2 exchange, used for the calculation, are given in parentheses. (After Lange 2000a)

Species (number of diel courses)	Metabolically active (% of time)	Carbon gain (% of time)	Carbon loss (% of time)
Cladonia convoluta (106)	65.6	26.1	39.5
Collema cristatum (116)	57.3	22.9	34.4
Squamarina lentigera (126)	49.5	21.2	28.3
Diploschistes muscorum (174)	49.1	15.7	33.4
Lecanora muralis (462)	35.2	16.5	18.7
Fulgensia fulgens (160)	34.9	13.0	21.9

differences in photosynthetic parameters such as moisture compensation point and water content optimums for photosynthesis (see Chap. 18).

As an example, annual dynamics of photosynthesis and respiration of *Co. cristatum* is described in detail. Distributed in four blocks to cover the main seasons, 113 continuous diel time courses were recorded (Fig. 20.7). Highest weekly maximal rates of net photosynthesis (NP) were attained in mid-summer when average temperatures and PPFDs were the highest. However, there were a few weeks in November and February during which net photosynthesis was higher than during some of the weeks in summer. Actual maximal NP in summer could attain about twice the winter rates. This was due to both more favorable environmental conditions and the seasonal acclimation of *Co. cristatum* photosynthetic rates, which were higher in summer than winter (measured under standard laboratory conditions). Overall, maximal NP of *Co. cristatum* was greater compared to other species from the same habitat, and this certainly reflects the high N content of the cyanolichen due to its N-fixing capability.

As was the case with momentary maximal rates of NP, maximum daily C gain was substantially higher during summer than winter, as photosynthesis was limited by low winter light. The highest daily photosynthetic gain was almost 1 gC m^{-2}. However, maximal daily C loss was also highest in summer, reaching 468.1 mgC m^{-2} day^{-1} during a wet and warm night in August. Nevertheless, maxima of diel C balance attained between 600 and 700 mgC m^{-2} day^{-1}. Thus, in the optimal case, daily net production of *Co. cristatum* was almost 0.4 % of its thallus C content.

Weekly integrals of C gain and balance (Fig. 20.7A,B) depended greatly on momentary weather conditions that tended to equalize the seasonal differences in lichen productivity. The 4-week period in late summer 1996 combined many completely dry days with days of high precipitation; this often supersaturated the lichen thallus and kept it moist during the relatively warm night periods. Even while short-term maxima of NP were high in these months, 2 weeks of negative C balance also occurred during this time (Fig. 20.7B). Combined, this time period had the lowest weekly average of C balance of all four measuring periods (0.19 gC m^{-2} $week^{-1}$). The weekly averages for photosynthetic C gain were similar in fall and early winter 1996. However, C balance was higher in late winter than in summer 1996, as well as in early spring 1997. High light intensities, in combination with extended favorable moisture conditions, promoted productivity of the lichen in early summer 1997, and this period had the highest weekly mean C balance (1.36 gC m^{-2} $week^{-1}$).

This case study illustrates that C acquisition by poikilohydric soil-crust organisms is highly variable, responds quickly to changes in the momentary microclimatic conditions, and does not necessarily follow seasonal trends. Hydration and light levels are the most important environmental factors. The

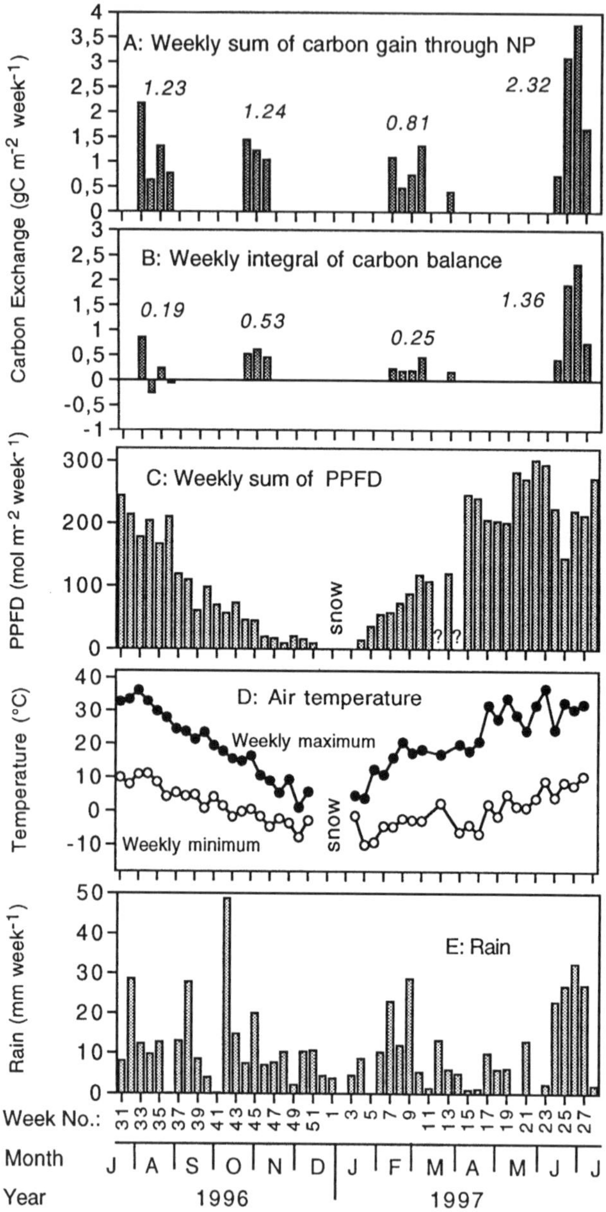

Fig. 20.7A–E. Time period July 1996 to July 1997 of field measurements with *Collema cristatum* (*abscissa* week numbers and months). **A, B** Weekly sum of carbon gain and integral of carbon balance, respectively, indicating the distribution of the four monitoring periods during the year. For the four seasonal periods of measurements the mean weekly totals (gC m^{-2} $week^{-1}$) are indicated by *numbers in italics*. **C** Weekly sums of photosynthetic active photon flux density (*PPFD*) incident on the lichen. No data are available during the winter period when the lichen was under snow; *question marks* indicate missing data (March/April 1997). **D** Weekly maxima and minima of air temperature near the lichen thallus. **E** Weekly rainfall totals (Wetterwarte Würzburg). (After Lange 2000a)

soil-crust habitat in the temperate local steppe formation represents an intermediate situation between the extremes of arid desert regions, where metabolic activity might be reduced to less than 10 % of total time (see Lange et al. 1994), and the boreal, arctic or alpine regions, where soil-crust organisms will experience longer periods of annual hydration or even extended snow cover. Comparison of the organisms shows that species-specific morphological and physiological features are important in determining photosynthetic C gain. It also appears that advantages and disadvantages bestowed by these different features were mutually balanced, thus allowing coexistence of heteromerous phycolichens and homoeomerous, gelatinous cyanolichens in the same temperate soil-crust habitats (see Lange 2000b). This might be different under other climatic conditions, where the different thallus structures and physiological features could lead to dominance of one or the other life form, and where their individual C input into the ecosystem might be altered.

20.3.2 Estimated Carbon Input

Integrated over all of the 113 diel time courses of actual measurements of the above case study, *Co. cristatum* had a total photosynthetic C gain of 22.48 gC m^{-2}. Of this, 61 % (13.77 gC m^{-2}) was lost through respiration. Assuming the 113 measured days were representative of the year, and taking into account that the lichen was frozen for ca. 30 days (during which C gains and losses are balanced; see Kappen 1993), C balance for the total course of the year is estimated at 25.8 gC m^{-2} a^{-1} or, related to thallus C content, 200 mgC (g C)$^{-1}$ a^{-1}. This translates to a potential annual production of about 20 %. Because of the extrapolation process, these numbers must be treated with caution. Nevertheless, they are a sound approximation for annual net C input by *Co. cristatum* under temperate soil-crust conditions. At the same location, CO_2 exchange of *L. muralis* was monitored continuously for longer than an entire year, and its integrated annual C gain was very similar (21.7 gC m^{-2} a^{-1}; O.L. Lange, unpubl.). These values estimate the C contribution of soil crusts to temperate ecosystems through C leakage and decomposition.

Other estimates of annual C contributions by soil-crust organisms are based on laboratory experiments or on shorter periods of field studies. Jeffries et al. (1993a,b) estimated CO_2 flux in *Scytonema*-dominated and *Microcoleus*-dominated crusts from a *Coleogyne* (blackbrush) community in southern Utah. Assuming 10 h of metabolic activity per day, 50 and 180 wet days per year, and adjusted for the relative field coverage of the two types, crust C inputs were estimated at 0.64 to 2.3 gC m^{-2} a^{-1}. Another laboratory estimate by Garcia-Pichel and Belnap (1996) used oxygen micro-profiles to estimate net photosynthesis and dark respiration of maximally wetted crusts from SE Utah during a winter season. They assumed soils were wet 100 h during one winter season, and an equal probability of rainfall during day and night. This resulted in an estimate of net C gain of 0.402 and 0.678 gC m^{-2} per winter season for the two crusts, both containing *Microcoleus vaginatus*, but with differing amounts of *Nostoc* and *Scytonema*.

Based on laboratory analyses, Klopatek (1992) estimated potential annual C contribution of lichen soil crusts (see Beymer and Klopatek 1991) in blackbrush and pinyon-juniper (Arizona, Utah) to be 12–22 and 20.5–37 gC m^{-2} a^{-1}, respectively. Lange et al. (1994) monitored CO_2 exchange of soil-crust samples covered by crustose lichens under natural conditions in a coastal fog zone of the Namib. Thirty-five complete diel courses with nocturnal fog (and/or dew) resulted in an average daily C gain of 126 mgC m^{-2} day^{-1}. Using the annual average of 250 foggy days reported for this part of Namibia, where rain precipitation is negligible and seasonal weather changes are small, the estimate of annual C gain for these lichen crusts is 32 gC m^{-2} a^{-1}.

Even when using different methods of extrapolation and estimation, reported estimates for crust production and thus for C input by biological soil crusts in their total ecosystems are remarkably similar. A lower range (0.4–2.3 gC m^{-2} a^{-1}) is reported for cyanobacterially dominated crusts, where the autotrophic biomass is at, or below, the soil surface, and higher estimates (12–37 gC m^{-2} a^{-1}) are observed in crusts where lichens and/or mosses on top of the soil surface contribute to production. This is in agreement with the differences in photosynthetic capacity of these two crust types (see Chap. 18).

Overall, vascular plant productivity in most arid and semiarid plant formations is low. These are the same ecosystems where biological soil crust cover is generally high. Under such circumstances, an additional photosynthetic C supply yielding up to several hundred kg ha^{-1}, and being fixed by very resistant crust organisms, represents a substantial contribution to the C budget of these ecosystems. In addition, much of the C contributed by crusts occurs in plant interspaces, where water and nutrient limitations otherwise restrict such inputs.

Acknowledgments. RDE was supported by the National Science Foundation (Award # 98-14510, 98-14358), the Department of Defense, and the Environmental Protection Agency. The work of OLL took place as a project of Sonderforschungsbereich 251 of the University of Würzburg.

References

Belnap J (1995) Surface disturbances: their role in accelerating desertification. Environ Monit Assess 37:39–57

Belnap J, Harper KT (1995) Influence of cryptobiotic soil crusts on elemental content of tissue of two desert seed plants. Arid Soil Res Rehabil 9:107–115

Beymer RJ, Klopatek JM (1991) Potential contribution of carbon by microphytic crusts in pinyon-juniper woodlands. Arid Soil Res Rehabil 5:187–198

Bolton H Jr, Smith JL, Wildung RE (1990) Nitrogen mineralization potentials of shrub-steppe soils with different disturbance histories. Soil Sci Soc Am J 54:887–891

Callison J, Brotherson JD (1985) Habitat relationships of the blackbrush community (*Coleogyne ramosissima*) of southwestern Utah. Great Basin Nat 45:321–326

Cameron RE, Fuller WH (1960) Nitrogen fixation by some algae in Arizona soils. Soil Sci Soc Proc 24:353–356

Evans RD, Belnap J (1999) Long-term consequences of disturbance on nitrogen dynamics in an arid ecosystem. Ecology 80:150–160

Evans RD, Ehleringer JR (1993) A break in the nitrogen cycle of aridlands? Evidence from $\delta^{15}N$ of soils. Oecologia 94:314–317

Evans RD, Ehleringer JR (1994) Water and nitrogen dynamics in an arid woodland. Oecologia 99:233–242

Evans RD, Johansen JR (1999) Microbiotic crusts and ecosystem processes. Crit Rev Plant Sci 18:183–225

Fletcher JE, Martin WP (1948) Some effects of algae and molds in the rain crust of desert soils. Ecology 29:95–100

Garcia-Pichel F, Belnap J (1996) Microenvironments and microscale productivity of cyanobacterial desert crusts. J Phycol 32:774–782

Harper KT, Marble JR (1988) A role for nonvascular plants in management of arid and semiarid rangelands. In: Tueller PT (ed) Vegetation science applications for rangeland analysis and management. Kluwer Academic Publishers, Boston, pp 135–169

Harper KT, Pendleton RL (1993) Cyanobacteria and cyanolichens: can they enhance availability of essential minerals for higher plants? Great Basin Nat 53:59–72

Haynes RJ, Sherlock RR (1986) Gaseous losses of nitrogen. In: Haynes RJ (ed) Mineral nitrogen in the plant-soil system. Academic Press, New York, pp 242–302

Heaton THE (1986) Isotopic studies of nitrogen pollution in the hydrosphere and atmosphere: a review. Chem Geol 59:87–102

Isichei AO (1980) Nitrogen fixation by blue-green algal soil crusts in Nigerian savanna. In: Rosswall T (ed) Nitrogen cycling in west African ecosystems. Royal Swedish Acad Sci, Stockholm, pp 191–198

Jackson RB, Caldwell MM (1993) The scale of nutrient heterogeneity around individual plants and its quantification with geostatistics. Ecology 74:612–614

Jeffries DL, Klopatek JM, Link SO, Bolton H (1992) Acetylene reduction of cryptogamic crusts from a blackbrush community as related to resaturation and dehydration. Soil Biol Biochem 24:1101–1105

Jeffries DL, Link SO, Klopatek JM (1993a) CO_2 fluxes of cryptogamic crusts. I. Response to resaturation. New Phytol 125:163–173

Jeffries DL, Link SO, Klopatek JM (1993b) CO_2 fluxes of cryptogamic crusts. II. Response to dehydration. New Phytol 125:391–396

Jones K, Stewart WDP (1969a) Nitrogen turnover in marine and brackish habitats. III. Production of extracellular N by *Calothrix*. J Mar Biol Assoc UK 49:475–488

Jones K, Stewart WDP (1969b) Nitrogen turnover in marine and brackish habitats. IV. Uptake of the extracellular products of the nitrogen fixing algae *Calothrix*. J Mar Biol Assoc UK 49:701–716

Kappen L (1993) Plant activity under snow and ice, with particular reference to lichens. Arct J Arct Inst North Am 46:297–302

Kershaw KA (1985) Physiological ecology of lichens. Cambridge University Press, New York

Kleiner EF, Harper KT (1972) Environment and community organization in grasslands of Canyonlands National Park. Ecology 53:299–309

Klopatek JM (1992) The use of cryptogamic crusts as indicators of disturbance in semiarid landscapes. In: McKenzie DH, Hyatt DE, MacDonald VJ (eds) Ecological indicators, 1. Elsevier, London, pp 773–786

Klubek B, Eberhardt PJ, Skujins J (1978) Ammonia volatilization from Great Basin Desert soils. In: West NE, Skujins JJ (eds) Nitrogen in desert ecosystems. Dowden, Hutchinson and Ross, Stroudsburg, pp 107–129

Lange OL (2000a) Photosynthetic performance of a gelatinous lichen under temperate habitat conditions: long-term monitoring of CO_2 exchange of *Collema cristatum*. Bibl Lichenol 75:307–332

Lange OL (2000b) Die Lebensbedingungen von Bodenkrusten-Organismen: Tagesverlauf der Photosynthese einheimischer Erdflechten. Hoppea: 61:423–443

Lange OL, Green TGA (1996) High thallus water content severely limits photosynthetic carbon gain of central European epilithic lichens under natural conditions. Oecologia 108:13–20

Lange OL, Meyer A, Zellner H, Heber U (1994) Photosynthesis and water relations of lichen soil crusts: field measurements in the coastal fog zone of the Namib Desert. Funct Ecol 8:253–264

Lange OL, Reichenberger H, Walz H (1997) Continuous monitoring of CO_2 exchange of lichens in the field: short-term enclosure with an automatically operating cuvette. Lichenologist 29:259–274

Liengen T (1999) Conversion factor between acetylene reduction and nitrogen fixation in free-living cyanobacteria from high arctic habitats. Can J Microbiol 45:223–229

MacGregor AN, Johnson DE (1971) Capacity of desert algal crusts to fix atmospheric nitrogen. Soil Sci Soc Am Proc 35:843–844

Mayland HF, McIntosh TH (1966) Availability of biologically fixed atmospheric nitrogen-15 to higher plants. Nature 209:421–422

Mayland HF, McIntosh TH, Fuller WH (1966) Fixation of isotopic nitrogen on a semiarid soil by algal crust organisms. Soil Sci Soc Am Proc 30:56–60

Nadelhoffer KJ, Fry B (1988) Controls on natural nitrogen-15 and carbon-13 abundances in forest soil organic matter. Soil Sci Soc Am J 52:1633–1640

Peterjohn WT, Schlesinger WH (1991) Factors controlling denitrification in a Chihuahuan Desert ecosystem. Soil Sci Soc Am J 55:1694–1701

Rogers SL, Burns RG (1994) Changes in aggregate stability, nutrient status, indigenous microbial populations, and seedling emergence, following inoculation of soil with *Nostoc muscorum*. Biol Fertil Soils 18:209–215

Rychert RC, Skujins J (1974) Nitrogen fixation by blue-green algae-lichen crusts in the Great Basin Desert. Soil Sci Soc Am Proc 38:768–771

Rychert R, Skujins J, Sorensen D, Porcella D (1978) Nitrogen fixation by lichens and free-living microorganisms in deserts. In: West NE, Skujins JJ (eds) Nitrogen in desert ecosystems. Dowden, Hutchinson and Ross, Stroudsburg, pp 20–30

Schlesinger WH, Peterjohn WT (1991) Processes controlling ammonia volatilization from Chihuahuan Desert soils. Soil Biol Biochem 23:637–642

Schlesinger WH, Raikes JA, Hartley AE, Cross AE (1996) On the spatial pattern of soil nutrients in desert ecosystems. Ecology 77:364–374

Shearer G, Kohl DH (1986) N_2-fixation in field settings: estimation based on natural ^{15}N abundance. Aust J Plant Phys 13:699–756

Stewart WDP (1967) Transfer of biologically fixed nitrogen in a sand dune slack region. Nature 214:603–604

West NE (1990) Structure and function of microphytic soil crusts in wildland ecosystems of arid to semi-arid regions. Adv Ecol Res 20:179–223

West NE, Skujins J (1977) The nitrogen cycle in North American cold-winter semi-desert ecosystems. Oecol Plant 12:45–53

Zaady E, Groffman PM, Shachak M (1996) Litter as a regulator of N and C dynamics in macrophytic patches in Negev Desert soils. Soil Biol Biochem 28:39–46

21 Influence of Biological Soil Crusts on Soil Environments and Vascular Plants

J. BELNAP, R. PRASSE, and K.T. HARPER

21.1 Introduction

The presence of biological soil crusts can significantly change the physical and chemical soil environment. Such alterations can affect the germination, survival, and nutritional status of vascular plants. Some landscape surveys have found a negative relationship between biological soil crusts and vascular plant cover (West 1990; Johansen 1993; Eldridge 1993). Based on this, some authors have concluded that crust organisms limit vascular plant cover (Savory 1988). However, it is also possible that soil crusts merely occupy all soil surfaces not first covered by vascular plants. In support of this, numerous authors have reported that vascular plant cover either shows no correlation with crust cover (Anderson et al. 1982; Beymer and Klopatek 1992; Jeffries and Klopatek 1987; Kleiner and Harper 1972, 1977) or that there is a positive correlation between vascular plant and crust cover (Carleton 1990; Dadlich et al. 1969; Graetz and Tongway 1986; Ladyman and Muldavin 1994; Lesica and Shelley 1992; Mücher et al. 1988).

Many studies have addressed aspects of the relationship between crusts and vascular plants. These studies have been done at sites with different soil textures, physical and chemical crusting, soil surface roughness and stability, climate regimes, species composition of the biological crusts, and vascular plant species (each with different strategies for germination and survival). It is essential that all these differences be accounted for when generalizing about the relationship between soil crusts and vascular plants. In this chapter, we will discuss how crusts modify soils in ways that can affect vascular plants. We will then address the effects of biological soil crusts on seed dispersal and germination, and how crusts affect vascular plant survival and nutritional status in different climate regimes. Finally, we will discuss the overall effects of crusts on plant community structure.

Ecological Studies, Vol. 150
J. Belnap and O.L. Lange (eds.) Biological Soil Crusts: Structure, Function, and Management

21.2 Modification of the Soil Environment by Biological Soil Crusts

The presence of biological soil crusts modifies soils in ways that can affect germination, emergence, and survival of vascular plants. Crust-caused soil alterations include changes in soil surface roughness, soil texture, thermal conditions, availability of nutrients, organic matter, and water. A large body of literature, cited below, shows that manipulation of each of these factors can influence vascular plants. As few experiments have directly linked crusts and vascular plant performance, it is hoped that the following discussion will help stimulate future research on this relationship.

21.2.1 Surface Roughness

Biological soil crusts influence the microtopography of soil surfaces (Chap. 1, Photos 30, 32, 48, 50; Chaps. 15, 16). In hot deserts that lack soil-freezing, changes in surface roughness are relatively small. Cyanobacterial crusts can smooth soil surfaces, while lichens and mosses growing above the soil surface increase roughness slightly. In contrast, soil surfaces in cool and cold deserts show large increases in soil-surface roughness due to frost-heaving and the differential erosion of crusted soils. Resulting soil pedicels can be up to 15 cm high (Belnap and Gardner 1993). These relatively large microtopographic features act to trap wind- and water-borne seeds, organic matter, soil fines, and water (Harper 1977; Harper and St. Clair 1985; Eckert et al. 1986; J. Belnap, unpubl. data).

21.2.2 Soil Texture and Chemistry

Crusted soil surfaces often have a greater silt/clay fraction than underlying soils, or adjacent uncrusted soil surfaces. Fine clay particles stick to the mucilaginous sheath material, especially when wet (Belnap and Gardner 1993; Verrecchia et al. 1995; see Chap. 15). These negatively charged clay particles bind positively charged plant macronutrients, thus increasing soil fertility (Black 1968). The presence of biological crusts can significantly alter soil pH and the amounts and availability of plant-essential nutrients and organic matter. Studies in cool deserts and laboratory settings show that crusts enhance the content of most bioessential elements including N, K, Ca, Mg, P, Fe, Mn, Cl, and S (Harper and Pendleton 1993; Mitchell and Fullen 1994; Belnap and Harper 1995; Harper and Belnap 2000).

Nitrogen. Up to 70 % of the nitrogen fixed by cyanobacteria and cyanolichens is released immediately into the surrounding soil environment, and is available to associated organisms including vascular plants, mosses, fungi, and other microbes (see Chap. 16). This has specifically been shown for *Collema*, *Microcoleus*, and *Nostoc–Microcoleus* soil crusts from SE Utah (F. Garcia-Pichel and J. Belnap, unpubl. data). Multiple studies have reported that the presence of soil crusts increases surrounding soil N by up to 200 % (Shields and Durrell 1964; Rogers and Burns 1994; DeFalco 1995; Harper and Belnap 2000). In addition, stable isotopes show that soil crusts can be the dominant source of N for desert soils and plants (Evans and Ehleringer 1993; Evans and Belnap 1999; Chap. 19, 20).

Carbon, C:N Ratios, and Organic Matter. All crust components secrete extracellular carbon (C) within minutes to a few days of C acquisition. In cyanobacteria, these secretions can represent up to 50 % of total fixed carbon (Fogg 1966; Lewin 1956). As a result, presence of soil-crust organisms increases soil polysaccharides and total carbon by up to 300 % (Rao and Burns 1990; Rogers and Burns 1994), benefiting heterotrophic microbes which are often carbon-limited. In addition, the presence of crusts can lower soil C:N ratios. This increases decomposition rates, making nutrients available faster to associated organisms (Kleiner and Harper 1972). Increases in soil organic matter are also well known to ameliorate compaction, reduce inorganic soil crusting, reduce nutrient leaching losses, and increase soil moisture retention (Tongway and Ludwig 1990; Evans and Young 1994).

Nutrient Binding. Microbial exopolymers not only add carbon to soils, they also modulate metal-ion concentrations at the microbial cell surface. The lipid, protein, and carbohydrate components combine to create a mosaic of many polyfunctional metal-binding sites that differ in affinity and specificity. Both cations and anions can be bound (Greene and Darnall 1990). Under natural conditions, most of the polymers act as polyanions that prevent excess quantities of highly charged molecules (such as heavy metals) from approaching the cell surface, while concentrating growth-promoting nutrients present at low concentrations in the surrounding environment (such as Na, K, Mg, Ca, Mn, Fe, Ni, Cu, and Zn) (Lange 1976; Geesey and Jang 1990). This binding occurs within minutes, and is largely a passive electrostatic interaction between cationic metals and anionic cell surface groups. Therefore, the organisms do not need to be alive or active (Gadd 1990a).

Binding capacity and affinity for one cation over another is determined by a variety of factors, including the microbial species, pH, temperature, cell densities, cell size, ion densities, and ion size. Different microbial species secrete polymers with different binding affinities, and selectivity coefficients may also be different, depending on whether polysaccharides are in gel or solution. Small-celled organisms with larger surface areas adsorb more than large filamentous species with less surface area (Geesey and Jang 1990).

External soil factors also influence the binding properties of microbial exopolymers. In soils above pH 7 (e.g., most desert soils), cation exchange capacity increases significantly for all microbial species. Binding of metal ions cause polymers to condense into gels, altering cation binding affinities. Polysaccharide production is stimulated by Mg, K, and Ca, which then results in greater binding of these nutrients. Affinities also fluctuate depending on concentrations of other cations present. Initial binding may also be followed by inorganic deposition of an increased amount of metal. Particulate material can bind to crustal organisms (e.g., sulfides of copper, Zn, Pb, zinc dust, magnesic oxide, and ferric hydroxide; Lange 1974; Gadd 1990a,b; McLean and Beveridge 1990; Pirszel et al. 1995). Most adsorbed metals stay on or within the extracellular sheath, and are not absorbed by the cells, thus remaining available to plants (Geesey and Jang 1990).

Metal Chelation. Cyanobacteria (including the common soil genera *Anabaena*, *Anacystis*, *Lyngbya*, and *Nostoc*) green algae, fungi, lichens, and bacteria also secrete powerful metal chelators such as siderochromes (Lange 1974; McLean and Beveridge 1990; Schelske et al. 1962). Chelating agents are organic compounds that form complexes with ions of polyvalent metals. Their presence maintains metals that would otherwise be precipitated in solution. They are effective in sequestering essential trace metals at exceedingly low ambient concentrations (Paerl 1988). This is especially important in soils with high pH (such as found in many deserts), as some elements form insoluble oxides or hydroxides at or above pH 7. Cyanobacteria also secrete peptide nitrogen and riboflavin. Together with siderochromes, these substances form complexes with tricalcium phosphate, Cu, Zn, Ni, and ferric iron, keeping them plant-available. These secreted chelators are water-soluble, and capable of being translocated in soil and water. Consequently, chelated compounds are made available to associated, nonchelating plants or microbes. Binding and chelation may be especially important for nitrogen fixers, as they require high amounts of otherwise unavailable Fe (Bose et al. 1971; Lange 1974; Geesey and Jang 1990; Gadd 1990a).

Other Plant Nutrients and Plant Factors. Cyanobacteria also secrete other compounds that can influence vascular plant growth. Glycollate is a common extracellular product. Secretion of glycollate is increased at low cyanobacterial cell density (as in soils) and when cyanobacteria go from low to high light intensities, as on a partly cloudy day. Glycollate stimulates uptake of phosphate in cyanobacteria (Fogg 1966). Cyanobacteria also secrete various vitamins (e.g., B_{12}), auxin-like substances, and other substances that promote growth and cell division in plant and animal tissue (Fogg 1966; Venkataraman and Neelakantan 1967).

Vesicular-Arbuscular Mycorrhizae. There is a strong positive correlation between mycorrhizal infections of some seed plants and crusted surfaces (Harper and Pendleton 1993; Pendleton and Warren 1995). In addition, consistent increases of biogenic mineral elements in crusted soils and associated plants that do not form complexes with chelating compounds suggest that the increased mycorrhizal fungi associated with crusts are also involved in increasing nutrient availability to plants (Harper and Pendleton 1993; Belnap and Harper 1995; Harper and Belnap 2000).

pH. The presence of photosynthetic crustal organisms significantly increases the pH of surrounding medium (see Chap. 16). Crustal organisms from SE Utah raise the soil pH from 8 to 10.5 (Garcia-Pichel and Belnap 1996). Endolithic cyanobacteria in Venezuela and South Africa raised the pH of surrounding solutions in rocks from about 8 to 10 as well (Büdel 2000). As seen above in the discussion on chelation, increases in pH decreases availability of some nutrients (e.g., P, Fe), CO_2 dissolution rates, carbonate formation rates, and solubilization of soil minerals (e.g., Si; Nobel 1991).

21.2.3 Thermal and Water Relations

The presence of biological soil crusts can increase soil surface temperatures by up to 14 °C relative to adjacent uncrusted soils in both temperate and polar deserts. This is expected to influence seed germination, plant growth, and nutrient uptake rates (Harper and Pendleton 1993; Belnap 1995; Gold and Bliss 1995).

Cyanobacteria and gelatinous lichens can absorb ten times or more their volume in water (Campbell 1979). While increased biomass of these organisms may increase the surface water-holding capacity of soils, this could be offset by increased loss of soil surface moisture due to higher surface temperatures in crusted than uncrusted soils (see Chap. 27). However, it has also been repeatedly observed in a cool desert setting that darker, more well-developed crusts retain surface moisture longer after rain events than bare sand or thin cyanobacterial crusts, regardless of air temperatures or insolation levels (J. Belnap, pers. obs.). While this may not affect germination of seeds lying on the crust surface, it may influence moisture status of seeds embedded in the crust. In hot deserts, surfaces dry very quickly when air temperatures are high; this increase in surface moisture may influence seed germination during hot periods. It is not known what effect crusts have on subsurface soil moisture. Three studies have directly addressed this question. Two studies found that the presence of crusts increased soil moisture (Brotherson and Rushforth 1983; D. George et al., unpubl.); one study found that crusts reduced soil moisture (Harper and Marble 1988). D. George et al. noted that the type of soil crust (cyanobacterial vs. lichen crust) influenced study results, as cyanobacterial crusts lost mosture quickly than lichen-moss crusts.

The presence of crusts can also influence water distribution on a landscape scale. In hot deserts, plant interspaces are covered by smooth crusts that generally reduce water infiltration. The resultant increase in runoff from crusted plant interspaces potentially provides increased water availability for nearby vascular plants (Tongway and Ludwig 1990; Cornet et al. 1992; Zaady and Shachak 1994; Eldridge et al. 2001; Chap. 22). In cool and cold deserts, crusts can enhance local water infiltration, thus reducing runoff on a landscape scale (Chap. 24).

21.3 Influence of Soil Crusts on Seed Dispersal and Germination

21.3.1 Seed Dispersal

Many factors influence movements of seeds once they are dispersed from the parent plant. As the establishment of perennial species is a fairly rare event, field studies concerning natural dispersal and emergence have generally addressed annual plant species. Regardless of seed size or shape, most seeds can be easily carried by wind or water across dry, smooth soil surfaces (Harper 1977). However, some seeds have mucilaginous coats or pappus hairs that can stick to wet soil surfaces, even when soil surfaces are smooth (Harper 1977; Gutterman 1994; Zaady et al. 1997). Smooth crust surfaces are most often found in hot deserts or soils recently colonized by cyanobacteria. Smooth surfaces are generally characterized by low plant litter, no frost-heaving, biological crusts heavily dominated by cyanobacteria, and some physical or chemical crusting. The effects of a smooth crustal surface on seed dispersal has been well illustrated by recent work at the Nizzana dunes in Israel. Using field experiments, Prasse (1999) showed that the probability of seed entrapment on these smooth cyanobacterial crusts is very low. Experimentally applied disturbances that roughened the smooth soil surface significantly increased seed entrapment. Seed-trapping was most effective when the disturbance was applied before main seed-dispersal times. Roughened plots that were covered to prevent seed input had no seedlings, indicating that no germinable seeds were available prior to surface roughening. This observation was supported by watering intact crusts in a greenhouse setting, where almost no germination was observed.

In contrast to these smooth soil surfaces found in hot deserts, surfaces in cool and cold deserts are characterized by lichen–moss crusts, plant litter, frost-heaved soils, and limited physical or chemical crusting. Many studies have shown these rough surfaces trap seeds. In addition, these surfaces trap organic matter, water and soil fines, thus increasing microsite fertility (Harper and St. Clair 1985; Eckert et al. 1986; Harper and Marble 1988; Mücher et al. 1988; Chap. 14). Using a portable wind tunnel to generate winds of known speed, J. Belnap (unpubl. data) showed that smoothing of crusted surfaces by disturbance results in less trapping of seeds and organic matter at all commonly-occurring wind speeds.

21.3.2 Vascular Plant Seed Germination and Establishment

Few studies have specifically addressed the effect of soil crusts on seed germination and plant establishment, and results have been variable. Most studies have used experimental sowing of selected species into crusted and uncrusted soils; thus, it should be kept in mind that seed placement may not mimic that of naturally dispersed seeds. Other studies have used comparisons of species composition and/or density of vascular plants in adjacent crusted and uncrusted soils. Interpreting results from these studies necessitates understanding the reasons for crust cover differences and the influence these factors may have on vascular plant germination and/or establishment.

Biological soil crusts do not present a physical barrier to root penetration for any plant species or soil type tested (Belnap and Gardner 1993; R.L. Pendleton, unpubl. data). In contrast, physical and chemical crusting often inhibits root penetration (see Chap. 15; Chartres 1992). Combinations of physical and biological crusts are expected to have intermediate effects on root penetration, depending on the level and type of crust development.

Biogeography. In cool deserts, crusts generally enhance germination and establishment of vascular plants. St. Clair et al. (1984) showed large increases in germination and establishment in the lab for the perennial grasses *Agropyron elongatum* (fivefold increase) and *Elymus junceus* (11-fold increase) on crusted southeastern Utah soils when compared to uncrusted soils. In the Great Basin, Eckert et al. (1986) showed that seedling germination was higher in crusts under shrub canopies and in adjacent crusted soils compared to the uncrusted plant interspace soils. Most seedlings were found at the base of crust pedicels. Schlatterer and Tisdale (1969) showed no effect of moss cover on the number of perennial grass seeds that germinated (*Stipa thurberiana, Sitanion hystrix, Agropyron spicatum*) when compared to artificial litter, although time to germination was slower. When compared to bare surfaces, germination was inhibited. However, seeds were placed under 2 cm of moss, rather than on top of the moss, as would occur in nature. In Idaho and southeastern Utah, Larsen (1995) and Howell (1998), respectively, showed that germination of native *Stipa* was not affected by biological soil crusts, while germination of the exotic annual grass *Bromus tectorum* was inhibited by intact crusts. *Bromus* germination was stimulated when crusts were broken but left in place. Thus, in cool deserts, seed germination is generally enhanced for perennial plants and decreased for exotic annual grasses.

In contrast with cool deserts, hot-desert species show a more ambivalent response to the presence of soil crusts. In Australia, Crisp (1975) found the same as Larsen (1995) did in cool deserts: germination and establishment of the native perennial grass *Stipa* was not inhibited on crusts, while germination and establishment of the annual exotic grass *Schismus* was inhibited. As with the annual *Bromus*, *Schismus* germination was stimulated when the crust was broken. In Arizona, McIlvanie (1942) found that two species of perennial warm-season grasses did not germinate on soils with 100 % moss cover in southern Arizona. In the laboratory, Zaady et al. (1997) found that cyanobacterial crusts in Israel inhibited germination of the annual plants *Plantago* and *Reboudia*, while germination of the annual *Carrichtera* was enhanced. Hacker (1984) reported reduced germination on "crusted" soils for two shrubs in Australia; however, he makes no mention of biological crusts, and it is likely that these were physical crusts, as they increased with grazing. In 1987, Hacker found more adult plants, but fewer seedlings, of *Selenothamnus* (a chenopod shrub) on crusted soils. He speculated that crusts may reduce germination of this species; however, he did not directly test this hypothesis.

Physical Influences on Seed Germination. Seeds require varying amounts of moisture to germinate. In the dry air of deserts, many species require some soil or plant cover to maintain sufficient moisture for germination. While the cracks and crannies in the soil surface are often sufficient for small seeds (Gutterman 1994; Chartres and Mücher 1989), most larger-seeded plants require additional cover by soil or litter (Evans and Young 1984). In deserts where litter is often limited, native seeds generally accomplish this by burial mechanisms such as hygrochasic awns or through rodent caching. However, some plant species lack specialized penetration structures (Boeken and Shachak 1994). These species are often annuals that evolved in regions where soil-surface disturbance is relatively high. Consequently, germination of these species is generally inhibited in areas where litter is sparse, soil crusts stabilize soils, and surface disturbance is low. This has been demonstrated for several annual *Bromus* species in the US and Israel (Evans and Young 1984; Larsen 1995; Howell 1998; Kaltenecker et al. 1999; Prasse 1999), *Schismus* in both Australia and Israel (Crisp 1975; Zaady et al. 1997), *Plantago* and *Reboudia* in Israel (Zaady et al. 1997) and *Salsola* in Australia and the US (West 1990). Increased spatial and temporal frequency of surface disturbance generally favors germination for these species. Unfortunately, this often results in exotic annual invasions of perennial plant-dominated cool desert communities (Larsen 1995). Increasing soil surface disturbance in ecosystems where annuals and disturbance are a significant component (e.g., Negev and Sonoran Deserts) leads to some increased plant productivity; however, seed availability and dispersal may quickly become limiting (Boeken et al. 1998).

Soil mobility is also an important control on seed germination. Seeds generally have optimal burial depths, and increased or decreased burial can prevent germination (Baskin and Baskin 1998). Because soil crusts enhance soil stability (see Part V), their presence or absence can differentially affect burial and subsequent germination of seeds.

Nutrient and pH Influence on Seed Germination. Laboratory studies show that N additions can stimulate seed germination in some plant species (Baskin and Baskin 1998). Consequently, germination of some species may be stimulated with the presence of crusts; however, we are not aware of any experiments which directly test this hypothesis. Similarly, extreme soil pH affects seed germination (Baskin and Baskin 1998). Thus, the large increase in soil pH seen on crusted soil surfaces may influence seed germination of some plant species. Again, no experiments have been done to directly investigate this.

Influence of Soil Crusts on Soil Thermal and Water Relations. We found no work that specifically addresses how the thermal and moisture environments of soil crusts affect seed germination. Mücher et al. (1988) found higher seed germination on crusted surfaces in dry woodlands of Australia, and attributed this to the higher soil moisture found on crusted surfaces; however, this was not directly measured. Gold and Bliss (1995) speculated that increased seed germination and plant biomass on crusted surfaces was due to substantial increases in soil surface and subsurface temperatures, as well as subsurface moisture. Again, this was not directly tested.

21.4 Influence of Soil Crusts on Vascular Plant Survival, Biomass, Fecundity and Nutrition

21.4.1 Plant Survival

Cool and Cold Deserts. Studies from cool and cold deserts have spanned many plant species and life-forms, and have included both coarse- and fine-textured soils. All studies have found that biological soil crusts either increase or do not affect vascular plant biomass and survival. Harper and St. Clair (1985) seeded two perennial grasses, two perennial forbs, two annual forbs, and one shrub in crusted and uncrusted plots at three cool desert sites around Utah (USA). Soils ranged from 33 to 66 % sand. After 1 year, crusted plots had 2.8 times as many seedlings as scalped plots, with all species enhanced except the two perennial grasses, which were not affected. After 3 years, crusted plots had 1.6 times as many seedlings. All the species that survived on both the crusted and uncrusted plots had enhanced survival on the crusted plots. In a separate study, St. Clair et al. (1984) reported that survival of three perennial grasses was enhanced on crusted soils relative to adjacent uncrusted soils in both the field (1.3–4 times) and the greenhouse (1.6–12-fold).

Lesica and Shelley (1992) found that while crusts had no effect on the survival of small *Arabis fecunda* individuals (a perennial forb), the presence of crusts enhanced the long-term survival of medium and larger individuals in Montana. Eckert et al. (1986) reported that crust removal by trampling decreased survival of perennial grass seedlings (*Poa sandbergii, Sitanion hystrix*) while increasing the emergence of the shrub *Artemisia tridentata* and of annual forbs. However, after long-term protection, crusted areas supported more of both the perennial grasses and the shrub. DeFalco (1995) reported a greater number of individuals of five annual grasses and forbs on crusted soils relative to adjacent uncrusted soils. Jeffries and Klopatek (1987) and Kleiner and Harper (1977) report that vascular plant species richness and cover was positively correlated with crusted soils, while Anderson et al. (1982) found no significant correlation between crusts and vascular plant cover. Meyer (1986) reported a positive correlation between vascular plant species richness and crust development on gypsiferous soils.

As with germination, Larsen (1995) showed that seedling density of native *Stipa* was not affected by biological soil crusts relative to uncrusted soils. However, in contrast to native annuals, density of the exotic annual grass *Bromus tectorum* was lower on intact crusts. *Bromus* density was greater on crusts that were broken, but left in place. Greater seedling success in crusted soils, when compared to uncrusted soils, has also been reported by Fritsch (1922), Booth (1941), Fletcher and Martin (1948), and Shields and Durrell (1964). The presence of a well-developed lichen-moss crust in arctic cold deserts supported a higher diversity of vascular plants. This was attributed to reduced cryoturbation and increased nutrient availability, organic matter, and soil temperature associated with the presence of soil crusts (Gold and Bliss 1995).

Hot Deserts. Eldridge and Greene (1994) found no relationship between Australian *Atriplex* shrub seedling establishment and crust cover. Graetz and Tongway (1986) found the cover of chenopod shrubs was positively correlated with crust cover, citing increased moisture and nutrients in the upper 10 cm as responsible. Survival of the native grass *Stipa* in Australia was not affected by the presence of soil crusts, while survival of the exotic *Schismus* was reduced on crusted sufaces (Crisp (1975). The annual grasses *Bromus* and *Schismus* have been found to have higher densities on broken crust in the cool deserts of the US and hot deserts of Australia and Israel (Crisp 1975; Larsen 1995; Prasse 1999). Prasse (1999) also found higher densities of all annuals on crusts that were broken. He also noted no effect of crusts on plant mortality. Studies in Israel showed that vascular plant densities correlated with surface stability (Danin 1978; Kadmon and Leschner 1995; Tielbörger 1997), although a direct connection with biological soil crusts has not been established.

21.4.2 Plant Biomass

Cool Deserts. Pendleton and Warren (1995) reported that one annual grass (*Bromus tectorum*), one perennial grass (*Sitanion hystrix*), two perennial forbs (*Gaillardia pulchella*, *Sphaeralcea munroana*) and one shrub (*Coleogyne ramosissima*) all showed greater shoot and total plant weight in crusted soils than adjacent uncrusted surfaces. All species except *Coleogyne* had greater root weight. All species showed smaller root/shoot ratio in crusted soils, indicating greater soil nutrients were available in the crusted soils. Schlatterer and Tisdale (1969) showed that the height and biomass of *Sitanion hystrix* were significantly increased, and biomass and height of *Stipa thurberiana* and *Agropyron spicatum* were not affected by moss cover when compared to artificial litter. Generally, height and biomass were greater with moss cover than no litter for all three species.

DeFalco (1995) reported greater biomass in individuals of five annual grasses and forbs on crusted soils relative to adjacent uncrusted soils. In another study, biomass of the native annual grass *Festuca octoflora* was twice that on crusted soils compared to uncrusted soils (Belnap 1995). Similar results were obtained by other authors (Brotherson and Rushforth 1983; Shields and Durrell 1964). In the Arctic, crusted sites had higher plant diversity and biomass. These sites also had increased soil nutrients, organic matter, and temperature, as well as reduced soil cryoturbation, due to the presence of crusts (Gold and Bliss 1995).

Hot deserts. In southern Arizona, McIlvanie (1942) found the biomass of four species increased 45–300 % when grown on crusted surfaces relative to uncrusted surfaces. However, it should be noted that he was also comparing grazed and protected soils. In Israel, vascular plant biomass was correlated with surface stability (Danin 1978; Kadmon and Leschner 1995; Tielbörger

1997). However, the role of biological soil crusts in these findings has not been directly tested.

21.4.3 Vascular Plant Fecundity

Only two studies have examined the effect of crusts on plant fecundity. Lesica and Shelley (1992) found that crusted soils had no effect on the fecundity of any size class of the perennial *Arabis fecunda* in Montana. In a hot desert (Israel) on a smooth cyanobacterial crust, Prasse (1999) found the annuals *Ifloga spicata* and *Senecio glaucus* had the same or higher capitules/plant where crusts were broken or removed compared to where crusts were left intact. Enhanced plant fecundity was attributed to greater water interception in areas where the crusts were broken.

21.4.4 Vascular Plant Nutrition

Alteration of soil-surface chemistry by biological soil crusts is significantly correlated with changes in the content of bioessential elements in the tissue of associated seed plants (Harper and Marble 1988; Belnap and Harper 1995; Harper and Pendleton 1993; Pendleton and Warren 1995; Harper and Belnap 2000). Findings from these studies are summarized for ten plant-essential elements in Table 21.1. All studies occurred with *Microcoleus-Nostoc-Collema* soil crusts on sandy, calcareous soils. Soil crusts enhanced N content of associated seed plants in all trials. In contrast, P in plant tissue from crusted areas was increased as often as reduced. The presence of soil crusts generally enhanced plant uptake of Mg, K, Cu, and Zn, while generally reducing Fe uptake. Ca and Mn in tissue of seed plants grown in crusted soils show no consistent differences from control plants.

Other researchers have found similar results. Schlatterer and Tisdale (1969) observed that N was higher in *Stipa thurberiana* (18 %), *Sitanion hystrix* (10 %), and *Agropyron spicatum* (19 %) under moss litter when compared to artificial litter. With no litter, only *Agropyron* had higher N (18 %). They found no difference in P among the treatments. DeFalco (1995) measured N and P in five annual forbs. Individual plants in crusted soils had lower concentrations of N; however, plant biomass was much higher on crusted soils. When plant concentrations were calculated on a m^2 basis, total N and P was much higher in combined plant tissue on the crusted soils. Individually, two of the species showed no difference between crusted and uncrusted soils (*Cryptantha pterocarya*, *Eriophyllum wallacei*). *Bromus rubens* and *Erodium cicutarium* showed higher N on crusted soils (sixfold and tenfold, respectively) and P (ninefold and tenfold, respectively). *Descurainia pinnata*, on the

Table 21.1. Elemental content of vascular plant tissue. (Harper and Belnap 2000)

Element	Substrate	Shrub		Forb				Grass
		CORA seedlings	*CORA* adults	*BRRA* adults	*LEMOJO* adults	*LEMOMO* adults	*STLO* adults	*STHY* adults
Ca, %	No Crust	0.88	3.13	2.98	0.62	1.17	1.16	1.42
	Crust	0.86	3.19	2.44*	0.64	1.41	1.80**	1.00**
K, %	No Crust	1.19	0.80	1.72	2.19	3.03	1.49	1.50
	Crust	0.86**	0.78	1.43**	2.26*	4.15*	2.06**	2.01*
Mg, %	No Crust	0.49	0.41	0.45	0.29	0.34	0.20	0.14
	Crust	0.46	0.43	0.34**	0.26*	0.43	0.25**	0.14
N, %	No Crust	5.60	0.98	NM	1.29	1.42	1.44	1.99
	Crust	5.83	1.07	1.24	1.41*	1.58	2.09**	2.30**
P, %	No Crust	0.55	0.11	0.35	0.07	0.25	0.28	0.24
	Crust	0.69*	0.11	0.23**	0.08*	0.28	0.27	0.13**
Cu, ppm	No Crust	13.80	4.00	4.80	8.60	3.08	5.80	8.60
	Crust	15.00*	4.00	3.80	8.80	6.33*	7.20*	9.10
Fe, ppm	No Crust	765.40	138.60	66.60	29.70	474.10	419.20	130.40
	Crust	646.90	84.60**	60.50	27.60	772.70	222.40*	117.90
Mn, ppm	No Crust	0.49	0.41	119.10	49.50	52.50.	28.80	24.20
	Crust	0.46	0.43	163.90**	47.10*	102.50	25.00	71.90**
Na, ppm	No Crust	61.50	8.40	28.50	51.30	161.00	NM	NM
	Crust	69.80	9.20	22.50	50.70	281.00*	NM	NM
Zn, ppm	No Crust	51.10	13.00	829.70	24.80	12.20	20.20	7.70
	Crust	74.10**	11.20	917.10	23.90	15.20	19.80	12.60*

CORA=*Coleogyne ramosissima*, *BRRA*=*Brassica rapa*, *LEMOJO*=*Lepidium montanum var. jonesii*, *LEMOMO*=*L. montanum* var. *montanum*, *STLO*=*Streptanthella longirostris*, *STHY*=*Stipa hymenoides*.
For adult *Coleogyne*, leaves and current year twigs were analyzed.
For *LEMOJO* and *STHY*, leaves were analyzed.
For *BRRA*, *LEMOMA*, and *STLO*, roots, stems and leaves were analyzed.
Crusted and uncrusted soils were separated by more than a 4.0 m linear distance. Sites were similar in soil type, aspect and vascular plant cover.
*=$P<0.05$, **=$P<0.01$. NM=not measured.

other hand, had lower N and P (3× and 10×, respectively). Cyanobacterial inoculations have long been known to increase N in rice plants (Venkataraman and Neelakantan 1967) and vegetable crops (Dadlich et al. 1969).

The negative effects of crusts on plant P and Fe may be related to competition between the soil-crust organisms and the roots of seed plants for bioessential elements. P and plant-available Fe are often in short supply in the calcareous, sandy soils common to many deserts (Black 1968). Thus, crustal microorganisms may compete with vascular plant roots for these elements. This competitive effect is expected to be most severe for small, short-lived (especially annual) plants that root heavily at the soil surface, which is the area most influenced by the biological soil-crust organisms. We have observed that many plants, including shrubs, produce hair-like feeder roots in the surface 2 cm of soil, especially in seasons of unusually abundant rainfall (Belnap and Harper 1995; Harper and Belnap 2001). Such roots could scarcely escape competition for bioessential elements with organisms of the microbiotic surface crusts.

Biological crusts can exert positive influence on uptake of essential elements in a variety of ways. Crusted soils are enriched by cyanobacterially-fixed N, and uptake of this N by associated organisms has been documented (see Sect. 21.2.2 above; Chap. 19). Soil content of K, Mn, Ca, Mg, and P are consistently shown to be greater where biological crusts occur. Although Fe and Zn soil concentrations are generally lower in crusted areas, plant tissue concentrations of Zn and, occasionally, Fe are elevated (Belnap and Harper 1995; Harper and Belnap 2000; Harper and Pendleton 1993). Thus, the biological crusts appear to somehow render those elements more available to plant roots. It is well known that both Zn and Fe, along with other elements such as Cu, Co, Mn, and Mo, can form stable complexes with chelating compounds (see Sect. 21.2.2).

Studies examining mycorrhizal infection rates show plants growing on biologically crusted surfaces consistently have higher infection rates than the same species growing on adjacent, uncrusted surfaces. This association may also partially explain higher nutrient concentrations in leaf tissue. Differences may also be partially ascribable to a thermal effect. Dark crusts are warmer than lighter uncrusted soils; therefore, uptake of nutrients would occur at a higher rate (Harper and Pendleton 1993).

Herbivores and other consumers may benefit indirectly from the enhanced nutrient status of these ecosystems (Belnap and Harper 1995). Populations of microarthropods are also correlated with the presence of well-developed soil crusts (see Chap. 14). Subsequent enhancement of decomposition rates may also contribute to greater availability of nutrients to associated vascular plants. Secreted subsurface carbon compounds may also influence water availability to vascular plant seedlings.

21.5 Conclusions

The effect of soil crusts on final seed placement once seeds are dispersed from the parent plant is fairly well understood. Seed entrapment is low in areas where soil surfaces are smooth (cyanobacterial crusts in hot deserts or soils with significant physical or chemical crusting that lack frost heaving), while entrapment is high in regions where crusts significantly increase surface roughness (well-developed crusts and frost-heaving).

The effect of biological soil crusts on seed germination is more complex. Germination of annuals and seeds that lack burial mechanisms appear most affected by biological soil crusts, especially in areas with smooth soil surfaces or for plant seeds that lack structures that facilitate soil-seed contact. Consequently, crusts may significantly affect plant germination and initial establishment in ecosystems where annuals or plants without burial mechanisms contribute a significant amount of the primary productivity, and where soil surfaces are smooth. This combination of characteristics is most often found in hot deserts. In contrast, crusts probably have much less influence on plant germination and initial establishment in communities dominated by perennial plants and/or seeds with burial mechanisms, and in areas with roughened soil surfaces that facilitate seed-soil contact. These characteristics are most often found in cool and cold deserts. However, it should be kept in mind that the ultimate determinant of perennial vascular plant density or biomass in any desert is seldom seed dispersal or germination per se. Instead, water and nutrients are generally the limiting factors (Mabbutt and Fanning 1987; Tongway and Ludwig 1990; Dunkerley and Brown 1995). Introduced annual grasses that lack burial mechanisms have shown inhibited germination on all intact crust types in both hot and cold deserts. This may result from their evolution in livestock-disturbed areas.

Once plants have become established, the effect of biological soil crusts on survival and biomass appears universally positive or without effect in cool and cold deserts. No study has documented a negative relationship between plant survival or biomass and the presence of crusted soils in these habitats. In hot deserts, studies are much more limited, and thus few generalizations can be made. Biomass of the four perennial plants tested in one study showed biomass was higher, or not affected, by the presence of crusts. However, studies of annual plants show a highly variable response.

To fully understand the relationship between soil crusts and vascular plants, we need studies that specifically address the influence of biological soil crusts as separate from the influences of inorganic crusts, soil types, and climate regimes. We need data on how variation in crustal species composition affects seed dispersal and germination, as the influence of cyanobacterial crusts on soil factors may be very different from moss-lichen crusts

(West 1990). And lastly, we need a better understanding of how crust-caused modifications of the soil environment affect vascular plant communities.

Acknowledgments. The contributions of Michelle Schmid, Sue Philllips, Katja Tielbörger (Potsdam) and Otto L. Lange are gratefully acknowledged.

References

Anderson DC, Harper KT, Holmgren RC (1982) Factors influencing development of cryptogamic soil crusts in Utah deserts. J Range Manage 35:180–185

Baskin CC, Baskin JM (1998) Seeds. Ecology, biogeography, and evolution of dormancy and germination. Academic Press, New York

Belnap J (1995) Surface disturbances: their role in accelerating desertification. Environ Mon Assess 37:39–57

Belnap J, Gardner JS (1993) Soil microstructure in soils of the Colorado Plateau: the role of the cyanobacterium *Microcoleus vaginatus*. Great Basin Nat 53:40–47

Belnap J, Harper KT (1995) Influence of cryptobiotic soil crusts on elemental content of tissue of two desert seed plants. Arid Soil Res Rehabil 9:107–115

Beymer RJ, Klopatek JM (1992) Effects of grazing on cryptogamic crusts in pinyon-juniper woodlands in Grand Canyon National Park. Am Midl Nat 127:139–148

Black CA (1968) Soil-plant relationships. John Wiley, New York

Boeken B, Shachak M (1994) Desert plant communities in human-made patches – implications for management. Ecol Appl 4:702–716

Boeken B, Lipchin C, Gutterman Y, van Rooyen N (1998) Annual plant community responses to density of small-scale soil disturbances in the Negev Desert of Israel. Oecologia 114:106–117

Booth WE (1941) Algae as pioneers in plant succession and their importance in erosion control. Ecology 22:38–46

Bose P, Nagpal US, Venkataraman GS, Goyal SK (1971) Solubilization of tricalcium phosphate by blue-green algae. Curr Sci 7:165–166

Brotherson JD, Rushforth SR (1983) Influence of cryptogamic crusts on moisture relationships of soils in Navajo National Monument, Arizona. Great Basin Nat 43:73–78

Büdel B (2000) Ecology and diversity of rock-inhabiting cyanobacteria in tropical regions. Europ J Phycol 34:361–370

Campbell SE (1979) Soil stabilization by a prokaryotic desert crust: implications for precambrian land biota. Origins Life 9:335–348

Carleton TJ (1990) Variation in terricolous bryophyte and macrolichen vegetation along primary gradients in Canadian boreal forests. J Veg Sci 1:585–594

Chartres CJ (1992) Soil crusting in Australia. In: Sumner ME, Stewart BA (eds) Soil crusting: chemical and physical processes. Lewis, Boca Raton, FL, pp 339–366

Chartres CJ, Mücher HJ (1989) The effects of fire on the surface properties and seed germination in two shallow monoliths from a rangeland soil subjected to simulated raindrop impact and water erosion. Earth Surf Proc Landforms 14:407–417

Cornet AF, Montana C, Delhoume JP, Lopez-Portillo J (1992) Water flows and the dynamics of desert vegetation stripes. In: Hansen AJ, di Castri F (eds) Landscape boundaries: consequences for biotic diversity and ecological flows. Springer, Berlin Heidelberg New York, pp 327–345

Crisp MD (1975) Long-term change in arid zone vegetation. Unpubl PhD University of Adelaide, Australia

Dadlich KS, Varma AK, Venkataraman GS (1969) The effect of *Calothrix* inoculation on vegetable crops. Plant Soil 31:377–379

Danin A (1976) Plant species diversity under desert conditions. I. Annual species diversity in the Dead Sea Valley. Oecologia 22:251–259

Danin A (1978) Plant species diversity and plant succession in a sandy area in the Northern Negev. Flora 167:409–422

DeFalco LA (1995) Influence of cryptobiotic crusts on winter annuals and foraging movements of the desert tortoise. Dept of Biol, Colorado State University, Fort Collins, CO

Dunkerley DL, Brown KJ (1995) Runoff and runon areas in a patterned chenopod shrubland, arid western New South Wales, Australia: characteristics and origin. J Arid Environ 30:41–55

Eckert RE, Peterson FF, Meurisse MS, Stephens JL (1986) Effects of soil-surface morphology on emergence and survival of seedlings in big sagebrush communities. J Range Manage 39:414–420

Eldridge DJ (1993) Cryptogams, vascular plants, and soil hydrological relations: some preliminary results from the semiarid woodlands of eastern Australia. Great Basin Nat 53:48–58

Eldridge DJ, Green RSB (1994) Microbiotic soil crusts: a review of their roles in soil and ecological processes in the rangelands of Australia. Aust J Soil Res 32:389–415

Eldridge DJ, Lepage M, Bryannah MA, Ouedraogo P (2001) Soil biota in banded landscapes. In: Tongway DJ, Valentin C, Seghieri J (eds) Banded vegetation patterning in arid and semiarid environments: ecological processes and consequences for management. Springer, Berlin Heidelber New York, pp 105–131

Evans RD, Belnap J (1999) Long-term consequences of disturbance on nitrogen dynamics in an arid ecosystem. Ecology 80:150–160

Evans RD, Ehleringer JR (1993) A break in the nitrogen cycle in arid lands? Evidence from ^{15}N of soils. Oecologia 94:314–317

Evans RA, Young JA (1984) Microsite requirements for downy brome (*Bromus tectorum*) infestation and control on sagebrush rangelands. Weed Sci 32:13–17

Fletcher JE, Martin WP (1948) Some effects of algae and molds in the rain-crust of desert soils. Ecology 29:95–100

Fogg GE (1966) The extracellular products of algae. Oceanogr Mar Biol Annu Rev 4: 195–212

Fritsch FE (1922) The moisture relations of terrestrial algae. Ann Bot 36:1–20

Gadd GM (1990a) Biosorption. Chem Ind 13:421–426

Gadd GM (1990b) Fungi and yeasts for metal accumulation. In: Ehrlich HL, Brierley CL (eds) Microbial mineral recovery. McGraw-Hill, New York, pp 249–275

Garcia-Pichel F, Belnap J (1996) Microenvironments and microscale productivity of cyanobacterial desert crusts. J Phycol 32:774–782

Geesey G, Jang L (1990) Extracellular polymers for metal binding. In: Ehrlich HL, Brierley CL (eds) Microbial mineral recovery. McGraw-Hill, New York, pp 223–247

Gold WG, Bliss LC (1995) Water limitations and plant community development in a polar desert. Ecology 76:1558–1568

Graetz RD, Tongway DJ (1986) Influence of grazing management on vegetation, soil structure and nutrient distribution and the infiltration of applied rainfall in a semi-arid chenopod shrubland. Aust J Ecol 11:347–360

Greene B, Darnall DW (1990) Microbial oxygenic photoautotrophs (cyanobacteria and algae) for metal-ion binding. In: Ehrlich HL, Brierley CL (eds) Microbial mineral recovery. McGraw-Hill, New York, pp 277–302

Gutterman Y (1994) Strategies of seed dispersal and germination in plants inhabiting deserts. Bot Rev 60:373–425

Hacker RB (1984) Vegetation dynamics in a grazed mulga shrubland community. I. The mid-storey shrubs. Aust J Bot 32:239–249

Harper JL (1977) Population biology of plants. Academic Press, New York

Harper KT, Belnap J (2001) The influence of biological soil crusts on mineral uptake by associated vascular plants. J Arid Environ 47:347–357

Harper KT, Marble JR (1988) A role for nonvascular plants in management of arid and semiarid rangelands. In: Tueller PT (ed) Vegetation science applications for rangeland analysis and management. Kluwer Academic Publishers, Dordrecht, pp 136–169

Harper KT, Pendleton RL (1993) Cyanobacteria and cyanolichens: can they enhance availability of essential minerals for higher plants? Great Basin Nat 53:59–72

Harper KT, St. Clair LL (1985) Cryptogamic soil crusts on arid and semiarid rangelands in Utah: effects on seedling establishment and soil stability. Dept Botany Range Sci, Brigham Young University, Provo, Utah

Howell W (1998) Germination and establishment of *Bromus tectorum* L. in relation to cation exchange capacity, seedbed, litter, soil cover and water. Unpubl Thesis, Prescott College, AZ

Jeffries DL, Klopatek JM (1987) Effects of grazing on the vegetation of the blackbrush association. J Range Manage 40:390

Johansen JR (1993) Cryptogamic crusts of semiarid and arid lands of North America. J Phycol 29:140–147

Kadmon R, Leschner H (1995) Ecology of linear dunes: effect of surface stability on the distribution and abundance of annual plants. Adv Geoecol 28:125–143

Kaltenecker JH, Wicklow-Howard M, Pellant M (1999) Biological soil crusts: natural barriers to *Bromus tectorum* L. establishment in the northern Great Basin, USA. In: Eldridge D, Freudenberger D (eds) Proc VI Int Rangeland Congress, Aitkenvale, Queensland, Australia, pp 109–111

Kleiner EF, Harper KT (1972) Environment and community organization in grasslands of Canyonlands National Park. Ecology 53:229–309

Kleiner EF, Harper KT (1977) Soil properties in relation to cryptogamic groundcover in Canyonlands National Park. J Range Manage 30:202–205

Ladyman JAR, Muldavin E (1994) A study of the terricolous cryptogam and other ground cover in low disturbance pinyon–juniper woodlands in New Mexico. Unpubl report to the USDA Forest Service, Albuquerque, NM

Lange W (1974) Chelating agents and blue-green algae. Can J Microbiol 20:1311–1321

Lange W (1976) Speculations on a possible essential function of the gelatinous sheath of blue-green algae. Canad J Microbiol 22:1181–1185

Larsen KD (1995) Effects of microbiotic crusts on the germination and establishment of three range grasses. Unpubl Thesis. Boise State University, Boise, ID

Lesica P, Shelly JS (1992) Effects of cryptogamic soil crust on the population dynamics of *Arabis fecunda* (Brassicaceae). Am Midl Nat 128:53–60

Lewin RA (1956) Extracellular polysaccharides of green algae. Can J Microbiol 2:665–673

Mabbutt JA, Fanning PC (1987) Vegetation banding in arid western Australia. J Arid Environ 12:41–59

McIlvanie SK (1942) Grass seedling establishment, and productivity-overgrazed vs. protected range soils. Ecology 23:228–231

McLean RJC, Beveridge TJ (1990) Metal-binding capacity of bacterial surfaces and their ability to form mineralized aggregates. In: Ehrlich HL, Brierley CL (eds) Microbial mineral recovery. McGraw-Hill, New York, pp 185–222

Meyer SE (1986) The ecology of gypsophile endemism in the eastern Mojave Desert. Ecology 67:1303–1313

Mitchell DJ, Fullen MA (1994) Soil-forming processes on reclaimed desertified land in north-central China. In: Millington AC, Pye K (eds) Environmental change in drylands: biogeographical and geomorphological perspectives. John Wiley, New York, pp 394–412

Mücher HJ, Chartres CJ, Tongway DJ, Greene RSB (1988) Micromorphology and significance of the surface crusts of soils in rangelands near Cobar, Australia. Geoderma 42:227–244

Nobel PS (1991) Physicochemical and environmental plant physiology. Academic Press, San Diego, CA

Paerl HW (1988) Growth and reproductive strategies of freshwater blue-green algae (cyanobacteria). In: Sandgren CD (ed) Freshwater blue-green algal ecology. Cambridge Univ Press, Cambridge, pp 261–315

Pendleton RL, Warren SD (1995) Effects of cryptobiotic soil crusts and VA mycorrhizal inoculation on growth of five rangeland plant species. In: West NE (ed) 5th Int Range Congress. Society for Range Management, Salt Lake City, UT, pp 436–437

Pirszel J, Pawlik B, Skowronski T (1995) Cation-exchange capacity of algae and cyanobacteria: a parameter of their metal sorption abilities. J Ind Microbiol 14:319–322

Prasse R (1999) Experimentelle Untersuchungen an Gefäßpflanzenpopulationen auf verschiedenen Geländeoberflächen in einem Sandwüstengebiet (Experimental studies with populations of vascular plants on different soil surfaces in a sand desert area). Universitätsverlag Rasch, Osnabrück, Germany

Rao DLN, Burns RG (1990) Use of blue-green algae and bryophyte biomass as a source of nitrogen for oil-seed rape. Biol Fertil Soils 10:61–64

Rogers SL, Burns RG (1994) Changes in aggregate stability, nutrient status, indigenous microbial populations, and seedling emergence, following inoculation of soil with *Nostoc muscorum*. Biol Fertil Soils 18:209–215

Savory A (1988) Holistic resource management. Island Press, Covelo, California

Schelske CL, Hooper FF, Haertl EJ (1962) Responses of a marl lake to chelated iron and fertilizer. Ecology 43:646–653

Schlatterer EF, Tisdale EW (1969) Effects of litter of *Artemisia, Chrysothamnus* and *Tortula* on germination and growth of three perennial grasses. Ecology 50:869–873

Shields LM, Durrell LW (1964) Algae in relation to soil fertility. Bot Rev 30:92–128

St Clair LL, Webb BL, Johansen JR, Nebeker GT (1984) Cryptogamic soil crusts: enhancement of seedling establishment in disturbed and undisturbed areas. Reclam Res 3:129–136

Tielbörger K (1997) The vegetation of linear desert dunes in the north-western Negev, Israel. Flora 192:261–278

Tongway DJ, Ludwig JA (1990) Vegetation and soil patterning in semi-arid mulga lands of eastern Australia. Aust J Ecol 15:23–34

Venkataraman GS, Neelakantan S (1967) Effect of the cellular constituents of the nitrogen-fixing blue-green alga *Cylindrospermum muscicola* on the root growth of rice seedlings. J Gen Appl Microbiol 13:53–61

Verrecchia E, Yair A, Kidron GJ, Verrecchia K (1995) Physical properties of the psammophile cryptogamic crust and their consequences to the water regime of sandy soils, north-western Negev Desert, Israel. J Arid Environ 29:427–437

West NE (1990) Structure and function of microphytic soil crusts in wildland ecosystems of arid to semi-arid regions. Adv Ecol Res 20:179–223

Zaady E, Shachak M (1994) Microphytic soil crust and ecosystem leakage in the Negev Desert. Am J Bot 81:109

Zaady E, Gutterman Y, Boeken B (1997) The germination of mucilaginous seeds of *Plantago coronopus*, *Reboudia pinnata*, and *Carrichtera annua* on cyanobacterial soil crust from the Negev Desert. Plant Soil 190:247–252

Part V

Soil Stability and Hydrology as Influenced by Soil Crusts

22 Effects of Biological Soil Crusts on Water Redistribution in the Negev Desert, Israel: a Case Study in Longitudinal Dunes

A. Yair

22.1 Introduction

The important role of biological crusts in arid ecosystems is widely recognized (see reviews in Isichei 1990; West 1990; Belnap 1994). However, their hydrological role is not clear. Studies from regions with soil freezing (and therefore pinnacled or rolling crusts; see Chap. 15) or fine-grained soils show that soil crusts can increase infiltration and soil moisture content (Booth 1941; Fletcher and Martin 1948; Loope and Gifford 1972; Brotherson et al. 1983; Eldridge and Tozer 1997; Perez 1997). Bare loamy soils are very sensitive to surface sealing by raindrop impact (McIntyre 1958; Morin and Benyamini 1977). Under such conditions, the cohesive and flexible biological elements absorb raindrop energy and prevent the rapid development of a rain crust conducive to runoff generation. Quite different conditions prevail in hot sandy deserts, which are characterized by extremely high infiltration rates. Here, smooth biological crusts generally reduce infiltration and generate runoff (Bond 1964; Roberts and Carson 1971; Dekker and Jungerius 1990; Yair 1990; Bisdom et al. 1993; Danin 1996; Kidron and Yair 1997). This phenomenon is explained by clogging of soil pores by microbially-produced polysaccharides (Avnimelech and Nevo 1964), the inherent water-repellent properties of some crusts, and surface-sealing processes caused by the combined swelling of microorganisms and soil fine particles when wetted (Campbell 1979; Wang et al. 1981; Verrecchia et al. 1995).

Due to the scarcity and short duration of rainstorms in arid areas, runoff generation plays an important role in arid ecosystems. Runoff generated at one site may be absorbed at a nearby area. Under such conditions, the spatial redistribution of water resources may affect plant and animal distribution, and may explain vegetation patterns within the ecosystem. The process of water redistribution caused by a biological soil crust is especially important in hot-desert sandy ecosystems. Sands devoid of biological crusts are known for their high infiltration capacity, and are able to absorb all rainwater at all rainfall intensities. The result is a uniform depth of water penetration and a

Ecological Studies, Vol. 150
J. Belnap and O.L. Lange (eds.) Biological Soil Crusts: Structure, Function, and Management

limited variability in edaphic conditions. The presence of a biological crust can redistribute surface water, thus increasing diversity of edaphic conditions in the system.

The present chapter has two aims: (1) to report the results of a systematic field monitoring of the frequency and magnitude of runoff generation from a biologically-crusted, sandy substratum, and (2) to analyze the role of surface runoff in the spatial redistribution of water resources in a longitudinal dune ecosystem.

22.2 Description of the Study Area

The Nizzana Research Site is located within the Hallamish dune field in the northwestern Negev Desert (Fig. 22.1), and is the eastern part of the extensive Sinai erg. It is characterized by west-east-trending, longitudinal dunes (15–20 m high), separated by 100–200-m wide interdune corridors (see also Chap. 31). Average annual rainfall is 95 mm, restricted to the winter months. Average monthly temperatures are 9 °C in January and 25 °C in August (Rosenan and Gilad 1985). Potential annual evaporation is approximately 2600 mm (Evenari 1980). The biological crust (see Chap. 1, Photos 49, 50) is dominated

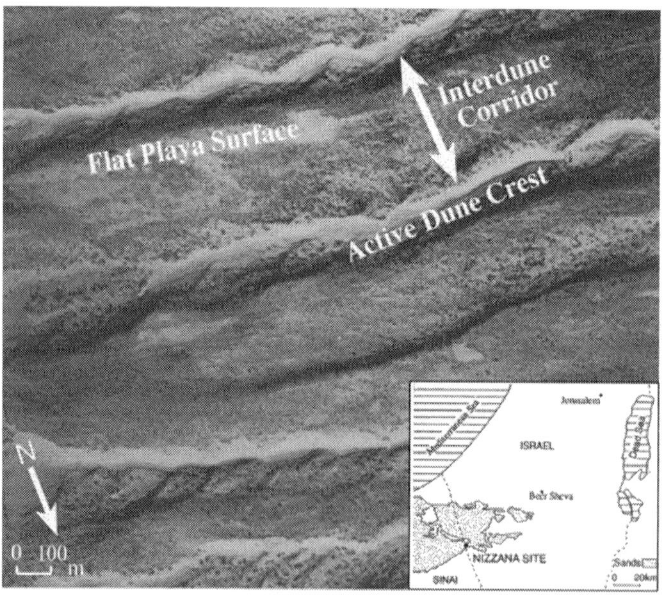

Fig. 22.1. General view of the longitudinal dunes and location map of Nizzana research site

Fig. 22.2. Dune crest with mobile sand (*right*) and crusted and vegetated area (*left*). (Photo A. Yair)

by cyanobacteria (*Microcoleus sociatus* and other *Microcoleus* spp., *Calothrix parietina, Nostoc microscopicum*) and also contains green algae and mosses (Lange et al. 1992; see also Chap. 8). Mosses are restricted to the wettest microenvironments found mostly at the concave base of the dune ridges. The degree of crust development and vascular plant cover increases from the upper to the lower part of the dune slope (Fig. 22.2). Crusts have greater chlorophyll *a* content on north-facing slopes (20–50 mg m^{-2}) than south-facing slopes (15–20 mg m^{-2}; Kidron and Yair 1997). The crust layer contains up to 50 % silt and clay as compared to 3–7 % in the underlying sand (Yair 1990). The dune soils are classified as unconsolidated arenosols, and interdune soils are characterized as sali-calcaric arenosols over the playa units (Blume et al. 1995).

22.3 Field Instrumentation and Methodology

Plots were constructed on north (N) and south (S) slopes (Fig. 22.3). Fig. 22.4 shows a typical experimental layout. Plots N1, N2, S1 and S3 were subdivided into upper, middle, and lower subplots (e.g., N1.1, N1.2, N1.3, respectively; Table 22.1). Plots N3, N4, S2, and S4 extended over the entire dune slope. Plots C1 and C4 covered a small stabilized dune in the interdune area. Miniplots NM2 and SM2 were located at the bottom of the north- and south-facing slopes, respectively. Rainfall was recorded with an electronic tipping-bucket rain recorder. Runoff was measured either with a V-notch weir and a pressure-gauge transducer or collected in large containers.

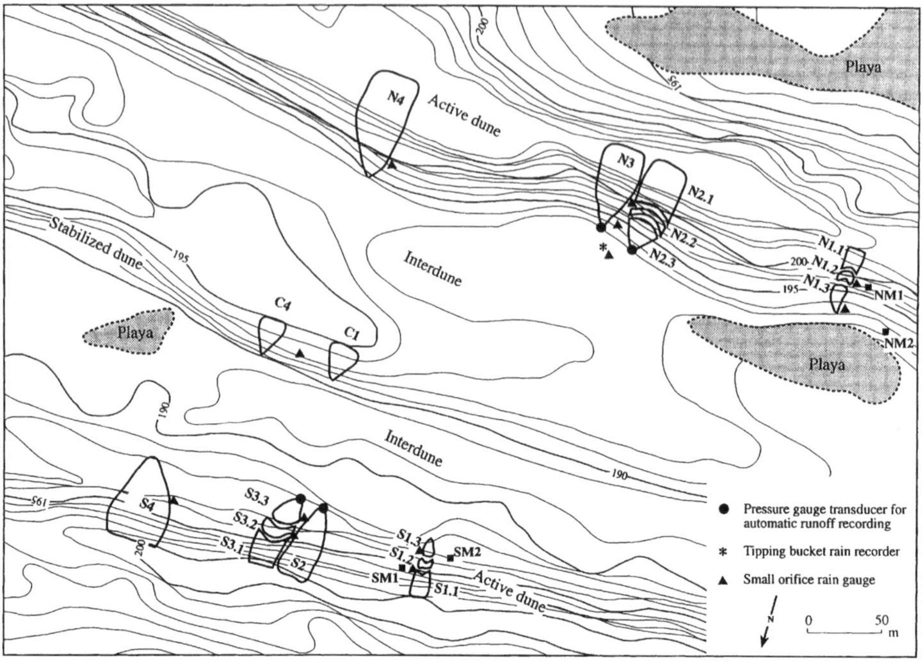

Fig. 22.3. Location of the runoff plots. (After Yair and Abeliovitch 1994; Kidron 1995)

Fig. 22.4. View of the runoff plots. (Photo A. Yair)

Table 22.1. Main characteristics of runoff plots. (After Yair and Abeliovitch 1994; Kidron 1995)

	Plot number	Plot characteristics	Plot location
Subdivided large plots	N1.1, N2.1, S1.1, S3.1	No crust, <5 % vegetation	Dune crest
	N1.2, N2.2, S1.2, S3.2	40–75 % patchy crust, 10–30 % vegetation	Mid-dune
	N1.3, N2.3, S1.3, S3.3	>96 % crust, 40–60 % vegetation	Dune bottom
Miniplots	NM2, SM2	100 % crust	Dune bottom
Whole slope plots	N3, N4, S2, S4	0–97 % crust, 5–60 % vegetation	All slope positions
Interdune plots	C1, C4	>95 % crust, 10–40 % vegetation	Interdune

22.4 Results

22.4.1 Rainfall

The monitoring period covered 1990–1994 (Table 22.2). Above-average rain fell in the first 2 years (119.7 and 131.4 mm, respectively), while the 2 following years were normal to very dry (85.3 and 46.9 mm, respectively). Seventy percent of the 10–20 storms recorded during 1990–94 had rain amounts below 5 mm, and 85 % had rainfall below 15 mm. Rainstorms in the range of 38 to 50 mm were recorded in the first 2 rainy years. Seventy to 80 % of annual rains fell at intensities below 12 mm h^{-1}. Annual rain amounts at intensities higher than 12 mm h^{-1} varied from 7.1 to 14.1 mm. These amounts are low, especially if one considers that they were recorded in several rain bursts during the same rainstorm.

22.4.2 Runoff

One to five annual rainstorms yielded runoff during the research period. Runoff developed only when rain intensities exceeded 9–12 mm h^{-1} under wet surface conditions (Kidron and Yair 1997). Data obtained highlight the role of the biological crust in runoff generation. No runoff was collected from the large subplots that drained the upper parts of the slopes, where crusts were absent. Negligible runoff was collected from subplots draining the mid-slope section, where crusts were patchy and thin and often covered by sand during sandstorms. Runoff frequency and magnitude were highest on the crusted and densely vegetated areas, located at the lower flanks of the north- and south-facing slopes. As expected, antecedent high moisture conditions resulted in higher runoff frequency and magnitude during rainy years (1990–1991 and 1991–1992) than the 2 following dry years (Table 22.2).

The small miniplots (2–4 m^2) at the base of the slopes generally had the highest runoff (m^{-2} encrusted area; Table 22.2), and thus were the most responsive to rainfall; however, total volumes collected from these plots were generally quite low, except for a few storms when rain amounts and intensities were high and soils were already wet from previous rainy spells. Higher runoff was always obtained from NM2 relative to SM2.

Table 22.2. Runoff (l m^{-2}) for encrusted area. (After Kidron 1995)

Plot type		*Plots with a recording device*					
Plot name		N3	N2.3	S2	S3.3		
Encrusted area (m^2)		478	307	411	198		
Rainfall year	Total rain (mm)						
1990–91	119.7	0.50	0.91	0.05	0.25		
1991–92	131.4	0.40	0.91	0.07	0.76		
1992–93	85.3	0.02	0.07	0.07	0.33		
1993–94	46.9	0.04	0.11	0.01	0.05		
		Plots with a runoff collector					
Plot name		N4	N1.3	S4	S1.4	C1	C4
Encrusted area (m^2)		1093	74	799	30	188	238
Rainfall year	Total rain (mm)						
1990–91	119.7	0.71	7.03	0.25	0.16	0.90	0.92
1991–92	131.4	0.60	7.70	0.50	0.16	1.86	2.20
1992–93	85.3	0.11	3.11	0.16	0.82	0.05	0.15
1993–94	46.9	0.03	1.22	0.01	0.49	0.02	0.02
		Miniplots					
Plot name		NM	SM				
Encrusted area (m^2)		3.6	3.9				
Rainfall year	Total rain (mm)						
1990–91	119.7	12.80	1.95				
1991–92	131.4	9.00	2.30				
1992–93	85.3	7.00	1.80				
1993–94	46.9	3.20	0.82				

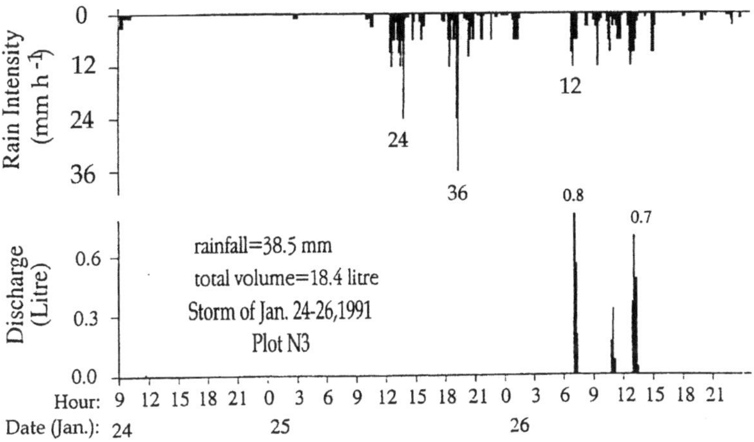

Fig. 22.5. Response of plot N3 to the storm of 24–26 January, 1991 (see text). (After Kidron and Yair 1997)

In the larger subplots and whole-slope plots, runoff volumes collected were generally low to very low, and often were an order of magnitude lower than runoff from the miniplots. A characteristic hydrological response to rainfall is presented in Fig. 22.5. This shows that rain was generally intermittent, with highly variable intensities. It also shows that runoff did not generally begin immediately after rainfall, even if rain intensity was high.

22.5 Discussion

Data analysis showed that variability in runoff volumes collected was not explained by total storm rain amount, storm peak rain intensity, or storm rain amount in excess of 12 mm h^{-1} (considered the threshold intensity for runoff generation in this study area; Yair 1990; Kidron and Yair 1997). This discussion will focus on three issues: (1) the conditions for runoff generation, (2) the extent of the contributing area, and (3) implications for the spatial redistribution of runoff.

22.5.1 Conditions for Runoff Generation

Figure 22.5 shows that rain intensities in excess of 12 mm h^{-1} at the beginning of a rainstorm are not sufficient for runoff generation. Instead, runoff begins when such intensity occurs after soils are wet at the end of the storm. This implies that surface sealing due to raindrop impact does not take place in the study area. Unlike other biological crusts reported in the literature, the crusted surfaces at Nizzana do not behave as hydrophobic, water-repellent soils (Kidron and Yair 1997). The biological soil crust at Nizzana behaves as a normal soil, capable of absorbing all rainwater at medium and high intensity rains when dry. Once wetted, infiltration decreases with time, due to swelling of fine-grained soil particles and cyanobacterial sheath material. Campbell (1979) and Wang et al. (1981) report cyanobacterial sheaths may absorb up to 12 times their dry weight and increase their volume up to 10 times. Once the crust is saturated, runoff occurs when rain intensities are higher than 12 mm h^{-1}. The fact that such intensities are not frequent, and last only for a few minutes, explains the limited amounts of runoff collected.

22.5.2 Extent of Contributing Area

The miniplots represent the units most responsive to rainfall. Despite their small dimensions, there is no correlation between total storm rainfall and runoff. The same lack of correlation was obtained for the larger plots (Kidron and Yair 1997). Such a result indicates the phenomenon of partial-area contribution; that is, areas contributing runoff were limited to the encrusted parts of the plots. One can gain an idea of the actual contributing area (in the large plots) by analyzing the ratio between peak flow rate and peak rain intensity in a given storm. Peak flow at the most extreme runoff event (22–23 March 1991) and the most responsive large plot (N2), was only 14.2 % of the causative rain burst (with an intensity of 83.5 mm h^{-1} lasting for 1 min). This rain-burst occurred at the very end of the storm when the biological crust had already reached saturation conditions. With a final infiltration rate of 12 mm h^{-1}, assuming that the whole crusted area of the plot had contributed runoff, peak flow value should have been 1.19 mm. The recorded value was, however, only 0.17 mm, indicating that only part of the area actually contributed runoff. The contributing area represented about 40–50 m^2 with a slope length of 7–10 m. The very steep rising and falling slopes of the hydrographs also indicate that only lower portions of the plot were contributing runoff.

The contributing areas were restricted to the lower part of the encrusted dune slope; this can be attributed to differences in infiltration rates within the large plots. It has already been mentioned that the thickness and spatial continuity of the crust decreased upslope, with the highest infiltration rate at the upper part of the encrusted slope, where the crust was thin and discontinuous. A smooth, thick, and continuous crust surface can be expected to enhance flow continuity in the downslope direction, while a patchy crust, with local sandy depressions, would impede this process. Such losses are minimized over a small area, thus explaining the high runoff rate per unit area recorded at the miniplots.

An interesting aspect is that runoff frequency and magnitude were highest on the lower part of the north-facing slope, where vegetation density is the highest. This is probably due to the fact that crust thickness, and its spatial continuity in the bare area between shrubs, was also highest there.

22.5.3 Implications for Spatial Runoff Water Redistribution

The data show that water runoff in crusted sandy soils is distributed differently from water runoff in uncrusted sandy soils. The area is composed of patches of runoff-contributing areas and run-on vegetated patches. Each shrub receives runoff from a limited contributing area extending for a short distance upslope. Some of this water flow would be expected to benefit the shrub (Eldridge et al. 2000). As each shrub can be regarded as a sink that interrupts the continuity of downslope water flow, the extent of the potential contributing area for each shrub is determined by the location of the neighboring shrubs. In view of short flow distances and total crust cover, data collected at the miniplots provide the best picture of water yield from a crusted surface at this site. Taking into account the density of the shrubs in the lower slope sections, each shrub has a potential contributing area of 4–6 m^2. Using the miniplot data (Table 22.2) each of the shrubs growing on the north-facing slope would theoretically have (in rainy years) on the order of 40 to 80 l of water available for collection. This amount is approximately equivalent to 40–90 % of the annual rainfall. Lower amounts (10 to 20 l) would be expected in drier years. Runoff contribution may be largest at the dune-interdune interface, due to the abrupt change in slope angle. The importance of runoff contribution to shrubs growing on south-facing slopes and at the mid-slope sections may be more limited due to their lower runoff frequency and magnitude.

However, altered spatial distribution of runoff does not explain the occurrence of narrow, densely vegetated belts found at slope concavities along the dune slope and at the base of dune slopes. In addition to surface runoff, subsurface flow is assumed to play an important role here. In the absence of a biological crust, all rainwater infiltrates at the upper part of the dune. However, due to the cross-bedding structure of the dune and localized differences in compaction and amounts of fine-grained material, infiltrated water does not percolate vertically but rather moves along flow lines parallel to the dune surface, thus concentrating locally at slope concavities. The phenomenon of subsurface flow was clearly observed and monitored at this site, where a neutron probe was used to detect water movement in 16 boreholes dug to a depth of 6 m (for more details see Yair et al. 1997). Slope concavities can therefore be regarded as special sites where both surface runoff and subsurface flow tend to concentrate, creating locally very wet environments that are rich in mosses and vascular plants.

22.6 Conclusions

Data obtained at the Nizzana site lead to three main conclusions:

1. Sandy areas, with well-developed dunes, are usually perceived as areas with very high infiltration rates (up to 300 mm h^{-1}) that prevent runoff generation. This is the case in moving unstable sands, where biological crusts cannot develop. However, stabilized or partly stabilized dunes, rich in fine-grained particles, allow the development of biological crusts. These crusts drastically reduce infiltration and enhance runoff generation. At the Nizzana Research Site, with an annual average rain amount of 95 mm, up to five runoff events were recorded in wet years, and one to two events in normal and dry rainfall years.
2. The biological crust plays an important role in the water regime in the area. It is responsible for runoff generation and nonuniform spatial distribution of rainwater and soil moisture. Runoff is higher on north-than south-facing slopes. On the north slopes in rainy years, runoff may locally increase water input to plants up to 80 % compared to direct rainfall. This process of water concentration may be responsible for the relatively high vegetal cover of the dunes that occurs despite the limited average annual rainfall, the long dry hot season, and a potential evaporation rate of the order of 2600 mm per year. Spatial differences in runoff may also explain the spatial differences in the density of the vegetal cover, which is less developed where the crust is patchy and thin. Thus, the crust plays a double role. On one hand, it limits water infiltration; but on the other hand, rainwater (minus infiltration) is transformed into surface runoff that can concentrate and infiltrate around the shrubs. Runoff water infiltration around shrubs may be enhanced by slowing water velocity and/or by the high water absorption capacity of the organic matter that accumulates below shrubs and by sand mounds, caused by bioturbation at their base. The process of water concentration is most efficient at slope concavities where surface and subsurface flows combine to create the wettest environment in the studied area.
3. Unlike many biological crusts described in the literature, the crust at the Nizzana dune field is not water-repellent. It can absorb large amounts of water when dry. Infiltration rate decreases upon wetting until saturation is reached at a rate of 9–12 mm h^{-1}. Runoff occurs once the crust is saturated.

Acknowledgments. The study was supported by the Arid Ecosystems Research Centre of the Hebrew University and a research grant by DISUM (German-Israeli Research Program). Warm thanks are due to Dr. G.J. Kidron and Mr. E. Sachs for their help in data collection and processing; to Mrs. M. Kidron of the Department of Geography of the Hebrew University for drawing the illustrations, and to Dr. J. Belnap for her fruitful comments.

References

Avnimelech Y, Nevo Z (1964) Biological clogging of sands. Soil Sci 98:222–226

Belnap J (1994) Potential role of cryptobiotic soil crusts in semiarid rangelands. Proceedings: Ecology and Management of annual range lands, US Department of Agriculture, Ogden, UT, pp179–185

Bisdom EBA, Dekker LW, Schoute JFT (1993) Water repellency of sieve fractions from sandy soils and relationships with organic material and soil structure. Geoderma 56:105–118

Blume HP, Yair A, Yaalon DH (1995) An initial study of pedogenic features along a transect across longitudinal dunes and interdune areas, Nizzana region, Israel. Adv Geoecol 28:51–64

Bond RD (1964) The influence of microflora on physical properties of sand. Effects associated with filamentous algae and fungi. Aust J Soil Res 2:123–131

Booth WE (1941) Algae as pioneers in plant succession and their importance in erosion control. Ecology 22:38–46

Brotherson JD, Rushforth SB, Johansen JR (1983) Influence of cryptogamic crusts on moisture relationships of soils in Navajo National Monument, Arizona. Great Basin Nat 43:73–78

Campbell SE (1979) Soil stabilization by prokaryotic desert crust. Implication for Precambrian land biota. Origins Life 9:335–348

Danin A (1996) Plants of desert dunes. Springer, Berlin Heidelberg New York

Dekker LW, Jungerius PD (1990) Water repellency in the dunes with special reference to the Netherlands dunes of the European Coasts. Catena, Suppl 18:173–183

Eldridge DE, Tozer ME (1997) A practical guide to soil lichens and bryophytes of Australia's dry country. Dept Land Water Conserv, Sydney

Eldridge DE, Zaady E, Shachak M (2000) Infiltration through three contrasting biological soil crusts in patterned landscapes in the Negev, Israel. Catena 40:323–336

Evenari M (1980) Ecology of the Negev Desert. A critical review of our knowledge. In: Shuval H (ed) Development in arid zone ecology and environmental quality. Balaban, Philadelphia, pp 1–33

Fletcher JE, Martin WP (1948) Some effects of algae and molds in the rain-crust of desert soil. Ecology 29:95–100

Isichei AO (1990) The role of algae and cyanobacteria in arid lands. A review. Arid Soil Res Rehabil 4:1–17

Kidron GJ (1995) The impact of a microbial crust upon rainfall-runoff and sediment yield relationships on longitudinal dune slopes, Nizzana, western Negev, Israel. PhD Thesis, Hebrew University of Jerusalem, Jerusalem (in Hebrew with English summary)

Kidron G J, Yair A (1997) Rainfall-runoff relationship over encrusted dune surfaces, Nizzana, western Negev, Israel. Earth Surface Processes Landforms 22:1169–1184

Lange OL, Kidron GJ, Büdel B, Meyer A, Kilian E, Abeliovitch A (1992) Taxonomic composition and photosynthetic characteristics of the biological soil crust covering sand dunes in the western Negev desert. Funct Ecol 6:519–527

Loope WL, Gifford GF (1972) Influence of a soil microfloral crust on selected properties of soils under pinyon-juniper in southeastern Utah. J Soil Water Conserv 28:27–52

McIntyre DS (1958) Soil splash and the formation of surface crusts by raindrop impact. Soil Sci 81:261–166

Morin J, Benyamini Y (1977) Rainfall infiltration into bare soils. Water Resour Res 13:813–817

Perez FL (1997) Microbiotic crusts in the high equatorial Andes, and their influence on Paramo soils. Catena 31:173–198

Roberts FG, Carson BA (1971) Water repellence in sandy soils of southwestern Australia. Aust J Soil Res 10:35–42

Rosenan N, Gilad M (1985) Atlas of Israel, meteorological data. Carta, Jerusalem

Verrecchia E, Yair A, Kidron GJ, Verrecchia K (1995) Physical properties of the psammophile cryptogamic crust and their consequences to the water regime of sandy soils, north western Negev, Israel. J Arid Environ 29:427–437

Wang F, Zhung Z, Hu Z (1981) Nitrogen fixation by an edible terrestrial blue-green algae. In: Gibson AH, Newton WE (eds) Current perspectives in nitrogen fixation. Elsevier, Amsterdam, pp 450–455

West NE (1990) Structure and function of microphytic soil crusts in wildland ecosystems of arid and semi-arid regions. Adv Ecol Res 20:179–223

Yair A (1990) Runoff generation in a sandy area – the Nizzana Sands, western Negev Israel. Earth Surface Processes Landforms 15:597–609

Yair A, Abeliovitch A (1994) Spatial variability of water resources in a sandy area. Final research report. DISUM research program in desert ecology. Hebrew University, Jerusalem

Yair A, Lavee H, Greitser N (1997) Spatial and temporal variabilty of water percolation and movement in a system of longitudinal dunes, Western Negev, Israel. Hydrol Process 11:43–58

23 Biological Soil Crusts and Water Relations in Australian Deserts

D.J. Eldridge

23.1 Introduction

In Australian rangelands, crusts comprise an assortment of lichens, bryophytes (mosses and liverworts), algae, cyanobacteria, and assorted bacteria and fungi (see Chap. 10). Whilst biological crusts in Australia are rather thin and occur often in association with physically or chemically crusted surfaces (Eldridge et al. 1995), they influence the flow of water through, and the movement of sediments over, the soil surface. Because of the marked differences in crust type and morphology at even small spatial scales, large differences in water flow through biological crusts often occur. It is these differences which make it difficult to generalize about the role of crusts in water flow in deserts.

In this chapter, the influence of biological soil crusts in hydrological processes will be discussed, drawing on published and unpublished literature from Australia, particularly the semiarid woodlands of eastern Australia. The influence of soil crusts on water infiltration and runoff, and soil erosion and deposition, will be discussed in relation to soil physical properties and type of biological soil crust.

23.2 Soil Hydrological Properties

The role of biological crusts in infiltration and runoff processes in the Australian context has received increasing attention over the past decade (Eldridge and Greene 1994b). Field observations of Australian biological soil crusts demonstrate the following: (1) biological soil crusts are extremely thin, usually extending less than a few millimeters below the ground surface (Chartres 1992); (2) except where thin cyanobacterial crusts occur on dune crests, they are almost always found in association with a range of chemical or physical crust types such as depositional and raindrop-impacted (structural) crusts; and (3) the visible components of biological soil crusts (i.e., the lichens

Ecological Studies, Vol. 150
J. Belnap and O.L. Lange (eds.) Biological Soil Crusts: Structure, Function, and Management

and bryophytes) rarely form a contiguous cover on the soil surface, even in areas where they are common (Eldridge 1996). In areas of almost continuous crust cover, such as on calcareous soils in southern Australia, the areas between lichen thalli or moss stems act to trap and transfer water to areas lower in the soil profile (Rogers 1982).

Physical or chemical soil crusts, which are frequently associated with biological soil crusts, are generally regarded as a degraded soil feature, as they restrict seed germination and provide a poor medium for plant growth. These crusts are known to reduce infiltration, a consequence of a breakdown in aggregate stability and a blocking of the matrix pores which conduct water through the profile (Romkens et al. 1990). Whilst infiltration and permeability of some of these physical and chemical crusts such as rain-impacted and depositional crusts are generally well understood, there is conflicting evidence on the role that biological soil crusts play in regulating the flow of water into the soil (Eldridge and Greene 1994b).

In Australia, researchers have generally employed two methods to investigate the impact of biological soil crusts on infiltration. Both of these techniques have limitations. Scalping experiments, where infiltration is compared before and after removal of the crust (e.g., Graetz and Tongway 1986), have generally been replaced by in situ measurements of crusts using disk permeameters (infiltrometers) and rainfall simulations, where surfaces with inherently different crust cover and composition are compared (e.g., Eldridge 1993a; Eldridge et al. 1997). Scalping has the disadvantage that the soil surface immediately under the crust is destroyed, and changes in infiltration can only be ascribed to a combination of removal of the crust and disturbance of the soil surface. The other technique of selecting areas of different cover presupposes that areas with reduced cover are not inherently different from areas supporting an extensive crust cover, and that bare "control areas" are devoid of cyanobacteria and other less visible crusts. More recent work has attempted to use algacides to eliminate the effect of cyanobacteria in the soil before infiltration measurements are made.

Research in Australia and other regions has demonstrated infiltration increasing on soils dominated by biological crusts when compared with crust-free soils (e.g., Fletcher and Martin 1948; Blackburn 1975; Greene and Tongway 1989; Eldridge 1993a). Other work (Loope and Gifford 1972; Brotherson and Rushforth 1983; Graetz and Tongway 1986; Eldridge et al. 1999) has demonstrated reduced infiltration with increasing biological soil-crust cover. Some studies from the semiarid woodlands of eastern Australia have also suggested that infiltration capacity can be independent of crust cover (Eldridge et al. 1997). Knowledge of the intimate relationships between biological soil crusts, and (1) soil physical and chemical properties and (2) different compositional factors in the crusts, leads to various hypotheses about how these crusts influence water flow. These are discussed below.

23.2.1 Soil Physicochemical Properties and Water Flow

There are divergent views on the roles of biological soil crusts in moderating water flow through soils. Experimental work on degraded red earth soils near Cobar led Eldridge (1993a) to propose that the importance of biological crusts in moderating water flow increased as the macroporosity of the soil declines. Macropores (large soil pores) are created by soil animals and plant roots, and conduct most of the infiltrating water. Overtrampling results in a breakdown in macroporosity and a decline in the productive potential of the soil. In these situations, even small increases in biological soil-crust cover result in markedly greater increases in infiltration through small pores remaining in the soil. It has been suggested that cyanobacteria and the hyphae of free-living fungi could enhance infiltration in situations where hyphae help to maintain continuous pores through the soil. Alternatively, they could reduce infiltration by one or more of the following mechanisms: (1) biological soil crusts could exclude water by occupying matrix pores, thereby creating a hydrophobic surface (Greene and Tongway 1989; Chartres 1992); (2) crusts could also exclude water, and thereby reduce infiltration, through the swelling of cyanobacteria in the crusts (see Chap. 21). Evidence of the hydrophobic nature of biological soil crusts comes from studies on a dunefield near Cobar in eastern Australia, where removal of the crust from the dune flank enhanced infiltration and delayed the onset of runoff by 1 h. When runoff finally occurred, it was significantly less than that on the undisturbed crust (R.S.B. Greene, unpubl. data).

Other hypotheses have been invoked that suggest that differences in infiltration may be due to different crust composition, or differences in the moisture status of the crusts. As much of the rain falls in small amounts, biological crusts may enhance infiltration by their effect on microtopography and thus surface detention (Eldridge and Greene 1994b). Equally however, crusts form close associations with surface soils, often cementing soil particles together and closing off pores at the surface. Electron microscopy studies have revealed that the upper 1–2-mm layer of biological crusts are typically non-porous compared with the soil below (Eldridge and Greene 1994a). Thus it is probable that, during most rainfall events, hydraulic penetration and water entry would be limited.

23.2.2 Crust Morphology and Water Flow

Depending on their component organisms, biological soil crusts in Australian deserts are morphologically and structurally diverse, ranging from black algal scums with little surface microrelief, to strongly patterned, thick, often-pedicelled crusts containing lichens and bryophytes. The interaction of widely differing morphologies with surface soils would be expected to produce a differential influence on soil hydrology (Eldridge and Rosentreter 1999).

During rainfall events, moss- and liverwort-dominated crusts would be expected to lengthen the time for runoff to occur by absorbing more surface water. Leaf architecture of mosses may also influence soil hydrological properties, as many mosses possess long tips or hair points on their leaves which channel water into the center of the plant. In a sagebrush shrubland near Boise, Idaho, cumulative infiltration through mats of the moss *Tortula ruralis* growing under sagebrush hummocks was significantly greater than that under shrubs in the absence of mosses (D.J. Eldridge and R.R. Rosentreter, unpubl. data). Apart from enhancing infiltration, moss leaves increase the surface area of the plant, protecting the soil against erosion, or helping to trap airborne dust and incorporating it into the upper soil horizons (Danin and Gaynor 1991). In Northern and Central Australia, where summer rainfall predominates and soil crusts are typically dominated by thallose liverworts (Rogers 1989; D.J. Eldridge and D.J. Albrecht, unpubl. data), crusts are likely to shed water once they have fully imbibed and have reached their interception capacity (Rogers 1989).

Apart from some of the foliose and fruticose lichens, most soil lichens in deserts (i.e., squamulose and crustose forms), do not alter their shape in response to rainfall, but merely imbibe water as the photobiont cells turn green and begin to photosynthesize. Eldridge and Koen (1998) showed that crusts from sites in poor rangeland health (i.e., sites showing evidence of active erosion, and supporting a sparse cover of generally poor-quality, weedy vascular plants) are dominated by cyanolichens such as *Collema coccophorum*, *Peltula patellata*, and *Heppia despreauxii*. *Collema* has an unstratified (homoiomerous) thallus, characterized by a uniform arrangement of the cyanobacterial cells amongst the fungal hyphae (see Eldridge et al. 1997). Species of this genus can absorb up to 13 times their own body weight in water (Galun et al. 1982), whereas most stratified lichens, characteristic of sites in moderate to good health, can absorb much less water (Blum 1973). Accumulations of cyanobacterial sheath material close to the soil surface can absorb up to eight times their weight in water (Wang et al. 1981), thereby prolonging the onset of ponding. This suggests that crust morphology, which is strongly associated with landscape condition (Eldridge and Rosentreter 1999), directly influences the fate of water on the soil surface in areas where crusts are a predominant cover component. West (1990) however contends that sponge-like lichens (e.g., *Collema*) could hold water on the surface for longer periods, thus increasing the possibility of evaporation. Therefore, the degree to which this intercepted water influences subsequent infiltration is not known. Clearly, more work is required to determine how and under what conditions these crust components influence water movement through arid soils.

23.2.3 Soil Stability

Numerous studies have indicated that biological soil crusts have the ability to enhance the stability of soil aggregates (Eldridge and Greene 1994b), and that aggregates from soil surfaces with biological crusts have greater stability than aggregates from bare surfaces (Greene and Tongway 1989; Greene et al. 1990; Eldridge and Robson 1996). Enhanced aggregation is generally greatest in the top few millimeters of the soil, and results from the combination of physical binding by larger plant structures (e.g., lichen rhizines, moss rhizoids and protonemata), microscopic fungal hyphae and cyanobacterial filaments, and organic gels and polysaccharides produced by cyanobacterial components.

Whilst little is known about the biodiversity of fungi and algae in Australia's arid soils (see Chap. 10), their impact on soil aggregation is well known. Gels secreted by biological crust organisms bind the organisms, their filaments, and surrounding soil particles into small, rounded aggregates (Greene et al. 1990). This aggregation converts highly erodible micro-aggregates (<0.25 mm diameter) into stable macroaggregates (>0.25 mm diameter; Tisdall and Oades 1982). This effect can also be obtained by inoculating soils with algae (Singh 1950; Bailey et al. 1973). The filamentous growth habit of cyanobacteria such as *Microcoleus vaginatus*, and production of a mucilaginous sheath or capsule around the filaments results in a binding together of surface soils (Bond and Harris 1964; Rogers 1989; Chap 15).

23.2.4 Soil Stability and Landscape Health

The strong links between soil structure and biological soil crusts have been incorporated into methodologies used to assess rangeland health (Tongway 1995; Eldridge and Rosentreter 1999). In landscapes where biological soil crusts are found, their presence is generally associated with stable landscapes in good health because of the processes they mediate. Healthy, stable landscape are defined as those with little or no evidence of active erosion, and supporting a diverse cover of nonvascular and vascular plants. The effect of biological soil crusts on soil aggregation means that surfaces with extensive crust cover are generally more productive and stable (in terms of erosion) than those with little or no cover. In the semiarid woodlands of eastern Australia, Greene and Tongway (1989) identified a patchwork pattern of noneroded (class 1) to seriously eroded (class 4) soil surfaces. The coverage by biological soil crusts ranged from "extensive" on class 1 to "not present" on class 4 soils. These differences in health and stability were also reflected in major differences in organic carbon in the top 20 mm of the soil, as well as in infiltration between the zones (Greene and Tongway 1989).

23.3 Erosion and Sediment Production

Australian studies have shown that the presence of biological soil crusts reduces erosion and sediment production (e.g., Mücher et al. 1988; Chartres and Mücher 1989; Greene et al. 1990; Kinnell et al. 1990; Eldridge 1993a,b; Eldridge and Greene 1994a; Eldridge and Kinnell 1997; Eldridge 1998). These studies have been undertaken predominantly in the semiarid woodlands of eastern Australia, using a combination of in situ rainfall simulation (using portable simulators), and intact monoliths using raindrop-impacted flow transport (Kinnell 1998).

23.3.1 Quantifying Water Erosion Using Simulated Rainfall

In average seasons, vascular plant cover is often sufficient to protect the soil against erosion. During drought, however, when the bulk of the vascular vegetation is absent, crusts provide vital soil protection (Eldridge et al. 1997; Eldridge and Tozer 1997). Using small cores of soil with varying crust-cover levels, Eldridge and Greene (1994b) showed that when crust cover was 25–50 %, soil erosion was reduced to very low levels (Fig. 23.1). Crust-cover levels of 25 % and lower resulted in dramatically higher levels of erosion (Eldridge and Greene 1994b). There was also higher sediment yield from surfaces with sparse crust cover, and most of the eroded sediment were fine particles, onto which nutrients such as nitrogen and phosphorus are likely to be adsorbed. Thus, both erosion and reductions in potential yield of vascular vegetation are likely to be much greater on surfaces of sparse crust cover.

Further studies under raindrop-impacted flow transport (Kinnell 1998) indicate that biological soil crusts protect the soil against water erosion under a range of management regimes. In the undisturbed state, erosion is less on biologically crusted soils compared with uncrusted soils (Kinnell et al. 1990; Eldridge and Kinnell 1997). Empirical data have shown that crusts provide greater soil protection, compared with crust-free soils, when a stable grass cover is removed (Chartres and Mücher 1989), or when soils are subjected to moderate trampling by sheep (Eldridge 1998). Compared with physical soil crusts (such as raindrop impact or depositional crusts), water erosion from crusted surfaces is likely to be up to five times less (Eldridge and Kinnell 1997).

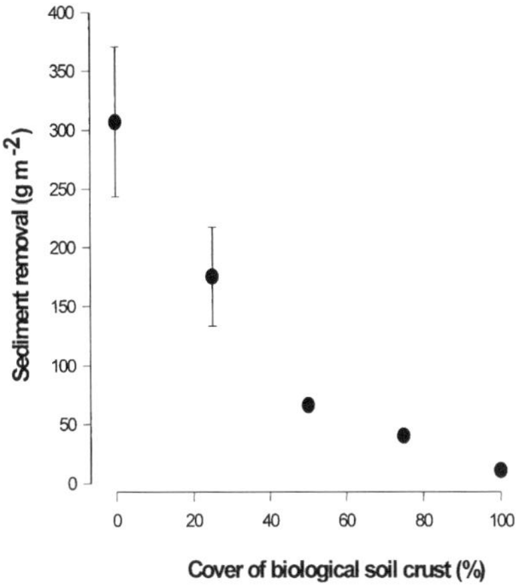

Fig. 23.1. Splash erosion (g m^{-2}) from soil cores of varying cover of biological soil crusts. (After Eldridge and Greene 1994b)

Laboratory simulations using intact soil crusts showed that trampling of biological soil crusts, combined with raindrop impact, results in significant sorting of material and removal of fine sediments (<250 µm) through erosion (Eldridge 1998). During these simulation experiments, the surface of severely trampled plots (equivalent to 1.6 sheep ha^{-1}) developed a thick depositional layer of quartz particles (>500 µm) which was confined to the micro-depressions between lichen and moss plants (Eldridge 1998; see Chap. 26). Severe trampling reduced the quality of the soil surface; the proportion of depressions declined from 32 to 7 %, and attached crust from 31 to 14 %. The majority of the impact of trampling occurred within the top few millimeters of the surface.

The impact of trampling on erosion is minimal provided that sufficient areas of crust remained on the surface to trap eroding soil particles. In the experiments described above, it was only when the trampling rate reached about 320 hoof impacts m^{-2} that the soil surface lost any structural integrity and large amounts of erosion resulted (Fig. 23.2, about 40 hoof impacts mono-

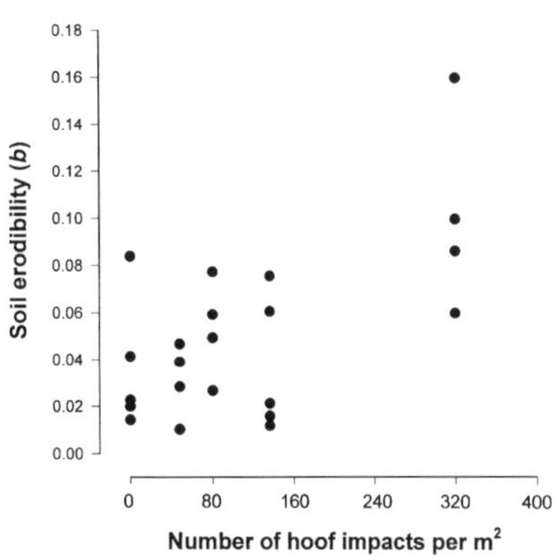

Fig. 23.2. The relationship between soil erodibility (relative units) and trampling rate. (After Eldridge 1998)

lith^{-1}). Trampling may reduce crust cover sufficiently to cause major differences in erosion or landscape stability between regions. However, this does not always mean that major processes are unaffected, as there may be a shift in the species composition of the crust with no appreciable change in crust cover.

23.3.2 Soil Biota as Moderators of Soil Erodibility

23.3.2.1 Mosses and Stratified Lichens

As discussed above, reduced erosion results from enhanced soil properties mediated by the actions of soil organisms in the crusts. Crust organisms can be arranged into broad categories according to their influence on water erosion (Eldridge and Rosentreter 1999). Morphological types which provide a continuous cover on the soil surface are likely to reduce raindrop impact, whilst those which enhance soil roughness and surface detention are likely to influence overland flow processes. Moss-dominated surfaces are the most effective, lichen-dominated surfaces of intermediate effectiveness and cyanobacterial-dominated surfaces least effective, at protecting the soil against raindrop impact (Tchoupopnou 1989; Eldridge and Rosentreter 1999). Mosses from the family Pottiaceae, which are common components of arid soils worldwide (see Chap. 1), have twisting leaves which enable them to "swim up" through layers of accreting sand and sediments (Scott 1982), effectively reducing raindrop impact and trapping sediments mobilized by overland flow (Danin and Gaynor 1991; Eldridge 1998).

Of the lichens, the squamulose (scaly or warty) types such as the widespread *Psora* spp. and *Endocarpon* spp. are effective trappers of water and soil particles. Their convex surfaces enhance both surface roughness and thus the sinuosity of flow, and thereby reduce the transportability of rainwater (West 1990). Similarly, some of the strap-like liverworts, such as *Riccia* spp. and *Asterella* spp., have the capacity to increase their surface area up to fivefold, creating effective barriers to raindrop action (Rogers 1994).

23.3.2.2 Cyanobacteria and Gelatinous Lichens

Although cyanobacteria are widespread in soils worldwide (see Chaps. 1, 12), their influence on erosion processes is more subtle than that of the lichens and mosses, and results indirectly from their effect on soil structure. Generally, along with the short-lived and structurally less complex gelatinous lichens such as *Collema coccophorum*, cyanobacteria are the morphological group least effective in protecting the surface against water erosion (Eldridge 1996). However, where dense mats of cyanobacterial filaments occur, they are often associated with areas of enhanced microtopography (Belnap and Gardner 1993). In some soils they may increase surface roughness by a few centimeters, creating small pools of ponded water on the surface, resulting in a higher hydraulic head, and thus enhancing infiltration.

Scott (1982) observed that crusted algal (i.e., cyanobacterial) surfaces had a greater degree of cracking than noncrusted surfaces, and proposed that this helps to prevent sand from engulfing the plants growing in the cracks. The popular perception is that crust organisms manipulate the soil surface environment to maximize the storage of water for their own use, whilst minimizing water flow within their immediate environment. Viewing the crust as a complex of organisms, it is in the crust's best interests to minimize redistribution to nearly vascular "neighbors" which compete directly for light and essential nutrients. It is conceivable that cyanobacteria could alter the balance between the water that is shed and the water that is trapped by regulating the production of polysaccharides and gels.

Acknowledgments. Much of this work was supported by the Department of Land and Water Conservation and the Land and Water Resources Research and Development Corporation. This is publication 99/26 of the Centre for Natural Resources, Department of Land and Water Conservation.

References

Bailey D, Mazurak PA, Rosowski JR (1973) Aggregation of soil particles by algae. J Phycol 9:99–101

Belnap J, Gardner JS (1993) Soil microstructure in soils of the Colorado Plateau: the role of the cyanobacterium *Microcoleus vaginatus*. Great Basin Nat 53:40–47

Blackburn WH (1975) Factors influencing infiltration rate and sediment production of semiarid rangelands in Nevada. Water Resour Res 11:929–937

Blum OB (1973) Water relations. In: Ahmadjian V, Hale ME (eds) The lichens. Academic Press, New York, pp 381–400

Bond RD, Harris JR (1964) The influence of the microflora on the physical properties of soils. I. Effects associated with filamentous algae and fungi. Aust J Soil Res 2:111–22

Brotherson JD, Rushforth SR (1983) Influence of cryptogamic crusts on moisture relationships of soils in Navajo National Monument, Arizona. Great Basin Nat 43: 73–79

Chartres CJ (1992) Soil crusting in Australia. In: Sumner ME, Stewart BA (eds) Soil crusting: chemical and physical processes. Lewis Publishers, Boca Raton, FL, pp 339–365

Chartres CJ, Mücher HJ (1989) The response of burnt and unburnt rangeland soil surfaces to simulated raindrop impact and water erosion. Earth Surface Processes Landforms 14:407–417

Danin A, Gaynor E (1991) Trapping of airborne dust by mosses in the Negev Desert, Israel. Earth Surface Processes Landforms 16:153–162
Eldridge DJ (1993a) Cryptogam cover and soil surface condition: effects on hydrology in a semi-arid woodland. Arid Soil Res Rehabil 7:203–217
Eldridge DJ (1993b) Cryptogams, vascular plants and soil hydrological relations: some preliminary results from the semi-arid woodlands of eastern Australia. Great Basin Nat 53:48–58
Eldridge DJ (1996) Dispersal of microphytes by water erosion in an Australian semi-arid woodland. Lichenologist 28:97–100
Eldridge DJ (1998) Trampling of microphytic crusts on calcareous soils and its impact on erosion under rain-impacted flow. Catena 33:221–239
Eldridge DJ, Greene RSB (1994a) Assessment of sediment yield from a semi-arid red earth soil with varying cryptogam cover. J Arid Environ 26:221–232
Eldridge DJ, Greene RSB (1994b) Microbiotic soil crusts: a review of their roles in soil and ecological processes in the rangelands of Australia. Aust J Soil Res 32:389–415
Eldridge DJ, Kinnell PIA (1997) Assessment of erosion rates from microphyte-dominated calcareous soils under rain-impacted flow. Aust J Soil Res 35:475–489
Eldridge DJ, Koen TB (1998) Cover and floristics of microphytic soil crusts in relation to indices of landscape health. Plant Ecol 137:101–114
Eldridge DJ, Robson AD (1996) Bladeploughing and exclosure influence soil properties in a semi-arid Australian woodland. J Range Manage 50:191–198
Eldridge DJ, Rosentreter RR (1999) Morphological groups: a framework for monitoring microphytic crusts in arid landscapes. J Arid Environ 41:11–25
Eldridge DJ, Tozer ME (1997) A practical guide to soil lichens and bryophytes of Australia's dry country. Dept Land Water Conserv, Sydney
Eldridge DJ, Tozer ME, Koen TB (1995) Composition of bryophytes in arid soil crusts: what can they tell us about rangeland health? Proc Vth Int Rangelands Congress, Salt Lake City, Utah
Eldridge DJ, Tozer ME, Slangen S (1997) Soil hydrology is independent of microphytic crust cover: further evidence from a wooded semiarid Australian rangeland. Arid Soil Res Rehabil 11:113–126
Eldridge DJ, Zaady E, Schachak M (1999) Infiltration through three contrasting biological soil crusts in patterned landscapes in the Negev, Israel. Catena 40:323–336
Fletcher JE, Martin WP (1948) Some effects of algae and moulds in the rain crust of desert soils. Ecology 29:95–100
Galun M, Bubrick P, Garty J (1982) Structural and metabolic diversity of two desert lichen populations. J Hattori Bot Lab 53:321–324
Graetz RD, Tongway DJ (1986) Influence of grazing management on vegetation, soil structure and nutrient distribution and the infiltration of applied rainfall in a semi-arid chenopod shrubland. Aust J Ecol 11:347–360
Greene RSB, Tongway DJ (1989) The importance of (surface) physical and chemical properties in determining soil surface condition of red earths in rangelands. Aust J Soil Res 27:213–225
Greene RSB, Chartres CJ, Hodgkinson KH (1990) The effect of fire on the soil in a degraded semi-arid woodland. I. Cryptogam cover and physical and micromorphological properties. Aust J Soil Res 28:755–775
Kinnell PIA (1998) Influence of surface zones with zero detachment on erosion by rain-impacted flow. Aust J Soil Res 36:327–333
Kinnell PIA, Chartres CJ, Watson CL (1990) The effects of fire on the soil in a degraded semi-arid woodland. II. Susceptibility of the soil to erosion by shallow rain-impacted flow. Aust J Soil Res 28:779–794

Loope WL, Gifford GF (1972) Influence of a soil microfloral crust on selected properties of soils under pinyon-juniper in southeastern Utah. J Soil Water Conserv 27:164–167

Mücher HJ, Chartres CJ, Tongway DJ, Greene RSB (1988) Micromorphology and significance of the surface crusts of soils in rangelands near Cobar, Australia. Geoderma 42:227–244

Rogers RW (1982) Lichens of arid Australia. J Hattori Bot Lab 53:351–355

Rogers RW (1989) Blue-green algae in southern Australian rangeland soils. Aust Range J 11:67–73

Rogers RW (1994) Zonation of the liverwort *Riccia* in a temporary watercourse in subtropical semi-arid Australia. Aust J Bot 42:659–662

Romkens MJM, Prasad SN, Whisle FD (1990) Surface sealing and infiltration. In: Anderson MG, Burt IP (eds) Process studies in hillslope hydrology. Wiley, New York, pp 127–172

Scott GAM (1982) Desert bryophytes. In: Smith AJE (ed) Bryophyte ecology. Chapman and Hall, London, pp 105–122

Singh RN (1950) Reclamation of "Usar" lands in India through blue-green algae. Nature 4191:325–326

Tchoupopnou E (1989) Experimental studies of rain splash erosion from soil microphytic crusts on Utah rangelands. MS Thesis, Utah State Univ, Logan, UT

Tisdall JM, Oades JM (1982) Organic matter and water stable aggregates in soils. J Soil Sci 33:141–163

Tongway DJ (1995) Monitoring soil productive potential. Environ Monit Assess 37:303–318

Wang F, Zhung Z, Hu Z (1981) Nitrogen fixation by an edible terrestrial blue-green algae. In: Gibson AH, Newton WE (eds) Current perspectives in nitrogen fixation. Elsevier-North Holland, Amsterdam, pp 455

West NE (1990) Structure and function of microphytic soil crusts in wildland ecosystems of arid and semi-arid regions. Adv Ecol Res 20: 179–223

24 Biological Soil Crusts and Hydrology in North American Deserts

S.D. Warren

24.1 Introduction

Precipitation that reaches the Earth's surface either remains on the surface or infiltrates into the soil profile. Water remaining on the surface evaporates or runs off into bodies of water. Water that enters the soil may be stored as groundwater, may move laterally through the soil to channels, bodies of water, or springs, or may be returned to the atmosphere through evaporation and/or transpiration. Numerous physical, biological, and chemical characteristics of the soil affect the partitioning of water between infiltration and runoff. Soil properties also govern the ultimate fate of infiltrated water and determine the erodibility of the soil when subjected to surface runoff. This chapter examines the effects of biological soil crusts on infiltration, runoff, soil moisture, and soil erosion by water in North America.

24.2 Infiltration and Runoff

Infiltration and runoff are closely related and inversely proportional, i.e., as more water infiltrates into the soil, less runs off. On loamy soils of the sagebrush steppe in the northern Great Basin Desert, infiltration rate is significantly related to soil surface morphology, with more rapid infiltration on the rougher surfaces of shrub coppices and slower infiltration on the flatter surfaces in the interspaces between the shrubs (Blackburn 1975; Pierson et al. 1994). Moss-dominated biological soil crusts are considerably more abundant on shrub coppices than in the interspaces (Eckert et al. 1978; Pierson et al. 1994), leading to the possible conclusion that the crusts govern the infiltration process in this area. While compelling, this conclusion may not be entirely justified based on these data alone. Although cover of biological soil crusts is positively correlated with infiltration rate in the sagebrush steppe (Spaeth et al. 1994), soil characteristics other than biological crusts also vary between

Ecological Studies, Vol. 150
J. Belnap and O.L. Lange (eds.) Biological Soil Crusts: Structure, Function, and Management

the shrub coppices and the interspaces, including greater litter cover under the shrubs (Eckert et al. 1978) and the presence of vesicular soil horizons in the interspaces (Blackburn 1975; Pierson et al. 1994), which are well known for impeding infiltration (Dobrowolski 1994).

In the pinyon-juniper woodlands of the Colorado Plateau, the infiltration rate of ponded water on intact moss-encrusted sandy soil was significantly faster than on adjacent areas where the crusts had been destroyed (Rushforth and Brotherson 1982; Brotherson and Rushforth 1983). Lichen and algal crusts slowed the infiltration rate of ponded water compared to disturbed biological crusts. However, when the infiltration characteristics were measured with simulated rainfall designed to approximate brief summer cloudbursts, the downslope spread of water on both types of biological crusts was significantly less, and the depth of water penetration was significantly greater than on the disturbed crust, implying significant enhancement of infiltration by the crust. The apparent discrepancy in the data was attributed to the rough surface topography associated with the biological crusts. Although some biological crusts reduced the rate of infiltration, the convoluted microtopography prevented runoff, thus retaining water on the surface for a longer period of time and facilitating greater net infiltration.

Elsewhere in pinyon-juniper woodlands of the Colorado Plateau, Loope and Gifford (1972) examined the effects of various levels of disturbance on infiltration into lichen-encrusted sandy loam soils. Treatments included an undisturbed relict area with intact crusts, a lightly grazed area with minimal disturbance, a previously grazed area with deteriorated crusts, an intensively trampled area with badly disturbed crusts, and an area subjected to mechanical brush removal causing the complete destruction of the biological crusts. When infiltration was measured by ponding water on the soil surface, no clear patterns were observed. However, when simulated rainfall was applied, sites with any degree of biological crust cover had significantly higher infiltration rates than sites where the crusts had been completely destroyed. The results suggest that the biological crusts protected the soil from raindrop impact and subsequent clogging of soil pores by detached soil particles.

Johansen et al. (1998) measured infiltration rates of biologically encrusted soils at sites in several North American deserts that had been subjected to various forms of disturbance. Plots on silty soils of a salt desert shrub site in the Great Basin were subjected to trampling and burning. Sandy sites in the Sonoran Desert and Colorado Plateau were trampled. Disturbances applied to a fine-textured soil in the Chihuahuan Desert included burning and tracked vehicle traffic. At all sites, infiltration rate was determined by comparing the rate of application of simulated rainfall with runoff from the experimental plots. Infiltration rates at the Sonoran Desert site exceeded the maximum rate of applied rainfall for all treatments, thus precluding comparisons between treatments. At all other locations, infiltration rates on disturbance treatments were consistently slower than on undisturbed controls, although the differences were not statistically significant for the trampled plots at the Great Basin site or the burned plots in the Chihuahuan Desert. While it is tempting to conclude that disturbances to the crust were responsible for the changes, reductions in infiltration rate can occur under trampling (Warren et al. 1986b), tracked vehicle traffic (Thurow et al. 1993), and burning (Hester et al. 1997), even where biological soil crusts are absent or poorly represented.

Research conducted at the same Great Basin site as in the preceding study attempted to separate the effects of disturbance to biological soil crusts from concomitant alterations to other soil properties (Dobrowolski and Williams 1994). Treatments included three levels of tracked vehicle traffic, careful removal (scalping) of the biological crust layer, scalping plus tracked vehicle traffic, shallow raking to disrupt the biological crust without affecting an underlying vesicular layer, deep raking to disturb both the biological crust and the vesicular layer, and scalping plus deep raking. Scalping had little impact on infiltration rate when compared to undisturbed soils, while shallow raking caused a small, temporary improvement in infiltration. All other treatments caused varied decreases in infiltration rates. Apparently, soil compaction and the presence of a vesicular horizon were of greater importance than biological crusts in determining infiltration at that site.

Two additional investigations in the Great Basin compared infiltration on intact biological soil crusts with scalped soils (Harper and St. Clair 1985; Harper and Marble 1988). In both cases, the infiltration rate was significantly more rapid on the intact crust than on the scalped soil. Harper and St. Clair (1985) also included a one-time trampling treatment and trampling plus the application of dilute chlorine bleach in an attempt to kill the crust organisms. Subsequent observations indicated that the treatments did not result in a complete kill of the crust organisms. This led to considerable within-treatment variability that precluded the detection of statistically significant differences between treatments. However, a regression analysis indicated that percent cover of the biological crust was significantly and positively correlated with infiltration rate.

An investigation on poor-condition rangeland at the northeastern edge of the Chihuahuan Desert near San Angelo, Texas, corroborates the results of the preceding studies, showing that an algal crust greatly reduced runoff (improved infiltration) compared to plots where the crust had been removed (Osborn 1952). Likewise, in the Sonoran Desert near Tucson, Arizona, scalping biological soil crusts increased the amount of runoff from two different soil types (Fletcher 1960).

Williams et al. (1995, 1996, 1999) investigated the effect of chemically killing a biological soil crust at a sandy loam site on the Colorado Plateau. Chemically killed plots were compared with undisturbed plots and scalped plots. Regardless of whether infiltration rate was measured under ponded water or under simulated rainfall, there were no differences between treatments, indicating that biological crusts at this site had less effect on the infiltration of water than other soil parameters.

24.2.1 Seasonal Variation

Based on data from the Sonoran Desert, Fletcher (1960) suggested that the relationship between biological soil crusts and soil hydrology may be seasonal. He noted that infiltration declined and runoff increased during extended periods of favorable moisture conditions that allowed biological crust organisms to produce abundant growth. When moist periods were followed by dry periods that allowed the crusts to shrink and crack, infiltration improved dramatically; but during extended dry seasons or drought, the crusts lost their ability to bind soil particles. Thus, the first rain of the season was marked by lower infiltration and higher runoff than subsequent rains due to the detachment of soil particles and clogging of soil pores. These observations are worthy of further exploration. However, it should be noted that significant seasonal differences in infiltration rate also occur in more temperate areas where biological soil crusts are poorly represented (Warren et al. 1986a). Such differences are most likely a response to seasonal fluctuations in soil aggregate stability (Blackburn and Pierson 1994).

24.2.2 Taxonomic and Successional Differences

As a general rule, cyanobacteria and algae tend to be early successional colonizers of disturbed lands in the deserts of North America, followed by lichens and mosses (Harper and Marble 1988). On a sandy soil in the pinyon-juniper woodlands of the Colorado Plateau, Brotherson and Rushforth (1983) compared infiltration of ponded water on crusts dominated by mosses versus crusts dominated by lichens and cyanobacteria. The rate of infiltration was significantly faster on moss-dominated crusts than on heavily disturbed crusts. Infiltration on lichen/cyanobacterial crusts was significantly slower than on the disturbed crust. The results are supported by another rather novel approach to evaluating the effect of biological crusts on infiltration in sandy soils of pinyon-juniper woodlands on the Colorado Plateau (Ladyman et al. 1993). These researchers compared the pH of the surface soil with the pH at a depth of 9–12 cm in both biologically encrusted and uncrusted areas. They reasoned that greater infiltration would lead to greater leaching of calcareous material and thus to higher pH at lower depths. Although the sample size was very small, the soil under a moss-dominated crust had higher pH at the greater depth, while the reverse was true for a cyanobacterial crust when compared to uncrusted areas, implying that the moss-dominated crust improved infiltration while the cyanobacterial crust limited infiltration (Ladyman and Muldavin 1996). Assuming that the moss-dominated crusts were more mature than the lichen/cyanobacterial crusts in these studies, the results suggest that later successional crusts may enhance infiltration while early successional crusts may deter infiltration. As noted in the next section, however, these results may be unique to sandy soils (see also Chap. 26).

24.2.3 Response by Soil Type

Infiltration of water is naturally more rapid in coarse-textured soils than in fine-textured soils. However, infiltration in fine-textured soils can be significantly enhanced by a well-aggregated granular structure. As biological soil crusts are well known for their ability to improve the aggregation of fine soil particles (Schulten 1985), it can logically be hypothesized that the presence of biological crusts on a fine-textured soil has greater potential to increase infiltration and reduce runoff than on a coarse-textured soil (see Chap. 22). Faust (1970) examined the effects of the presence of a biological soil crust on a sandy loam soil (coarse texture) and a clay loam soil (fine texture) in the Sonoran Desert. At a simulated rainfall intensity of 2.5 cm h^{-1}, infiltration rate did not differ between biologically crusted and uncrusted conditions on the coarse-textured soil. However, on the clay loam soil, infiltration was improved by 40 % and runoff declined by 65 % when a biological crust was present. Also in the Sonoran Desert, Fletcher (1960) detected minimal differences in runoff between crusted and uncrusted conditions on a gravelly soil, while on a finer-textured alluvial soil, the presence of a biological soil crust reduced runoff by over 40 %.

24.2.4 Effect of Rainfall Intensity

To date, only one study in North America has evaluated the effect of rainfall intensity on the relationship between biological soil crusts and infiltration. Faust (1970) applied rainfall intensities of 2.5 and 5 cm h^{-1} to a sandy loam soil and a clay loam soil. The higher intensity rainfall caused at least a slight reduction in infiltration rate for both soils, regardless of the presence of a crust. The difference was most pronounced on the biologically encrusted clay loam soil where the higher rainfall intensity caused a 45 % reduction in infiltration rate. These results suggest that intense rainstorms have the potential to disrupt biological soil crusts and break down soil aggregates, leading to the detachment of soil particles that subsequently clog soil pores and reduce infiltration particularly on fine-textured soil. It should be noted that the crusts in this study were in an early stage of development and were dominated by cyanobacteria. It is possible that the soil hydrologic response might have differed if the crusts had been more mature (see Sect. 24.2.2).

24.3 Soil Moisture

Water is the primary factor limiting the productivity of desert ecosystems. The conservation of water that enters the soil is paramount to the survival of desert biota. Near the southern edge of the Chihuahuan Desert, in the state of San Luis Potosí, Mexico, soil moisture was measured over a period of 1 year in areas of intact crusts and in areas where the crusts were disturbed by grazing (Meyer and García-Moya 1989). In two grassland communities, soil moisture to a depth of 20 cm was greater in grazed than in ungrazed areas, although the differences were significant in only one community. Because the grasses were rooted at this depth, the authors attributed the differences in soil moisture to a decline in water consumption by grazed plants, not to changes in the biological crust cover. In contrast, in a shrubland where woody plants use moisture from a greater depth, soil moisture to a depth of 45 cm was diminished by grazing, presumably due to greater evaporative loss where the biological crust was destroyed.

During a very dry period in the southern Great Plains, Booth (1941) was unable to detect differences in soil moisture in the surface 2.5 cm below biologically crusted and physically crusted soil. However, 2 days following a light rain, moisture beneath the biological crust averaged 8.9 % compared to 1.3 % beneath the physical crust. Rushforth and Brotherson (1982) reported 75 % more moisture under a biological crust than for a disturbed surface in the Colorado Plateau region of the Great Basin. Actual moisture content, and the depth and dates of sampling were not reported. Elsewhere in the Great Basin, evaporative loss over a 7-day period from the surface 7.5 cm of a soil profile was almost 20 % higher from a heavily crusted soil than from a nearby scalped soil (Harper and Marble 1988). The difference was significant only where biological crusts covered 85 % of the soil surface; there was no significant difference where the crust covered 61 % of the surface. The results were attributed to darkening of the surface by the cyanobacteria and lichens, resulting in greater absorption of solar radiation and higher surface temperatures. Again, the actual moisture content and dates of sampling were not reported. In each of the latter two studies, it would be enlightening to know if similar patterns existed throughout the year.

24.4 Soil Erosion by Water

Soil erosion occurs as the result of detachment and transport of individual soil particles or small aggregates. The energy of falling raindrops or water moving across the soil surface may cause detachment. Coarse-textured soils and soils with large, water-stable aggregates are less susceptible to detachment and transport than finer and less-aggregated soils. Biological soil crusts contribute to the stability of the soil surface in the face of the erosive forces of raindrops and runoff. Where larger crust organisms such as mosses, liverworts, and lichens cover the soil, raindrops are unable to directly impact the soil surface and detach soil particles. Depending on the type of biological crust, various combinations of fungal hyphae, cyanobacterial filaments, and rooting structures of lichens and mosses may create a mesh throughout the surface layer of the soil, physically entwining soil particles and aggregrates (Fletcher and Martin 1948; Went and Stark 1968; Schulten 1985; Belnap and Gardner 1993; see Chap. 1). Some bacteria, fungi, and cyanobacteria exude extracellular polysaccharide compounds that glue soil particles together (Tisdale and Oades 1982; Schulten 1985). In addition, many biological crusts are associated with a rough surface microtopography (Eckert et al. 1978; Belnap 1995) that serves to diminish the energy of surface runoff, thus reducing detachment and transport of sediment (Blackburn 1975).

McCalla (1946) first showed that soil aggregates formed in association with moss, algae, and fungi were more resistant to disintegration by falling raindrops than aggregates formed without the benefit of crust organisms. Subsequently, Osborn (1952) determined that the presence of an algal crust on a degraded rangeland in the southern Great Plains significantly reduced detachment and splash erosion by raindrops. Greater rainfall intensity was required to initiate splash erosion on moss- and lichen-dominated crusts than on algal crusts or bare soil in a salt desert shrub community of the Great Basin, and all forms of biological crusts significantly reduced the height of splashed sediment and presumably the total sediment movement (Tchoupopnou 1989). Booth (1941) and Fletcher and Martin (1948) observed that biological soil crusts minimized detachment of soil particles by overland flow of runoff.

While not separating the processes of detachment and transport, additional studies confirm that biological soil crusts effectively reduce soil erosion. Faust (1970, 1971) inoculated bare soils with naturally occurring cyanobacteria. Five months later, the amount of sediment produced under simulated rainfall was significantly less on the early successional crust than on adjacent uninoculated soils. Other researchers have compared erosion between intact crusts and crusts that have been disturbed by various mechanical methods, including vehicular traffic, livestock trampling, human trampling, scalping, brush removal, burning, and raking (Booth 1941; Loope and Gifford 1972; Dogan 1975; Rushforth and Brotherson 1982; Brotherson and Rushforth 1983; Harper and St. Clair 1985; Dobrowolski and Williams 1994; Williams et al. 1995, 1996; Johansen et al. 1998). In all cases, erosion from the disturbed crusts was greater than from intact crusts. Similar results were obtained when crusts were chemically killed, but not mechanically disturbed (Dogan 1975; Harper and St. Clair 1985; Williams et al. 1995, 1996). In the case of Harper and St. Clair (1985), the differences were not statistically different, due in large part to within-treatment variability from uneven crust mortality. However, a regression analysis indicated that soil loss was inversely correlated with crust cover.

24.5 Conclusions

Within North America, studies examining the effects of biological soil crusts on infiltration and runoff have produced mixed results. At many locations, the effects of soil properties unrelated to the presence of biological soil crusts appear to far outweigh contributions by the crusts, thus limiting interpretations of the hydrologic role of biological crusts. In addition, where intact crusts have been compared to disturbed crusts, it is often difficult to separate the impact of the disturbance to the crust from the disturbance to the soil itself. Nevertheless, the data tend to indicate that to the degree that biological soil crusts contribute to the formation or stability of a roughened soil surface, they also contribute to greater net infiltration and less runoff. Various taxonomic and successional groups of crust organisms may have different effects on infiltration. In sandy soils, crusts dominated by cyanobacteria may have a tendency to impede infiltration, while moss-dominated crusts appear to enhance infiltration and reduce runoff. To the degree that biological crust organisms contribute to the formation of water-stable aggregates in fine-textured soils, it is likely that they improve the ability of water to infiltrate. The ability of biological soil crusts to improve infiltration is minimal in coarse-textured soils that are already characterized by high porosity and rapid infiltration. Rainfall intensity appears to affect the ability of crusts to regulate infiltration. At high rainfall intensities, biological soil crusts dominated by algae and cyanobacteria may lose their effectiveness, probably due to the breakdown of soil aggregates resulting from raindrop impact. Vegetative structures of lichens and mosses may provide greater protection from detachment of soil particles and subsequent clogging of soil pores.

As with infiltration, studies examining the role of biological soil crusts relative to soil moisture have been inconclusive. Intuitively, where the crusts enhance infiltration, they will enhance soil moisture; the opposite will be true where the crusts impede infiltration. Once water has entered the soil, the bulk of the evidence tends to indicate that biological soil crusts conserve soil moisture. However, further research should examine seasonal variability in the relationship.

Without exception, all studies conducted in arid and semiarid regions of North America indicate that biological soil crusts are effective in reducing soil erosion by water, both from raindrop impact and surface runoff.

References

Belnap J (1995) Surface disturbances: their role in accelerating desertification. Environ Monit Assess 37:39–57

Belnap J, Gardner JS (1993) Soil microstructure in soils of the Colorado Plateau: the role of the cyanobacterium *Microcoleus vaginatus*. Great Basin Nat 53:40–47

Blackburn WH (1975) Factors influencing infiltration and sediment production of semiarid rangelands in Nevada. Water Resour Res 11:929–937

Blackburn WH, Pierson FB Jr (1994) Sources of variation in interrill erosion on rangelands. In: Blackburn WH, Pierson FB Jr, Schuman GE, Zartman R (eds) Variability in rangeland water erosion processes. Soil Sci Soc Am Spec Publ 38, Madison, Wisconsin, pp 1–9

Booth WE (1941) Algae as pioneers in plant succession and their importance in erosion control. Ecology 22:38–46

Brotherson JD, Rushforth SR (1983) Influence of cryptogamic crusts on moisture relationships of soils in Navajo National Monument, Arizona. Great Basin Nat 43:73–78

Dobrowolski JP (1994) In situ estimation of effective hydraulic conductivity to improve erosion modeling for rangeland conditions. In: Blackburn WH, Pierson FB Jr, Schuman GE, Zartman R (eds) Variability of rangeland water erosion processes. Soil Sci Soc Am Spec Publ 38, Madison, Wisconsin, pp 83–91

Dobrowolski JP, Williams JD (1994) Effects of disturbance by tracked vehicles on wind and water erosion. Report to US Dep Agric, For Serv, Shrub Sci Lab, Provo, Utah. Utah State Univ, Logan, Utah

Dogan A (1975) Some effects of microflora on surface runoff quality. PhD Dissertation, Utah State Univ, Logan, Utah

Eckert RE Jr, Wood MK, Blackburn WH, Peterson FF, Stephens JL, Meurisse MS (1978) Effects of surface-soil morphology on improvement and management of some arid and semiarid rangelands. In: Hyder DN (ed) Proc 1st Int Rangeland Congress, Soc Range Manage, Denver, Colorado, pp 299–302

Faust WF (1970) The effect of algal-mold crusts on the hydrologic processes of infiltration, runoff, and soil erosion under simulated conditions. MS Thesis, Univ Arizona, Tucson, Arizona

Faust WF (1971) Blue-green algal effects on some hydrologic processes at the soil surface. In: Hydrology and water resources in Arizona and the southwest, Proc Arizona Sect, Am Water Resour Assoc, and Hydrol Sect, Arizona Acad Sci, Tempe, Arizona, pp 99–105

Fletcher JE (1960) Some effects of plant growth on infiltration in the southwest. In: Gardner JL (arranger) Symposium – Water yield in relation to environment in the southwestern United States. SW Rocky Mt Div Am Assoc Adv Sci, pp 51–63

Fletcher JE, Martin WP (1948) Some effects of algae and molds in the rain-crust of desert soils. Ecology 29:95–100

Harper KT, Marble JR (1988) A role for nonvascular plants in management of arid and semiarid rangelands. In: Tueller PT (ed) Vegetation science applications for rangeland analysis and management. Kluwer Academic Publishers, Dordrecht, pp 135–169

Harper KT, St Clair LL (1985) Cryptogamic soil crusts on arid and semiarid rangelands in Utah: effects on seedling establishment and soil stability. Report to US Dep Inter, Bur Land Manage, Denver, Colorado. Brigham Young Univ, Provo, Utah

Hester JW, Thurow TL, Taylor CA Jr (1997) Hydrologic characteristics of vegetation types as affected by prescribed burning. J Range Manage 50:199–204

Johansen JR, St Clair LL, Evans RD, Flechtner VR, Balczon J, Webb BL (1998) Resilience of microbiotic species to military training pressures: natural and stimulated recovery following disturbance. Report to US Army Const Eng Res Lab, Champaign, Illinois. John Carroll Univ, University Heights, Ohio

Ladyman JAR, Muldavin E (1996) Terrestrial cryptogams of pinyon-juniper woodlands in the southwestern United States: a review. US Dep Agric, For Serv Gen Tech Rep RM-GTR-280, Fort Collins, Colorado

Ladyman JAR, Muldavin E, Fletcher R (1993) Pattern and relationships of terrestrial cryptogam cover in two piñon-juniper communities in New Mexico. In: Aldon EF, Shaw DW (coords) Managing piñon-juniper ecosystems for sustainability and social needs. US Dep Agric, For Serv Gen Tech Rep RM-GTR-236, Fort Collins, Colorado, pp 97–104

Loope WL, Gifford GF (1972) Influence of a soil microfloral crust on select properties of soils under pinyon-juniper in southeastern Utah. J Soil Water Conserv 27:164–167

McCalla TM (1946) Influence of some microbial groups on stabilizing soil structure against falling water drops. Soil Sci Soc Am Proc 11:260–263

Meyer SE, García-Moya E (1989) Plant community patterns and soil moisture regime in gypsum grasslands of north central Mexico. J Arid Environ 16:147–155

Osborn B (1952) Range soil conditions influence water intake. J Soil Water Conserv 7: 128–132

Pierson FB Jr, Van Vactor SS, Blackburn WH, Wood JC (1994) Incorporating small scale spatial variability into predictions of hydrologic response on sagebrush rangelands. In: Blackburn WH, Pierson FB Jr, Schuman GE, Zartman R (eds) Variability in rangeland water erosion processes. Soil Sci Soc Am Spec Publ 38, Madison, Wisconsin, pp 23–34

Rushforth SR, Brotherson JD (1982) Cryptogamic soil crusts in the deserts of North America. Am Biol Teach 44:472–475

Schulten JA (1985) Soil aggregation by cryptogams of a sand prairie. Am J Bot 72: 1657–1661

Spaeth KE, Weltz MA, Fox HD, Pierson FB Jr (1994) Spatial pattern analysis of sagebrush vegetation and potential influences on hydrology and erosion. In:Blackburn WH, Pierson FB Jr, Schuman GE, Zartman R (eds) Variability in rangeland water erosion processes. Soil Sci Soc Am Spec Publ 38, Madison, Wisconsin, pp 35–50

Tchoupopnou E (1989) Splash from microphytic soil crusts following simulated rain. MS Thesis, Utah State Univ, Logan, Utah

Thurow TL, Warren SD, Carlson DH (1993) Tracked vehicle traffic effects on the hydrologic characterisitics of central Texas rangeland. Trans Am Soc Agric Eng 36:1645–1650

Tisdall JM, Oades JM (1982) Organic matter and water-stable aggregates in soils. J Soil Sci 33:141–163

Warren SD, Blackburn WH, Taylor CA Jr (1986a) Effects of season and stage of rotation cycle on hydrologic condition of rangeland under intensive rotation grazing. J Range Manage 39:486–491

Warren SD, Thurow TL, Blackburn WH, Garza NE (1986b) The influence of livestock trampling under intensive rotation grazing on soil hydrologic characteristics. J Range Manage 39:491–495

Went FW, Stark N (1968) The biological and mechanical role of soil fungi. Proc Natl Acad Sci 60:497–504

Williams JD, Dobrowolski JP, West NE (1995) Microphytic crust influence on interrill erosion and infiltration capacity. Trans Am Soc Agric Eng 38:139–146

Williams JD, Dobrowolski JP, West NE (1996) Microphytic-crust influence on soil stability and hydrologic properties of a sandy-loam soil in Capitol Reef National Park, Utah, USA. In: West NE (ed) Rangelands in a sustainable biosphere. Proc 5th Int Rangeland Congr, Soc Range Manage, Denver, Colorado, pp 601–602

Williams JD, Dobrowolski JP, West NE (1999) Microbiotic crust influence on unsaturated hydraulic conductivity. Arid Soil Res Rehabil 13:145–154

25 Biological Soil Crusts and Wind Erosion

J. Belnap

25.1 Introduction

Wind is an important erosive force in deserts, where limited cover of vascular plant material offers little soil-surface protection. Dust deposition by wind often exceeds that of fluvial deposition in these drier regions (Goudie 1978; Williams et al. 1995). Sediment production from soil surfaces occurs when wind forces exceed soil threshold friction velocities (TFV: the wind velocity needed to detach particles from soil surfaces). Decreased TFVs are directly associated with increased sediment movement. Soil fine particles are preferentially lost over larger sand particles (Leys 1990; Williams et al. 1995). Increased sediment movement can result in many direct and indirect problems for semiarid and arid ecosystems. Soils weather slowly from parent rock in deserts, often taking 5000–10 000 years (Webb and Wilshire 1983). Much of the soil fine material found in these regions is from atmospheric deposition (Danin and Yaalon 1982). Soils from the Colorado Plateau deserts show that most fine particles are derived from surrounding igneous mountains. Current deposition rates in these regions are low and when soils are disturbed, loss rates may far exceed deposition (Gillette et al. 1980; Offer et al. 1992; Belnap and Gillette 1997,1998; Reynolds et al. 1998).

Most of the soil photosynthetic productivity and nitrogen (N) fixation in desert soils is concentrated within 3 mm of the surface (Garcia-Pichel and Belnap 1996; see Chap. 16). Thus, only a little soil loss can significantly reduce C and N inputs from these organisms (see Chaps. 18, 19). In addition, the top few mm of soil contain a much higher percentage of soil fine particles than underlying soils (Danin and Ganor 1991; Verrecchia et al. 1995). Loss of soil fines can reduce site productivity, as plant-essential nutrients are often bound to these particles. Burial of nearby biological soil crusts from windblown sediments generally means death for the photosynthetic components of the soil crusts, further reducing fertility. Reduced fertility of systems is one of the most definitive, and problematic, aspects of desertification. Worldwide increases in windborne sediments are amply documented (Goudie 1978; Kovda 1980; Tsoar and Pye 1987).

Ecological Studies, Vol. 150
J. Belnap and O.L. Lange (eds.) Biological Soil Crusts: Structure, Function, and Management

Both plant and soil characteristics influence wind erosion. In deserts, vascular plants are generally short with sparsely vegetated stems, and very large spaces occur between individuals. Plant litter cover is also very low. Thus, plant materials in semiarid and arid regions offer limited protection to soils from wind erosion. In such environments, biological soil crusts can play a critical role in soil stabilization.

25.2 The Influence of Undisturbed Biological Soil Crusts on Wind Erosion

Soil characteristics that reduce wind erosion include rock cover, high salt or calcium carbonate content, high clay/silt content, physical crusts, and extensive biological soil-crust cover. Where rocks form a pavement, soils are completely armored from wind erosion. However, many desert soils are not covered by pavement, and any soil between rocks is still vulnerable to wind erosion (Gillette et al. 1980; Musick 1998). Undisturbed, fine-textured desert soils have many of the above characteristics and are most often stable until disturbed (Gillette et al. 1980). When disturbed, however, the physical crust crumbles into tiny particles and most soil-surface protection is lost until the next rain storm reforms the crust. However, when biological crusts are disturbed, large aggregrates are left, bound together by unbroken filaments (Gillette et al. 1980; J.E. Herrick and J. Belnap, unpubl.). When undisturbed, coarse-textured soils are inherently more erodible than fine-textured soils, as they have less salt, clay and other characteristics that lead to significant surface physical crusting. Therefore, coarse soils are generally dependent on rocks and biological soil crusts for surface protection (Williams et al. 1995; Leys and Eldridge 1998). When different surfaces are tested under naturally occurring wind speeds, ranking of the inherent resistance of soils to erosion is: desert pavements=lichen/moss biological soil crusts > playa centers > playa edges > cyanobacterial soil crusts >disturbed clay soils > alluvial and aeolian sandy soils > sand dunes = disturbed sandy or silty soils (Gillette et al. 1980; Belnap and Gillette 1997, 1998).

Many studies worldwide have shown that biological soil crusts can be a critical factor in reducing soil erosion by wind (Dulieu et al. 1977; Van den Ancker et al. 1985; Tsoar and Møller 1986; Danin et al. 1989; Pluis 1994; Williams et al. 1995; Belnap and Gillette 1997, 1998; Marticorena et al. 1997; Leys and Eldridge 1998). Polysaccharides extruded by cyanobacteria and microfungi entrap and bind soil particles together, creating larger soil aggregates. Polysaccharides also then link these larger aggregates together (Belnap and Gardner 1993). These larger, linked aggregates require greater wind velo-

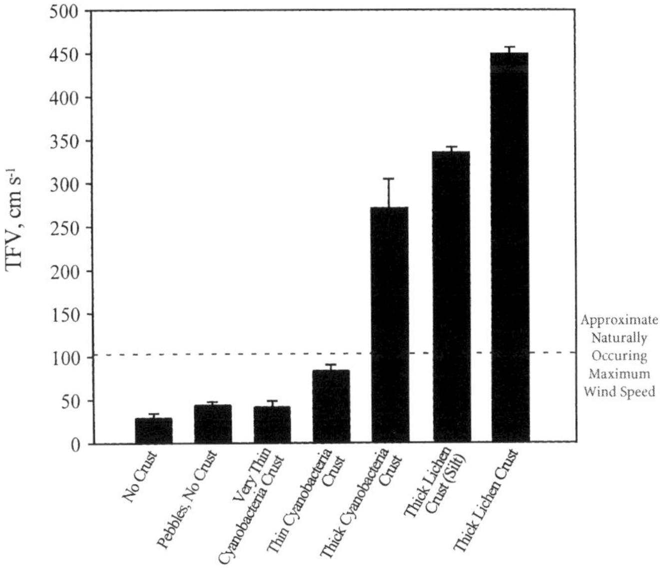

Fig. 25.1. Threshold friction velocities (*TFV*) of soil surfaces with different levels of biological soil-crust development. (After Belnap and Gilette 1998)

city to move than single grains (Gillette et al. 1980; Marticorena et al. 1997); this can be seen vividly with SEM photos (see Chap. 15; Van den Ancker et al. 1985; Danin et al. 1989; Chartes 1992; Belnap and Gardner 1993; Eldridge and Greene 1994). In laboratory wind-tunnel trials, cyanobacteria appear better able to protect soil surfaces than green algae. Cyanobacteria are generally longer and larger than green algae, and thus better able to connect soil particles together (McKenna-Neuman et al. 1996).

Soil resistance to wind erosion increases with biological crust development on all substrates tested (Fig. 25.1). While any level of biological crust development reduces soil loss by wind, both silt and sandy soils with the greatest erosion resistance are those with the most well-developed biological crusts. Even undisturbed, chemically-killed crusts (leaving the polysaccharide material intact) continue to protect the soil surface from erosion, at least temporarily (Williams et al. 1995). In most cases, well-developed crusts can withstand winds well above those recorded at field sites, and so offer complete protection from wind erosion, even in coarse soils (Leys 1990; Williams et al. 1995; Belnap and Gillette 1997,1998; Leys and Eldridge 1998).

25.3 Crust Disturbance and Wind Erosion

Biological soil crusts are highly susceptible to disturbance, especially in soils with low aggregate stability such as sands (see Chap. 27; Webb and Wilshire 1983; Belnap and Gardner 1993). Cyanobacterial filaments, lichens, and

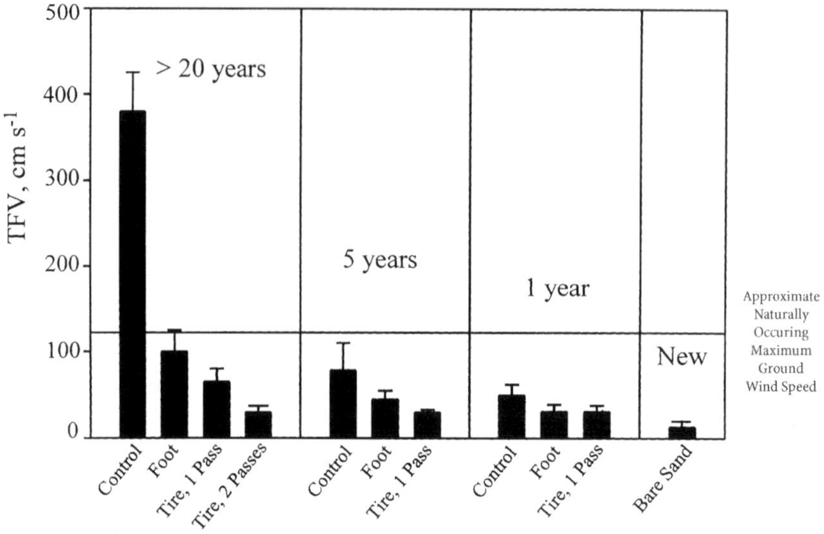

Fig. 25.2. Threshold friction velocities (*TFV*) and sediment production of control (undisturbed) and vehicle-disturbed (two passes) surfaces in Joshua Tree National Park, CA. Seven of eight control sites have TFVs above maximum field-recorded wind speeds; seven of eight disturbed sites have TFVs below recorded wind speeds. Sediment production is increased 2–36× with disturbance. (J. Belnap, unpubl.)

mosses are brittle when dry, and crush easily when subjected to compressional or shear forces incurred by activities such as trampling or vehicular traffic. Such disturbances break apart soil aggregates and the connections between them. Thus, TFV decreases significantly when the soil crusts are disturbed. Belnap and Gillette (1997) showed soil surface TFV in a sandy soil decreased by 73–92 % when moderate disturbance was applied (Fig. 25.2). Leys and Eldridge (1998) reported that moderate disturbance reduced soil TFV by 57 % in a loamy soil and by 40 % in sandy soils. Williams et al. (1995) found removing crust from silty soils decreased TFV by 50 %. Two passes with a vehicle decreased TFV in silty soils up to 96 % in coarse sandy and silty soils (J. Belnap, unpubl.). Belnap and Gillette (1998) applied the same moderate disturbance to a range of adjacent soil types. Physically-crusted playa soils with no biological crust showed the greatest impact with disturbance, as crushing of the mineral crust left the surface with virtually no erosion protection. Silty and coarse-structured soils covered with varying types of biological soil crusts showed an intermediate response, with gravel soils being the least compromised (Fig. 25.1). Severe disturbances show greater decreases in TFV than moderate disturbances (Leys and Eldridge 1998).

As TFV decreases, sediment production increases; thus, sediment loss is increased on disturbed surfaces. In Australia, Leys (1990) showed a fivefold increase in sediment production when a crusted surface was disturbed. Leys

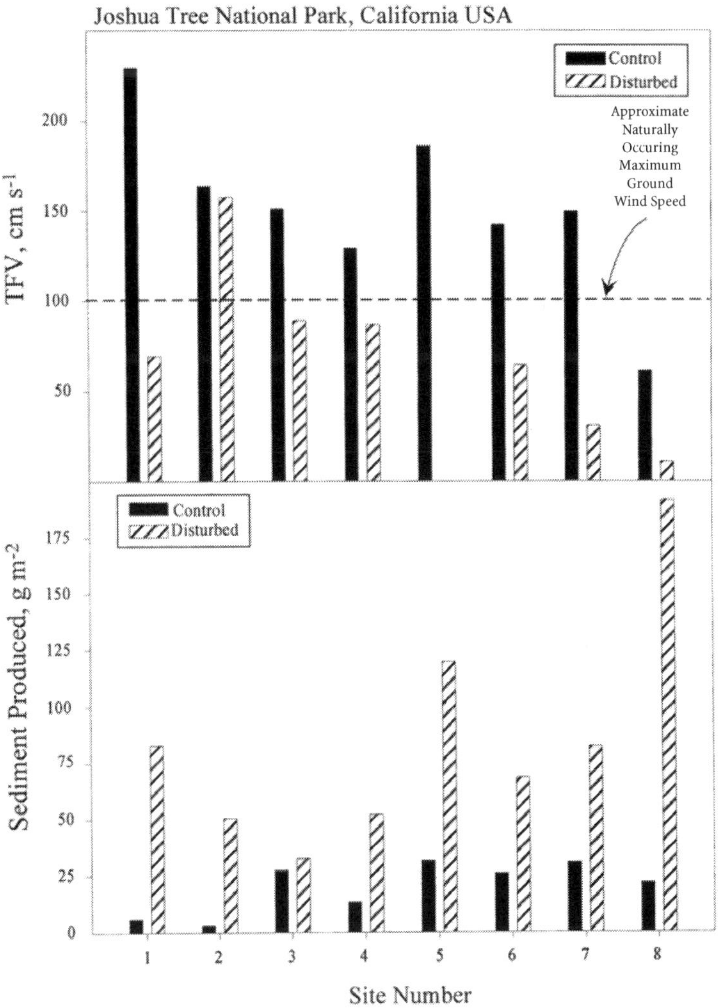

Fig. 25.3. Threshold friction velocities (*TFV*) of soil surfaces. Experimental disturbances were applied to four surfaces in SE Utah with different disturbance histories: undisturbed (>20 years), lichen-moss soil crusts; crust vehicle-disturbed 5 years previously (currently cyanobacterially dominated); crust vehicle-disturbed 1 year previously (currently cyanobacterially dominated) and newly disturbed lichen-moss crust. These soils were then walked and driven over, except for controls. All differences within a soil type are significantly different ($p<0.05$). TFVs of undisturbed crusts are above maximum field-recorded wind speeds, while those for all other surfaces are below recorded wind speeds, and thus subject to wind erosion. (After Belnap and Gilette 1997)

and Eldridge (1998) showed a fivefold increase in sediment loss from a loamy soil and a fourfold increase from sandy soils when soil crusts were severely disturbed by raking (however, due to greater erodibility in sandy soil, much more total sediment was lost from the sandy site). Williams et al. 1995 showed a fivefold increase in sediment loss when crusts were scalped from a silty soil. Figure 25.2 shows 2- to 35-fold increases in sediment production at high spring wind speeds when crust-covered sandy loam soils in the Mojave Desert were disturbed by vehicle tracks. As disturbance increases in severity, more aggregates are crushed, resulting in greater soil loss with greater disturbance levels (Leys and Eldridge 1998).

Disturbance type can affect wind erodibility. Hoof and human footprints have less effect on soil TFV than vehicle tracks in sandy and some silty soils (Belnap and Gillette 1997, 1998). In other silty soils, hoof prints have more impact than vehicle tracks (Belnap and Gillette 1998). This response is probably dependent on the depth of the print in a given soil, as deeper, discrete prints would have greater protection from wind than shallow prints or continuous vehicle tracks.

The condition of the biological soil crust prior to disturbance can affect the wind erodibility of a surface after the new disturbance. Disturbance in well-developed, undisturbed lichen-moss soil crusts have less effect on TFVs than disturbance to crusts that are recovering from impacts (Fig 25.3; Belnap and Gillette 1997). Footprints and one pass with a vehicle in an undisturbed lichen-moss crust have less impact than the same disturbance on a crust disturbed 5 years ago, 1 year ago, or just recently, as undisturbed crusts have a buildup of soil-binding polysaccharide material both at and below the soil surface that is lacking in crusts that have recently been disturbed. Nearby disturbances can also have a detrimental effect on the stability of soil surfaces. Sediment produced from such disturbances can "sandblast" adjacent crusts and erode them at a much faster than normal rate. Thus, a small disturbance can trigger much larger impacts (Belnap and Gillette 1997; Evans and Johansen 1999).

25.4 Recovery of Wind Resistance

Because crustal organisms are only metabolically active when wet, re-establishment time is slow in arid systems. Recovery is faster under plant canopies than shrub interspaces (see Chap. 27), but soils in plant interspaces are more vulnerable to wind erosion. Experiments show that sandy soils can remain susceptible to monthly maximum winds for at least 5 years after disturbance, and to annual maximum winds for at least 10 years after disturbance (Fig. 25.2).

Although silty alluvial and sandy soils showed a similar response to new vehicle disturbance (83 vs. 74% TFV decline), silty soils show a much faster recovery rate than sandy soils. Silty soils are able to form a thin physical crust after even a light rain that increases TFV by 65%. In contrast, sandy soils with no physical crusting showed little TFV recovery after 5 years (Belnap and Gillette 1997).

25.5 Conclusions

All studies show that biological soil crusts play a significant role in reducing soil loss by wind. Degree of crustal development parallels amount of protection from wind erosion. As coarse soils are inherently more erodible than silt soils and do not generally form physical crusts, the protection offered by biological soil crusts is especially important in coarse soils. Disturbance to crusts results in decreased TFV and increased sediment production. This response is related to the severity of disturbance; to the biomass, species composition, and physical structure of the soil crusts; and to soil characteristics such as texture, physical crusting, and vesicular horizons. However, with the limited number of data sets available, it is not yet possible to develop generalizations predicting the relative importance of these factors on a regional or global basis.

Recovery time of wind resistance is much faster in soils with inherent aggregate stability, such as clays and gypsum. Recovery is also much faster in soils where physical crusts reform after rain events. Thus, coarse soils have less inherent resistance to wind erosion, and take longer to redevelop protection from wind erosion once disturbed. This suggests that, while biological soil crusts play an important role in preventing wind erosion in all soil types, the protection offered by biological soil crusts is especially critical in coarse soils.

References

Belnap J, Gardner JS (1993) Soil microstructure of the Colorado Plateau: the role of the cyanobacterium *Microcoleus vaginatus*. Great Basin Nat 53:40–47

Belnap J, Gillette DA (1997) Disturbance of biological soil crusts: impacts on potential wind erodibility of sandy desert soils in southeastern Utah. Land Degrad Dev 8:355–362

Belnap J, Gillette DA (1998) Vulnerability of desert biological soil crusts to wind erosion: the influence of crust development, soil texture, and disturbance. J Arid Environ 39:133–142

Chartes CJ (1992) Soil crusting in Australia. In: Summer ME, Stewart BA (eds) Soil crusting: chemical and physical processes. Lewis Publisher, Boca Raton, FL, pp 339–366

Danin A, Ganor E (1991) Trapping of airborne dust by mosses in the Negev Desert, Israel. Earth Surface Processes Landforms 16:153–162

Danin A, Yaalon DH (1982) Silt plus clay sedimentation and decalcification during plant succession in sands of the mediterranean coastal plain of Israel. Isrl J Earth Sci 31:101–109

Danin A, Bar-Or Y, Dor I, Yisraeli T (1989) The role of cyanobacteria in stabilization of sand dunes in southern Israel. Ecol Medit 15:55–64

Dulieu D, Gaston A, Darley J (1977) La dégradation des pâturages de la région N'Djamena (République du Tchad) en relation avec la présence de cyanophycées psammophiles – étude préliminaire. Rev Elev Méd Vét Pays Trop 30:181–190

Eldridge DJ, Green RSB (1994) Microbiotic soil crusts: a review of their roles in soil and ecological processes in the rangelands of Australia. Aust J Soil Res 32:389–415

Evans RD, Johansen JR (1999) Microbiotic crusts and ecosystem processes. Crit Rev Plant Sci 18:183–203

Garcia-Pichel F, Belnap J (1996) Microenvironments and microscale productivity of cyanobacterial desert crusts. J Phycol 32:774–782

Gillette DA, Adams J, Endo A, Smith D, Kihl R (1980) Threshold velocities for input of soil particles into the air by desert soils. J Geophys Res 85(C10):5621–5630

Goudie AS (1978) Dust storms and their geomorphological implications. J Arid Environ 1:291–310

Kovda V (1980) Land aridization and drought control. Westview Press, Boulder, CO

Leys JN (1990) Soil crusts, their effect on wind erosion. Soil Conservation Service of NSW Res Note 1/90

Leys JF, Eldridge DJ (1998) Influence of cryptogamic crust disturbance to wind erosion on sand and loam rangeland soils. Earth Surface Processes Landforms 23:963–974

Marticorena B, Bergametti G, Gillette D, Belnap J (1997) Factors controlling threshold friction velocity in semiarid and arid areas of the United States. J Geophys Res 102(D19):23,277–23,287

McKenna-Neuman C, Maxwell CD, Boulton JW (1996) Wind transport of sand surfaces crusted with photoautotrophic microorganisms. Catena 27:229–247

Musick HB (1998) Field monitoring of vegetation characteristics related to surface changes in the Yuma Desert, Arizona, and at the Jornada Experimental Range in the Chihuahuan Desert, New Mexico. US Geological Survey, Publication 1598-D

Offer ZY, Goossens D, Shachak M (1992) Aeolian deposition of nitrogen to sandy and loessial ecosystems in the Negev Desert. J Arid Environ 23:355–363

Pluis JLA (1994) Algal crust formation in the inland dune area, Laarder Wasmeer, The Netherlands. Vegetatio 113:41–51

Reynolds R, Belnap J, Reheis M, Mazza N (1998) Eolian dust on the Colorado Plateau – magnetic and geochemical evidence from sediment in potholes and biologic soil crust. In: Proc Conf on Dust Aerosols, Loess, Soils, and Global Change, Seattle, WA

Tsoar H, Møller JT (1986) The role of vegetation in the formation of linear sand dunes. In: Nickling WG (ed) Aeolian geomorphology. Allen and Unwin, Boston, MA, pp 75–95

Tsoar H, Pye K (1987) Dust transport and the question of desert loess formation. Sedimentology 34:139–153

Van den Ancker JAM, Jungerius PD, Mur LR (1985) The role of algae in the stabilization of coastal dune blow-outs. Earth Surface Processes Landforms 10:189–192

Verrecchia E, Yair A, Kidron GJ, Verrecchia K (1995) Physical properties of the psammophile cryptogamic crust and their consequences to the water regime of sandy soils, north-western Negev Desert, Israel. J Arid Environ 29:427–437

Webb RH, Wilshire HG (eds) (1983) Environmental effects of off-road vehicles: impacts and management in arid regions. Springer, Berlin Heidelberg New York

Williams JD, Dobrowolski JP, West NE, Gillette DA (1995) Microphytic crust influence on wind erosion. Trans Am Soc Agric Eng 38:131–137

26 Synopsis: Influence of Biological Soil Crusts on Arid Land Hydrology and Soil Stability

S.D. Warren

26.1 Introduction

Chapters 22–24 of this Volume review the relationships between biological soil crusts and soil hydrology in the deserts of Israel, Australia, and North America, respectively. It is readily apparent that there are contradictions regarding the effects of biological soil crusts on arid land hydrology and soil stability. A logical approach to understanding some of the commonalities and disparities is to seek an understanding of how the crusts affect the processes of infiltration, runoff and erosion. Infiltration and runoff are closely related and inversely proportional, i.e., as more water infiltrates into the soil, less is available for surface runoff. This chapter discusses the effects of biological soil crusts on parameters known to control the partitioning of surface water between infiltration and runoff, as well as parameters related to soil erosion. These include soil physical and chemical properties, surface cover and roughness, below-ground biomass, soil hydrophobicity, and antecedent soil moisture content.

26.2 Soil Properties

Sandy soils are generally characterized by large pores, rapid infiltration, and low water-holding capacity. Finer-textured soils often have higher water-holding capacity and more total pore space, but the pores are narrow and the transmission of water through them is slower, being regulated by capillary movement rather than gravitational flow. The hydraulic conductivity of a clay soil may be as much as 3 orders of magnitude slower than sand, and the addition of as little as 10–20 % fine particles to sand can reduce its hydraulic conductivity by 1 order of magnitude (Rawls et al. 1983). Biological soil crusts on sandy soils are known to accumulate windblown silt and clay particles at the surface (e.g., Danin and Ganor 1991; Verrecchia et al. 1995; Malam Issa et

Ecological Studies, Vol. 150
J. Belnap and O.L. Lange (eds.) Biological Soil Crusts: Structure, Function, and Management

al. 1999; Chap. 15). Clay particles in particular may be tightly bound to the sticky sheaths of some cyanobacteria (e.g., Campbell et al. 1989; Danin et al. 1989; Belnap and Gardner 1993; see also Chaps. 1, 15). In addition, fungal hyphae, cyanobacterial filaments, and rooting structures of lichens, mosses, and liverworts may occupy soil pores. Upon wetting, the sheaths of some cyanobacteria absorb up to eight times their weight in water (Wang et al. 1981; Belnap and Gardner 1993) and swell to several times their original volume (Campbell 1979), further blocking available pore space near the soil surface (Verrecchia et al. 1995; Kidron and Yair 1997; Kidron et al. 1999). Hence, it may be expected that, other factors being equal, the presence of biological soil crusts on very sandy soils will reduce infiltration rates relative to uncrusted soils of similar texture. Table 26.1 presents the results of published studies that have compared the infiltration rates of biologically crusted versus uncrusted or less-crusted soils. As postulated, it appears evident that the presence of biological crusts on soils with greater than 80 % sand-sized particles tends to cause a reduction in infiltration rate. However, it should be noted that all of the cited studies occurred in regions where soils do not freeze. In regions where frost-heaved crusts result in high surface roughness, it is possible that infiltration is increased in sandy soils, although this has not been experimentally tested.

As the clay content of a soil increases, the ratio of micro- to macropores also increases, the effective porosity declines, the rate of infiltration is slowed, and the role of biological soil crusts relative to infiltration changes dramatically. Biological soil crusts contribute significant amounts of organic carbon to arid soils (Beymer and Klopatek 1991; Jeffries et al. 1993). The contribution may be in the form of decaying organic matter (Danin and Ganor 1991) or polysaccharide secretions (Rogers and Burns 1994; Mazor et al. 1996). Organic matter is effective in aggregate formation when decomposed by soil microbes into various metabolites (Burns and Davies 1986); polysaccharide secretions function as gluing agents and contribute directly to the formation of aggregates (Rogers and Burns 1994; de Cano et al. 1997; Zulpa de Caire et al. 1997). Fungal hyphae and cyanobacterial filaments contribute further to the stability of soil aggregates through physical entanglement of soil particles (Durrell and Shields 1961; Bond and Harris 1964a; Schulten 1985). A well-aggregated, granular structure increases the proportion of larger pores, thus increasing the infiltration rate of water. By increasing the size and stability of soil aggregates, the presence of biological crusts on soils with a significant percentage of clay-sized particles should enhance infiltration relative to uncrusted soils of similar texture. An examination of Table 26.1 indicates that this is true more often than not where the clay content of arid soils exceeds approximately 15 %. In the only documented study where the presence of a biological crust on soil with at least 15 % clay resulted in a decline in the rate of infiltration, the authors noted the presence of abundant faunal channels

Table 26.1. A comparison of the relative infiltration rate of biologically crusted versus uncrusted or less-crusted soils as a function of soil particle size distribution. Infiltration rates on biological soil crusts are reported as higher (+), lower (-) or not significantly different (ns) than infiltration on the uncrusted or less-crusted counterpart

% Sand	% Silt	% Clay	Response	Reference
100	0	0	–	Mazor et al. (1996)
98	1	1	–	Kidron et al. (1999)
90	9	1	–	Yair (1990)
90	9	1	–	Kidron and Yair (1997)
90	9	1	–	Eldridge et al. (2000)
90	5	5	–	Graetz and Tongway (1986)
86	10	4	–	Dulieu et al. (1977)
83	11	6	–	Zombre et al. (1996)
81	7	12	ns	Eldridge et al. (1997)
66	28	6	ns	Williams et al. (1995, 1999)
66	11	23	–	Greene et al. (1990)
65	25	10	+	Osborn (1952)
65	16	19	+	Greene and Tongway (1989)
64	11	25	ns	Eldridge (1993)[a]
63	14	23	+	Eldridge (1993)[b]
62	29	8	ns	Faust (1970)
59	27	14	–	Eldridge et al. (2000)
56	17	27	+	Seghieri et al. (1997)
55	35	10	+	Loope and Gifford (1972)
50	40	10	ns	Dobrowolski (1994)
37	33	30	+	Faust (1970)
36	49	15	ns	Dobrowolski and Williams (1994)
28	47	25	+	Harper and St. Clair (1985)
14	74	12	–	Abaturov (1993)

[a] Protected site in good or better condition in a nature preserve.
[b] Degraded site subjected to grazing.

just beneath the surface that facilitated rapid infiltration once the crust was removed (Greene et al. 1990).

Arid soils with a high percentage of silt are prone to the development of vesicular horizons characterized by numerous air-filled voids (Miller 1971) and are well-known for restricting infiltration (Blackburn 1975). Where vesicular horizons are present, the potential effects of biological soil crusts on infiltration may be largely negated (Dobrowolski 1994). Unfortunately, few studies relative to infiltration rates on biologically encrusted soils have discussed the presence or absence of a vesicular horizon. Nevertheless, in the majority of the studies summarized in Table 26.1 where the combined silt and clay content of the soil exceeded 20 % and the presence of a biological crust had no significant impact on infiltration, the ratio of silt to clay was at least 2:1, indicating a likelihood of a vesicular horizon.

Eldridge (1993) and Eldridge et al. (1997) suggest that the relationship between biological soil crusts and infiltration may depend on the condition of the site. They found that the presence of biological soil crusts was positively correlated with infiltration at degraded sites, but had little effect on infiltration at sites in good condition. Their data suggest that infiltration occurred mainly through macropores at good condition sites and through micropores at degraded sites. On poor-condition sites, the occurrence of biological crusts probably correlates well with soil aggregate development, thus having a significant influence on infiltration. Conversely, on good-condition sites where a well-aggregated structure is rather ubiquitous, other factors such as litter, plant roots, and invertebrate activity may be better correlated with infiltration.

Many of the same soil parameters that control infiltration also affect soil erodibility. Less energy is required to detach and suspend fine soil particles than coarse soil particles or stable soil aggregates. Thus, to the degree that biological soil crusts contribute to the aggregation of soil particles, they contribute to soil stability. As expected, the scientific literature points to a near consensus that biological soil crusts deter soil erosion (see Chaps. 23, 24). Exceptions occur on sand dunes where biological soil crusts collect fine soil particles that subsequently impede infiltration and increase runoff. Runoff accumulation in downslope positions can generate sufficient erosive energy to incise the crust and create gullies (Talbot and Williams 1978; Jungerius and van der Meulen 1988).

26.3 Surface Cover

On sandy soils, where biological soil crusts reduce infiltration, infiltration is negatively correlated with the percent cover of the crusts (Kidron 1995). Crusts dominated by cyanobacteria cause greater reductions than crusts dominated by mosses (Kidron 1995). When finer-textured soils are subjected to raindrop impact, soil aggregates at the surface may break down, and individual soil particles or microaggregates may be washed into soil pores, where they impede infiltration. To the degree that biological crusts provide surface cover on finer-textured soils, they reduce the kinetic energy of falling raindrops and help maintain high infiltration rates while limiting soil erosion (Eldridge and Greene 1994). Given the small size of cyanobacteria and algae, plus the fact that they live within the surface few millimeters of the soil rather than on the surface, they are less effective in preventing detachment of soil particles, while mosses and lichens growing on the surface are more effective

(Tchoupopnou 1989). Despite their ability to mitigate the impact of falling raindrops, the vegetative structures of crustose and squamulose lichen species tend to grow flush with the soil surface, thus blocking access to soil pores. Where a significant portion of the soil surface is covered by these lichens, infiltration may be limited and runoff may increase (Alexander and Calvo 1990). However, because their presence also limits detachment and transport of soil particles by raindrops and runoff, there is generally a net decline in erosion when compared to similar areas without lichens (Alexander and Calvo 1990).

During dry periods, many mosses and liverworts shrivel, fold, or curl, thus giving the appearance of minimal hydrologic benefit. However, rehydration may occur very rapidly when moisture is present, even during small precipitation events. Upon rehydration, the organisms swell, unfold, and uncurl, increasing their surface area up to fourfold and covering a greater portion of the soil surface (Scott 1982; Rogers 1994; Eldridge and Kinnell 1997).

26.4 Surface Roughness

When the intensity of rainfall exceeds the rate of infiltration, water remains on the soil surface. Surface water either runs off or is stored in small depressions, where it may subsequently infiltrate. Biological soil crusts contribute to surface roughness, thus providing surface detention, increasing net infiltration, and reducing net runoff. Lichens, mosses, and liverworts are generally larger than algae and cyanobacteria, and create greater surface roughness (Cameron and Blank 1966; Danin and Barbour 1982; Danin et al. 1998). In regions without frost-heaving, crusts dominated by algae and cyanobacteria are often difficult to discern in the field, but frequently contribute to a rugose microtopography (see Chap. 1, Photos 49, 55; Chap. 15). In regions where soils freeze, frost-heaving of crusted surfaces creates pinnacles up to 15 cm high, creating a highly roughened surface (Photos 31–33). On fine-textured soils prone to polygonal surface cracking, the shrinking and swelling of cyanobacterial polysaccharides may increase the density of polygons (Danin et al. 1998; Kidron et al. 1999). Cyanobacteria and algae also contribute to differential drying between the upper and lower surfaces of the crust, leading to upward warping or curling of the polygon margins (Durrell and Shields 1961; Cameron and Blank 1966; Danin et al. 1998). Surface cracking facilitates infiltration into an otherwise less-permeable surface (Kidron et al. 1999). In addition, the roughness of the cracked surface lessens the erosive energy of surface runoff and allows more time for infiltration to occur.

26.5 Below-Ground Biomass

Below-ground biomass plays an important role in soil stability. Rooting structures serve as physical barriers to soil detachment by overland flow of water. When dead, they add organic matter to the soil, thus enhancing soil aggregate formation and stability. Mosses, fungi, and many lichens produce root-like protonemata, hyphae and rhizines, respectively (see Chap. 1). Many algae and cyanobacteria are filamentous, functioning much like the roots of higher plants in terms of binding soil particles. In some desert ecosystems the organic mass of algal crusts may exceed that of vascular plants (Rodin and Bazilevich 1967; Kononova 1975). Even where the below-ground biomass of biological soil crusts is proportionately small, the contribution to soil stability may be disproportionately large. Because soil erosion takes place at the soil surface, and because the rooting structures and filaments of biological soil-crust organisms are located in the surface few millimeters, their role in soil stabilization may be more important than that of the roots of higher plants.

26.6 Soil Hydrophobicity

Soil hydrophobicity or water repellency occurs where the cohesive forces between water molecules are stronger than the adhesive forces between the water molecules and the soil surface, resulting in a high contact angle of the water on the surface (Wessel 1988) and beading of water droplets on the surface. A distinction should be made between hydrophobic soils and those with inherently low permeability. Low permeability may be caused by low total porosity, small pore size, pore discontinuity, or physical obstructions. Hydrophobicity is generally related to the coating of soil particles with an organic film that actually repels water (Roberts and Carbon 1972). The phenomenon is most commonly reported on sandy soils. Factors and processes contributing to soil hydrophobicity are poorly understood. Nevertheless, soil hydrophobicity has been attributed to chemical metabolites of some fungi (Bond and Harris 1964a,b; Savage et al. 1969) and polysaccharide secretions of some algae and cyanobacteria (Rietveld 1978; Mazor et al. 1996; Kidron et al. 1999). Water repellency appears to occur only when soils are dry (Bond and Harris 1964b; Rutin 1983; Jungerius and de Jong 1989). Repellency tends to increase when biological soil crusts experience long periods of favorable moisture followed by drying (Rutin 1983; Kidron et al. 1999), but disappears after the application of as little as a few millimeters of water (Rietveld 1978; Kidron et al. 1999). Therefore, hydrophobicity induced by biological soil crusts is likely to significantly affect infiltration and erosion only during very light precipitation events or during the first few minutes of heavier events.

26.7 Antecedent Soil Moisture

With few exceptions (e.g., Harper and Marble 1988), it is generally reported that biological soil crusts serve as a surface mulch and conserve soil moisture (Fritsch 1922; Mücher et al. 1988; Belnap et al. 1996; Mazor et al. 1996; Perez 1997; Seghieri et al. 1997; Malam Issa et al. 1999; see also Chap. 21). Antecedent soil moisture, or soil moisture at the beginning of a rainfall event, has little impact on steady-state infiltration rate. However, as a general rule, the greater the moisture content of the soil at the beginning of a rainfall event, the slower the initial rate of infiltration and the greater the amount of total runoff. Because precipitation is scant and soils tend to dry between the sporadic rainfall events in arid environments, it is unlikely that the temporary retention of soil moisture by biological soil crusts has a major effect on infiltration. Nevertheless, the preservation of soil moisture, albeit temporary, may have a positive effect on the germination and establishment of vascular plants (Tsoar and Møller 1986) which, in turn, enhance infiltration. Hence, while the increased antecedent soil moisture may have a negative effect on infiltration at the scale of a small plot, the overall effect at the landscape scale may be positive.

26.8 Conclusions

Although the literature is replete with apparent contradictions regarding the role of biological crusts in soil hydrology and erosion, an evaluation of the effects of the crusts on the various processes and parameters controlling infiltration and erosion is enlightening. Where the sand content of a soil exceeds 80 % and soils are not frost-heaved, the effect of biological crusts on infiltration is generally negative. This may result from a reduction in the effective porosity of sandy soils and/or organically induced hydrophobicity. Porosity of sandy soils may be reduced as the crusts accumulate fine soil particles or where polysaccharide secretions, rooting structures, or cyanobacterial/algal filaments clog the soil pores. Lichens that grow flush with the soil surface may also block access to the soil pores. The effect of soil crusts on frost-heaved, sandy soils has not been experimentally tested.

Where the sand content of the soil is less than 80 %, the effect of biological crusts on infiltration tends to reverse. Soil-crust organisms contribute to the aggregation of fine soil particles into stable soil aggregates. Well-aggregated soils are characterized by higher effective porosity, higher rates of infiltration, and minimal erosion when compared to poorly aggregated soils of similar texture. This generalization may not hold true where the silt content of the soil is high; desert soils with a high silt content are prone to the development of vesicular horizons characterized by discontinuous pores and slow infiltration. Vesicular horizons may mask any potential effects of biological soil crusts.

The correlation of biological soil crusts with favorable soil hydrologic conditions may be most pronounced on degraded sites. While biological crusts contribute to the development of a well-aggregated, porous soil structure, other factors may become equally or more important in maintaining that structure once it has been created.

The presence of biological soil crusts contributes cover and roughness to the soil surface. Surface cover reduces the energy of falling raindrops, thus minimizing detachment of soil particles and subsequent clogging of soil pores. Surface roughness slows surface runoff and provides detention storage, thereby allowing more time for water to infiltrate and reducing soil erosion.

Filamentous cyanobacteria and algae, as well as the rooting structures of lichens, mosses, and liverworts, are concentrated in the surface few millimeters of the soil where erosion takes place. As such, they may be disproportionately more important than the root mass of higher plants with regard to soil stability in some desert ecosystems.

With rare exceptions, the presence of biological soil crusts tends to conserve soil moisture. While the effect of moisture conservation has little direct impact on infiltration and erosion, the potential to enhance the germination and growth of higher plants may indirectly contribute to the stability of desert landscapes.

Acknowledgments. Financial support for this work was provided by Gwyn Howard, US Army Construction Engineering Research Laboratories, Champaign, Illinois.

References

Abaturov BD (1993) Alteration of small relief forms of the hydrophysical properties of heavy loam soils in the semidesert zone by grazing. Euras Soil Sci 25:17–28

Alexander RW, Calvo A (1990) The influence of lichens on slope processes in some Spanish badlands. In: Thornes JB (ed) Vegetation and erosion. John Wiley, Chichester, UK, pp 385–398

Belnap J, Gardner JS (1993) Soil microstructure in soils of the Colorado Plateau: the role of the cyanobacterium *Microcoleus vaginatus*. Great Basin Nat 53:40–47

Belnap J, Sanford RL, Lungu L (1996) Biological soil crusts: ecological roles and response to fire in miombo woodlands of Zimbabwe. Trans Zimbabwe Sci Assoc 70:14–20

Beymer RJ, Klopatek JM (1991) Potential contribution of carbon by microphytic crusts in pinyon-juniper woodland. Arid Soil Res Rehabil 5:187–198

Blackburn WH (1975) Factors influencing infiltration and sediment production of semiarid rangelands in Nevada. Water Resour Res 11:929–937

Bond RD, Harris JR (1964a) The influence of the microflora on physical properties of soils. I. Effects associated with filamentous algae and fungi. Aust J Soil Res 2:111–122

Bond RD, Harris JR (1964b) The influence of the microflora on physical properties of soils. II. Field studies on water-repellent sands. Aust J Soil Res 2:123–131

Burns RG, Davies JA (1986) The microbiology of soil structure. In: Lopez-Real JM, Hodges RD (eds) The role of microorganisms in a sustainable agriculture. AB Academic Publ, Great Britain, pp 9–27

Cameron RE, Blank GB (1966) Desert algae: soil crusts and diaphanous substrata as algal habitats. Nat Aeronaut Space Adm, Jet Propul Lab Tech Rep 32–971. California Inst Tech, Pasadena, California

Campbell SE (1979) Soil stabilization by prokaryotic desert crust: implications for Precambrian land biota. Origins Life 9:335–348

Campbell SE, Seeler JS, Golubic S (1989) Desert crust formation and soil stabilization. Arid Soil Res Rehabil 3:217–228

Danin A, Barbour MG (1982) Microsuccession of cryptogams and phanerogams in the Dead Sea area, Israel. Flora 172:173–179

Danin A, Ganor E (1991) Trapping of airborne dust by mosses in the Negev Desert, Israel. Earth Surface Processes Landforms 16:153–162

Danin A, Bar-Or Y, Dor I, Yisraeli T (1989) The role of cyanobacteria in stabilization of sand dunes in southern Israel. Ecol Mediterr 15:55–64

Danin A, Dor I, Sandler A, Amit R (1998) Desert crust morphology and its relations to microbiotic succession at Mt. Sedom, Israel. J Arid Environ 38:161–174

de Cano MS, de Mulé MCZ, de Caire GZ, Palma RM, Colombo K (1997) Aggregation of soil particles by *Nostoc muscorum* Ag. (Cyanobacteria). Phyton 60:33–38

Dobrowolski JP (1994) In situ estimation of effective hydraulic conductivity to improve erosion modeling for rangeland conditions. In: Blackburn WH, Schuman GE, Pierson FB Jr, Zartman R (eds) Variability in rangeland water erosion processes. Soil Sci Soc Am Spec Publ 38, Madison, Wisconsin, pp 83–91

Dobrowolski JP, Williams JD (1994) Effects of disturbance by tracked vehicles on wind and water erosion. Report to US Dep Agric, For Serv, Shrub Sci Lab, Provo, Utah. Utah State Univ, Logan, Utah

Dulieu D, Gaston A, Darley J (1977) La dégradation des pâturages de la région de N'Djamena (République du Tchad) en relation avec la présence de cyanophycées psammophiles – étude préliminaire. Rev Elev Méd Vét Pays Trop 30:181–190

Durrell LW, Shields LM (1961) Characteristics of soil algae relating to crust formation. Trans Am Microsc Soc 80:73–79

Eldridge DJ (1993) Cryptogam cover and soil surface condition: effects on hydrology on a semiarid woodland soil. Arid Soil Res Rehabil 7:203–217

Eldridge DJ, Greene RSB (1994) Assessment of sediment yield by splash erosion on a semi-arid soil with varying cryptogam cover. J Arid Environ 26:221–232

Eldridge DJ, Kinnell PIA (1997) Assessment of erosion rates from microphyte-dominated calcareous soils under rain-impacted flow. Aust J Soil Res 35:475–489

Eldridge DJ, Tozer ME, Slangen S (1997) Soil hydrology is independent of microphytic crust cover: further evidence from a wooded semiarid Australian rangeland. Arid Soil Res Rehabil 11:113–126

Eldridge DJ, Zaady E, Shachak M (2000) Infiltration through three contrasting biological soil crusts in patterned landscapes in the Negev, Israel. Catena 40:323–336

Faust WF (1970) The effect of algal-mold crusts on the hydrologic processes of infiltration, runoff, and soil erosion under simulated conditions. MS Thesis, Univ Arizona, Tucson, Arizona

Fritsch FE (1922) The terrestrial algae. J Ecol 10:220–236

Graetz RD, Tongway DJ (1986) Influence of grazing management on vegetation, soil structure and nutrient distribution and the infiltration of applied rainfall in a semi-arid chenopod shrubland. Aust J Ecol 11:347–360

Greene RSB, Tongway DJ (1989) The significance of (surface) physical and chemical properties in determining soil surface condition of red earths in rangelands. Aust J Soil Res 27:213–225

Greene RSB, Chartres RSB, Hodgkinson KC (1990) The effects of fire on soil in a degraded semi-arid woodland. I. Cryptogamic cover and physical and micro-morphological properties. Aust J Soil Res 28:755–777

Harper KT, Marble JR (1988) A role for nonvascular plants in management of arid and semiarid rangelands. In: Tueller PT (ed) Vegetation science applications for rangeland analysis and management. Kluwer Academic Publishers, Dordrecht, pp 135–169

Harper KT, St Clair LL (1985) Cryptogamic soil crusts on arid and semiarid rangelands in Utah: effects on seedling establishment and soil stability. Report to US Dep Inter, Bur Land Manage, Denver, Colorado. Brigham Young Univ, Provo, Utah

Jeffries DL, Link SO, Klopatek JM (1993) CO_2 fluxes of cryptogamic crusts. I. Response to resaturation. New Phytol 125:163–173

Jungerius PD, de Jong JH (1989) Variability of water repellence in the dunes along the Dutch coast. Catena 16:491–497

Jungerius PD, van der Meulen F (1988) Erosion processes in a dune landscape along the Dutch coast. Catena 15:217–228

Kidron GJ (1995) The impact of microbial crust upon rainfall-runoff-sediment yield relationships on longitudinal dune slopes, Nizzana, western Negev Desert, Israel. PhD Thesis, Hebrew University of Jerusalem, Jerusalem

Kidron GJ, Yair A (1997) Rainfall-runoff relationships over encrusted dune surfaces, Nizzana, western Negev, Israel. Earth Surface Processes Landforms 22:1169–1184

Kidron GJ, Yaalon DH, Vonshak A (1999) Two causes for runoff initiation on microbiotic crusts: hydrophobicity and pore clogging. Soil Sci 164:18–27

Kononova MM (1975) Humus of virgin and cultivated soils. In: Gieseking JE (ed) Soil components, 1. Organic components. Springer, Berlin Heidelberg New York, pp 475–526

Loope WL, Gifford GF (1972) Influence of a soil microfloral crust on select properties of soils under pinyon-juniper in southeastern Utah. J Soil Water Conserv 27:164–167

Malam Issa O, Trichet J, Défarge C, Couté A, Valentin C (1999) Morphology and microstructure of microbiotic soil crusts on a tiger bush sequence (Niger, Sahel). Catena 37:175–196

Mazor G, Kidron GJ, Vonshak A, Abeliovich A (1996) The role of cyanobacterial exopolysaccharides in structuring desert microbial crusts. FEMS Microbiol Ecol 21:121–130

Miller DE (1971) Formation of vesicular structure in soil. Soil Sci Soc Am Proc 35:635–637

Mücher HJ, Chartres CJ, Tongway DJ, Greene RSB (1988) Micromorphology and significance of the surface crusts of soils in rangelands near Cobar, Australia. Geoderma 42:227–244

Osborn B (1952) Range soil conditions influence water intake. J Soil Water Conserv 7:128–132

Perez FL (1997) Microbiotic crusts in the high equatorial Andes, and their influence on paramo soils. Catena 31:173–198

Rawls WJ, Brakensiek DL, Miller N (1983) Green-Ampt infiltration parameters from soils data. J Hydraul Eng 109:62–70

Rietveld J (1978) Soil non wettability and its relevance as a contributing factor to surface runoff on sandy soils in Mali. Report of the project Production Primaire au Sahel, Wageningen

Roberts FJ, Carbon BA (1972) Water repellence in sandy soils of south-western Australia. II. Some chemical characteristics of the hydrophobic skins. Aust J Soil Res 10:35–42

Rodin LE, Bazilevich NI (1967) Production and mineral cycling in terrestrial vegetation. Oliver and Boyd, Edinburgh

Rogers RW (1994) Zonation of the liverwort *Riccia* in a temporary watercourse in subtropical, semi-arid Australia. Aust J Bot 42:659–662

Rogers SL, Burns RG (1994) Changes in aggregate stability, nutrient status, indigenous microbial populations, and seedling emergence, following inoculation of soil with *Nostoc muscorum*. Biol Fertil Soils 18:209–215

Rutin J (1983) Erosional processes on a coastal sand dune, De Blink, Noordwijkerhout. Physical Geography and Soils Laboratory Publ 35. Univ Amsterdam, Amsterdam

Savage SM, Martin JP, Letey J (1969) Contribution of some soil fungi to natural and heat-induced water repellency in sand. Soil Sci Soc Am Proc 33:405–409

Schulten JA (1985) Soil aggregation by cryptogams of a sand prairie. Am J Bot 72:1657–1661

Scott GAM (1982) Desert bryophytes. In: Smith AJE (ed) Bryophyte ecology. Chapman and Hall, London, pp 105–122

Seghieri J, Galle S, Rajot JL, Ehrmann M (1997) Relationships between soil moisture and growth of herbaceous plants in a natural vegetation mosaic in Niger. J Arid Environ 36:87–102

Talbot MR, Williams MAJ (1978) Erosion of fixed dunes in the Sahel, central Niger. Earth Surface Processes 3:107–113

Tchoupopnou E (1989) Splash from microphytic soil crusts following simulated rain. MS Thesis, Utah State Univ, Logan; Utah

Tsoar H, Møller JT (1986) The role of vegetation in the formation of linear sand dunes. In: Nickling WG (ed) Aeolian geomorphology. Allen & Unwin, Boston, pp 75–95

Verrecchia E, Yair A, Kidron GJ, Verrecchia K (1995) Physical properties of the psammophile cryptogamic crust and their consequences to the water regime of sandy soils, north-western Negev Desert, Israel. J Arid Environ 29:427–437

Wang F, Zhung Z, Hu Z (1981) Nitrogen fixation by an edible terrestrial blue-green alga. In: Gibson AH, Newton WE (eds) Current perspective in nitrogen fixation. Elsevier, Amsterdam, p 455

Wessel AT (1988) On using the effective contact angle and the water drop penetration time for classification of water repellency in dune soils. Earth Surface Processes Landforms 13:555–561

Williams JD, Dobrowolski JP, West NE (1995) Microphytic crust influence on interrill erosion and infiltration capacity. Trans Am Soc Agric Eng 38:139–146

Williams JD, Dobrowolski JP, West NE (1999) Microbiotic crust influence on unsaturated hydraulic conductivity. Arid Soil Res Rehabil 13:145–154

Yair A (1990) Runoff generation in a sandy area – the Nizzana sands, western Negev, Israel. Earth Surface Processes Landforms 15:597–609

Zombré PN, Pallo F, Mulders MA, Thiombiano L, Stroosnijder L, Hoog-Moed WB, Tromp M, Kaboré BR, Van Baren JHV (1996) Excursion to the Kaya Region, Burkina Faso. In: Escadafal R, Mulders MA, Thiombiano L (eds) Surveillance des sols dans l'environnement par télédétection et systèmes d'information géographiques. Institut Français de Recherche Scientifique pour le Développement en Coopération, Paris, pp 569–592

Zulpa de Caire G, Storni de Cano M, Zaccaro de Mulé MC, Palma RM, Colombo K (1997) Exopolysaccharide of *Nostoc muscorum* (Cyanobacteria) in the aggregation of soil particles. J Appl Phycol 9:249–253

Part VI

Disturbance to Biological Soil Crusts: Resistance, Resilience, and Restoration

27 Disturbance and Recovery of Biological Soil Crusts

J. BELNAP and D. ELDRIDGE

27.1 Introduction

Disturbance can profoundly affect the cover, species composition, and the physiological functioning of biological soil crusts. The disturbances we discuss include air pollution; exposure to oil, herbicides, and pesticides; invasion by annual exotic weeds; mechanical disturbances such as human and livestock trampling (see Chap. 29), off-road driving, mining, and hiking; and, briefly, wildfire (for extensive discussion, see Chap. 28). Studies on disturbance have generally been limited to the western US and Australia, with some additional work done in China, Israel, South Africa, and Zimbabwe.

Methods of assessing impacts of, and recovery from, disturbance have been highly variable in the past. Generally, measurements have been limited to visual estimates of crust cover. However, Belnap (1993) showed that visual assessment can accurately assess only moss and lichen cover, and cannot be used to measure recovery of cyanobacterial biomass, soil stability, and/or physiological functioning of crustal organisms. In addition, some studies have only considered total crust cover but have not delimited the relative cover of cyanobacteria, mosses, and lichens. The relationships between total crust cover and impacts of disturbance can be weak, as cyanobacterial cover generally increases, while moss and lichen cover decreases, after disturbance. This often makes total crust cover a poor measure of the dynamics of soil-crust recovery. Differentiating between crustal components is also important because alteration of species composition can heavily influence ecological functioning of the crusts (Eldridge 1998).

Comparing recovery rates from different studies can be problematic, as factors known to control recovery rates (such as site stability and precipitation following disturbance) are often not reported. More importantly, severity of disturbance is seldom quantified. Studies generally report disturbance levels as "light", "moderate", or "heavy" without any definition of these categories; thus, what is "moderate" in one study may be considered "heavy" in another. As studies cover a large range of climatic zones, soil types, levels of

Ecological Studies, Vol. 150
J. Belnap and O.L. Lange (eds.) Biological Soil Crusts: Structure, Function, and Management

disturbance, and ways to calculate recovery, and because there has been no standard for measuring crust recovery, it is not surprising that recovery rates in the literature have ranged widely (2 years in cool deserts to over 3000 years in very dry deserts), and either show no pattern or results often appear contradictory (Anderson et al. 1982; Callison et al. 1985; Jeffries and Klopatek 1987; Cole 1990; Belnap 1995,1996; Belnap and Warren 1998).

27.2 Disturbance Effects

The impact of a given disturbance depends on the soil, plant, and climate characteristics of the site, and the severity, frequency, timing, and type of the disturbance (Fig. 27.1). While most disturbances result in similar types of impacts, severity can vary widely. For instance, vehicles and trampling exert both compressional and shear forces; however, these forces are much greater for vehicles than trampling. In addition, vehicles often turn soils over and bury crustal organisms, while trampling tends to only compress the surface. Vehicle tracks often channel water and thus slow or prevent recovery (Webb and Wilshire 1983). Intensifying physical impacts (such as high-intensity, short-duration grazing) is deleterious to the cover and species richness of biological soil crusts (Johansen 1993). Disturbance that removes or kills crustal organisms results in greater impact and slower recovery than disturbance that leaves crushed crust material in place (see Sect. 27.3.3). As lichens and mosses are less tolerant of disturbance than cyanobacteria, frequent disturbance can maintain the biological soil crust at a low successional stage (e.g., cyanobacterially dominated; Harper and Marble 1988).

Timing of disturbance can affect the degree to which the cover and species richness of the crust is reduced. The intrinsic strength of soil varies with moisture content (Fig. 27.2). Soils with little tendency to form aggregates, such as sands, are more susceptible to compressional stresses when dry. Fine-textured soils or those with inorganic crusts are more vulnerable to compressional disturbance when wet (Webb and Wilshire 1983). Crust components are brittle when dry, and the connections they make between soil particles are easily crushed. Thus, compressional disturbances can severely affect the ability of crusts to stabilize soils, especially in dry sandy and silty soils. For instance, grazing in the early wet season (winter) has been shown to have less impact on crust cover and species composition than late winter or spring grazing. As crustal species are only metabolically active when wet and are brittle when dry, disturbance in dry seasons is more destructive, and organisms are less able to recover, than when disturbed in wet seasons (Harper and Marble 1988).

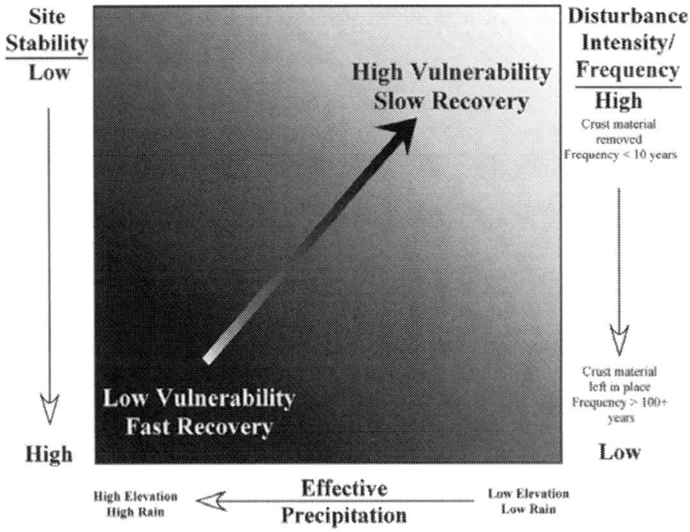

Factors Determining Site Stability

	Low Stability	High Stability
Soil Texture/Age	Coarse/Young	Fine/Old
Rock/Gravel Cover	Rolling	Embedded
Soil Depth	Deep	Shallow
Plant Spacing	Sparse	Dense

Fig. 27.1. Vulnerability and recoverability of crusts depends on gradients of site stability, effective precipitation, and disturbance regimes. *Top panel* Crusts at sites with the greatest stability (defined in *bottom panel*), greatest effective precipitation, and lowest disturbance frequency or intensity will be less impacted (*dark shading*) than sites with lower stability, less effective precipitation, and higher disturbance frequency or intensity (*light shading*). Similarly, recovery time is faster (*dark shading*) in areas of low vulnerability, and slower (*light shading*) where vulnerability is higher. *Bottom panel* Factors influencing site stability

		Soil Texture		
		Sand	Clay	Silt
Soil Moisture Content	Frozen	Low	Low	Low
	Wet	Medium-High	High	Low
	Moist	Medium-High	Medium	Medium
	Dry	Very High	Low	Very High

Fig. 27.2. Vulnerability of crusts to disturbance varies according to soil texture and moisture. Note crusts on sandy soils are more susceptible to disturbance when dry than when wet, while crusts on clay soils have the opposite response

27.2.1 Disturbance Effects on Species Composition of Biological Soil Crusts

Overall, disturbance generally results in loss of species diversity, biomass, and surface cover of cyanobacteria, lichens, and mosses. The more severe the disturbance, the greater the loss. Thus, after severe disturbance, the crust community with multiple species of cyanobacteria, lichens, and mosses is often simplified to a community dominated by one or a few species of cyanobacteria.

Air Pollution. A few studies have addressed the impact of air pollutants on soil lichen communities in desert environments. No differences have been found in species composition near pollution sources when compared to control sites. It is reasoned that lichens with thalli closely appressed to the soil surface, as found in most desert soil-crust lichens, are less susceptible to damage by air pollutants than lichens whose tissue is more exposed to air. In addition, most desert soils are very alkaline, and thus thought to buffer acidic pollutants (Nash and Sommerfeld 1981).

Oil Spills, Insecticides, and Herbicides. We know of no study that directly addresses the effects of oil, oil dispersants, insecticides, or herbicides on species composition of intact soil crusts. However, much work has been done on individual species of cyanobacteria, green algae, and mosses isolated from soil crusts. Tested species were differentially affected, both positively and negatively, depending on the compound and the species tested. Thus, exposure to these agents could either kill or potentially alter species composition of crusts (Metting 1981; Youtie et al. 1999).

Annual Plant Invasion. Invasion of exotic annual plants into perennial plant communities can impose a long-term threat to biological soil-crust communities because the crust-dominated interspace between perennial plants is often heavily invaded. Surveys in invaded communities show that rich perennial moss/lichen communities are quickly replaced with only a few species of annual mosses and cyanobacteria (Kaltenecker 1997; Belnap and Phillips 2001). The mechanism by which this shift occurs is not known, but probably results from increased plant material shading the soil surface, increased soil turnover by elevated rodent populations, and/or increased fire frequency.

Fire. The effects of fire are covered extensively in Chapter 28. In summary, the main effects of fire are loss of cover, biomass, and species diversity of soil crusts (Johansen et al. 1993). Damage to, and recovery of, soil crusts depends on the prefire composition and structure of the vascular plant community and distribution of fuel, fire intensity, and fire frequency. Hotter and more frequent fires result in more damage to the crusts. Frequent fires prevent the recovery of lichens and mosses, resulting in crusts containing only a few species of cyanobacteria.

Mechanical Disturbance. Mechanical disturbance includes activities such as vehicle traffic (military and recreational), trampling by livestock (Chap. 29) and people, and land-clearing (such as mining). Such uses are intensifying exponentially in arid and semiarid areas of the world. Effects of mechanical disturbance are especially noticeable at sites with highly erodible soils and large topographic relief (Harper and Marble 1988).

Over 30 studies on four continents document that livestock grazing, vehicles (both recreational and military), and human trampling dramatically reduce lichen-moss cover and species richness of crusts. Resistance to disturbance generally decreases in the following order: cyanobacteria > gelatinous lichens > squamulose, crustose lichens; mosses > foliose lichens (reviewed in Harper and Marble 1988; West 1990; Johansen 1993; Eldridge and Greene 1994; Ladyman and Muldavin 1996). Cyanobacteria, being relatively mobile, are the most resistant to disturbance. Lichens use a combination of thallus structures, pigments, water storage capacities, tolerance of inundation, and/or an ability to fix atmospheric nitrogen (N) to increase resistance to disturbance. For example, Rogers and Lange (1971) showed that the lichens *Collema coccophorum* and *Heppia lutosa* (*H. despreauxii*) were the lichen species least affected by sheep trampling around a watering point. These lichens are able to fix atmospheric N and to store greater amounts of water than phycolichens (Galun et al. 1982). In a comparison of species inside and outside a grazing exclosure, cover of most lichen species was significantly reduced with grazing (Table 27.1). In addition, most soil-surface disturbance compacts soils. Compaction influences soil water and nutrient-holding capacity, which can lead to changes in the species composition of soil-crust communities. These subtle compositional changes often occur before changes in cover are apparent (Eldridge 1996).

Burial. Disruption of crusts often destabilizes underlying soils, leaving adjacent crusts vulnerable to burial by wind and water-moved sediments. Large filamentous cyanobacteria can respond to limited burial by moving up to 5 mm every 24 h if wet. Mosses can also cope with limited burial (Fig. 15.2). However, when burial is too deep, these organisms will die. Burial often kills other photosynthetic components of the crust, including lichens, green algae, and smaller cyanobacteria (Campbell 1979), and thus generally results in a greatly simplified crustal community.

Table 27.1. Frequency (%) of lichens and a liverwort on loamy and sandy soils inside (ungrazed) and outside (grazed) exclosures at Koonamore Vegetation Reserve, South Australia in May 1972. Species extirpated by grazing (frequency = 0) were not evaluated statistically. Frequencies of all other species were significantly reduced with grazing (p<0.10). (After Crisp 1975)

Species	Type	Loamy soil		Sandy soil	
		Ungrazed	Grazed	Ungrazed	Grazed
Acarospora smaragdula	Crustose	2.2	0.4	0.2	0.0
Aspicilia calcarea	Crustose	27.4	18.0	11.3	2.1
Aspicilia calcarea	Fruticose	12.1	0.7	1.0	0.0
Collema coccophorum	Squamulose	46.7	40.2	36.4	17.4
Fulgensia subbracteata	Crustose	16.4	4.6	3.1	0.1
Psora decipiens	Squamulose	32.6	12.7	5.0	0.7
Eremastrella crystallifera	Squamulose	9.5	0.4	1.6	0.2
Toninia sedifolia	Squamulose	9.4	2.8	3.0	0.2
Riccia lamellosa	Liverwort	3.7	0.3	0.4	0.0

27.2.2 Disturbance Effects on Nutrient Inputs and Retention

Carbon inputs. Cyanobacterial crusts near coal-fired power plants have greater chlorophyll *a* per unit soil surface area than crusts away from the plant. This is partially a result of fertilization by nitrogen (N) and sulfur compounds from the effluents (Sheridan 1979; Belnap 1991). Photosynthesis-inhibiting herbicides show significant impact on *Nostoc* growth and N fixation (Gadkari 1988). In general, herbicides inhibit growth and reproduction in culture (Metting 1981). However, effects appear to be more pronounced in the laboratory than in the field, and may be transitory (Prasad et al. 1984).

Alteration in species composition of the crust will affect the total carbon (C) fixation of crusts, as lichens and mosses fix more C per unit soil surface area than cyanobacteria (Phillips and Belnap 1998). As much of the C fixed by crustal species is released into the surrounding soils (Lewin 1956), reduction in crust cover is expected to reduce soil C available for often carbon-limited microbial populations. This, in turn, may affect decomposition rates and soil-nutrient levels (Paul and Clark 1996).

Nitrogen Inputs. Effluents from power plants have been shown to decrease N fixation in *Collema* and *Microcoleus-Nostoc-Scytonema*-dominated crusts (Sheridan 1979; Belnap 1991). Atmospheric N deposition may offset reduced N inputs from crusts; alternatively, anthropogenic N deposition may aggravate N loss through increased ammonia volatilization and denitrification. In addition, biological soil crusts release ammonia in the soils, while anthropogenic N deposition contains large amounts of nitrate (F. Garcia-Pichel et al., unpubl. data). Microbial and vascular plant species differentially use ammonia and nitrate; thus, deposition of additional nitrates may alter ecosystem dynamics (Binkley et al. 1997).

Free-living or lichenized *Nostoc* show stimulation of N fixation at low concentrations or short exposure to arsenic, nickel, lead, palladium, and zinc. However, longer-term exposure to cadmium, lead, and zinc inhibit fixation (Henriksson and DaSilva 1978). The impact of insecticides, herbicides, phenolics, and oil compounds on N fixation is highly variable, as can be expected, given the vast array of these compounds. Exposure to crude oil and oil dispersants decreases N fixation in *Nostoc* (Marowitch et al. 1988). Nitrogen fixation is significantly inhibited in *Nostoc* by many insecticides, herbicides, and phenolic compounds tested (Megharaj et al. 1988; Bhunia et al. 1991).

Mechanical disturbance can result in large decreases in soil N through a combination of reduced input and elevated losses (Peterjohn and Schlesinger 1990; Evans and Belnap 1999). In all soils tested, disturbance by vehicles, human foot traffic, mountain bikes, and raking immediately reduces N input from crusts (25–40 % on silty soils; 76–89 % on sandy soils). Over time, nitrogenase activity can drop to 80–100 % of controls, due to subsequent death of buried material (Belnap et al. 1994; Belnap 1995, 1996).

Changes in species composition also affect N inputs, as cyanobacteria fix an order of magnitude less N than the equivalent soil surface area of cyanolichens such as *Collema* (see Chap. 16). Thus, the shift from a lichen crust to a cyanobacteria crust can result in less N entering the ecosystem, as has been shown in multiple studies. Jeffries et al. (1992) showed that heavy grazing reduced N fixation in sandy soils by 95 %. Terry and Burns (1987) measured a 64 % reduction of N fixation in a silty loam soil in burned areas, 85–94 % reduction in grazed areas, and 99 % reduction in a tilled area. *Collema* cover was reduced 50–80 % in grazed areas relative to adjacent ungrazed areas (Brotherson et al. 1983; Johansen and St. Clair 1986); thus, N inputs would be expected to be reduced accordingly. Evans and Belnap (1999) showed N fixation in an area released from grazing for 25 years was still 2.5 times less than an adjacent, never-grazed area due to a reduction in *Collema* cover.

Decreased N inputs from crusts can have long-term impacts on soil N levels (see Chap. 20). Jeffries (1989) found 50 % less N in grazed soils compared to adjacent ungrazed soils. Evans and Belnap (1999) found a 42 % decrease in soil N and 34 % decrease in plant tissue N when comparing a previously grazed to an adjacent ungrazed area. In the same area, stable N isotopes showed that both soil and plants in the grazed area contained less newly fixed N than the ungrazed area, and N mineralization potential decreased almost 80 % (Rimer and Evans 1997). This has large implications for ecosystems that are dependent on soil crusts for N inputs, such as found on the Colorado Plateau (Evans and Ehleringer 1993).

27.2.3 Disturbance and Vascular Plants

As discussed extensively in Chapter 21, crusts can affect vascular plant germination, survival, and nutrition. Lichen-moss crusts in areas with frost-heaving (as in cool and cold deserts) increase perennial vascular plant seed entrapment, germination, establishment, survival, biomass, and nutritional status (Belnap and Harper 1995). In these deserts, disturbance that flattens frost-heaved surfaces and/or eliminates mosses and lichens can have a negative effect on some or all of these characteristics. However, germination and survival of some exotic annual grasses can be enhanced with soil-crust disturbance.

In hot deserts, seed entrapment, biomass, and fecundity of annual plants is generally increased when smooth cyanobacterial crusts are disturbed (Prasse 1999), though there are some exceptions (Gutterman 1994). Disturbance of smooth cyanobacterial crusts enhances germination of some annual species, while decreasing germination of other annual species (Zaady et al. 1997). Rugose crusts in hot deserts have received little study. We found no work that addressed the effects of disturbance on seed entrapment in rugose crusts in hot deserts. Effects on germination in rugose crusts appear to be species-dependent. In contrast to smooth crusts, disturbance to hot-desert rugose crusts lessens survival and biomass of perennial plants (McIlvanie 1942; Crisp 1975).

Biological soil crusts have been shown to influence availability of many plant-essential nutrients, and the loss of crusts can result in less plant-available Mg, K, Fe, Ca, P, Mn, and S (Chap. 21). Spatial distribution of nutrients is also affected with disturbance, as lichen-moss crusts are more patchily distributed across the landscape than cyanobacterial crusts (Klopatek 1992).

27.2.4 Disturbance and Surface Albedo

Trampling of dark-crusted surfaces exposes underlying, lighter soils, thus increasing albedo (reflectance) by up to 50 % at most wavelengths measured. This represents an energy loss from the soil surface of approximately 40 W m^{-2}, and can reduce soil temperatures by up to 14 °C (Belnap 1995). Such changes in reflectance are especially evident with remote platforms, as can be seen in the Negev Desert, along the US-Mexico border and in Australia (see Chap. 31). Much of this difference is attributed to loss of biological soil crusts (Graetz and Tongway 1986). These large-scale changes in surface color may lead to changes in regional climate patterns (Sagan et al. 1979).

27.2.5 Disturbance and Soil Hydrology

The effect of crust disturbance on soil hydrology is very site-specific (see Part V). Water infiltration, runoff, and resultant soil moisture are influenced by soil surface roughness, soil texture, microtopography, surface albedo and temperature, vegetative cover (both vascular and nonvascular), and climatic conditions. All these factors can be altered when soil crusts are disturbed. In general, disturbance flattens rolling and pinnacled crusts, decreasing water infiltration and increasing runoff. Disturbance to smooth and rugose lichen- or cyanobacterial-crusted surfaces in hot deserts can increase overall infiltration rates (Greene et al. 1990; Eldridge et al. 2000). However, this increase in infiltration may negatively affect nearby vascular plants that depend on interspace runoff for survival in hyperarid regions (see Chaps. 21, 22). Alteration of soil structure as a result of soil compaction can also reduce infiltration rates. Available soil moisture is a complex interaction of many soil and plant characteristics, such as infiltration, soil moisture-holding capacity, albedo, and plant root density. Only three studies were found that addressed the combined effects of crusts on available soil moisture, with disparate results. Two studies found increased soil moisture under intact crusts, while the other found soil moisture increased under disturbed crusts (reviewed in Harper and Marble 1988; D. George, unpubl.).

27.2.6 Disturbance and Wind and Water Erosion

Compressional disturbance damages soil-stabilizing cyanobacterial sheath material both on the surface and at depth. The damage at depth is nonrepairable, as living cyanobacteria are not present to regenerate their sheaths. As more than 75 % of photosynthetic biomass and productivity are from organisms living in the top 3 mm of soils, disturbance that results in even small soil losses can dramatically reduce site fertility and further reduce soil-surface stability (Garcia-Pichel and Belnap 1996).

Water Erosion. Globally, studies consistently show that crust cover significantly decreases water erosion of both coarse-textured and fine-textured soils (see Chaps. 22–26). Continuous tracks have a greater impact than non-connected disturbances (such as hoof prints) due to increased volume and velocity of water flow along the tracks (Webb and Wilshire 1983). Crushing of cyanobacterial connections between particles results in smaller, unconnected soil aggregates, which are then more easily moved by water. In addition, surfaces flattened by compressional disturbances are less able to reduce water velocity, thereby allowing movement of larger soil particles.

Wind Erosion. All studies of wind erosion indicate that disturbed soils are more susceptible to wind erosion than undisturbed soils. Soil particle movement is initiated at lower wind speeds when crusts are crushed or absent. Resistance to wind erosion varies with the development of biological soil crusts. Crusts with lichens and mosses, on both silt and sandy soils, have 2–130 times the resistance to soil erosion than less well-developed crusts or bare soil. Vehicle tracks result in greater damage than hoof prints on a given soil type. Decreased resistance of soils to wind erosion is directly associated with increased sediment movement. Experimental disturbances applied to crusts increased sediment production up to 35 times compared to adjacent undisturbed crusts (Belnap and Gillette 1997, 1998; J. Belnap, unpubl. data). Nearby soil crusts are often buried by blowing sediment, resulting in the death of photosynthetic organisms (for extensive discussion, see Chap. 25).

27.3 What Ecosystem Characteristics Confer Resistance to Soil-Surface Disturbance?

Biological soil crusts occur in many different habitats, ranging from subhumid areas that evolved with large herds of hooved animals (e.g., the Great Plains of the US, the Serengeti Plain in Africa, and the central Asian steppes) to arid and semiarid areas where limited food and water prehistorically supported few grazing animals. In habitats that evolved with high numbers of grazers, crusts are highly resistant to trampling, and grasses have characteristics that increase their tolerance of grazing (e.g., tillering, secondary compounds, and high tissue silica content). In contrast, most arid and semiarid desert areas have grasses with very low tolerance of grazing, and crusts intolerant of trampling, as even limited disturbance results in severe damage or extirpation of most crust species present (Martin 1975; Stebbins 1981; Mack and Thompson 1982; Belnap 1995, 1996).

Comparing regions with different evolutionary histories can aid in identifying ecosystem characteristics that support trampling-resistant crusts. For this purpose, crusts in four regions were compared: two regions that evolved with large hooved herds (central Asian steppes, Serengeti Plains); one region with herds of nonhooved animals (southeastern Australia) and one region without any large herds (Colorado Plateau). Crusts along rainfall gradients (200–400 mm) in these ecosystems were compared (J. Belnap, unpubl.). The cyanobacterium *Microcoleus* sp. was present on all soils, regardless of grazing pressure or precipitation. More-developed soil crusts showed increases in the biomass of *Microcoleus*, cover of the lichens *Collema* sp. and *Catapyrenium* sp. and biomass of surface-dwelling colonies of *Nostoc commune*. Several factors were present when more-developed soil crusts were observed in areas of heavy grazing pressure: (1) soils were highly weathered, fine-textured to loamy (<50 % sand), moderately to highly calcareous, had high water-holding capacity, high bulk density (and thus did not churn easily), and some physical crusting. If sandy, soils were highly calcareous and very shallow; (2) sites had high effective rainfall, due to (a) greater total amount of rainfall, (b) cool-season rainfall, and/or (c) soils where water ponding occurred; (3) crust species present were those known to be resistant to trampling; (4) live and dead grass clumps were closely spaced (10 cm) or soils were very rocky, which protected much of the soil from direct hoof contact; and (5) animals did not stay in one area for more than a day, but kept moving through the landscape. Two important features were noted. First, high lichen diversity (>three species) was found only where grazing pressure was absent or minimal, or grazing pressure was due to animals without hooves (kangaroos). Secondly, where hooved animals were continually present, only cyanobacterial soil crusts were found.

27.4 Recovery of Biological Soil Crusts

27.4.1 Natural Recovery Rates

Recovery rates are dependent on many factors, including the type, severity, and extent of disturbance; structure of the vascular plant community; condition of adjoining substrates; availability of inoculation material; and climate during and after disturbance (Fig. 27.1). As discussed in the Introduction (Sect. 27.1), comparing reported recovery rates is problematic, as studies include a wide range of climatic zones, soil types, and levels of disturbance, and these factors are seldom quantified. Thus, most studies can only be used to define the general range of recovery rates.

All published estimates of crust recovery time are based on linear extrapolations of observed recovery rates (e.g., Anderson et al. 1982; Callison et al. 1985; Jeffries and Klopatek 1987; Cole 1990; Belnap 1995; Eldridge 1996). It is not known under what conditions this method is valid. On the Colorado Plateau, scalped plots were reassessed 2–5 and 10–14 years after disturbance. After 2–5 years, cyanobacterial cover was predicted to recover in 45–110 years, whereas it recovered within 14–34 years; thus, extrapolations based on 2–5 years greatly overestimated recovery time. Estimate of moss recovery after the first sample time was 400 years; after the second sampling time, it was 42 years. As with cyanobacteria, extrapolations based on early assessments overestimated recovery time. In contrast, *Collema* recovery after

3 years was estimated at 85 years; after 14 years, estimates for full recovery were 50 years, a fairly close match. *Collema* recovery at another site was highly variable: after 2 years, recovery time was estimated at 487 years; after 11 years, estimates at plots within the site ranged from 40 to 766 years. Thus, original estimates were both under- and over-optimistic, depending on individual plot characteristics. Sites with more shade and less sandy soils were faster to recover than the original estimates predicted, while sunnier sites with sandier soils were slower to recover than the original estimates (J. Belnap, unpubl.).

Sequence of Species Appearance. Large filamentous cyanobacteria such as *Microcoleus* spp. are generally the first photosynthetic crustal organisms to appear (Fig. 27.3) after disturbance, especially on unstable sandy soils (Ashley and Rushforth 1984; Belnap 1995). This is probably due to their mobility across soil surfaces and their ability to easily disperse via the air (Schlichting 1969). These large cyanobacteria are then followed by smaller cyanobacteria and green algae. Rarely, desert soils are slightly acidic, and here green algae can be the first to appear (Johansen 1993).

Lichens and mosses require stable soil surfaces for growth. This can be provided by a physical crust or by colonization of large, filamentous cyanobacteria. Once soils are stabilized, gelatinous N-fixing lichens (e.g., *Collema* spp.) are generally the first lichens to appear in disturbed areas, followed by early successional phycolichens and mosses (Johansen et al. 1984; Belnap 1993). Early-colonizing lichen species often disperse photobiont-containing tissue (e.g., isidia, soredia, thallus fragments, or algal-covered spores), as opposed to bare spores, thus precluding the need to find photobionts on site. In addition, cyanolichens may be the first to colonize alkaline desert soils, as the cyanobiont *Nostoc* is much more numerous than any of the green-algal phycobionts required by phycolichens.

Colonization times vary widely for mid- and late-successional crust species, depending on site characteristics. Little is known about growth rates once colonization occurs. *Psora crenata* requires 55–100 years to regain full size in Australia (240 mm annual rain; Eldridge and Ferris 2000). Similar recovery times have been observed for *Diploschistes* in southern Idaho (R. Rosentreter, unpubl.).

Soil Texture. Soil texture influences recovery rates. In general, when climate and disturbance characteristics are similiar, recovery of all crust components is faster in fine-textured soils that form physical crusts compared to coarse-textured soils that do not form physical crusts when climate and disturbance characteristics are similar (Fig. 27.2; Table 27.2; reviewed in Harper and Marble 1988; Johansen 1993; Ladyman and Muldavin 1996). This is true both for disturbances severe enough to remove crusts and those crushed in place. Recovery of wind resistance is also more rapid in fine-textured soils, due to

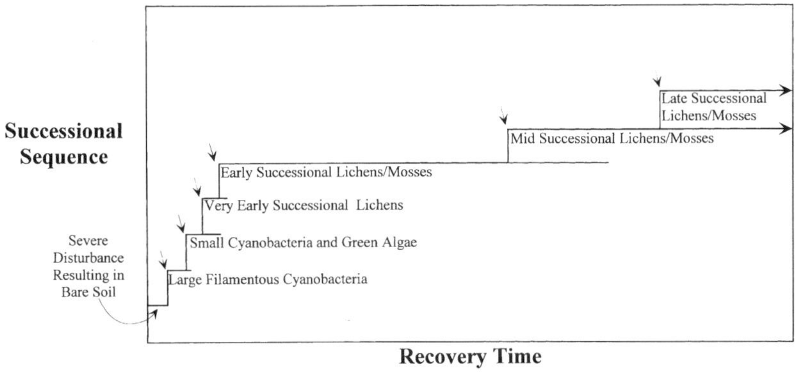

Desert Type	Years to Recovery: Very Early Successional	Early Successional	Mid Successional	Late Successional
Lower Mojave (<1500 ,100 mm)	1200 →	3800		
High Mojave (1500 m, 200 mm)	200 →	800	NK	NK
Colorado Plateau (1500 m, 200 mm)	50 →	500	NK	NK
Northern Great Basin (1000 m, 350 mm)	20 →	25 →	60 →	125

Fig. 27.3. Colonization sequence and estimated recovery times for crustal species in the western US. *Top panel* Arrows indicate colonization events; *length of line* indicates relative time for recovery of each successional group. Species indicative of successional groups include: large filamentous cyanobacteria: *Microcoleus* spp.; small cyanobacteria: *Nostoc* spp.; very early-successional: *Collema* spp.; early-successional: *Catapyrenium* spp., *Pterygoneurum* spp.; mid-successional: *Psora* spp., *Fulgensia* spp. *Tortula* spp, *Bryum* spp.; late-successional: *Acarospora* spp., *Pannaria* spp. *Bottom panel* Relative recovery rates in sandy soils with complete crust removal in different climates. Elevation and annual precipitation of characteristic regions are in parentheses. Reported estimates are averages as sites show considerable variation in recovery times. Recovery rates of mid- and late-successional species are not known (*NK*) in drier deserts, where slow recovery times have precluded estimates. Estimates are based on published rates from studies cited in text

formation of chemical crusts after rainfall. While silty and sandy soils show a similar reduction in wind resistance to vehicle disturbance (83 and 74 %, respectively), silty soils show a 50 % recovery of wind resistance after a single large rain event. Sandy soils can take up to 10 years for a similar recovery (Belnap and Gillette 1997; J.E. Herrick et al., unpubl. data).

Climate Regimes. Crustal organisms are metabolically only active when wet; thus, recovery is faster in regions with greater effective precipitation (Figs. 27.1, 27.3; Table 27.2; Harper and Marble 1988; Johansen et al. 1993). Table 27.2 shows that cool-desert sites at higher latitudes (and thus lower annual temperatures) have quicker recovery for all crustal components than hot-desert sites with the same precipitation but higher annual temperatures, when disturbed similarly. Sites with lower annual precipitation and higher annual temperatures are the slowest to recover.

Table 27.2. Estimated recovery times of dated disturbances of known severity in the Mojave, Sonoran, Chihuahuan and Colorado Plateau deserts of the western US (J. Belnap et al., unpubl.)

Desert type	Disturbance severity	Soil texture	Annual rainfall range (mm)	Elevation range (m)	Years since disturbance	Estimated recovery times (years)				No. of lichen species	
						Cyanobacteria	*Collema*	Other lichens	Moss	Disturbed	Control
Coastal	Removed	Fine	179	100	5		>20–200	>200		1	5
Cool	Removed	Coarse	220–230	1500–1800	10–14	14–34	50–400	>200–875	82–>766	1	4–5
		Fine	230	1600	14	9	>20–240	>200	>140	1	3–5
	Crushed	Coarse	220–230	1500–1800	6–13	7-14	10–71		26–39	1	1–4
		Fine	215	1550	2	2	6	3	4	1	2
Hot	Removed	Coarse	119–230	271–1730	50–80	<81–100	217–1250	580–3700	<81	2	2–7
	Crushed	Coarse	74–248	500–950	2–20	3	10–90	>9		1	1–2
		Fine	248–256	500–1300	2	2–3	4–15	>4		1	2

The "<" indicates that the component was fully recovered when the site was sampled. If a crust component was observed in the control plots but not found in the disturbed plots, a value of 0.01 % relative cover was used to calculate a minimum recovery time; these calculated values are preceeded by a ">". The last two columns indicate the number of lichen species present in the disturbed areas relative to the number of species in the control areas.

Type, Frequency, Severity, and Size of Disturbance. We found no work that addressed recovery rates of soil-crust lichens or cyanobacteria from air pollution. Recovery from fire is specifically covered in Chapter 28. In general, crusts are highly susceptible to damage by hot fires; thus, recovery will depend on the size and intensity of fires. As noted previously in Section 27.2 above, most compressional disturbances have similar types of impacts. However, severity of mechanical disturbance can vary widely with disturbance type. Thus, on similar soils, vehicle tracks generally have longer recovery times than disturbances that do not churn the soil or make continuous tracks (Wilshire 1983; Belnap 1996). Repeated disturbance will generally keep crusts at an early-successional stage (e.g., cyanobacteria-dominated) by preventing colonization of lichens or mosses (Fig. 27.3; Belnap 1996).

Severity of disturbance can control recovery rates. In the western US, recovery of adjacent sites within a given soil and climate type is generally faster when disturbance is not severe enough to remove all crust material (Table 27.2). In addition, because recolonization of disturbances occurs mostly by material washing or blowing in from adjacent, less-disturbed areas, the rate of lichen recovery is much slower in areas with higher internal surface areas relative to perimeter area (Eldridge and Ferris 2000; J. Belnap, unpubl. data).

Condition of Adjoining Substrate. In general, recovery is slower if soils adjacent to disturbed areas are destabilized. Detached sediment can both bury adjacent crusts and/or provide material for sandblasting nearby surfaces, thus increasing wind erosion rates and reducing recovery rates (Belnap 1995; McKenna-Neuman et al. 1996; Leys and Eldridge 1998).

Vascular Plant Structure. Crusts recover more quickly under shrub canopies than adjacent plant interspaces. This is probably due to greater soil moisture and fertility under shrub canopies. In the Mojave Desert, lichen recovery 50 years after total removal was 36 % complete under shrub canopies and only 4 % complete in the shrub interspaces (Belnap and Warren 1998). A similar pattern was seen in plots on the Colorado Plateau, where lichen recovery was two to three times faster under shrubs than in adjacent interspaces (J. Belnap, unpubl. data).

The growth of biological soil crusts and vascular plants can be an interactive process (Danin et al. 1989). Eldridge and Greene (1994) present a conceptual model in which crust development at sites is strongly linked to the restoration of vital soil processes. As primary colonizers of recovering sites, biological soil crusts enhance microsite fertility and provide safe sites for seeds (see Chap. 21). In turn, established vascular plants help stabilize soils, provide shade, and reduce wind speeds at the soil surface, providing conditions conducive to further diversification of the biological soil crusts (Wood et al. 1982; Danin et al. 1989).

Nitrogen and Carbon Fixation. Recovery of N and C fixation rates after disturbance is dependent on the species composition, biomass, and physical structure of the crust. Thus, recovery of N fixation is dependent on recovery rates of cyanobacteria and cyanolichens (see Sect. 27.3.1). Nitrogen fixation in crusts can require anaerobic microzones in the soils, which may be dependent on the buildup of cyanobacterial biomass (Belnap 1996). Recovery of C fixation is mainly dependent on moss and lichen cover. As recovery is faster with higher available moisture, N and C fixation rates are expected to recover faster in cool deserts than in hot deserts, other factors being equal.

Lichen cover may not fully recover after disturbance, as conditions which allowed its initial establishment may no longer be present. For example, 30 years after cessation of grazing, soil space previously occupied by lichens was occupied by mosses in SE Utah, and thus lichens are not necessarily expected to return (Evans and Belnap 1999). Consequently, soil N, plant N, and N fixation are still lower in the previously grazed areas when compared to an ungrazed area. Reduced N inputs are expected as long as lichen recovery does not occur.

Surface Albedo. Restoration of normal surface albedos and temperatures will depend on the restoration of predisturbance cover of cyanobacteria, lichens, and mosses. While cyanobacteria form a darkish matrix in which other components are embedded, most mosses and lichens are much darker (Belnap 1993). Consequently, recovery of surface albedos will be limited by factors that control recovery of lichens and mosses.

27.4.2 Enhancing Recovery with Inoculation

Inoculants can be used to speed soil-crust recovery (Lewin 1977; Tiedemann et al. 1980; Ashley and Rushforth 1984; St. Clair et al. 1986; Belnap 1993; Buttars et al. 1998). Applied inoculants have ranged from commercially available preparations (Lewin 1977) to laboratory-prepared material (Buttars et al. 1998) to material transported from nearby areas (Belnap 1993). In all cases, inoculation enhanced recovery rates of cyanobacterial and lichen biomass and species diversity. For example, 3 years after scalping and inoculating plots with nearby crumbled crusts (10:1 removed to inoculated material), cyanobacterial biomass in inoculated plots was almost twice that of uninoculated plots (0.14 vs. 0.077), lichen cover was 43.3 versus 3.6% (respectively) and lichen species were 4.3 versus 1.8 species (respectively). The entire surface area of the inoculated plots appeared covered by cyanobacteria, and pedicellation had begun (Belnap 1993). In a laboratory study, Buttars et al. (1998) isolated, cultured, and then made alginate pellets of *Microcoleus vaginatus*. These pellets were then placed on petri dishes. After 3 months, inoculated dishes had higher filament density and N fixation rates.

27.5 Conclusions

Disturbance can directly and indirectly affect many aspects of the structure and function of biological crust communities. Moderate to severe disturbance generally results in the simplification of the crust community from one containing multiple species of cyanobacteria, lichens, and mosses to a community consisting of only a few species of cyanobacteria. This alteration of species composition reduces C and N inputs from the crusts. Disturbance is also generally accompanied by loss of surface roughness, which affects local hydrologic regimes and vascular plant seed entrapment. Compressional disturbances break the cyanobacterial connections to soil particles, thus increasing vulnerability of the underlying soils to wind and water erosion. As fine soil particles are preferentially lost, increased erosion reduces site fertility. Disturbance also increases soil albedo and reduces soil temperatures. This, in turn, affects the soil microbial and vascular plant communities.

Some generalizations can be made regarding recovery rates for crusts. Recovery rates are faster for infrequent or minor disturbances where crustal material is crushed in place compared to severe or repeated disturbances that result in complete removal of crustal material. Recovery is faster on fine-textured compared to sandy soils. Recovery is also faster in regions with greater effective precipitation. Recovery is different among crust functional groups, with rates increasing in the following order: cyanobacteria > gelatinous cyanolichens > annual mosses, perennial phycolichens with dispersal mechanisms that avoid early growth stages > perennial phycolichens and mosses without specialized dispersal mechanisms. Recovery of physiological functioning is dependent on the recovery of specific species, e.g., N inputs are dependent on recovery rates for cyanobacteria and cyanolichens. Complete recovery to predisturbance states may not be possible, due to altered edaphic or climate regimes, or lack of source material for colonization.

Acknowledgments. We thank Sue Phillips, Tonya Troxler, Roger Rosentreter, Julie Kaltenecker, and Michelle Schmid for useful comments and graphics.

References

Anderson DC, Harper KT, Rushforth SR (1982) Recovery of cryptogamic soil crusts from grazing on Utah winter ranges. J Range Manage 35:355–359

Ashley J, Rushforth SR (1984) Growth of soil algae on top soil and processed oil shale from the Uintah Basin, Utah, USA. Reclam Reveg Res 3:49–63

Belnap J (1991) Effects of wet and dry pollutants on the physiology and elemental accumulation of cryptogamic crusts and selected lichens of the Colorado Plateau. National Park Service, Air Quality Division Final Report, Denver, CO

Belnap J (1993) Recovery rates of cryptobiotic crusts: inoculant use and assessment methods. Great Basin Nat 53:89–95

Belnap J (1995) Surface disturbances: their role in accelerating desertification. Environ Mon Assess 37:39–57

Belnap J (1996) Soil surface disturbances in cold deserts: effects on nitrogenase activity in cyanobacterial-lichen soil crusts. Biol Fertil Soils 23:362–367

Belnap J, Gillette DA (1997) Disturbance of biological soil crusts: impacts on potential wind erodibility of sandy desert soils in southeastern Utah. Land Degrad Dev 8:355–362

Belnap J, Gillette DA (1998) Vulnerability of desert biological soil crusts to wind erosion: the influence of crust development, soil texture, and disturbance. J Arid Environ 39: 133–142

Belnap J, Harper KT (1995) Influence of cryptobiotic soil crusts on elemental content of tissue of two desert seed plants. Arid Soil Res Rehabil 9:107–115

Belnap J, Phillips S (2001) Soil biota in an ungrazed grassland: response to annual grass (*Bromus tectorum*) invasion. Ecol Appl 11:1261–1275

Belnap J, Warren S (1998) Measuring restoration success: a lesson from Patton's tank tracks. Ecol Bull 79:33

Belnap J, Harper KT, Warren SD (1994) Surface disturbance of cryptobiotic soil crusts: nitrogenase activity, chlorophyll content, and chlorophyll degradation. Arid Soil Res Rehabil 8:1–8

Bhunia AK, Basu NK, Roy D, Chakrabarti A, Banerjee SK (1991) Growth, chlorophyll *a* content, nitrogen-fixing ability, and certain metabolic activities of *Nostoc muscorum*: effect of methylparathion and benthiocarb. Bull Environ Contam Toxicol 47:43–50

Binkley D, Giardiana C, Dockersmith I, Morse K, Scruggs M, Tonnessen K (1997) Status of air quality and related values in Class I National Parks and Monuments of the Colorado Plateau. NPS-AQ D-358

Brotherson JD, Rushforth SR, Johansen JR (1983) Effects of long-term grazing on cryptogam crust cover in Navajo National Monument, Arizona. J Range Manage 36:579–581

Buttars SM, St Clair LL, Johansen JR, Sray JC, Payne MC, Webb BL, Terry RE, Pendleton BK, Warren SD (1998) Pelletized cyanobacterial soil amendments: laboratory testing for survival, escapability, and nitrogen fixation. Arid Soil Res Rehabil 12:165–178

Callison J, Brotherson JD, Bowns JE (1985) The effects of fire on the blackbrush (*Coleogyne ramosissima*) community of southwestern Utah. J Range Manage 38: 535–538

Campbell SE (1979) Soil stabilization by a prokaryotic desert crust: implications for Precambrian land biota. Origins Life 9:335–348

Cole DN (1990) Trampling disturbance and recovery of cryptogamic soil crusts in Grand Canyon National Park. Great Basin Nat 50:321–325

Crisp MD (1975) Long-term change in arid-zone vegetation at Koonamore, South Australia. PhD Dissertation, University of Adelaide, Adelaide

Danin A, Bar-Or Y, Dor I, Yisraeli T (1989) The role of cyanobacteria in stabilization of sand dunes in southern Israel. Ecol Mediterr 15:55–64

Eldridge DJ (1996) Distribution and floristics of terricolous lichens in soil crusts in arid and semi-arid New South Wales, Australia. Aust J Bot 44:581–599

Eldridge DJ (1998) Trampling of microphytic crusts on calcareous soils and its impact on erosion under rain-impacted flow. Catena 33:221–239

Eldridge DJ, Ferris J (2000) Recovery of populations of the soil lichen *Psora crenata* after disturbance in arid South Australia. Rangel J 21:194–198

Eldridge DJ, Greene RSB (1994) Microbiotic soil crusts: a review of their roles in soil and ecological processes in the rangelands of Australia. Aust J Soil Res 389–415

Eldridge DJ, Koen TB (1998) Cover and floristics of microphytic soil crusts in relation to indices of landscape health. Plant Ecol 137:101–114

Eldridge DJ, Zaady E, Shachak M (2000) Infiltration through three contrasting biological soil crusts in patterned landscapes in the Negev, Israel. Catena 40:323–336

Evans RD, Belnap J (1999) Long-term consequences of disturbance on nitrogen dynamics in an arid ecosystem. Ecology 80:150–160

Evans RD, Ehleringer JR (1993) A break in the nitrogen cycle in arid lands? Evidence from $\delta^{15}N$ of soils. Oecologia 94:314–317

Gadkari D (1988) Effect of some photosynthesis-inhibiting herbicides on growth and nitrogenase activity of a new isolate of cyanobacteria, *Nostoc* G3. J Basic Microb 28:419–426

Galun M, Burbrick P, Garty J (1982) Structural and metabolic diversity of two desert lichen populations. J Hattori Bot Lab 53:321–324

Garcia-Pichel F, Belnap J (1996) Microenvironments and microscale productivity of cyanobacterial desert crusts. J Phycol 32:774–782

Graetz RD, Tongway DJ (1986) Influence of grazing management on vegetation, soil structure and nutrient distribution and the infiltration of applied rainfall in a semi-arid chenopod shrubland. Aust J Ecol 11:347–360

Greene RSB, Chartes CJ, Hodgkinson KC (1990) The effects of fire on the soil in a degraded semi-arid woodland. I. Cryptogam cover and physical and micromorphological properties. Aust J Soil Res 28:755–777

Gutterman Y (1994) Strategies of seed dispersal and germination in plants inhabiting deserts. Bot Rev 60:373–425

Harper KT, Marble JR (1988) A role for nonvascular plants in management of arid and semi-arid rangelands. In: Tueller PT (ed) Vegetation science applications for rangeland analysis and management. Kluwer Academic Publishers, Dordrecht, pp 136–169

Henriksson LE, DaSilva EJ (1978) Effects of some inorganic elements on nitrogen-fixation in blue-green algae and some ecological aspects of pollution. Z Allg Mikrobiol 18:487–494

Jeffries DL (1989) The vegetation, soil, and cryptogamic crusts of blackbrush communities in the Kapairowits Basin. PhD Dissertation, Dept Botany, Arizona State University, Tempe, AZ

Jeffries DL, Klopatek JM (1987) Effects of grazing on the vegetation of the blackbrush association. J Range Manage 40:390

Jeffries DL, Klopatek JM, Link SO, Bolton H Jr (1992) Acetylene reduction by cryptogamic crusts from a blackbrush community as related to resaturation and dehydration. Soil Biol Biochem 24:1101–1105

Johansen JR (1993) Cryptogamic crusts of semiarid and arid lands of North America. J Phycol 29:140–147

Johansen JR, St Clair LL (1986) Cryptogamic soil crusts: recovery from grazing near Camp Floyd State Park, Utah, USA. Great Basin Nat 46:632–640

Johansen JR, St Clair LL, Webb, BL, Nebeker GT (1984) Recovery patterns of cryptogamic soil crusts in desert rangelands following fire disturbance. Bryologist 87: 238–243

Johansen JR, Ashley J, Rayburn WR (1993) Effects of rangefire on soil algal crusts in semiarid shrub-steppe of the lower Columbia Basin and their subsequent recovery. Great Basin Nat 53:73–88

Kaltenecker JH (1997) The recovery of microbiotic crusts following post-fire rehabilitation on rangelands of the western Snake River Plain. MS Thesis, Boise State University, Boise

Klopatek JM (1992) Cryptogamic crusts as potential indicators of disturbance in semi-arid landscapes. In: McKenzie DH, Hyatt DE, McDonald VJ (eds) Ecological indicators, vol. 1. Elsevier, Brisbane, UK, pp 773–786

Ladyman JAR, Muldavin E (1996) Terrestrial cryptogams of pinyon-juniper woodlands in the southwestern United States: a review. USDA Forest Service, Rocky Mountain Forest and Range Experiment Station, General Technical Report, RM-GTR-280

Lewin RA (1956) Extracellular polysaccharides of green algae. Can J Microbiol 2:665–673

Lewin RA (1977) The use of algae as soil conditioners. Scripps Inst Oceanogr Trans 3:33–35

Leys JF, Eldridge DJ (1998) Influence of cryptogamic crust disturbances to wind erosion on sand and loam rangeland soils. Earth Surface Processes Landforms 23:963–974

Mack RN, Thompson JN (1982) Evolution in steppe with few large, hooved mammals. Am Nat 119:757–773

Martin PS (1975) Vanishings, and future of the prairie. Geosci Man 10:39–49

Marowitch J, Dale MRT, Hoddinott J (1988) The effect of crude oil and oil spill chemicals on nitrogen fixation in the cyanobacteria *Nostoc* sp. Environ Pollut 51:75–83

McIlvanie SK (1942) Grass seedling establishment and productivity-overgrazed vs. protected range soils. Ecology 23:228–231

McKenna-Neuman C, Maxwell CD, Boulton JW (1996) Wind transport of sand surfaces crusted with photoautotrophic microorganisms. Catena 27:229–247

Megharaj M, Venkateswarlu K, Rao AS (1988) Effect of insecticides and phenolics on nitrogen fixation by *Nostoc linckia*. Bull Environ Contam Toxicol 41:277–281

Metting FB (1981) The systematics and ecology of soil algae. Bot Rev 47:195–312

Nash TH III, Sommerfeld MR (1981) Elemental concentrations in lichens in the area of the Four Corners Power Plant, New Mexico. Environ Exp Bot 21:153–162

Paul EA, Clark FE (1996) Soil microbiology and biochemistry. Academic Press, London

Peterjohn WT, Schlesinger WH (1990) Nitrogen loss from deserts in the southwestern United States. Biogeochemistry 10:67–79

Phillips SL, Belnap J (1998) Shifting carbon dynamics due to the effects of *Bromus tectorum* invasion on biological soil crusts. Ecol Bull 79:205

Prasad RBV, Dhanaraj PS, Narayana Rao VVS (1984) Effects of insecticides on soil microorganisms. In: Lal R (ed) Insecticide microbiology. Springer, Berlin Heidelberg New York, pp 169–202

Prasse R (1999) Experimentelle Untersuchungen an Gefäßpflanzenpopulationen auf verschiedenen Geländeoberflächen in einem Sandwüstengebiet (Experimental studies with populations of vascular plants on different soil surfaces in a sand desert area). Universitätsverlag Rasch, Osnabrück, Germany

Rimer RL, RD Evans (1997) The impact of disturbance and an invasive grass species on N mineralization in a cold-desert system (abs). Ecol Bull 78:303

Rogers RW, Lange RT (1971) Lichen populations on arid soil crusts around sheep watering places in South Australia. Oikos 22:93–100

Sagan C, Toon OB, Pollack JB (1979) Anthropogenic albedo changes and the earth's climate. Science 206:1363–1368

Schlichting HE Jr (1969) The importance of airborne algae and protozoa. J Air Pollut Control Assoc 19:946–951

Sheridan RP (1979) Impact of emissions from coal-fired electricity generating facilities on N_2-fixing lichens. Bryologist 82:54–58

St Clair LL, Johansen JR, Webb BL (1986) Rapid stabilization of fire-disturbed sites using a soil crust slurry: inoculation studies. Reclam Reveg Res 4:261–269

Stebbins GL (1981) Coevolution of grasses and herbivores. Ann Mo Bot Gard 68:75–86

Terry RE, Burns SJ (1987) Nitrogen fixation in cryptogamic soil crusts as affected by disturbance. US Dept of Agriculture, Forest Service, Intermountain Research Station, Reno, NV, pp 369–372

Tiedemann AR, Lopushinsky W, Larsen HJ Jr (1980) Plant and soil responses to a commercial blue-green algae inoculant. Soil Biol Biochem 12:471–475

Webb RH, Wilshire HG (1983) Environmental effects of off-road vehicles: impacts and management in arid regions. Springer Series on Environmental Management. Springer, Berlin Heidelberg New York

West NE (1990) Structure and function of microphytic soil crusts in wildland ecosystems of arid to semi-arid regions. Adv Ecol Res 20:179–223

Wilshire HG (1983) The impact of vehicles on desert soil stabilizers. In: Webb RH, Wilshire HG (eds) Environmental effects of off-road vehicles. Springer, Berlin Heidelberg New York, pp 31–50

Wood MK, Blackburn WH, Eckert RE Jr, Peterson FF (1982) Influence of crusting soil surfaces on emergence and establishment of crested wheatgrass, squirreltail, thurber needlegrass, and fourwing saltbush. J Range Manage 35:282–287

Youtie B, Ponzetti J, Salzer D (1999) Fire and herbicides for exotic annual grass control: effects of native plants and microbiotic soil organisms. In: Eldridge D, Freudenberger D (eds) People and rangelands: building the future. Proc 6th Int Rangelands Congress, Townsville, Australia, vol 2, pp 590–591

Zaady E, Gutterman Y, Boeken B (1997) The germination of mucilaginous seeds of *Plantago coronopus*, *Reboudia pinnata*, and *Carrichtera annua* on cyanobacterial soil crusts from the Negev Desert. Plant Soil 190:247–252

28 Impacts of Fire on Biological Soil Crusts

J.R. Johansen

28.1 Introduction

Rangefire is an important intermittent disturbance in many semiarid and arid lands. Limited interspace vegetation in hot deserts restricts the occurrence of fire, while shrub-steppe communities, with abundant bunch grasses between shrubs, have historically been prone to fire. The frequency of fire has increased notably with the invasion of exotic annual grasses, which are capable of carrying fire over thousands of acres in a single fire event. In particular, *Bromus tectorum*, or cheatgrass, has increased the frequency of rangefire in the shrub-steppes of North America (Billings 1994). Many native shrubs are not fire-resistant, and increased fire frequency could conceivably convert shrub-steppe to grasslands of annual and perennial grasses.

Biological soil crusts are thought to slow the invasion of annual grasses such as cheatgrass (Kaltenecker et al. 1999). Thus, undisturbed areas with good crust cover may not experience the same increase in fire frequency that overgrazed and otherwise disturbed areas experience. However, once the crust is damaged and cheatgrass invasion occurs, recovery of the biological soil crust may never reach predisturbance levels, due to increased frequency of fire. Thus, the invasion of exotic annual grasses may permanently alter the biological soil community.

This chapter reports the effects of fire on biological soil crusts and describes recovery following fire. The research in this area is limited, and so the existing studies will be discussed in some detail. Suggestions for further work conclude the chapter.

28.2 Initial Effects of Fire on Biological Soil Crusts

Deserts and semiarid lands differ with respect to both frequency and severity of rangefires. Hot, dry deserts frequently do not possess sufficient intershrub vegetation to sustain rangefire. Thus, natural rangefire generally plays

Ecological Studies, Vol. 150
J. Belnap and O.L. Lange (eds.) Biological Soil Crusts: Structure, Function, and Management

a minor role in the Mojave and Sonoran Deserts of the western United States, unless invaded by exotic plants. The Chihuahuan Desert is hot, but subject to monsoonal rains from the Gulf of Mexico, and so in some instances will catch fire during lightning storms. The Colorado Plateau generally has sparse vegetation on the shallow pockets of sandy soil in slickrock areas, but can support fire in bunchgrass communities invaded by exotic annual grasses. The rangelands most prone to wildfire in the United States are the shrub-steppe communities of the Basin and Range Province, which includes both the Great Basin (Utah, Nevada, Oregon, and Idaho) and the Columbia Basin (Washington and Oregon). These basin areas historically have been used for cattle ranching, and frequently are invaded by exotic annual grasses such as *Bromus tectorum.* The annual grasses carry fire well, and with the widespread expansion of these exotic grasses, the frequency of fire has increased in the shrub-steppe (Billings 1994). It is these latter rangelands that have received the most study with regard to biological soil crusts and fire.

This chapter will first review decreases in biomass of algae, mosses, and lichens associated with fires in the Basin and Range Province, the Chihuahuan Desert, the miombo woodland of Zimbabwe, and Australian eucalypt woodlands. Following this, the compositional changes will be discussed. The chapter will end with changes in ecosystem function associated with damage to the biological soil community following rangefire.

28.2.1 Decreases in Biomass and Cover

All four studies conducted in the Basin and Range Province show significant reductions in algal biomass following rangefire. At a sagebrush (*Artemisia tridentata*)-dominated site in Wasatch County, Utah, Johansen et al. (1982) found significantly lower algal frequency and diatom density in burned plots than in the unburned shrubland on the other side of the fireline. These results are interesting, given the fact that this study was conducted 3 years after the fire event, and no biological soil crusts were present at the site at that time. The site was subject to cattle grazing, and lacked the vesicular A-horizon typical of Great Basin soils.

Similar results were found in the well-crusted, shrink-swell soils of Camp Floyd, Utah County, Utah, 3 years after an extensive fire. The site was dominated by greasewood (*Sarcobatus vermiculatus*), shadscale (*Atriplex confertifolia*), and rabbitbrush (*Chrysothamnus nauseosus*). The biological soil crust was subtended by a thick vesicular layer and, when dry, formed a distinctive polygonal microtopography (Johansen 1993, Figs. 4, 5). Johansen et al. (1984) observed significantly lower algal frequency and diatom density in the burned area than in an adjacent unburned island. Mean frequencies of both

Microcoleus vaginatus and green coccoid algae were about one quarter those found in unburned plots. Visible algal cover was less than half that seen in undisturbed plots. Diatom densities were also significantly lower (1.58×10^6 cells g^{-1} compared to 2.55×10^6 cells g^{-1}). Lichens were reduced to an even greater extent, with lichen cover falling from 6.2 % in the unburned area to 0.1 % in the burned plots. Bryophytes were completely absent in the burned plots, while the unburned area had moss cover of 13.5 %. This led to a much lighter coloration of the soil in the burned area (Johansen 1993, Fig. 4) than in the unburned area (Johansen 1993, Fig. 5).

Kasper (1994) applied both trampling and burning treatments to 108 plots, 1 m^2 in a well-crusted soil at Dugway Proving Ground, Tooele County, Utah. As at Camp Floyd, the crusts were polygonal when dry and subtended by a vesicular A-horizon. Total crust cover was 42 %. The site was dominated by gray molly (*Kochia americana*), shadscale, and greasewood. Selected plots were burned and trampled in the fall and sampled in the spring. Burning had the most damaging effect on the crust, with a postfire mean of 6.29 μg chl *a* g^{-1} soil, compared to 7.24 μg chl *a* g^{-1} soil in the trampled plots and 8.02 μg chl *a* g^{-1} soil in the undisturbed plots ($p<0.0001$). Epifluorescence direct counts revealed significantly lower biovolume ($p=0.0067$) and density ($p=0.0031$) of *Microcoleus vaginatus* in burned plots. Other algae showed no significant differences, although coccoid eukaryotic algae and *Nostoc* species tended towards lower densities in burned plots. Diatom densities tended to be actually higher in burned plots than in either undisturbed or trampled plots. Five years after disturbance, lichen and bryophyte cover were still significantly lower in burned plots, with 4 % lichen cover and 0 % moss cover (Johansen et al. 1998).

The fourth study in the Basin and Range Province was conducted in the Columbia Basin on the Arid Lands Ecology Reserve in eastern Washington. Big sagebrush and various bunchgrasses dominated this area before a 200 000 acre fire destroyed most of the woody vegetation on the reserve. The biological soil crust was dominated by mosses and coccoid chlorophytes, with limited lichen and cyanobacterial crust cover, and was subtended by a vesicular layer. An unburned island served as a control site, similar to the study at Camp Floyd. One month after the wildfire, density of chlorophyte algae in burned plots was one third that in unburned plots ($p<0.001$, Johansen et al. 1993). Frequency of cyanobacteria and living diatoms was lower in the burned area (Fig. 28.1; $p<0.05$). An interesting finding in this study was the depression of seasonal changes in algal abundance in the burned areas. Unburned plots showed peak chlorophyte density in the winter months, with minor peaks during autumnal rains, while the burned plots showed no seasonal response in the first year of the study (Fig. 28.2). It was interesting to note that algal propagules were present in the burned area even 1 month after the fire; they simply did not grow to any extent during the first 17 months. All mosses and

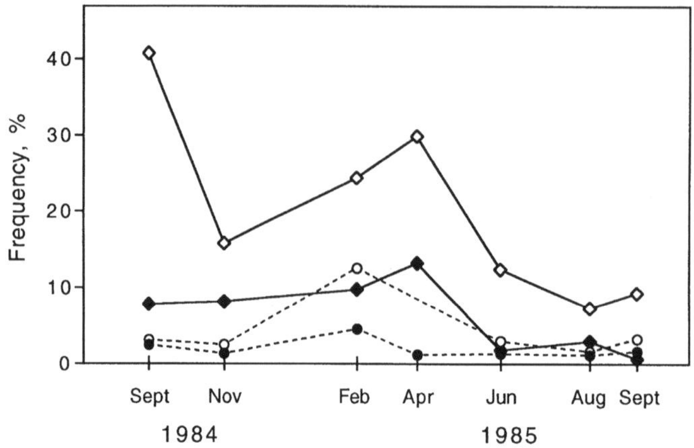

Fig. 28.1. Mean percent frequency of cyanobacteria and diatoms in moistened soils from adjacent burned and unburned areas of the Arid Land Ecology Reserve in Washington. *Diamonds* Cyanobacteria, *circles* diatoms; unburned plots *hollow symbols*; burned plots *solid symbols*. Rangefire occurred in August, 1984. (After Johansen et al. 1993)

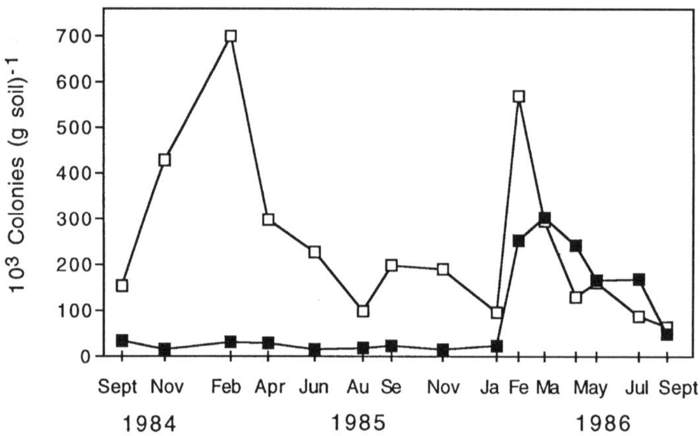

Fig. 28.2. Mean density of algae (all taxa) on solidified Bristol's medium from adjacent burned and unburned areas of the Arid Land Ecology Reserve in Washington. Unburned plots *hollow squares*; burned plots *solid squares*. Vertical axis in units of 1000 colony forming units g^{-1} dry weight soil. Rangefire occurred in August, 1984. (After Johansen et al. 1993)

lichens were destroyed in heavily burned plots, while mosses survived the fire in grassy intershrub areas.

Only one study of rangefire effects on biological soil crusts has been undertaken in the Chihuahuan Desert (Johansen et al. 1998). Burned and unburned study areas were established in Pendajo Wash, Otero County, New Mexico, on fine-textured soils. Creosote (*Larrea divaricata*) and burrograss (*Scleropogon brevifolius*) were the dominant vascular plants prior to the fire. The rangefire was patchy in its effects, with intensely burned plots (dense creosote woodland), lightly burned plots (creosote present, but less densely arranged), and unburned plots. The intensely burned plots had significantly ($p<0.0001$) lower biological soil biomass (1.55 µg chl *a* g^{-1} soil) than was observed in lightly burned (6.10 µg chl *a* g^{-1} soil) or unburned (6.24 µg chl *a* g^{-1} soil) plots. Epifluorescence direct counts revealed significantly lower biovolume ($p=0.0011$) and density ($p=0.0001$) of *Microcoleus vaginatus* in burned plots than in undisturbed plots. Biovolumes for cyanobacterial trichomes 6–8 µm in diameter and 10–12 µm in diameter were also significantly lower. Densities of filamentous cyanobacteria in all four size classes, 1–3 µm (mostly *Schizothrix* and *Leptolyngbya* species), 4–5 µm (mostly *Microcoleus vaginatus*), 6–8 µm (mostly *Oscillatoria* and *Phormidium* species), and 10–12 µm (mostly Scytonemataceae), were all significantly lower in the burned area. Other algae showed no significant differences, although coccoid eukaryotic algae and diatom species had lower densities in burned plots. Visual cover of cyanobacterial crust dropped from 40.5 to 4.4 %, while lichen cover dropped from 23.2 to 0.8 %, both changes being statistically significant.

Effects of fire on biological soil crusts in the miombo savanna woodland of Zimbabwe were studied by Belnap et al. (1996). This tropical area of high rainfall (1 m a^{-1}) supports small trees (up to 9 m), with well-defined openings. The study at the Grasslands Research Station used plots receiving a prescribed burning regimen since 1953. Fires were set at 2-year and 4-year intervals, with and without mattocking. An ungrazed control that had not been burned for at least 45 years was established in an adjacent woodland. The results of this study were very different from those obtained in studies in North America. Mean crust cover was highest in plots receiving the regular mattocked burning every 4 years (48 %), intermediate in control plots (34 %), and lowest in plots burned every 2 years (22 and 17 % in unmattocked and mattocked plots, respectively). Although variability in the data prevented cover data from yielding significant differences, these results were bolstered by the biomass data. Chlorophyll *a* was significantly higher in 4-year plots than in 2-year and control plots, being actually lowest in control plots. This is probably due to faster recovery of the cyanobacterial community in this warm, relatively moist ecosystem, and to the fact that the biological crust community declines somewhere between 4 and 45 years due to increasing vascular plant cover. Mosses were absent from all burned plots, but were present in the control area.

The Australian eucalypt woodlands also demonstrate resilience to fire not observed in North American arid environments. Greene et al. (1990) observed a significant drop in soil-crust cover in controlled fires in an eucalypt woodland in New South Wales. Plots were covered with wheat straw before burning to carry fire through the intershrub/tree spaces. However, sites burned every 4 years had similar soil-crust cover to those burned 9 years previously and over 60 years previously. Eldridge and Bradstock (1994) likewise found reductions in soil-crust cover from 35 to 5 % 2 weeks after the fire, but sites quickly recovered. At 16 years, soil-crust cover reached a peak in their study areas, with a decline in abundance after that time. Thus, it appears that in woodlands with moderate rainfall recovery is rapid, and mature woodland actually has lower cover than sites with some history of fire disturbance (Greene et al. 1990; Eldridge and Bradstock 1994; Belnap et al. 1996).

28.2.2 Compositional Changes

In the studies examining the effects of fire on algal species composition, it has been surprising to see how few compositional changes occur in response to fire. Following a fire, most algal taxa are present, although in reduced numbers. Three years after a rangefire near Wallsburg, Utah, the living algal floras in burned and unburned areas were highly similar according to the Spearman rank correlation coefficient (Johansen et al. 1982). Diatom floras were even more similar, with a significant ($p<0.0002$) Spearman rank correlation coefficient and a Ruzicka's similarity of 84.9 %. The major differences in community composition at this site were the reduction in frequency of *Microcoleus vaginatus* and *Phormidium tenue* and increase in density of *Hantzschia amphioxys*. The increase in *Hantzschia* was most evident in the living algal flora, where it increased from 28 to 50 % frequency. Study of the subfossil diatom community showed a similar trend, but a lesser percent increase, most likely due to the long residence time of diatom frustules in the soil after cell death.

Three years after the rangefire at Camp Floyd State Park, Utah, algal species richness in the burned area was only slightly less than in the unburned area (6 and 9 taxa, respectively), while Shannon diversity was also similar (1.5 and 2.1, respectively). *Nostoc* species were rare in the unburned area, but totally absent in the burned plots (Johansen et al. 1984). Two years later, after unusually wet weather, recovery of species composition was evident, as well as marked increases in species richness and diversity (23 and 19 taxa, Shannon diversity 3.4 and 3.1, in burned and unburned areas, respectively). The subfossil diatom community also showed a high degree of similarity between burned and unburned areas in both study years. *Hantzschia* was denser in the unburned area 3 years after the fire, but denser in the burned plots 5 years after the fire, although in neither year were these differences significant (Johansen et al. 1984).

The study conducted at the Arid Land Ecology Reserve in the Columbia Basin is the only study that examined the effect of fire on algal species composition immediately following fire (Johansen et al. 1993). The surprising outcome in this study was that most species present in the unburned area

Table 28.1. Comparison of species richness and Sorenson's Similarity Index in burned and unburned plots at the ALE Reserve, Washington (from data reported in Johansen et al. 1993). The fire occurred in late August, 1984. Data are based on algal species recovered from agar enrichment plates (all algal divisions) as well as from the Moistened Soil Method (MSM – cyanobacteria and diatoms only)

Month	Plate data richness burned	Plate data richness unburned	Plate data Sorenson index (%)	MSM data richness burned	MSM data richness unburned	MSM data Sorenson index (%)
Sept. 1984				10	16	76.9 %
Nov. 1984	18	23	68.3 %	14	13	81.5 %
Feb. 1985	38	43	66.7 %	13	13	84.6 %
Apr. 1985	35	37	58.3 %	11	14	72.0 %
Jun. 1985	29	33	54.8 %	8	12	80.0 %
Aug. 1985	23	39	61.3 %	10	10	50.0 %
Sept. 1985	12	34	47.8 %	8	12	30.0 %
Entire year	59	66	86.4 %	24	24	79.2 %

were also present in the burned area even in the first year (Table 28.1). The similarity between the lists of taxa from the burned and unburned plots varied from 47.8 to 68.3 % during the first year, with overall similarity 86.4 %. The moistened-soil method produced a wider range of similarities (30.0 to 84.6 %), although the overall similarity for the entire year was also high (79.2 %). I suspect that even with the large sample size (15 per treatment), the differences in richness may be primarily due to inadequate sampling, i.e., with enough examination it is likely that all taxa could be found in all treatments in all months; the rare taxa simply are not observed very often in any site. Despite the similarity between treatments, some taxa did seem to be adversely affected by the rangefire, as evidenced either by their consistent absence in burned plots or reduced frequency in burned plots. *Klebsormidium flaccidum, Scytonema schmidtii, Nostoc commune,* and *Nostoc paludosum* all showed clear reductions in presence or frequency in burned areas. It is probable that the high algal species diversity in the burned area may be due to the presence of a few lightly burned plots. Intensely burned plots had much fewer taxa.

Lichens and mosses show much more striking compositional changes. At Camp Floyd, traces of only *Collema tenax* survived the fire. The other four lichen species and all moss species present in the area before the fire were extirpated (Johansen et al. 1984). The lichens at the ALE site were decimated by the fire, and the mosses were killed in all but a few lightly burned plots (Johansen et al. 1993). Mosses were absent from all burning treatments in the miombo woodland (Belnap et al. 1996). Lichens and mosses were both reduced in burned eucalypt woodlands, as evidenced by the loss of soil-crust cover and the decrease in lichen and moss components within the crust that remained (Eldridge and Bradstock 1994).

In summary, species composition changes are evident in all taxonomic groups studied thus far. Lichens and mosses are most seriously hurt by fire, and the crust remaining after fire is generally dominated by the algal components. Green algae, which have not been demonstrated to be important in ecosystem function, seem to be most resistant to fire. Cyanobacteria are reduced in numbers by fire, and some filamentous taxa appear to be more seriously damaged than others, including the heterocystous genera *Nostoc* and *Scytonema*.

28.2.3 Changes in Ecosystem Function

Nitrogen fixation has been demonstrated to decline significantly in crusted areas submitted to trampling or scalping disturbance (Jeffries et al. 1992; Belnap 1995, 1996; Evans and Johansen 1999; see Chap. 19). The effects of rangefire on nitrogen fixation have been much less studied. Five years after the controlled burn at Dugway Proving Ground, Utah, we found that the rate of nitrogen fixation was depressed in the burned plots (1.610 mmol N_2 m^{-2} h^{-1}) in comparison to the rate observed in unburned plots (5.150 mmol N_2 m^{-2} h^{-1}). However, the difference was not significant (Johansen et al. 1998). Similarly, 2 years after a fire in a creosote community in Fort Bliss, New Mexico, nitrogen fixation was depressed in the burned area (850 mmol N_2 m^{-2} h^{-1} burned area, 1260 mmol N_2 m^{-2} h^{-1} unburned control site). Again, the difference was not significant (Johansen et al. 1998). Certainly, more studies are needed to determine whether or not significant reductions in nitrogen fixation rates are detectable after rangefire. The patchiness of nitrogen fixation activity even in undisturbed sites may have led to these inconclusive results.

Our studies at both Dugway Proving Ground and Fort Bliss sites indicate a significant effect of fire on runoff and sedimentation (Johansen et al. 1998). Rainfall simulation was used to examine the following parameters at both sites: time to first runoff, total runoff, and total sediment production per trial. At Dugway, burned plots showed these significant effects: less time before runoff, more total runoff, and more total sediment production. Fort Bliss showed identical trends, with only total sediment showing significance. At Fort Bliss, infiltration was significantly less in burned plots (3.77 cm) than in unburned controls (7.21 cm).

Greene et al. (1990) found that repeated rangefire caused complete destruction of the soil crust in an Australian semiarid woodland, and that the number of fires that a site experienced was negatively correlated with the saturated infiltration rate. However, repeated fires did not appear to have an impact on unsaturated infiltration. More detailed studies of the effects of biological soil crust on infiltration have revealed an interesting and complex relationship. Eldridge (1993) and Eldridge et al. (1997) found that degree of crust cover in undisturbed areas had little effect on infiltration in Mungo National Park or Yathong Nature Reserve. However, trampling disturbance caused reductions in biological soil cover, aggregate stability, soil organic matter, and sorptivity. The conclusion drawn was that crusts have little effect on infiltration, but their presence is often tied to better physical and chemical properties in soil, which are tied to better infiltration. With rangefire, the effects on soil physical and chemical properties are different than the effects of livestock trampling, and thus it is difficult to know whether or not destruction of the crust by fire will necessarily lead to changes in infiltration. Rangefire has been observed to cause increases in erosion, and these increases have been attributed to the loss of biological soil crust (St. Clair et al. 1986; Chartres and Mücher 1989). However, given the changes in soil physical properties that may occur with rangefire, an effort should be made to separate the effects of crust loss from the effects of soil changes. The loss of the protective vascular plant overstory also undoubtedly has an effect on soil erosion rates following a fire, as shown by Greene et al. (1990) and Kinnell et al. (1990).

In summary, it appears that rangefire can cause a loss in nitrogen fixation rates, a loss in infiltration rate, and increased soil erosion, but the studies indicating these effects are few, and the effects have not always been significant. More studies on the effects of soil-crust loss due to rangefire on ecosystem processes are sorely needed.

28.3 Recovery

Many studies of recovery after fire have been undertaken years after the fire occurred (Johansen et al. 1982, 1984; Callison et al. 1985; Greene et al. 1990; Eldridge and Bradstock 1994). Recovery of the biological soil-crust community following rangefire is somewhat difficult to study when the sites are not chosen immediately following the fire. Natural fires are very uneven in their intensity, leaving a mosaic of biological soil-crust patches, some of which are practically unharmed, while others are thoroughly incinerated (Johansen et al. 1993). This patchiness may not be evident after several years, and thus collecting representative samples can be difficult. Relatively small patches not only persist through time, but also can serve as source sites for algal, bryophyte, and lichen colonizers, and the recovery period can be affected by the proximity of propagules (Johansen et al. 1984). Other unmeasured disturbances (such as trampling by cattle) can also interfere with biological soil-crust recovery, and cause old burn sites to appear as if recovery is slower than it might be without these confounding effects.

Despite these limitations, some trends have still emerged from the recovery studies undertaken. Recovery of the algal community can occur fairly rapidly. At the controlled burn at Dugway Proving Ground in the Great Basin, we found that burned sites showed full recovery of photosynthetic biomass (as estimated by chlorophyll *a*) 5 years after the fire (Johansen et al. 1998). At Camp Floyd State Park, another Great Basin site, recovery as measured by frequency counts in moistened soils also took 5 years (Johansen et al. 1984). In this latter study, restoration of species composition and diversity accompanied the recovery of algal densities. In both cases, unusually wet winters preceded the year of recovery. It seems likely that recovery of the algal components of the biological soil crust is not dependent on a linear passage of time, but rather that recovery is stalled until proper climatic conditions allow recovery to occur.

Recovery of algal density at the Arid Land Ecology Reserve in Washington was even more rapid. In the second winter following the fire, the green algal densities in the burned area jumped from low levels in the unburned areas to densities equal to, or even higher than, the unburned area at the same time (Johansen et al. 1993). Again, recovery was stalled until appropriate climatic conditions (winter snow and spring precipitation), when sudden and complete recovery occurred (Fig. 28.2). However, since species-specific data were not collected in the second year, the degree of recovery in species composition and diversity is unknown.

The earliest species to recover are the green coccoid algae, which have more rapid rates of growth than the cyanobacteria. Diatoms are also quick to recover, with *Hantzschia amphioxys* even showing elevated abundance in recently burned sites (Johansen et al. 1982, 1984, 1998). Early cyanobacterial species to show recovery include members of the morphospecies complex circumscribed by *Schizothrix calcicola* sensu Drouet (Johansen et al. 1982, 1984, as *Phormidium tenue* and *P. minnesotense*, respectively). *Microcoleus vaginatus* soon becomes the dominant cyanobacterium, particularly following a wet winter (Johansen et al. 1984, 1993).

The next floristic relay includes the weedy mosses, such as *Pterygoneurum ovatum*, and the less abundant heterocystous cyanobacterial species of *Nostoc*, *Tolypothrix*, and *Anabaena* (Johansen et al. 1984). Lichens with vegetative diaspores also invade at this time, particularly *Collema tenax* (Johansen et al. 1984). The earliest reported invasion of lichens and mosses into a burned site is 2 years, for the fire that occurred near Mills, Utah (West and Hassan 1985).

Although long-term burns have been less studied, it appears that the moss community, in general, recovers faster than the lichen community. Eldridge and Bradstock (1994) found that mosses reached their peak abundance in 4–7 years, while lichens dominated during years 13–35. Despite these trends in recovery, some sites do not recover. Callison et al. (1985) did not observe recovery of the biological soil crust in some blackbrush communities burned 37 years ago.

28.4 Conclusions

Rangefire has been demonstrated to be very damaging to biological soil crusts in a variety of arid and semiarid lands. Damage is usually patchy in shrublands, due to the patchiness of shrub cover. Grasses generally do not burn hot enough to eradicate crust organisms, and recovery in such areas can be fairly rapid.

Destruction of biological soil crusts due to fire reduces nitrogen fixation rates and infiltration rates, and increases soil erosion. Photoautotrophic biomass can recover fairly quickly (as short as 2 years) in shrub-steppe regions, but can take much longer in warmer, drier regions. Green algae and diatoms appear to recover early (2–5 years), followed by cyanobacteria, particularly members of the Oscillatoriales, such as *Microcoleus vaginatus* and the *Schizothrix/Leptolyngbya* taxonomic complex (4–6 years). Heterocystous cyanobacteria, such as *Nostoc* and *Tolypothrix*, are slower to recover. Mosses can take longer than the algae (3–8 years), but can come back fairly rapidly in moister, cooler climates such as those found in the northern Great Basin and Columbia Basin of North America. Lichens are the slowest component to recover, generally taking decades to develop significant cover. In tropical woodlands of Australia and Zimbabwe, where vascular vegetation is high, fire can actually stimulate biological soil-crust development, possibly through the flush of nutrients that occurs following fire, as well as the greater amount of light reaching the soil surface.

Our understanding of the effects of fire on biological soil crusts is limited to the few studies discussed in this chapter. The differences between regions noted in these few studies indicate that a clear picture of fire effects is not yet available, and other deserts worldwide need study. While fire is often not common in deserts due to low amount of plant cover between shrubs, invasion of annual grasses such as *Bromus tectorum* and *Schismus barbatus* may increase the frequency of fire. Studies of midterm recovery (5–20 years) are especially needed, as well as more long-term recovery studies in which the initial effects of the fire are documented. Establishment of permanent, long-term study sites in a number of regions where fire disturbance is common would greatly benefit our understanding of recovery. Long-term studies would be especially valuable in documenting the onset and progress of biological soil-crust decline with the increase in vascular plant cover that has been observed in woodland sites.

Probably the most pressing need (with regard to understanding the fire ecology of biological soil crusts) is a better picture of the effects of fire on ecosystem function. Very few studies have been undertaken in this area, and those that have been done indicate that the effects and recovery times may be quite different from effects and recovery times from trampling disturbances. While recovery of photosynthetic biomass can be rapid in some regions, there is no empirical evidence that the soil is as consolidated as it was before the fire. It seems likely that the buildup of sheath material and fungi to preburn levels may take many more years, and thus recovery of stabilization effects may be similarly delayed. The effects of biological soil crusts on hydrology are very complex, and need further study in undisturbed, trampled, and burned areas throughout the world where crusts occur. Recovery of ecosystem function most likely takes much longer than simple recovery of photoautotrophic biomass, algal density, and percent cover of crusted soil.

References

Belnap J (1995) Surface disturbances: their role in accelerating desertification. Environ Monit Assess 37:39–57

Belnap J (1996) Soil surface disturbances in cold deserts: effects on nitrogenase activity in cyanobacterial-lichen soil crusts. Biol Fertil Soils 23:362–367

Belnap J, Sanford RL, Lungu L (1996) Biological soil crusts: ecological roles and response to fire in Miombo woodlands of Zimbabwe. Trans Zimbabwe Sci Assoc 70:14–20

Billings WD (1994) Ecological impacts of cheatgrass and resultant fire on ecosystems in the western Great Basin. In: Monson SB, Kitchen SG (eds) Ecology and management of annual rangelands. USDA Techn Report Ser, INT-GTR-313

Callison J, Brotherson JD, Bowns JE (1985) The effects of fire on the blackbrush (*Coleogyne ramosissima*) community of southwestern Utah. J Range Manage 38: 535–538

Chartres CJ, Mücher HJ (1989) The response of burned and unburned rangeland soil surfaces to simulated raindrop impact and water erosion. Earth Surface Processes Landforms 14:407–417

Eldridge DJ (1993) Cryptogam cover and soil surface condition: effects on hydrology on a semiarid woodland soil. Arid Soil Res Rehabil 7:203–217

Eldridge DJ, Bradstock RA (1994) The effect of time since fire on the cover and composition of cryptogamic soil crusts on *Eucalyptus* shrubland soil. Cunninghamia 3:521–527

Eldridge DJ, Tozer ME, Slangen S (1997) Soil hydrology is independent of microphytic crust cover: further evidence from a wooded semiarid Australian rangeland. Arid Soil Res Rehabil 11:113–126

Evans RD, Johansen JR (1999) Microbiotic crusts and ecosystem processes. Crit Rev Plant Sci 18:183–225

Greene RSB, Chartres CJ, Hodgkinson KC (1990) The effects of fire on the soil in a degraded semi-arid woodland. I. Cryptogam cover and physical and micromorphological properties. Aust J Soil Res 28:755–777

Jeffries DL, Klopatek JM, Link SO, Bolton H (1992) Acetylene reduction of cryptogamic crusts from a blackbrush community as related to resaturation and dehydration. Soil Biol Biochem 24:1101–1105

Johansen JR (1993) Cryptogamic crusts of semiarid and arid lands of North America. J Phycol 29:140–147

Johansen JR, Javakul A, Rushforth SR (1982) Effects of burning on the algal communities of a high desert soil near Wallsburg, Utah. J Range Manage 35:598–600

Johansen JR, St Clair LL, Webb BL, Nebeker GT (1984) Recovery patterns of cryptogamic soil crusts in desert rangelands following fire disturbance. Bryologist 87:238–243

Johansen JR, Ashley J, Rayburn WR (1993) Effects of rangefire on soil algal crusts in semiarid shrub-steppe of the lower Columbia Basin and their subsequent recovery. Great Basin Nat 53:73–88

Johansen JR, St Clair LL, Evans RD, Flechtner VR, Balczon J, Webb BL (1998) Resilience of biological soil species to military training pressures: natural and stimulated recovery following disturbance. Final Report for Contract DACA88-95-C-0015, submitted to the US Army Construction Engineering Res. Lab, Champaign, Illinois

Kaltenecker JH, Wicklow-Howard M, Pellant M (1999) Biological soil crusts: natural barriers to *Bromus tectorum* L. establishment in the northern Great Basin, USA. In: Eldridge D, Freudenberger D (eds) Proc VI Int Rangeland Congress, Aitkenvale, Queensland, Australia, pp 109–111

Kasper KA (1994) The effects of rangefire and trampling upon algal soil crusts of a greasewood community in the West Desert of Utah. MS Thesis, John Carroll University, University Heights, Ohio

Kinnell PIA, Chartres CJ,Watson CL (1990) The effect of fire on the soil in a degraded semi-arid woodland. II. Susceptibility of the soil to erosion by shallow rain-impacted flow. Aust J Soil Res 28:779–794

St Clair LL, Johansen JR, Webb BL (1986) Rapid stabilization of fire-disturbed sites using a soil crust slurry: inoculation studies. Reclam Reveg Res 4:261–269

West NE, Hassan MA (1985) Recovery of sagebrush-grass vegetation following wildfire. J Range Manage 38:131–134

Part VII

Monitoring and Management of Biological Soil Crusts

29 Biological Soil Crusts and Livestock in Arid Ecosystems: Are They Compatible?

S.D. WARREN and D.J. ELDRIDGE

29.1 Introduction

Approximately 35 % of the Earth's land surface can be classified as arid or semiarid. Most plant communities in arid and semiarid areas evolved under the influence of grazing and/or browsing by native ungulates (marsupials in Australia), reaching a quasi-stable successional equilibrium based on the nature, frequency, and intensity of the various disturbance regimes to which they were naturally subjected. For example, prior to the arrival of European man, vast herds of ungulates, primarily bison, roamed the Great Plains region of North America east of the Rocky Mountains (England and DeVos 1969); ungulates were comparatively rare in the more arid regions west of the Rocky Mountains (Grayson 1993; Daubenmire 1988). The evolutionary implications of these diverse scenarios are reflected in the contemporary vascular plant and biological soil-crust communities (Mack and Thompson 1982). The Great Plains east of the Rocky Mountains are dominated by rhizomatous, sod-forming grasses; biological crusts occupy only a relatively small portion of the soil surface. In contrast, lands west of the Rocky Mountains are dominated by scattered bunch grasses and shrubs; the interspaces between the vascular plants are frequently covered with well-developed biological crusts.

Alterations to a natural disturbance regime, such as the introduction of domestic livestock, may significantly alter a plant community. The effects of livestock on vascular plants have been widely studied. However, comparatively little research has been focused on the relationship between biological crusts and livestock impacts. This chapter evaluates the limited extant literature in an effort to draw conclusions and formulate relevant hypotheses.

Ecological Studies, Vol. 150
J. Belnap and O.L. Lange (eds.) Biological Soil Crusts: Structure, Function, and Management

29.2 Physical Impacts of Ungulates

29.2.1 Primary Impacts

Trampling is the most common disturbance caused by grazing animals to biological soil crusts. A thorough review of the literature reveals an overwhelming consensus that trampling is detrimental to biological soil crusts in arid regions. Numerous studies in Australia, North America, central Asia, and the Middle East reveal marked reductions in the cover, frequency, biomass, and diversity of the crusts in actively grazed areas when compared with long-term exclosures or relic areas where domestic livestock are absent (Poulton 1955; Crisp 1975; Phillips et al. 1977; Nechaeva 1981; Otterman 1981; Anderson et al. 1982a, b; Brotherson et al. 1983; Graetz and Tongway 1986; Hacker 1987; Jeffries and Klopatek 1987; Marble and Harper 1989; Meyer and García-Moya 1989; Beymer and Klopatek 1992; McCune and Rosentreter 1992; Abaturov 1993; Eldridge 1993; Karnieli and Tsoar 1995; Tsoar and Karnieli 1996; Eldridge and Robson 1997; Memmott et al. 1998; Kaltenecker et al. 1999a,b). These comparative differences may persist for many years, even where grazing is terminated (Kleiner and Harper 1972, 1977; Antonova 1981; Nechaeva 1981; Kleiner 1982, 1983; Johansen and St. Clair 1986; Jeffries and Klopatek 1987; Evans and Belnap 1999). Andrew and Lange (1986) located an area in arid South Australia sufficiently distant from a developed water source that it was rarely traversed by livestock. A watering trough was subsequently installed and sheep introduced, allowing the authors to compare the biological soil crusts before and after significant livestock trampling. The lichen-dominated crust cover began to decline immediately after stocking. Within 2 years, most of the crust had disappeared inside a 50-m radius of the trough where trampling was greatest. Similar results were observed at other watering points in Australia (Rogers and Lange 1971; Graetz and Ludwig 1978; Hodgins and Rogers 1997) and Turkmenistan (Nechaeva 1981), although the latter studies did not have the benefit of monitoring prior to livestock use.

29.2.2 Secondary Impacts

Within Australia and North America, there is a growing consensus that the presence of biological soil crusts in arid ecosystems is an indicator of healthy, stable landscapes (e.g., Tongway and Smith 1989; Klopatek 1992). Indeed, it is well-established that biological crusts contribute to a stable soil surface that resists erosion by water (see Chaps. 23, 24, 26). As the presence of biological soil crusts declines in response to disturbance by domestic livestock, their ability to stabilize the soil surface also diminishes. Loope and Gifford (1972) reported higher water erosion losses from rangelands where the crusts had been disturbed by trampling. Rushforth and Brotherson (1982) attributed catastrophic gully erosion near Navajo National Monument, Arizona, USA, to the introduction of livestock and subsequent destruction of the biological soil crust. The disruption of biological soil crusts by overgrazing was also identified as the cause of severe gully erosion on the Navajo Indian Reservation, Arizona and an area within the Uintah Basin of Utah, USA (Johansen 1986). In a controlled study, Eldridge (1998b) determined that heavy trampling resulted in a marked reduction in biological soil crusts and a dramatic increase in erosion.

Intact biological soil crusts also stabilize soil surfaces against the erosive forces of wind (see Chap. 25). Gillette and Dobrowolski (1993) point to the disruption of biological soil crusts by livestock as a significant contributor to wind erosion in Tadzhikistan. Daubenmire (1988) drew a similar conclusion for the steppe region of the northwestern United States. A weak relationship between trampling intensity and dust emissions was detected in Australia at varying distances from a sheep watering trough (Andrew and Lange 1986). In the northern Chihuahuan Desert of North America, Belnap and Gillette (1998) measured significant reductions in the force required to detach soil particles (threshold friction velocity) from biologically crusted soils when disturbed by simulated livestock trampling. The threshold friction velocity declined an average of 50 and 60 % on sandy and silty soils, respectively, and fell below maximum wind speeds recorded at the site, thus indicating susceptibility to wind erosion. In Australia, Leys and Eldridge (1998) subjected biologically crusted soils to three levels of simulated livestock trampling. Where extensive disturbance destroyed the biological soil crust, wind erosion exceeded maximum acceptable levels by as much as 160 and 670 % on loamy and sandy soils, respectively.

The consequences of wind and water erosion can be particularly severe and long-lasting in arid areas where soil nutrients tend to accumulate in the surface layers. The loss of this nutrient-rich layer can significantly reduce the productivity of the soil (Charley and Cowling 1968). Given that biological soil crusts are often a significant source of fixed nitrogen and carbon in arid ecosystems (see Chap. 20), disruption can reduce the ability of the crusts to function in the role of nutrient cycling. Nitrogen fixation rates at two undisturbed cold-desert shrub sites in Utah, USA, were generally higher than at adjacent sites where livestock grazing was permitted (Terry and Burns 1987). In blackbrush and pinyon-juniper communities at the ecotone between the Great Basin and Mojave Deserts of North America, the contribution of nitrogen was reduced by as much as 2 orders of magnitude where livestock trampling severely damaged the biological soil crusts (Klopatek 1992). Lower levels of nitrogen were detected in both plants and soil on the Colorado Plateau of North America even after 32 years following the removal of livestock (Evans and Belnap 1999). Reductions in carbon cycling can also be expected where livestock trampling disrupts biological soil crusts (Klopatek 1992; Jeffries et al. 1993). Even light grazing pressure can reduce carbon input by almost half (Jeffries et al. 1993). This break in the nitrogen and carbon cycles can lead to heterogenous distribution of these elements, with depletion occurring primarily in the interspaces between scattered shrubs and grasses where trampling is most evident (Evans and Ehleringer 1993).

Disturbance of biological soil crusts by livestock may also contribute to invasion by undesirable plants. Mack (1981) attributes the catastrophic invasion of *Bromus tectorum*, a weedy annual grass, in North America to the destruction of fragile biological soil crusts coincident with the introduction of ranching on arid rangelands. *B. tectorum* establishes more readily on disturbed than on undisturbed crusts (Larsen 1995; Kaltenecker et al. 1999a), while the germination of native perennial grasses may be unaffected by disturbance (Larsen 1995). Once established, *B. tectorum* is an aggressive competitor for light, moisture, and soil nutrients, thus effectively precluding the reestablishment of biological crust organisms, even when grazing is removed (Rickard and Vaughan 1988). In Australia, Crisp (1975) and Scarlett (1994) observed that the presence of a well-developed biological crust precluded the invasion of rangelands by several species of weeds, while trampling facilitated their invasion and rapid establishment.

29.3 Factors Influencing the Impact of Livestock on Biological Soil Crusts

29.3.1 Trampling Intensity

The preceding section provides conclusive evidence that livestock trampling is detrimental to the existence and function of biological soil crusts in arid regions. Empirical data suggest that the degree of damage is related to the intensity of disturbance. For example, there was little or no difference in crust cover in a lightly grazed area of the northeastern Mojave Desert, compared to ungrazed exclosures (Anderson 1994). Similarly, total cover of biological soil crusts was not significantly different between an ungrazed relic area and a lightly grazed area near the Utah/Arizona border in the southwestern United States, whereas heavy grazing resulted in a decline of an order of magnitude (Jeffries and Klopatek 1987). A significant reduction in crust cover was recorded on the Colorado Plateau near the Grand Canyon National Park, Arizona, in an area that was subjected only to occasional trespass grazing, although a large part of the reduction appeared to be attributable to long-term impacts of heavy grazing that had occurred 55 years earlier (Beymer and Klopatek 1992). Continued moderate to heavy grazing in the same area resulted in a further 50 % reduction in crust cover. Scarlett (1994) reported that light stocking with sheep resulted in only minor damage to the biological soil crusts of Australia, while heavy stocking eliminated them altogether. Nevertheless, lower cover of lichens was evident following 32 years of recovery in an area of the Colorado Plateau in Utah previously subjected to only light grazing (Evans and Belnap 1999).

Another approach to investigating the effects of trampling intensity is to measure crust cover at increasing distances from water sources. Livestock tend to congregate at water sources and range radially for forage; trampling intensity declines with distance from the water source. This piosphere concept was investigated at several locations in Australia (Rogers and Lange 1971; Andrew and Lange 1986; Hodgins and Rogers 1997; Page 1997). At all locations except western Queensland (Page 1997), the cover of biological soil crusts was correlated with distance from the water source, thus indicating that crust cover is inversely proportional to trampling intensity.

29.3.2 Seasonal Effects

Marble and Harper (1989) suggested that the timing of livestock impact in relation to the season of year may affect the degree of damage to biologically encrusted soil. In their study on rangelands in southwestern Utah, USA, heavy grazing in early winter had no significant impact on crust cover compared to an ungrazed control. When grazing was conducted at the same intensity but distributed between early and late winter, the cover and species richness of the crust was significantly reduced. Because winter and early summer tend to be dry in that portion of the Great Basin Desert, the authors concluded that grazing during dry periods is more detrimental to the biological crust than grazing during moister periods. Further north in the cold desert of south-central Idaho, Memmott et al. (1998) compared an ungrazed area with areas grazed in the spring, summer, and winter. Areas grazed during the spring or summer experienced significant declines in biological crust cover, while the area grazed only during the winter was not significantly different from the ungrazed area. The authors concluded that biological crusts were protected by frozen soil conditions and snow cover during the winter.

Lusby (1979) monitored vegetation and sediment yield from a watershed in the Colorado Plateau region of the southern Great Basin Desert over a period of 20 years. He discovered that sediment yield was reduced 38 % when the period of livestock use was reduced from winter through late spring to early and midwinter only, even when the livestock utilization rate was kept approximately equal. Lusby did not specifically monitor biological soil crusts, but because no changes in vascular plant cover were detected, Marble and Harper (1989) concluded that the differences were most likely attributable to greater levels of damage to the crusts in the late spring when conditions were drier. A subsequent evaluation of the site (K.T. Harper and J. Belnap, unpubl.) showed that while there were still no changes in vascular plant cover, soil-crust cover was significantly reduced in grazed paddocks when compared to the ungrazed paddocks, with the greatest reduction seen in paddocks subjected to grazing through late spring. Others have also suggested that biological crusts are most susceptible when dry (Anderson et al. 1982a; Eldridge and Kinnell 1997). Among the various crust components, lichens (Pegau 1970; Oksanen 1978) and cyanobacteria (Campbell et al. 1989; Belnap and Gardner 1993; Belnap et al. 1994) seem to be particularly vulnerable to damage when dry.

29.3.3 Soil Substrate

To date, no research has specifically addressed the issue of the potential effects of soil substrate on the response of biological crusts to livestock impact. As all studies have shown a negative impact regardless of soil type, the issue may seem a moot point. However, there is a general consensus that biological soil crusts develop most readily on fine- to medium-textured soils (e.g., Rogers 1972; Anderson et al. 1982a). Due to the inherent instability of sandy soils, biological crusts are often absent or develop only after the moving sands have been stabilized by vascular plants (Danin et al. 1989; Gillette and Dobrowolski 1993). Even after establishment, biologically encrusted soils are susceptible to destabilization by saltating sand particles (Pluis and van Boxel 1993; McKenna Neuman et al. 1996). Hence, it seems reasonable to hypothesize that, under similar levels of livestock disturbance, biological crusts on sandy soils may be more susceptible to rapid degeneration than crusts on finer-textured soils.

29.3.4 Taxonomic or Morphological Groups

At a gross resolution, West (1990) suggested that mosses tend to be most susceptible to physical damage, followed by lichens, and then microscopic organisms. The biological crust components of Great Basin and Colorado Plateau cold-desert shrublands exhibit that pattern when subjected to grazing (Anderson et al. 1982a,b; Brotherson et al. 1983). Among the microscopic organisms, diatoms suffer more than the cyanobacterial/algal component (Anderson et al. 1982a). Crusts dominated by filamentous cyanobacteria and algae are more stable than those dominated by nonfilamentous forms

(McKenna Neuman et al. 1996). Susceptibility to impact seems to be inversely related to the ability of the various components to recover. The cyanobacterial/algal component recovers more rapidly, followed by lichens and mosses (Johansen and St. Clair 1986; Belnap 1993, 1994).

Biological crusts in grazed areas exhibit less species diversity than in undisturbed areas (Kleiner and Harper 1972), implying that some species are more susceptible than others. Likewise, the species richness of biological crusts decreases along a gradient of increasing disturbance surrounding livestock water sources in Australia (Hodgins and Rogers 1997). Among the lichens, some species seem more tolerant of trampling than others (Rogers and Lange 1971). This may be due to their respective morphologies, as foliose and fruticose forms with elevated vegetative structures seem to be more susceptible than crustose and squamulose forms that tend to grow flat against the soil surface (Crisp 1975; Eldridge 1998b; Eldridge and Rosentreter 1999). Gelatinous lichens appear to be least susceptible to trampling (Eldridge and Rosentreter 1999). Given the variable responses of species, the presence and abundance of individual species or morphological groups of species may be better indicators of range condition and soil stability than total crust cover (Eldridge and Koen 1998; Eldridge and Rosentreter 1999; see also Chap. 32).

29.3.5 Climatic Zones

Algae and cyanobacteria are important early successional species in nearly all climatic zones (e.g., Booth 1941; Roger and Reynaud 1982; Ellis-Evans and Walton 1990). The contribution of these and other soil-crust organisms to later successional stages is maximized in arid regions (e.g., Metting 1991) where they are highly susceptible to disturbance, but declines in mesic climates where their abundance relative to vascular plants generally heralds earlier successional conditions (e.g., Hunt et al. 1979). For example, soil-crust organisms are always present in the temperate rangelands of New South Wales, Australia, but increase in density when the vascular plant cover declines in response to grazing (Eldridge 1998a). Likewise, several cyanobacteria increased in abundance in response to heavy grazing in a semiarid Sahelian grassland near N'Djamena, Republic of Chad (Dulieu et al. 1977). Further south, in the semiarid grasslands of south-central Africa, Walker (1979) noted the occurrence of an algal crust only in areas that were badly deteriorated as a result of overgrazing. Booth (1941) likewise reported an increase in the presence of algal crusts in the temperate central plains where the natural vegetation had been severely disturbed. In the northern Great Plains, the clubmoss *Selaginella densa* may increase in response to grazing (Smoliak 1965), and most lichen species disappear when grazing is discontinued

(Looman 1964). In all of the aforementioned studies, the greater abundance of biological crust organisms in grazed areas was attributed to the elimination of competition from vascular plants; their abundance was not a positive response to the physical impacts of trampling.

Based on his observations of the presence of biologically encrusted soil in Israel despite historic grazing pressure, West (1990) criticized the suggestion that livestock cause extensive, long-term loss of biological crusts in arid climates. Notably, however, much of the Negev Desert of Israel, as well as portions of other coastal deserts such as the Atacama, the Baja California, the Namib, and the western edge of the Sahara in Mauritania, represent a unique phenomenon. While correctly classified as arid zones based on annual precipitation, such coastal deserts experience periods of fog, dew, and high relative humidity (Reitan and Green 1968; Walter 1986; Jürgens and Niebel-Lohmann 1995). Biological crust organisms are capable of utilizing these sources of atmospheric moisture for photosynthesis and growth (Thomson and Iltis 1968; Lange et al. 1994). Depending on the location, these supplemental sources of moisture may be available on as many as 200 days a^{-1} and can represent an annual equivalent of 150 mm of additional moisture (Reitan and Green 1968). While coastal deserts are highly susceptible to grazing impacts (Tsoar 1990), recovery following disturbance can be relatively rapid (Danin 1987; Tsoar and Karnieli 1996) compared to inland deserts that do not experience favorable atmospheric moisture conditions with the same degree of frequency.

29.4 Are the Impacts of Livestock on Biological Soil Crusts Good or Bad?

Throughout this chapter, emphasis has been on the negative impacts of livestock trampling on biological soil crusts in arid regions. An apparent exception has been reported from central Asia southeast of the Caspian Sea near the southern edge of the Turkestan Desert (Antonova 1981; Nechaeva 1981). Severe grazing by camels and sheep led to the destruction of both the vascular plant community and the biological soil crust. The net result was a dwindling forage resource and mobilization of the sandy soils. When rested for up to 7 years, the cover of both the vascular plants and biological crusts increased. Thereafter, the biological crust component continued to increase at the apparent expense of the vascular plant community, suggesting that some degree of crust disturbance may be beneficial in maintaining the forage resource. Abaturov (1993) and Miroshnichenko (1996) reported similar phenomena in semiarid areas of the northern Caspian region. There is evidence that the semiarid fringes of those central Asian deserts evolved in association with congregating ungulates that were in a continual state of migration in the face of erratic rainfall and periodic drought (e.g., Butzer 1965). While excessive grazing disrupts the evolutionary balance between ungulates, vascular plants, and biological soil crusts, total removal of livestock impact may also constitute a disturbance and can result in a less productive community greatly different from that which evolved in the presence of episodic grazing (Milchunas et al. 1990).

An alternative point of view was also offered by Savory (1988), who recommended intensive episodic grazing at heavy stocking rates as a mechanism of destroying biological soil crusts and improving the productivity of vascular plant communities. He contended that the crusts represent a low-successional phenomenon that impedes water infiltration into the soil and restricts germination of vascular plants. Noting the periodic intense impacts caused by the animals as they migrated across the tropical African savanna and the apparent benefits to the ecosystem that had evolved under such a disturbance regime, he patterned a high-intensity rotational grazing strategy for domestic livestock to mimic this natural process. It should be noted that the biological crusts observed by Savory were undoubtedly of the early-successional variety observed by Walker (1979) in other semiarid areas of southern Africa where grasslands evolved under alternating episodes of intense grazing followed by rest (Petrides 1974). In that ecosystem, livestock trampling may indeed disrupt the crusts and favor the germination and establishment of vascular plants (Savory 1988). However, the application of such a grazing strategy in ecosystems that evolved in the absence of a similar disturbance regime is unwise (Platou and Tueller 1985) and, as noted throughout this chapter, the consequences to biological soil crusts and to the physical and ecological processes which they govern in arid ecosystems can be catastrophic.

29.5 Conclusions and Recommended Grazing Management Strategies

From the foregoing sections, several reasonable conclusions can be formulated:

1. Biological soil crusts are a natural late-successional feature of most arid ecosystems; on a continuum to more mesic environments, the occurrence of biological crust organisms often signals a degraded or early-successional status.

2. Biological soil crusts of most arid ecosystems evolved in the absence of large herds of grazing/browsing ungulates.
3. Livestock trampling is physically detrimental to biological soil crusts, although some crust species may respond positively to the concomitant reduction of vascular plant cover, particularly in more mesic climates.
4. Disturbance of biological soil crusts in arid regions may result in accelerated erosion by wind and water, reductions in nutrient cycling and soil nutrient status, and invasion of undesirable plants.
5. Damage to biological soil crusts by livestock trampling is proportional to the intensity of impact, determined in large part by stocking rate and distance to water sources.
6. Biological soil crusts appear to be most susceptible to damage during the driest seasons.
7. For any given level of livestock impact, biological soil crusts on sandy soil may suffer greater damage than crusts on finer-textured soils.
8. Ranked in order of susceptibility, mosses > foliose/fruticose lichens > squamulose/crustose lichens > gelatinous lichens > diatoms > non-filamentous algae/cyanobacteria.
9. Some individual soil crust species are more tolerant of trampling than others.
10. Biological crust cover and/or relative abundance of individual species may be reliable indicators of range condition.

While some of these conclusions are founded in a small but significant body of scientific literature, others should be considered hypotheses that warrant additional research. Nevertheless, given the current state of knowledge and the importance of ecosystem management, it is imperative that range managers in arid areas seek to develop grazing management strategies that are compatible with the maintenance of viable biological soil crusts. Mack and Thompson (1982) concluded that ungulates are simply incompatible with ecosystems where biological soil crusts predominate later-successional stages. While this perspective has merit in most arid areas, crusts in some semiarid areas may thrive under limited disturbance. Stocking rates and grazing systems should mimic the natural disturbance regimes in which the ecosystems evolved. In most cases, this will be limited to a conservative, low-risk stocking rate at appropriate times of the year. In addition, strategies should be developed to distribute livestock evenly across the landscape in order to avoid concentrated zones of impact. Such strategies may include the development of multiple water sources and feeding locations where supplementation is required. When possible, livestock usage should be minimized during dry seasons of the year. Destocking will most likely be required during periods of prolonged drought. Alternatively, the use of "sacrifice paddocks" may be warranted where drought feeding can occur on a limited area without causing widespread damage (Eldridge 1998a). Monitoring programs (see Chap. 32) should include cover and floristics of biological soil crusts as indicators of range condition and health. Where monitoring programs reveal a decline in indicator species, stocking rates should be reduced and prolonged rest may be necessary to bring the range condition back to a healthy, productive level.

Acknowledgments. Financial support for this chapter was provided by Gwyn Howard, US Army Construction Engineering Research Laboratories, Champaign, Illinois. Kathryn Northcut, William Doe and Robert Brozka provided valuable editorial comments.

References

Abaturov BD (1993) Alteration of small relief forms of the hydrophysical properties of heavy loam soils in the semidesert zone by grazing. Euras Soil Sci 25:17–28

Anderson DC, Harper KT, Holmgren RC (1982a) Factors influencing development of cryptogamic soil crusts in Utah deserts. J Range Manage 35:180–185

Anderson DC, Harper KT, Rushforth SR (1982b) Recovery of cryptogamic soil crusts from grazing on Utah winter ranges. J Range Manage 35:355–359

Anderson JL (1994) Exclosures on the Beaver Dam slope in Arizona and Utah. Rangelands 16:184–188

Andrew MH, Lange RT (1986) Development of a new piosphere in arid chenopod shrubland grazed by sheep. 1. Changes to the soil surface. Aust J Ecol 11:395–409

Antonova KG (1981) Influence of reservation management upon range production. In: Rangeland ecology, management and productivity. Collection of instructional materials of the international training course, vol 2. UN Environ Programme, USSR Comm, Cent Int Proj, Moscow, pp 291–310

Belnap J (1993) Recovery rates of cryptobiotic crusts: inoculant use and assessment methods. Great Basin Nat 53:89–95

Belnap J (1994) Potential role of cryptobiotic soil crusts in semiarid rangelands. In: Monsen SB, Kitchen SG (compilers) Proceedings – ecology and management of annual rangelands. US Dep Agric, For Serv Gen Tech Rep INT-GTR-313, Ogden, Utah, pp 179–185

Belnap J, Gardner JS (1993) Soil microstructure in soils of the Colorado Plateau: the role of the cyanobacterium *Microcoleus vaginatus*. Great Basin Nat 53:40–47

Belnap J, Gillette DA (1998) Vulnerability of desert biological soil crusts to wind erosion: the influences of crust development, soil texture, and disturbance. J Arid Environ 39:133–142

Belnap J, Harper KT, Warren SD (1994) Surface disturbance of cryptobiotic soil crusts: nitrogenase activity, chlorophyll content and chlorophyll degradation. Arid Soil Res Rehabil 8:1–8

Beymer RJ, Klopatek JM (1992) Effects of grazing on cryptogamic crusts in pinyon-juniper woodlands in Grand Canyon National Park. Am Midl Nat 127:139–148

Booth WE (1941) Revegetation of abandoned fields in Kansas and Oklahoma. Am J Bot 28:415–422

Brotherson JD, Rushforth SR, Johansen JR (1983) Effects of long-term grazing on cryptogam crust cover in Navajo National Monument, Ariz. J Range Manage 36:579–581

Butzer KW (1965) Physical conditions in eastern Europe, western Asia and Egypt before the period of agricultural and urban settlement, vol I, chap 2. University Press, Cambridge

Campbell SE, Seeler JS, Golubic S (1989) Desert crust formation and soil stabilization. Arid Soil Res Rehabil 3:217–228

Charley JL, Cowling SW (1968) Changes in soil nutrient status resulting from overgrazing and their consequences in plant communities of semi-arid areas. Proc Ecol Soc Aust 3:28–38

Crisp MD (1975) Long-term change in arid zone vegetation. PhD Diss, Univ Adelaide, Adelaide, Australia

Danin A (1987) Impact of man on the biological components of desert ecosystems in Israel. Isr J Bot 36:46

Danin A, Bar-Or Y, Dor I, Yisraeli T (1989) The role of cyanobacteria in stabilization of sand dunes in Southern Israel. Ecol Mediterr 15:55–64

Daubenmire R (1988) Steppe vegetation of Washington. Washington State Univ Coop Ext Publ EB1446, Pullman, Washington

Dulieu D, Gaston A, Darley J (1977) La dégradation des pâturages de la région de N'Djamena (République du Tchad) en relation avec la présence de cyanophycées psammophiles – étude préliminaire. Rev Elev Méd Vét Pays Trop 30:181–190

Eldridge DJ (1993) Cryptogam cover and soil surface condition: effects on hydrology on a semiarid woodland soil. Arid Soil Res Rehabil 7:203–217

Eldridge DJ (1998a) Lichens and mosses and liverworts? Biological soil crusts – living protectors of the soil. Nat Res Manage 1:19–24

Eldridge DJ (1998b) Trampling of microphytic crusts on calcareous soils, and its impact on erosion under rain-impacted flow. Catena 33:221–239

Eldridge DJ, Kinnell PIA (1997) Assessment of erosion rates from microphyte-dominated calcareous soils under rain-impacted flow. Aust J Soil Res 35:475–489

Eldridge DJ, Koen TB (1998) Cover and floristics of microphytic soil crusts in relation to indices of landscape health. Plant Ecol 137:101–114

Eldridge DJ, Robson AD (1997) Bladeploughing and exclosure influence soil properties in a semi-arid Australian woodland. J Range Manage 50:191–198

Eldridge DJ, Rosentreter R (1999) Morphological groups: a framework for monitoring microphytic crusts in arid landscapes. J Arid Environ 41:11–25

Ellis-Evans JC, Walton D (1990) The process of colonization in Antarctic terrestrial and freshwater ecosystems. Proc NIPR Symp Polar Biol 3:151–163

England RC, DeVos A (1969) Influence of animals on pristine conditions on the Canadian grasslands. J Range Manage 22:87–94

Evans RD, Belnap J (1999) Long-term consequences of disturbance on nitrogen dynamics in an arid ecosystem. Ecology 80:150–160

Evans RD, Ehleringer JR (1993) A break in the nitrogen cycle in aridlands? Evidence from δ^{15}N of soils. Oecologia 94:314–317

Gillette DA, Dobrowolski JP (1993) Soil crust formation by dust deposition at Shaartuz, Tadzhik, S.S.R. Atmos Environ 27A:2519–2525

Graetz RD, Ludwig JA (1978) A method for the analysis of piosphere data applicable to range assessment. Aust Rangl J 1:126–136

Graetz RD, Tongway D.J. (1986) Influence of grazing management on vegetation, soil structure and nutrient distribution and the infiltration of applied rainfall in a semi-arid chenopod shrubland. Aust J Ecol 11:347–360

Grayson DK (1993) The desert's past – a natural prehistory of the Great Basin. Smithsonian Inst Press, Washington, DC, USA

Hacker RB (1987) Species responses to grazing and environmental factors in an arid halophytic shrubland community. Aust J Bot 35:135–150

Hodgins IW, Rogers RW (1997) Correlations of stocking with the cryptogamic soil crust of a semi-arid rangeland in southwest Queensland. Aust J Ecol 22:425–431

Hunt ME, Floyd GL, Stout BB (1979) Soil algae in field and forest environments. Ecology 60:362–375

Jeffries DL, Klopatek JM (1987) Effects of grazing on the vegetation of blackbrush association. J Range Manage 40:390–392

Jeffries DL, Link SO, Klopatek JM (1993) CO_2 fluxes of cryptogamic crusts. I. Response to resaturation. New Phytol 125:163–173

Johansen JR (1986) Importance of cryptogamic soil crusts to arid rangelands: implications for short duration grazing. In: Tiedemann JA (ed) Short duration grazing. Washington State Univ, Pullman, Washington, pp 127–136

Johansen JR, St. Clair LL (1986) Cryptogamic soil crusts: recovery from grazing near Camp Floyd State Park, Utah, USA. Great Basin Nat 46:632–640

Jürgens N, Niebel-Lohmann A (1995) Geobotanical observations on lichen fields of the southern Namib Desert. Mitt Inst Allg Bot Hamburg 25:135–156

Kaltenecker JH, Wicklow-Howard MC, Pellant M (1999a) Biological soil crusts: natural barriers to *Bromus tectorum* L. establishment in the northern Great Basin. In: Eldridge DJ, Freundberger D (eds) People and rangelands building the future, VI Int Rangel Congr, Aitkenvale, Queensland, Australia, pp 109–111

Kaltenecker JH, Wicklow-Howard MC, Rosentreter R (1999b) Biological soil crusts in three sagebrush communities recovering from a century of livestock trampling. In: McArthur ED, Ostler WK, Wambolt CL (compilers), Proceedings: shrubland ecotones. US Dep Agric, For Serv, Rocky Mt Res Sta Proc RMRS-P-11, pp 222–226

Karnieli A, Tsoar H (1995) Spectral reflectance of biogenic crust developed on desert dune sand along the Israel-Egypt border. Int J Remote Sens 16:369–374

Kleiner EF (1982) Eleven year vegetational comparison in an arid grassland. Proc Am Philos Soc 126:520–526

Kleiner EF (1983) Successional trends in an ungrazed, arid grassland over a decade. J Range Manage 36:114–118

Kleiner EF, Harper KT (1972) Environment and community organization in grasslands of Canyonlands National Park. Ecology 53:299–309

Kleiner EF, Harper KT (1977) Soil properties in relation to cryptogamic groundcover in Canyonlands National Park. J Range Manage 30:202–205

Klopatek JM (1992) Cryptogamic crusts as potential indicators of disturbances in semi-arid landscapes. In: McKenzie DH, Hyatt DE, McDonald VJ (eds) Ecological indicators, vol 1, Elsevier Applied Science, London, pp 773–786

Lange OL, Meyer A, Zellner H, Heber U (1994) Photosynthesis and water relations of lichen soil crusts: field measurements in the coastal fog zone of the Namib Desert. Funct Ecol 8:253–264

Larsen KD (1995) Effects of microbiotic crusts on the germination and establishment of three range grasses. MS Thesis, Boise State Univ, Idaho

Leys JF, Eldridge DJ (1998) Influence of cryptogamic crust disturbance to wind erosion on sand and loam rangeland soils. Earth Surface Processes Landforms 23:963–974.

Looman J (1964) Ecology of lichen and bryophyte communities in Saskatchewan. Ecology 45:481–491

Loope WL, Gifford GF (1972) Influence of a soil microfloral crust on select properties of soils under pinyon-juniper in southeastern Utah. J Soil Water Conserv 27:164–167

Lusby GC (1979) Effects of grazing on runoff and sediment yield from Badger Wash in western Colorado, 1953–1973. Geol Surv Water-Supply Pap 1532–I

Mack RN (1981) Invasion of *Bromus tectorum* L. into western North America: an ecological chronicle. Agro-Ecosystems 7:145–165

Mack RN, Thompson JN (1982) Evolution in steppe with few large, hooved mammals. Am Nat 119:757–773

Marble JR, Harper KT (1989) Effect of timing of grazing on soil-surface cryptogamic communities in a Great Basin low-shrub desert: a preliminary report. Great Basin Nat 49:104–107

McCune B, Rosentreter R (1992) *Texosporium sancti-jacobi*, a rare North American lichen. Bryologist 95:329–333

McKenna Neuman C, Maxwell CD, Boulton JW (1996) Wind transport of sand surfaces crusted with photoautotrophic microorganisms. Catena 27:229–247

Memmott KL, Anderson VJ, Monsen SB (1998) Seasonal grazing impact on cryptogamic crusts in a cold desert ecosystem. J Range Manage 51:547–550

Metting B (1991) Biological surface features of semiarid lands and deserts. In: Skujins J (ed) Semiarid lands and deserts. Marcel Dekker, New York, pp 257–293

Meyer SE, García-Moya E (1989) Plant community patterns and soil moisture regime in gypsum grasslands of north central Mexico. J Arid Environ 16:147–155

Milchunas DG, Lauenroth WK, Chapman PL, Kazempour MK (1990) Community attributes along a perturbation gradient in a shortgrass steppe. J Veg Sci 1:375–384

Miroshnichenko Y (1996) Desertification of pastures or why NW Caspian is believed to be a steppe and not a desert. In: West NE (ed), Rangelands in a sustainable biosphere. Proc 5th Int Rangel Congr Soc Range Manage, Denver, Colorado, pp 337

Nechaeva NT (1981) Indicators of range degradation. In: Rangeland ecology, management and productivity. Collection of instructional materials of the international training course, vol 1. UN Environ Prog, USSR Comm, Cent Int Proj, Moscow, pp 98–123

Oksanen L (1978) Lichen grounds of Finnmarksvidda, northern Norway, in relation to summer and winter grazing by reindeer. Rep Kevo Subarct Res Stn 14:64–71

Otterman J (1981) Satellite and field studies of man's impact on the surface in arid regions. Tellus 33:68–77

Page M (1997) Vegetation dynamics in the mulgalands: a Currawinya National Park case study. PhD Thesis, University of Queensland, Gatton College, Australia

Pegau RE (1970) Effect of reindeer trampling and grazing on lichens. J Range Manage 23:95–97

Petrides GA (1974) The overgrazing cycle as a characteristic of tropical savannas and grasslands in Africa. In: Structure, functioning and management of ecosystems, Proc 1st Int Congr Ecol, Centre for Agricultural Publishing and Documentation, Wageningen, The Netherlands

Phillips AM, Green LT, Ruffner GA (1977) Investigations of feral burro impact on plant communities, Grand Canyon, Arizona. In: Carothers SW (submitter) Biology and ecology of feral burros '*Equus asinus*' at Grand Canyon National Park, Arizona. Report to US Dep Inter, Nat Park Serv, Grand Canyon Nat Park, Arizona. Harold S Colton Res Cent, Mus North Arizona, Flagstaff, Arizona

Platou KA, Tueller PT (1985) Evolutionary implications for grazing management systems. Rangelands 7:57–61

Pluis JLA, van Boxel JH (1993) Wind velocity and algal crusts in dune blowouts. Catena 20:581–594

Poulton CE (1955) Ecology of the non-forested vegetation in Umatilla and Morrow counties, Oregon. PhD Thesis, State College of Washington, Pullman, Washington

Reitan CR, Green CR (1968) Weather and climate of desert environments. In: McGinnies WG, Goldman BJ, Paylore P (eds) Deserts of the world – an appraisal of research into their physical and biological environments. University of Arizona Press, Tucson, Arizona, pp 21–92

Rickard WH, Vaughan BE (1988) Plant community characteristics and responses. In: Rickard WH, Rogers LE, Vaughan BE, Liebetrau SF (eds) Shrub-steppe – balance and change in a semi-arid terrestrial ecosystem. Elsevier, Amsterdam, pp 109–179

Roger PA, Reynaud PA (1982) Free-living blue-green algae in tropical soils. In: Dommergues YR, Diem HG (eds) Microbiology of tropical soils and plant productivity. Martinus Nijhoff/Dr W Junk, The Hague, pp 147–168

Rogers RW (1972) Soil surface lichens in arid and subarid south-eastern Australia. III. The relationship between distribution and environment. Aust J Bot 20:301–316

Rogers RW, Lange RT (1971) Lichen populations on arid soil crusts around sheep watering places in South Australia. Oikos 22:93–100

Rushforth SR, Brotherson JD (1982) Cryptogamic soil crusts in the deserts of North America. Am Biol Teach 44:472–475

Savory A (1988) Holistic resource management. Island Press, Covelo, California

Scarlett N (1994) Soil crusts, germination and weeds – issues to consider. Victorian Nat 111:125–130

Smoliak S (1965) A comparison of ungrazed and lightly grazed *Stipa-Bouteloua* prairie in southeastern Alberta. Can J Plant Sci 45:270–275

Terry RE, Burns SJ (1987) Nitrogen fixation in cryptogamic soil crusts as affected by disturbance. In: Everett RL (ed) Proceedings pinyon-juniper conference. US Dep Agric, For Serv Gen Tech Rep INT-215, pp 369–372

Thomson JW, Iltis HH (1968) A fog-induced lichen community in the coastal desert of southern Peru. Bryologist 71:31–34

Tongway DJ, Smith EL (1989) Soil surface features as indicators of rangeland site productivity. Aust Rangel J 11:15–20

Tsoar H (1990) The ecological background, deterioration and reclamation of desert dune sand. Agric Ecosyst Environ 33:147–170

Tsoar H, Karnieli A (1996) What determines the spectral reflectance of the Negev-Sinai sand dunes. Int J Remote Sens 17:513–525

Walker BH (1979) Game ranching in Africa. In: Walker BH (ed) Management of semi-arid ecosystems. Elsevier, Amsterdam, pp 55–81

Walter H (1986) The Namib Desert. In: Evenari M, Noy-Meir I, Goodall DW (eds) Ecosystems of the world 12B, Hot deserts and arid shrublands, B. Elsevier, Amsterdam, pp 245–282

West NE (1990) Structure and function of microphytic soil crusts in wildland ecosystems of arid and semi-arid regions. Adv Ecol Res 20:179–223

30 Global Change and the Future of Biological Soil Crusts

R.D. Evans, J. Belnap, F. Garcia-Pichel, and S.L. Phillips

30.1 Introduction

Global change caused by anthropogenic activities may significantly modify the abundance, structure, and function of biological soil crusts. The components of global change can be grouped into (1) those that impact discreet sites but are occurring on a global scale, such as land-use change and invasive species, and (2) those that alter atmospheric chemistry resulting in changes in CO_2 concentration, UV radiation, temperature, and precipitation (Vitousek 1994). In this chapter we will examine how land-use change, invasive species, elevated atmospheric CO_2, increased UV radiation, and climate change may affect biological soil crusts and the ecosystems which they inhabit.

30.2 Land-Use Change

One of the most significant impacts of anthropogenic activity is land-use change; Kates et al. (1990) estimates that almost 50 % of the ice-free terrestrial land surface has been transformed or utilized by humans. The direct impact of humans on disturbance regimes and local biogeochemical cycles will most likely override any direct effects of changes in atmospheric chemistry (Walker and Steffen 1997). Land-use change is widespread in the arid and semiarid regions where most biological soil crusts are found. These regions are home to 35 % of the world's population (Brooks and Pokshishevsky 1986) and use is rapidly increasing due to increased need for food production in areas formerly regarded as unsuitable for this purpose, water development, and increased population pressures (Brown et al. 1995). Recreational use of arid lands is also increasing dramatically in developed regions such as the western United States.

Surface disturbance associated with land-use change will decrease the abundance of biological soil crusts and alter species composition. Almost all studies of the impact of disturbance on biological soil crusts have found

Ecological Studies, Vol. 150
J. Belnap and O.L. Lange (eds.) Biological Soil Crusts: Structure, Function, and Management

reduced cover and biomass (see reviews by Harper and Marble 1988; West 1990; Evans and Johansen 1999; Chap. 27). With disturbance, lichens and mosses are replaced by cyanobacteria. Recovery of the lichen and moss component can take decades to centuries (Belnap 1993; Chaps. 27, 29). This decrease in crust cover and change in species composition will have direct effects on rates of soil erosion and ecosystem nitrogen (N) and carbon (C) cycles.

30.2.1 Erosion

Biological soil crusts often form a continuous cover in plant interspaces and, during drought, they are the only organisms capable of stabilizing the soil surface (Eldridge and Greene 1994). Loss of crust cover significantly increases water erosion of both coarse-textured and fine-textured soils (Chap. 26). This is due to several factors including: (1) decreased protection of soils from raindrop impact due to loss of lichens and mosses; (2) disruption of cyanobacterial connections that aggregate soil particles, creating smaller soil aggregates that are more easily moved by water; and (3) increased velocity of water movement on compressed, flattened surfaces, that allows movement of larger soil particles (Harper and Marble 1988).

Similarly, loss of crust cover increases soil loss from wind erosion (Chap. 25). Soil particle movement is initiated at lower wind speeds when crusts are disturbed or absent. Lichen-moss crusts on both silt and sandy soils have 2–130 times more resistance to soil erosion than cyanobacteria-dominated crusts on the same soils, and sediment production is increased up to 35 times when well-developed crusts are disturbed (Belnap and Gillette 1997, 1998).

30.2.2 Nitrogen and Carbon Cycles

Surface disturbance can decrease N_2-fixation by 60 to 100 % following disturbance (Jeffries et al. 1992; Belnap 1995), but it is important to emphasize that crusts vary in their ability to fix N (Chap. 20). The change in the cover and species composition of biological soil crusts will have a direct impact on ecosystem N cycles (Fig. 30.1), as lichens such as *Collema* and late-successional cyanobacteria can fix more N per unit surface area than early-successional cyanobacterial or moss-dominated soil crusts (Evans and Belnap 1999; Chap. 19). This has large implications for ecosystems where N_2-fixation by the biological soil crust is the dominant source of N input (Evans and Ehleringer 1993). The reduction of N input from crusts virtually eliminates new N input into the ecosystem (Fig. 30.1). The loss of N input, coupled with continued loss of N from gaseous emissions and erosion, causes an overall

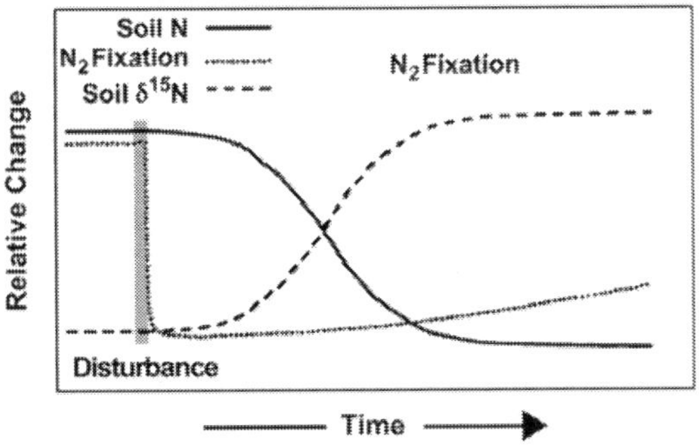

Fig. 30.1. Predicted changes in N_2-fixation, soil N content, and soil and plant $\delta^{15}N$ following surface disturbance caused by land-use change

decrease in the soil N content and an increase in soil and plant $\delta^{15}N$ values (Ehleringer et al. 1998; Evans and Johansen 1999). This pattern was observed by Evans and Belnap (1999), who found that a disturbed grassland with moss-cyanobacterial crusts had 40 % lower soil N than a nearby, undisturbed site with lichen-moss crusts. Soil $\delta^{15}N$ values often increase when rates of N loss exceed those of input (Evans and Ehleringer 1993), and values were 1.5 ‰ greater in the disturbed grassland. Potential net mineralization was reduced by 80 %, so disturbance of the crust will most likely have a direct impact on nutrient uptake by higher plants. In the long-term this may favor species such as C_4 grasses that are more efficient in their N use, perennial species that are capable of using N sources deeper in the soil, or higher plants that are capable of N_2-fixation (Schlesinger et al. 1990; Evans and Ehleringer 1994; Ehleringer et al. 1998; Evans and Belnap 1999). Decreased soil N will also impact soil food webs (Chap. 14).

Recovery of N_2-fixation can be slow; Evans and Belnap (1999) observed that rates of N_2-fixation in an undisturbed grassland were 250 % greater than those in an adjacent site that had been disturbed 30 years ago. This long-term decrease in N_2-fixation can be attributed to (1) elimination of species capable of N_2-fixation, (2) the long recovery rates for the lichen component of the biological soil crust (Belnap 1995), and (3) loss of anaerobic zones necessary for N_2-fixation by nonheterocystic bacteria (Steppe et al. 1996).

Change in crust species composition will also affect C cycles. Studies on the Colorado Plateau show that when soils were moist and temperatures adequate for optimal photosynthesis, soil respiration and net photosynthesis were most correlated with the composition of the biological soil crust. Areas covered primarily by cyanobacteria had the lowest rates, while rates were greatest in areas with a high cover of the soil lichen *Collema tenax* (Phillips and Belnap 1998). Carbon fixed by crust species can be released into the surrounding soils (Lewin 1956); thus, reduction in crust cover or biomass, or a shift to cyanobacterially dominated crusts, is expected to reduce soil C available for often C-limited microbial populations. This, in turn, will affect decomposition rates and soil nutrient levels (Paul and Clark 1996).

30.3 Invasive Species

The introduction of exotic species is one of the most significant components of global change (D'Antonio and Vitousek 1992), and the water-limited, nutrient-poor ecosystems characterized by biological soil crusts are especially susceptible (Ehleringer et al. 1998). The consequences of biological invasions on biological soil crusts has been best studied in arid and semiarid regions of North America, where exotic annual grasses comprise 50 to 85 % of the vascular plant cover in over 60 % of the region (EPA-EMAP, unpubl. report). The most aggressive invader in US cold-desert ecosystems is the annual grass *Bromus tectorum*. *Bromus* was first observed in this region in the 1890s and had spread to over 40 000 000 ha by 1930 (Mack 1981). Invasion in US hot-desert ecosystems occurs primarily by *Bromus*, *Schismus*, and African C_4 grasses that are replacing both native bunchgrasses and winter annuals (Williams and Baruch 2000). Invasion of these ecosystems will have direct consequences for biological soil crusts because of altered vegetation structure and fire frequency (Whisenant 1990; Evans et al. 2001).

Invasive plants often occupy the previously crust-dominated interspace between perennial species, so the initial effect of invasion is reduced light availability at the soil surface. Photosynthesis by biological soil crusts is a direct function of light intensity (Chap. 18) and crusts can become light-limited when shaded. Due to this and perhaps other factors, moss-lichen communities are replaced by annual mosses and cyanobacteria after invasion (Kaltenecker 1997; J. Belnap, unpubl.). A recent invasion in southeastern Utah by *Bromus* resulted in a 50 % loss of the nitrogen-fixing lichen *Collema* and replacement by mosses and cyanobacteria within 3 years (Phillips and Belnap 1998). As seen with surface disturbance (see Sect. 30.2.2), the replacement of lichen-moss crusts by cyanobacterially dominated ones will decrease both C and N inputs into the soils, impacting both vascular plants and soil food webs.

Communities once characterized by large, intercanopy spaces with biological soil crusts that now have a continuous cover of exotic annuals are also much more susceptible to fire (Whisenant 1990). Sites dominated by *Bromus* are 10 to 500 times more likely to burn than pristine sites (Knapp 1996), and fire interval can decrease from >100 years to <5 years (Whisenant 1990). The increase in fire frequency and hotter fires reduce cover, biomass, and species diversity of biological soil crusts and further enhances establishment of invasive annual grasses (Chap. 28; Johansen et al. 1993; Belnap et al. 1996). Frequent fires also prevent the recovery of lichens and mosses, leaving only a few species of cyanobacteria dominating what were once lichen and moss-rich crusts (Whisenant 1990; Eldridge and Bradstock 1994).

30.4 Elevated CO_2

Current projections predict a doubling of atmospheric CO_2 concentration by the end of the 21st century due to anthropogenic activity. Many studies have focused on the response of higher plants to elevated CO_2, but little is known about potential responses of biological soil crusts. Our knowledge will increase dramatically as long-term studies are completed in Free Air CO_2 Enrichment Systems (FACE) and in natural systems such as CO_2 springs and vents.

Photosynthesis by autotrophic organisms in biological soil crusts is often limited by CO_2 (Chap. 18). For example, under controlled conditions of optimal light and temperature and with minimal thallus diffusion resistance, net photosynthesis of three soil-crust lichens (*Fulgensia fulgens, Diploschistes muscorum, D. scruposus*) was between 70 and 80 % of their maximal rates at ambient (350 ppm) external CO_2 partial pressure (Lange et al. 1999). Higher thallus water contents limit CO_2 diffusion. In lichens such as *F. fulgens*, photosynthetic rates become much more depressed at ambient CO_2 concentrations, and CO_2 uptake increases in proportion to external CO_2 partial pressure over the range of several 1000 ppm. Studies using gas exchange systems that increase CO_2 concentrations over a period of hours or days show that higher CO_2 concentrations result in greater rates of net photosynthesis. An initial prediction is that elevated CO_2 will increase primary production of crust communities. However, when higher plants are exposed to elevated CO_2 concentrations over longer time periods (weeks to years), substantial downregulation of photosynthetic activity can decrease photosynthetic capacity (Moore et al. 1999). Balaguer et al. (1999) measured photosynthesis of the green-algal lichen *Parmelia caperata* growing by a natural CO_2 spring (>729 ppm CO_2) and nearby locations with normal CO_2. No differences were observed between the two populations in the photosynthetic response to light and CO_2, indicating that downregulation of photosynthesis did not occur. Biomass was less under elevated CO_2, but this may have been due to secondary effects associated with the CO_2 spring. More work is needed before generalizations can be developed on potential responses of other crust organisms.

Under elevated CO_2, species-specific responses may alter competitive relationships among soil-crust species. Elevated CO_2 may cause a shift from N-fixing cyanolichens to phycolichens. Cyanobacteria possess a CO_2 concentrating mechanism (CCM) that lowers their CO_2 compensation points, making them "C_4-like" in their photosynthetic metabolism (Badger et al. 1993; Green and Lange 1994; Máguas et al. 1995); thus, increased CO_2 may not alter extensively C acquisition. In contrast, the more "C_3-like" green algae in phycolichens may benefit from increased CO_2 because the CCM is absent or mildly expressed (Badger et al. 1993). Heterotrophic nitrogen-fixing organisms may also be favored if elevated CO_2 increases C available in the soil. Skujins (1981) concluded that heterotrophic, free-living nitrogen-fixing organisms are of minor importance in arid soils because of C limitations. Recent research has demonstrated that these organisms are common inside *Microcoleus* sheaths (Steppe et al. 1996); however, the few studies done indicate that N fixation rates are low (Wullstein 1989). Zaady et al. (1998) observed that rates of heterotrophic N_2-fixation were greatest under plant canopies, and attributed this to greater C availability.

The greatest impact of elevated CO_2 on soil crusts may be an increase in the cover and altered species composition of higher plant communities. Large increases in net primary productivity are predicted to occur in arid ecosystems from enhanced water availability (Melillo et al. 1993). Elevated CO_2 may also shift the competitive balance among higher plants, favoring invasive annuals such as *Bromus tectorum* (Smith et al. 1987; Sage 1996). Early results from FACE experiments support this hypothesis; *Bromus* cover and biomass has exhibited a substantial increase under elevated CO_2 in the Mojave Desert (Smith et al. 2000). The increase in plant cover and shift towards invasive annual grasses will shade soil crusts and increase fire susceptibility, possibly changing species composition and decreasing rates of N_2-fixation and photosynthesis (see Sect. 30.3).

30.5 Ozone Depletion and the Increase in Short-Wave Ultraviolet B (UVB)

The progressive destruction of the stratospheric ozone layer has been conclusively linked to the anthropogenic emissions of chlorofluorocarbons (CFC; Jagger 1985). Preindustrial levels of stratospheric ozone absorbed solar radiation in the UV-C and far UV-B ranges so that the Earth's surface received virtually no radiation below 290 nm. These wavelengths are extremely harmful, as they interact directly with DNA and proteins. The thinning of the ozone layer has resulted in more, and shorter, wavelengths of UV-B reaching the Earth's surface and is expected to eventually affect the entire globe (reviewed in Muller et al. 1997).

Forecasting the biological effects of ozone depletion is difficult, as many effects involve a nonlinear dependence on exposure time and wavelength. Although skin cancer doses have increased substantially (Madronich et al. 1998), effects of dose increases on other biological processes (e.g., photosynthesis) may be quite different. Most research has focused on oceanic phytoplankton (Smith et al. 1992) and higher plants (Caldwell et al. 1995), with results generally showing a significant reduction of primary productivity. However, models and results have been controversial, since possible organismal acclimation and foodweb effects have not always been taken into account (Bothwell et al. 1994). Behavior and characteristics of biological crust organisms indicates that solar ultraviolet directly influences crust ecology. Particular aspects of these communities that make them especially sensitive to the effect of ozone depletion are discussed below.

30.5.1 Timing of Exposure and Optical Microhabitats

The interaction between exposure to UV radiation and other stress factors (e.g., desiccation) has synergistic biological consequences (Garcia-Pichel and Castenholz 1991; Potts 1994). Most solar radiation occurs when crusts are dormant and dry; however, repair of cellular damage can only occur when crusts are wet. Thus, the ratio of dry to wet periods ("metabolically timed exposure"; Castenholz and Garcia-Pichel 2000) may be a key parameter in adaptations of desert crust biota to the harsh soil-surface environment.

Soils have two zones in terms of UV radiation: the soil surface, where local scalar irradiance can be actually higher than incident light, and the subsurface, where refuge from the short wavelengths can be found (Chap. 16). Thus, UV exposure is greater for surface-dwelling crust species (*Scytonema, Nostoc*, lichens, and mosses) than subsurface species (*Microcoleus* and other large filamentous cyanobacteria). Lichens and cyanobacteria can also have internal zonation, as UV is much more quickly attenuated when passing through thalli or polysaccharide material than longer wavelengths. Wavelengths <450 nm are reduced by over 90 % in the center of *Collema coccophorum* lobes and *Nostoc commune* colonies, while amounts of longer wavelengths are still sufficient for photosynthesis (Dodds 1989; Büdel et al. 1997).

30.5.2 UV-Protective Mechanisms

Many crust species produce high levels of sunscreen pigments, such as scytonemin in cyanobacteria (Garcia-Pichel and Castenholz 1991; Büdel et al. 1997), melanins in black fungi (Prota 1988), phenolic compounds in lichens (Rikkinen 1995) and anthocyanins in mosses (Post 1990). Mycosporine-like amino acids appear common in all cyanobacteria, fungi, and lichens. Up to 15 % of cyanobacterial dry biomass can be scytonemin alone (Bouillant et al. 1981; Garcia-Pichel and Castenholz 1993; Böhm et al. 1995; Büdel et al. 1997). Lichen and cyanobacterial sunscreens are concentrated in UV-exposed tissue, while shaded parts contain fewer pigments (Büdel et al. 1997). Extracellular polysaccharides may also offer UV protection (Ehling-Schulz and Scherer 1999).

Behavioral and morphological adaptations are also important in avoiding UV exposure. Cyanobacterial species are often layered, with heavily pigmented species (e.g., *Scytonema, Nostoc, Calothrix*) on the surface and relatively less pigmented filamentous species (e.g., *Microcoleus*) gliding to the surface only when soils are wet. Mosses and lichens often roll up sensitive surfaces when dry, and we have observed large concentrations of UV-absorbing substances on the back of *Tortula ruralis* leaves. The presence of hairs and sunken capsules may also reduce UV exposure (Frey and Kürschner 1991).

30.5.3 Sensitivity of Photosynthetic Soil-Crust Organisms

Studies with terrestrial cyanobacteria show that photosynthesis, respiration, N_2-fixation, and nutrient uptake are strongly affected by UV-B radiation (reviewed in Castenholz and Garcia-Pichel 2000). UV-B exposure increases synthesis of phenolic compounds in terricolous lichens (Bachereau and Asta 1997). Anthraquinones are found in *Caloplaca* and *Fulgensia,* and xanthones are found in the outer cortex of *Lecanora, Lecidea,* and *Buellia* (Rundel 1978). Enhanced UV decreases photosynthetic performance in moss (Rader and Bellish 1997; Gehrke 1999) and three types of desert soil crusts (J. Belnap et al., unpubl.). In *Nostoc commune,* UV exposure increased production of carotenoids, extracellular polysaccharides (containing mycosporine), and scytonemin, and affected the induction or repression of a large number of proteins (Ehling-Schulz and Scherer 1999).

Even with these adaptations, solar radiation is still a source of significant mortality for soil-crust species. In a cold-desert ecosystem, cyanobacteria showed significant mortality over the summer, despite an increase in pigment production. Mortality was higher in surface-dwelling species (*Nostoc* and *Scytonema*; M.A. Bowker et al., unpubl.) than subsurface *Microcoleus vaginatus.* Similar differential mortality during dry times has been noted in cyanobacterial populations on intertidal rocks (Nagarkar and Williams 1999).

30.6 Climate Changes

Biological activity in arid and semiarid ecosystems is determined primarily by pulses of precipitation (Ehleringer et al. 1998). Because soil crusts are metabolically active only when wet, and their physiological functions are highly responsive to temperature, changes in the timing or amount of precipitation may have profound consequences for their species composition and roles in ecosystems.

An example of how precipitation changes may affect soil crusts in the western US is illustrated here. Sea-surface temperatures in the Pacific Ocean have a major influence over the seasonal occurrence and intensity of precipitation events in western North America. Predicted increases in sea temperatures are expected to result in greater precipitation during the summer monsoon season and drier-than-normal winters (Cayan and Webb 1992; Cayan 1996). Increase in summer rain, with a concomitant decrease in winter rain, can be expected to affect biological soil crusts in regions that receive significant amounts of summer moisture (Chihuahuan, Sonoran, and Colorado Plateau deserts). High summer temperatures will dry soils faster than low winter temperatures. Combined with less winter rain, soils will be wet for a shorter time during the year, resulting in less overall crust activity time. N_2-fixation rates are greatest during prolonged wet, cool periods and minimal-to-none during hot periods (Chap. 19). Thus, a reduction in winter moisture will result in less N fixation, and less N input to soils. In addition, short wet-dry cycles, such as occur with summer precipitation, generally result in net C losses and negligible N_2-fixation rates in Colorado Plateau soil crusts (Jeffries et al. 1993a,b). C and N losses may result in increased mortality or extirpation of some crust species. Such an effect may explain why lichen and moss diversity decreases sharply in regions of summer rainfall on several continents when compared to winter rainfall areas of equal precipitation (US, Australia, and central Asia; see Part I). The two lichen genera most consistently seen in summer rainfall areas are *Collema* and *Catapyrenium* (see Part I), although cover of these species is far lower than in areas with rainfall at cooler temperatures.

In contrast to warmer deserts, increasing summer rainfall in the colder deserts (e.g., northern Great Basin, the Mongolian steppes, or the Arctic), would be expected to greatly benefit most crust species, as rainfall would be occurring when annual temperatures are optimal for C- and N_2-fixation. Accordingly, biomass, cover, and species richness of the crusts would be expected to increase in these regions.

30.7 Conclusions

The loss of key functional groups can result in permanent changes in ecosystem structure and function (Chapin et al. 1996). Biological soil crusts are characteristic of extreme environments with low biodiversity (Chap. 1). Soil-crust formers are often the only organisms capable of N_2-fixation and stabilizing the soil surface, and may provide significant C input into ecosystems. The largest threat to biological soil crusts is the changes caused by land-use change and invasive species. The reduction or loss of lichens and mosses, coupled with the slow recovery rates, will cause immediate decreases in soil stability and in C and N influx into the ecosystem. In the long term, more gradual changes in cover and species composition may also occur in response to elevated CO_2 and increased UV radiation. The organisms within the crust are highly specialized and it is likely that other, more resistant, organisms will not replace them. Therefore the change in the abundance, species composition, and function of biological soil crusts caused by global change will have a direct effect on the stability, biodiversity, and biogeochemistry of the ecosystems where they are present.

Acknowledgments. RDE was supported by awards from the National Science Foundation (Award no. 98-14510, 98-14358), the Department of Defense, and the Environmental Protection Agency. J.B. acknowledges the assistance of Michelle Schmid.

References

Bachereau F, Asta J (1997) Effects of solar ultraviolet radiation at high altitude on the physiology and the biochemistry of a terricolous lichen [*Cetraria islandica* (L.) Ach.]. Symbiosis 23:197–217

Badger MR, Pfanz H, Büdel B, Heber U, Lange OL (1993) Evidence for the functioning of photosynthetic CO_2-concentrating mechanisms in lichens containing green algal and cyanobacterial photobionts. Planta 191:57–70

Balaguer L, Manrique E, de los Rios A, Ascaso C, Palmqvist K, Fordham M, Barnes JD (1999) Long-term responses of the green-algal lichen *Parmelia caperata* to natural CO_2 enrichment. Oecologia 119:166–174

Belnap J (1993) Recovery rates of cryptobiotic crusts: inoculant use and assessment methods. Great Basin Nat 53:89–95

Belnap J (1995) Surface disturbances: their role in accelerating desertification. Environ Monit Assess 37:39–57

Belnap J, Gillette DA (1997) Disturbance of biological soil crusts: impacts on potential wind erodibility of sandy desert soils in southeastern Utah, USA. Land Degrad Dev 8:355–362

Belnap J, Gillette DA (1998) Vulnerability of desert biological soil crusts to wind erosion: the influence of crust development, soil texture, and disturbance. J Arid Environ 39:133–142

Belnap J, Sanford RL, Lungu L (1996) Biological soil crusts: ecological roles and response to fire in Miombo woodlands of Zimbabwe. Trans Zimbabwe Sci Assoc 70:14–20

Böhm GA, Pfleiderer W, Böger P, Scherer S (1995) Structure of a novel oligosaccharide-mycosporine-amino-acid ultraviolet A/B sunscreen pigment from the terrestrial cyanobacterium *Nostoc commune*. J Biol Chem 270:8536–8539

Bothwell ML, Sherbot DM, Pollock CJ (1994) Ecosystem response to solar ultraviolet B radiation: influence of trophic level interactions. Science 265:97–100

Bouillant ML, Pittet JL, Bernillon J, Favre-Bonvin J, Arpin N (1981) Mycosporins from *Ascochyta pisi, Cladosporium herbarum, Septoria nodorum*. Phytochemistry 20:2705–2707

Brooks SI, Pokshishevsky VV (1986) Population problems in the arid and semiarid regions of the world. In: Gerasimov IP (ed) Arid land development and the combat against desertification: an integrated approach, USSR Commission for the United Nations Environment Program, Moscow

Brown LR, Denniston D, Flavin C, French H, Kane H, Lenssen N, Renner M, Roodman D, Ryan M, Sachs A, Starke L, Weber P (eds) (1995) State of the World. WW Norton, New York, NY

Büdel B, Karsten U, Garcia-Pichel F (1997) UV sunscreen compounds, scytonemin and mycosporine like amino acids in exposed cyanobacterial lichens. Oecologia 112: 165–172

Caldwell MM, Tekamura AH, Tevini M, Bornman JF, Björn LO, Kulandaivelu G (1995) Effects of increased solar ultraviolet radiation on terrestrial plants. Ambio 24: 163–173

Castenholz RW, Garcia-Pichel F (2000) Cyanobacterial responses to UV-radiation. In: Whitton BA, Potts M (eds) Ecology of cyanobacteria: their diversity in time and space. Kluwer Academic Publishers, Dordrecht, pp 591–611

Cayan DR (1996) Interannual climate variability and snowpack in the western United States. J Climate 9:928–948

Cayan DR, Webb RH (1992) El Nino/southern oscillation and streamflow in the western United States. In: Diaz HF, Markgraf V (eds) El Nino historical and paleoclimatic aspects of the southern oscillation. Cambridge University Press, Cambridge, pp 29–69

Chapin FS III, Torn MS, Tateno M (1996) Principles of ecosystem sustainability. Am Nat 148:1016–1037

D'Antonio CM, Vitousek PM (1992) Biological invasions by exotic grasses, the grass/fire cycle, and global change. Annu Rev Ecol Syst 23:63–87

Dodds WK (1989) Microscale vertical profiles of N_2 fixation, photosynthesis, O_2, chlorophyll *a* and light in a cyanobacterial assemblage. Appl Environ Microbiol 55:882–886

Ehleringer JR, Evans RD, Williams D (1998) Assessing sensitivity to change in desert ecosystems – a stable isotope approach. In: Griffiths H (ed) Stable isotopes. BIOS Scientific Publishers, Oxford, pp 223–237

Ehling-Schulz M, Scherer S (1999) UV protection in cyanobacteria. Eur J Phycol 34:329-338

Eldridge DJ, Bradstock RA (1994) The effect of time since fire on the cover and composition of cryptogamic soil crusts on *Eucalyptus* shrubland soil. Cunninghamia 3:521–527

Eldridge DJ, Greene RSB (1994) Assessment of sediment yield by splash erosion on a semi-arid soil with varying cryptogam cover. J Arid Environ 26:221–232

Evans RD, Belnap J (1999) Long-term consequences of disturbance on nitrogen dynamics in an arid ecosystem. Ecology 80:150–160

Evans RD, Ehleringer JR (1993) A break in the nitrogen cycle of aridlands? Evidence from $\delta^{15}N$ of soils. Oecologia 94:314–317

Evans RD, Ehleringer JR (1994) Water and nitrogen dynamics in an arid woodland. Oecologia 99:233–242

Evans RD, Johansen JR (1999) Microbiotic crusts and ecosystem processes. Crit Rev Plant Sci 18:183–225

Evans RD, Rimer R, Sperry L, Belnap J (2001) Exotic plant invasion alters nitrogen dynamics in an arid grassland. Ecol Applic 11:1301–1310

Frey W, Kürschner H (1991) Morphological and anatomical adaptation of the species in terrestrial bryophyte communities along an ecological transect in the Judean Desert, Israel (in German). Bot Jahrb Syst 112:529–552

Garcia-Pichel F, Castenholz RW (1991) Characterization and biological implications of scytonemin, a cyanobacterial sheath pigment. J Phycol 27:395–409

Garcia-Pichel F, Castenholz RW (1993) Occurrence of UV-absorbing, mycosporine-like compounds among cyanobacterial isolates and an estimate of their screening capacity. Appl Environ Microbiol 59:163–169

Gehrke C (1999) Impacts of enhanced ultraviolet-B radiation on mosses in a sub-arctic heath ecosystem. Ecology 80:1844–1851

Green TGA, Lange OL (1994) Photosynthesis in poikilohydric plants: a comparison of lichens and bryophytes. In: Schulze E-D, Caldwell M (eds) Ecophysiology of photosynthesis. Springer, Berlin Heidelberg New York, pp 319–341

Harper KT, Marble JR (1988) A role for nonvascular plants in management of arid and semiarid rangelands. In: Tueller PT (ed) Vegetation science applications for rangeland analysis and management. Kluwer Academic Publishers, Boston, pp 135–169

Jagger J (1985) Solar-UV actions on living cells. Praeger, New York

Jeffries DL, Klopatek JM, Link SO, Bolton H (1992) Acetylene reduction of cryptogamic crusts from a blackbrush community as related to resaturation and dehydration. Soil Biol Biochem 24:1101–1105

Jeffries DL, Link SO, Klopatek JM (1993a) CO_2 fluxes of cryptogamic crusts. I. response to resaturation. New Phytol 125:163–173

Jeffries DL, Link SO, Klopatek JM (1993b) CO_2 fluxes of cryptogamic crusts. II. response to dehydration. New Phytol 125:391–396

Johansen JR, Ashley J, Rayburn WR (1993) Effects of rangefire on soil algal crusts in semiarid shrub-steppe of the lower Columbia Basin and their subsequent recovery. Great Basin Nat 53:73–88

Kaltenecker JH (1997) The recovery of microbiotic crusts following post-fire rehabilitation on rangelands of the Western Snake River Plain. Unpubl MSc thesis, Boise State University Idaho, Boise

Kates RW, Turner BL, Clark WC (1990) The great transformation. In: Turner BL (ed) The Earth as transformed by human action: global and regional changes in the biosphere over the past 300 years. Cambridge University Press, Cambridge, New York, pp 1–17

Knapp PA (1996) Cheatgrass (*Bromus tectorum* L.) dominance in the Great Basin Desert - history, persistence, and influences to human activities. Global Environ Change 6:37–52

Lange OL, Green TGA, Reichenberger H (1999) The response of lichen photosynthesis to external CO_2 concentration and its interaction with thallus water status. J Plant Physiol 154:157–166

Lewin RA (1956) Extracellular polysaccharides of green algae. Can J Microbiol 2:665–673

Mack RN (1981) Invasion of *Bromus tectorum* L. into western North America: an ecological chronicle. Agro-Ecosystems 7:145–165

Madronich S, MacKenzie R, BJorn L, Caldwell MM (1998) Changes in biologically active ultraviolet radiation reaching the Earth's surface. J Photochem Photobiol B 46:5–19

Máguas C, Griffiths H, Broadmeadow MSJ (1995) Gas exchange and C isotope discrimination in lichens: evidence for interactions between CO_2-concentrating mechanisms and diffusion limitation. Planta 196:95–102

Melillo JM, McGuire AD, Kicklighter DW, Moore B III, Vorosmarty CJ, Schloss AL (1993) Global climate change and terrestrial net primary production. Nature 363:234–240

Moore BD, Cheng S-H, Sims D, Seemann JR (1999) The biochemical and molecular basis for photosynthetic acclimation to elevated atmospheric CO_2. Plant Cell Environ 22:567–582

Muller R, Crutzen PJ, Gross JU, Bruhl C, Russell JM, Gernandt H, McKenna, DS, Tuck AF (1997) Severe chemical ozone loss in the Arctic during the winter of 1995-1996. Nature 389:709–712

Nagarkar S, Williams GA (1999) Spatial and temporal variation of cyanobacteria-dominated epilithic communities on a tropical shore in Hong Kong. Phycologia 38:385–393

Paul EA, Clark FE (1996) Soil microbiology and biochemistry. Academic Press, New York
Phillips SL, Belnap J (1998) Shifting carbon dynamics due to the effects of *Bromus tectorum* invasion on biological soil crusts (abs). Ecol Bull 79:205
Post A (1990) Photoprotective pigment as an adaptive strategy in the Antarctic moss *Ceratodon purpureus*. Polar Biol 10:241–246
Potts M (1994) Desiccation tolerance of prokaryotes. Microb Rev 58:755–805
Prota G (1988) Some new aspects of eumelanin chemistry. In: Baynara JT (ed) Advances in pigment cell research. Alan R Liss, New York, pp 101–127
Rader RB, Bellish TA (1997) Short-term effect of ambient and enhanced UVB on moss (*Fontinalis neomexicana*) in a mountain stream. J Freshwater Ecol 12:395–403
Rikkinen J (1995) What's behind the pretty colours? A study on the photobiology of lichens. Bryobrothera 4:1–239
Rundel, P (1978) Ecological relationships of desert fog zone lichens. Bryologist 81:277-293
Sage RF (1996) Modification of fire disturbance by elevated CO_2. In: Körner C, Bazzaz FA (eds) Carbon dioxide, populations and communities. Academic Press, San Diego, pp 231–249
Schlesinger WH, Reynolds JF, Cunningham GL, Heunneke LF, Jarrell WM, Virginia RA, Whitford WG (1990) Biological feedbacks in global desertification. Science 247:1043–1048
Skujins J (1981) Nitrogen cycling in arid ecosystems. Ecol Bull 33:447–491
Smith RC, Prézelin BB, Baker KS, Bidigare RR, Boucher NP, Cooley T, Karentz D, MacIntyre S, Matlick HA, Menzies D, Onrusek M, Wan Z, Waters KJ (1992) Ozone depletion: ultraviolet radiation and phytoplankton biology in Antarctic waters. Science 255:952–959
Smith SD, Strain BR, Sharkey TD (1987) Effects of CO_2 enrichment on four Great Basin grasses. Funct Ecol 1:139–143
Smith SD, Huxman TE, Zitzer SF, Charlet TN, Housman DC, Coleman JS, Fenstermaker LK, Seemann JR, Nowak RS (2000) Elevated CO_2 increases productivity and invasive species success in an arid ecosystem. Nature 408:79–82
Steppe TF, Olson JB, Paerl HW, Litaker RW, Belnap J (1996) Consortial N_2 fixation: a strategy for meeting nitrogen requirements of marine and terrestrial cyanobacterial mats. FEMS Microbiol Ecol 21:149–156
Vitousek PM (1994) Beyond global warming: ecology and global change. Ecology 75:1861–1876
Walker B, Steffen W (eds) (1997) The terrestrial biosphere and global change: implications for natural and managed ecosystems. IGBP Science No 1. Stockholm, Sweden
West NE (1990) Structure and function of microphytic soil crusts in wildland ecosystems of arid to semi-arid regions. Adv Ecol Res 20:179–223
Whisenant SG (1990) Changing fire frequencies on Idaho's Snake River Plains: ecological and management implications. In: McArthur ED, Romney EM, Smith SD, Tueller PT (eds) Proceedings, symposium on *Bromus* invasion. USDA GTR-INT-276
Williams DG, Baruch Z (2000) African grass invasion in the Americas: ecosystem consequences and the role of ecophysiology. Biol Invasions 2:123–140
Wullstein LH (1989) Evaluation and significance of associative dinitrogen fixation for arid soil rehabilitation. Arid Soil Res Rehabil 3:259–265
Zaady E, Groffman P, Shachak M (1998) Nitrogen fixation in macro and microphytic patches in the Negev Desert. Soil Biol Bichem 30:449–454

31 Remote Sensing of Biological Soil Crusts

A. Karnieli, R.F. Kokaly, N.E. West, and R.N. Clark

31.1 Introduction

The ability of remote sensing to detect and map the distribution of biological soil crusts offers the opportunity to extend site-specific ecological studies of crusts to a regional scale, thus reducing the time and costs associated with ground surveys. However, despite the global extent of soil crusts and the expanding interest in their ecological roles, there have been relatively few studies published on the use of remote sensing to detect and map their distributions. Wessels and Van Vuuren (1986) were the first to use satellite imagery to detect and map biological soil crusts. Their study of the Namib Desert of SW Africa used Landsat TM to discriminate lichen-covered areas from bare ground and vegetated surfaces. Subsequently, only a relatively small number of publications have either presented the spectral properties of biological soil crusts or applied remote sensing to map them (Clark et al. 1993a,b,c; Kokaly et al. 1994; O'Neill 1994; Karnieli and Tsoar 1995; Karnieli and Sarafis 1996; Karnieli et al. 1996, 1999; Tsoar and Karnieli 1996; Tromp and Steenis 1996; Karnieli 1997).

The objective of this chapter is to present the spectral characteristics of biological soil crusts and explore the methods and issues in using remote sensing for their identification. As the reflectance spectra of soil crusts vary among geographical regions, this chapter presents results from three study sites: the Negev Desert in Israel, the semiarid Colorado Plateau of southern Utah, and the rangelands of southeastern Idaho.

31.2 Brief Review of Remote Sensing

Remote sensing is used to detect or infer the properties of a substance without being in direct physical contact. Remote sensing instruments measure the electromagnetic energy upwelling from a surface at various parts of the wavelength spectrum. This energy might be an inherent emission from an object due to its temperature, reflected from a natural source such as the sun,

Ecological Studies, Vol. 150
J. Belnap and O.L. Lange (eds.) Biological Soil Crusts: Structure, Function, and Management

or reflected from an artificial source such as radar. Past studies of biological soil crusts interpreted the reflected sunlight from surfaces to discriminate soil crusts from vegetation and other surface cover. Reflected solar radiation is divided into three main regions: the visible (VIS, 400–700 nm), the near infrared (NIR, 700–1000 nm), and the short-wave infrared (SWIR, 1000–2500 nm). The VIS region may be further subdivided into three parts: the blue (B, 400–500 nm), the green (G, 500–600 nm), and the red (R, 600–700 nm). These spectral regions are delineated in Fig. 31.1.

The relative amounts of energy reflected from surfaces vary as a function of wavelength. Thus, water can be distinguished from other materials by its low reflectance in the VIS and near-zero reflectance in the NIR and SWIR. In contrast, bare soil usually has a high reflectance in the NIR and SWIR. The most important factor affecting soil reflectance is the soil mineralogy (e.g., iron oxides, clay minerals, and carbonates), with additional effects due to soil-water content, organic matter content, soil texture, and soil roughness (Stoner and Baumgardner 1981). Characteristically, healthy green vegetation shows an extreme difference in reflectance between the red (very low reflectance) and NIR (relatively high reflectance) regions of the electromagnetic spectrum. Figure 31.1 shows a plot of the reflectance of soil and vegetation, which illustrates the disparate reflectance properties of green plants in the red and NIR. The presence of chlorophyll in green plants causes the strong absorption of light in the blue and red wavelengths and the plot of reflectance shows an absorption feature at 680 nm (a local minimum in reflectance).

The percent cover and biophysical condition of vegetation have been monitored from satellite remote sensing images by utilizing channels in the

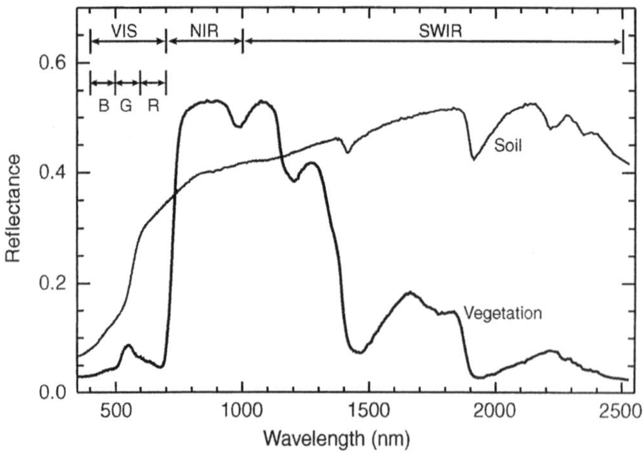

Fig. 31.1. Spectral reflectance measurements of soil and vegetation showing regions of the electromagnetic spectrum referred to in remote sensing of soil crusts

red and NIR combined in spectral vegetation indices (VIs) (e.g., Bannari et al. 1995). The most widely used index is the Normalized Difference Vegetation Index: NDVI = [NIR-(NIR-R)/(NIR + R], where R and NIR are the radiance or reflectance in the red and near infrared spectral bands, respectively (Rouse et al. 1974). NDVI values range from −1 to +1, where denser and/or healthier vegetation cover has higher positive values. However, in arid and semiarid regions, exposed soils have a significant effect on NDVI values and, thus, limit their use (Huete et al. 1984). Data from satellite-borne sensors used to generate vegetation indices usually consist of measurements in a few channels with coarse spectral resolution (100–200 nm). These instruments are referred to as broad-band, multispectral sensors (e.g., Landsat TM and MSS).

Imaging spectroscopy (also called hyperspectral remote sensing) measures the reflectance of solar radiation in hundreds of continuous narrow-band channels (for example, the Airborne Visible and Infra-Red Imaging Spectrometer (AVIRIS) operated by NASA/JPL; Green et al. 1998). The advantage offered by spectroscopic remote sensing is the ability to resolve and identify absorption features in a reflectance spectrum that arise from chemical bonds present in the surface materials. Imaging spectrometers resolve the shape and wavelength position of the absorption features rather than just sensing the differential reflectance level in two channels. For example, Fig. 31.1 compares spectral curves of vegetation as measured by a spectrometer to the overall level of reflectance in the red and NIR as detected by a broad-band sensor. Among the expanding number of applications, imaging spectroscopy has been used for geologic mapping (Clark et al. 1993a,b,c; Swayze 1997; Clark 1999), environmental studies (Swayze et al. 2000), vegetation cover identification (Kokaly et al. 1998; Martin et al. 1998; Roberts et al. 1998), and vegetation biochemical composition (Kokaly and Clark 1999). For detailed discussions of spectroscopic approaches to remote sensing see Clark (1999), and Clark et al. (2002).

31.3 The Negev Desert (Israel) Experience

Extensive studies of biological soil crusts have been conducted in Sede Halamish dune field along the Israel/Egypt political border, about 50 km south of the Mediterranean coastline (34°23'E/30°57'N) (Karnieli and Tsoar 1995; Karnieli 1997; Karnieli et al., 1999). The area is characterized by sandy, linear dunes, including the Nizzana Research Site (see Chap. 22 and Chap. 1, Photo 49). Average annual rainfall is 100 mm and is restricted almost entirely to the winter period (October–March). Sparse vascular vegetation is about 14 % perennial cover and 0–27 % annual cover (depending on the season), which is dominated by *Artemisia monosperma*. Most of the basal dune and interdune corridors are covered by cyanobacterially-dominated soil crusts associated with fine-grained soils that are 15–20% silt and clay (Danin et al. 1989; Lange et al. 1992; see Chap. 8). Five biological soil-crust communities (A–E) were defined in this area (Kidron 1995; Karnieli et al. 1999). Community A, with three cyanobacterial species (Table 31.1), extends over the lower, south-facing dune slopes, and the interdune area. Communities B to E occupy the north-facing slopes and support five additional cyanobacterial species, as well as green algae and mosses. Chlorophyll content, organic matter, polysaccharides, protein, crust thickness, and crust darkness increase from community A to E (Table 31.2).

Table 31.1. Microphytic communities at the Sede Halamish site. (After Karnieli et al. 1999)

Species	Community				
	A	B	C	D	E
Microcoleus sp.	*****	*****	****	****	***
Phormidium sp.	***	**	**	*	*
Scytonema sp.	*	**	***	****	***
Nostoc microscopicum	*	*	**	***	***
Oscillatoria sp.	s	*	**	***	***
Schizothrix friesii	s	*	s	***	***
Chroococcidiopsis sp.	s	s	*	*	*
Green algae	s	s	*	**	***
Mosses	5.5×10^2	8.3×10^2	3.3×10^3	4.7×10^4	7.8×10^5

Asterisks represent the relative abundance of the different microphytic communities where one asterisk stands for low, five for high, and "s" stands for scarce, except for moss cover which is given in individual counts m^{-2}.

Table 31.2. Physiological and soil parameters of microphytic communities at the Sede Halamish site. (After Karnieli et al. 1999)

Community	Chlorophyll content (g m^{-2})	Organic matter (%)	Polysacch. content (g m^{-2})	Protein content (g m^{-2})	Crust thickness (mm)	Soil color Dry	Soil color Wet
A	0.017	0.5	4.17	3.72	1.1	10YR6/4	10YR5/4
B	0.021	0.6	7.71	5.29	1.5	10YR6/4	10YR5/3
C	0.029	0.7	8.27	5.51	2.0	10YR6/4	10YR4/4
D	0.041	0.9	14.35	9.31	2.8	10YR6/3	10YR4/3
E	0.053	2.6	30.60	28.37	10.0	10YR6/3	10YR3/3

All values increase from community A to E as a function of the relative abundance and distribution of the microphytes. Community E, which consists mostly of mosses, has relatively higher values.

31.3.1 The Visible Region

Spectra for bare dune sand, cyanobacterial crust, and vascular plants (measured by a Li-Cor 1800 spectrometer) are shown in Fig. 31.2. The shape of the vascular plant spectrum is typical for plants, and has lower reflectance than the sand and crust spectra. Biological soil crust reflectance is generally lower than that of bare sand. Both curves are typical for soil, but the slight flattening between 600 to 700 nm in the soil crust curve is due to absorption by photosynthetic pigments. Relative to the soil-crust spectrum, the active dune sand spectrum is higher in the red and NIR bands, lower in the blue band, and intermixed in the green bands. This spectral characteristic has been observed in several other dune fields, such as Egypt (Jacobberger 1989), Burkina Faso (Tromp and Steenis 1996), Niger (Houssa et al. 1996), Senegal (A. Karnieli and C. Glaesser, unpubl.), and Turkmenistan (A. Karnieli, unpubl.). Karnieli and Sarafis (1996) suggest that the blue absorption feature is due to phycobilin pigments unique to cyanobacteria, or possibly to the biological crusts covering Fe^{3+}-bearing minerals of the bare soil, which absorb strongly in UV and blue regions.

Karnieli (1997) used the spectral signature of biological soil crusts in the Negev to develop a spectral Crust Index (CI), which is the normalized difference between red and blue spectral values [CI = 1-(RED – BLUE)/(RED + BLUE)]. In a sand-dune environment, the CI can distinguish different lithologic/morphologic units such as active sands, crusted interdune areas, and playas. A hyperspectral airborne DAIS image of the Sede Halamish sand field is shown in Fig. 31.3. A color composite image (Fig. 31.3A) is compared to the CI image (Fig. 31.3B). In the CI image (Fig. 31.3B), the wadi bed (upper right) and playa areas are dark, interdune crusted areas are gray, and mobile sand

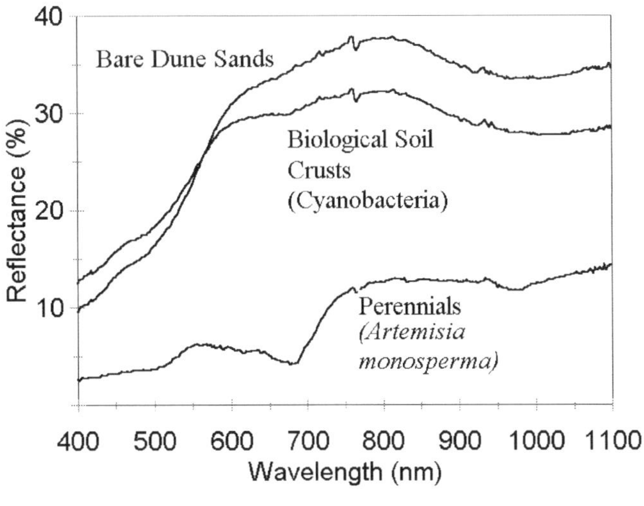

Fig. 31.2. Spectral reflectance field measurements of bare dune sands, biogenic soil crusts (cyanobacteria), and higher plants (*Artemisia monosperma*), Sede Halamish dune fields. (After Karnieli and Tsoar 1995)

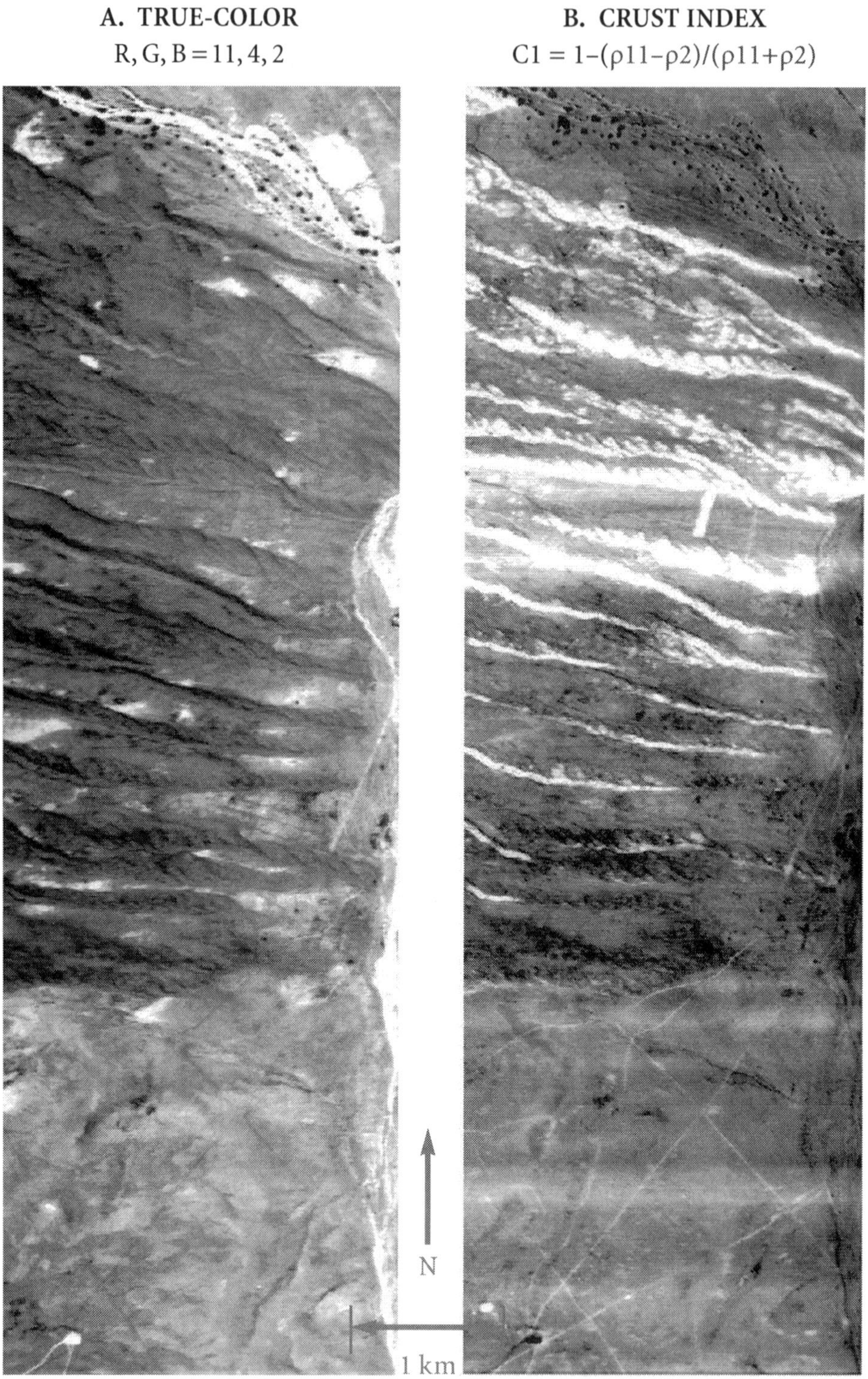

Fig. 31.3A–B. The dune study area in northwest Negev Desert, based on the DAIS airborne hyperspectral image spectrometer. **A** True-color composite. **B** Crust Index. (E. Ben-Dor and A. Karnieli, unpubl.)

dunes are white. Linear roads crossing the dunes (lower part, Fig. 31.3b) have destroyed the biological soil crusts (gray) and, thus, are clearly seen as white. The CI image is more sensitive to crusts features than the color composite image, and can be used with data from any sensor that measures blue and red bands. There are many sources for such data, including color aerial photographs and Landsat-TM images.

31.3.2 Effect of Wetting

Biological soil crusts are dormant when dry. A few minutes after wetting, respiration and, if light is available, photosynthetic activity begins (see Chap. 19). Laboratory spectral measurements (using a Li-Cor 1800 spectrometer) of the five biological soil crust communities show that dry (Fig. 31.4 top) and wet (Fig. 31.4 center, bottom) soil-crust spectra are different. All these curves have similar spectral features, but differ in the overall magnitude of reflectance and the depth of the pigment absorption zone. Community A, with the fewest species, is characterized by very slight absorption features in the red and is very much like a typical bare soil spectrum. As the number of species and the biomass increases, absorption features in the red increase, and overall reflectance (due to crust darkening) decreases. Vascular vegetation reflectance (Fig 31.2) is similar. However, in the crust spectra (Fig. 31.4) the absence of a pronounced peak at 550 nm and the gentle increase in slope around 700 nm indicate that these spectra are not due to vascular vegetation. Figure 31.4 (center and bottom) shows reflectance from the same communities 1 day and 7 days after wetting, respectively. With increased time since wetting, reflectance decreases and stronger absorption in the red region is observed.

When crusts were dry, NDVI values were low (0.06), a level also typical of dry soils and rocks (Karnieli et al. 1996). NDVI values increased in all five communities with wetting (Table 31.3). Because NVDI values for dune sand change little with wetting, the increase in crust NDVI values is unlikely to be due to an increase in soil moisture; more likely, it is caused by hydration of crust organisms. The biggest change in NVDI occurred 1 day after wetting; after 7 days, only slight increases in NVDI values were noted.

The chlorophyll content of dry crusts was relatively low, but much higher values were measured under the "wet+7 day" conditions (Table 31.3). In a different study, high correlation was found between chlorophyll content of wet crusts and NDVI values of saline sandy soils near the Dead Sea (Fig. 31.5). However, Table 31.3 indicates that this correlation may be more driven by dry versus hydrated crusts, rather than total chlorophyll content. In this study, correlations between NDVI values and organic matter, polysaccharide content, protein contents, crust thickness, and species numbers were observed (Table 31.2, 3).

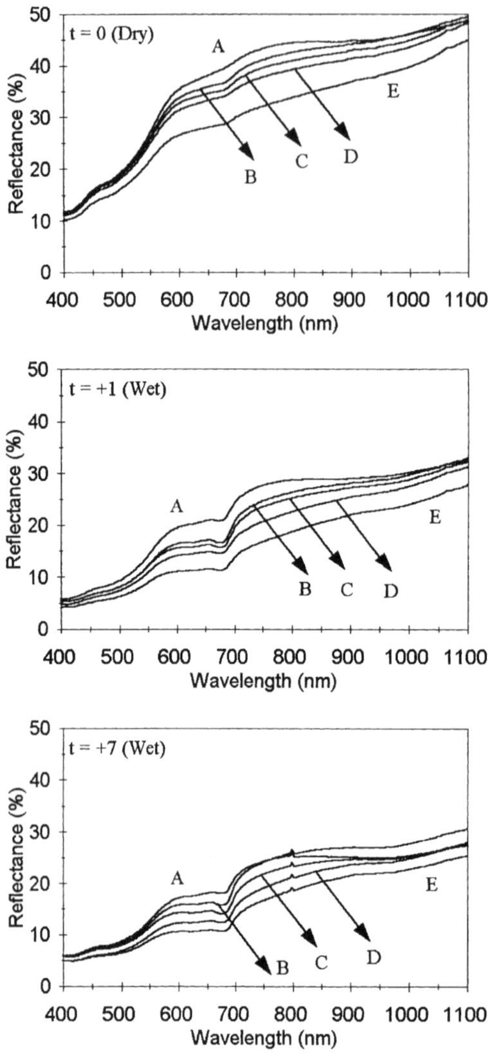

Fig. 31.4. Spectral response of different microphytic communities (*A, B, C, D, E*) in the sand dune study site (Sede Halamish) under different moisture conditions. *Top* Dry samples; *center* samples after wetting and 1 day of incubation; *bottom* samples after wetting and 7 days of incubation. Note that the spectral signals decrease from community A to E as a function of the relative abundance and distribution of the microphytes, and also due to the wetting schedule. (After Karnieli et al. 1999)

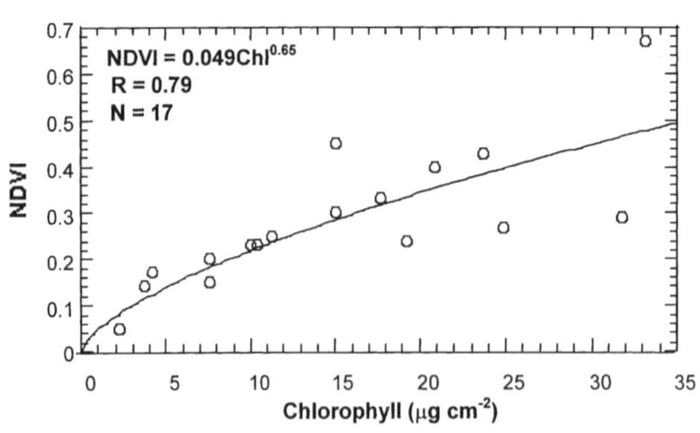

Fig. 31.5. Correlation between chlorophyll content and NDVI, near the Dead Sea

Table 31.3. NDVI and chlorophyll (in g m^{-2}) values for the sand dune and microphytic communities at the Sede Halamish site in different moisture conditions. (After Karnieli et al. 1999)

	Dry conditions		Wet+1 conditions		Wet+7 conditions	
Community	NDVI	Chlorophyll	NDVI	Chlorophyll	NDVI	Chlorophyll
Sand	0.07	N/A	0.08	N/A	0.08	N/A
A	0.06	0.017	0.14	N/A	0.15	0.076
B	0.08	0.021	0.22	N/A	0.23	0.100
D	0.08	0.029	0.23	N/A	0.24	0.193
D	0.08	0.041	0.23	N/A	0.27	0.249
E	0.10	0.051	0.27	N/A	0.29	0.318

N/A means data not available. Note that both NDVI and chlorophyll values increase from community A to E with respect to the relative abundance and distribution of the microphytes, and also due to the wetting schedule; see text.

31.3.4 The SWIR Region

Figure 31.6 shows the spectra of an active dune sand and three crust communities in the visible, NIR, and the SWIR regions (400–2500 nm), as obtained with a GER Mark V spectrometer. Iron oxide and clay mineral features are evident at 860 and 2200 nm, respectively, and evidence for carbonates is seen at 2330 nm. These findings were verified by X-ray analysis: about 95 % of the sample was quartz (which does not have a diagnostic absorption in the 400–2500-nm range), with the remaining 5 % being calcite, feldspar, and clay minerals. The two other distinctive absorption features at about 1400 and 1900 nm are related to structural water.

Spectra of communities A, C, and E have a shape similar to that of sand, but reflectance levels are lower. In the SWIR, there are several broad and shallow absorption features related to fresh organic matter (especially in community E): 1720 nm (cellulose, lignin, starch, pectin); 2180 nm (starch, lignin, wax, tannins); and 2309 nm (humic acid, wax, starch) (Ben-Dor et al. 1997). Organic matter absorption features occur at 1447 and 1920 nm, but are masked by water absorption bands (Elvidge 1990; Ben-Dor et al. 1997). Significant spectral differences occur as crusts become more developed (Fig. 31.6). Iron oxide and clay mineral absorption features at 860 nm and 2200 nm in the spectra diminish gradually from the sand to community E, while the chlorophyll absorption dip at 680 nm becomes deeper. As more dense biogenic crust forms, the iron oxide and clay mineral absorption features are masked, while the organic matter features become dominant. The water absorption features at 1447 and 1920 nm are very sensitive to external conditions (e.g., relative

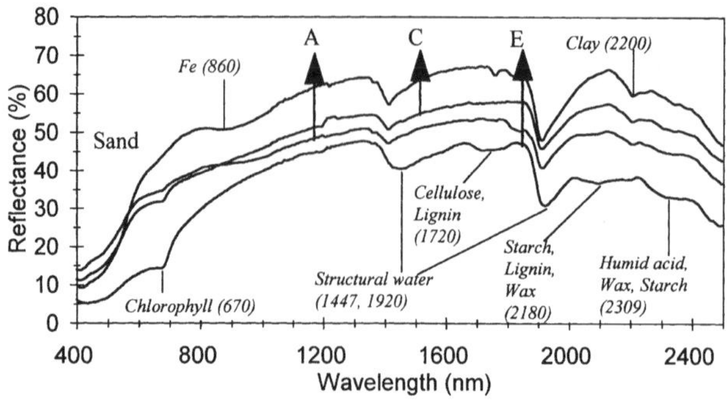

Fig. 31.6. Spectra of active sand dune and microphytic communities *A*, *C*, and *E*. *Lines* point to spectral signatures for different compounds and the *numbers* indicate wavelength in nanometers. (After Karnieli et al. 1999)

humidity, temperature, etc.) that were probably not constant at the time of measurement.

The community E spectrum (Fig. 31.6) shows a remarkable similarity to an organic matter spectrum measured by Ben-Dor et al. (1997). The similarities at 1720, 2180, and 2309 nm indicate that the same functional groups are active in both materials. The protein and polysaccharide contents in community E are much higher than in the other crust communities (Table 31.2), and the absorption features are much stronger (Fig. 31.6). Therefore, it and other related features of the biological soil crusts in sample E are probably also included in the organic matter component.

Because the spectra of crust and organic matter can be very similar, precautions must be taken in data analyses, and the 680-nm chlorophyll absorption should be examined at multiple dates. Decreasing absorption at 680 nm over time is due to loss of chlorophyll in decomposing organic matter (Ben-Dor et al. 1997), while residual peaks or persistent absorption may be attributable to the living crust organisms.

In summary, remote sensing and spectral analyses of biological soil crusts in the Negev Desert show that (1) the NDVI is strongly related to hydrated chlorophyll content. When biological soil crusts are wet, their NDVI value can reach up to 0.30 due to photosynthetic activity. The spectral signal of wet crusts might be misinterpreted as higher vegetation; (2) phycobilins, unique pigments in the biological soil crusts, contribute to higher reflectance in the blue region relative to the sand substrate. Consequently, the Crust Index (CI) can be useful for detecting and mapping soil crusts using remote sensing imagery; (3) the absorption features representing the biological soil crusts are chlorophyll (at 680 nm) and organic matter (at 1720, 2180, and 2309 nm). These spectral features increase progressively from the sand spectrum to the highest biomass biogenic crust spectrum. Water absorption features at 1400 and 1900 nm become weaker and shift towards the longer wavelengths.

31.4 The Colorado Plateau Region of Southwest United States

The Colorado Plateau is a semiarid region encompassing southern Utah, northern Arizona, western Colorado, and northwestern New Mexico. Arches and Canyonlands National Parks (NP) are large parks located in southern Utah. The most common vascular plant communities in these parks are pinyon-juniper (*Pinus edulis, Juniperus osteosperma*), saltbush (*Atriplex corrugata*), mixed grassland (*Stipa comata, Sporobolus cryptandrus, Hilaria jamesii*), or blackbrush (*Coleogyne ramosissima*) (Allan 1977). Biological soil crusts in this area are dominated by the cyanobacterium *Microcoleus vaginatus* and also contain lichens, green algae, and mosses. Soil crusts can represent up to 70 % of the living cover (Belnap 1990). Crusts range from smooth, light-colored crusts (*Microcoleus*-only) to pinnacled, dark-colored crusts (*Microcoleus, Scytonema, Nostoc,* lichens, and mosses; see Photos 31–33).

31.4.1 Laboratory Reflectance Spectra of Biological Soil Crusts

Laboratory reflectance spectra were measured for samples of biological soil crusts collected from Arches and Canyonlands using a modified Beckman 5270 spectrometer (Fig. 31.7). Samples A1 and A2 were collected from sandy soils in Arches NP, while A3 was from gypsiferous soils. A1 had no mosses or lichens on the surface, while A2 and A3 had both mosses and lichens. N1 and M1 were collected nearby from Canyonlands NP. N1 is from sandy soils; M1 is from soil developed on the clay-containing Moenkopi formation. Each of these samples of well-developed soil crusts shows an absorption feature at approximately 0.68 μm due to chlorophyll *a* present in crustal organisms. The spectra of soil crusts from the Colorado Plateau (Fig 31.7) are similar to spectra of soil crusts in the Negev Desert (Fig. 31.4).

Although the absorption feature at 0.68 μm is similar between samples, the spectra vary in the 2.0 μm region. In some samples, such as A3 and M1, clay mineral absorption features at 2.1 μm are apparent. The spectra show the strong wing of an absorption feature centered in the ultraviolet and extending through 1.30 μm, giving all the spectra in this region (except M1) a concave shape. The shapes of the spectra from 0.35–1.40 μm arise from both the iron mineralogy and the organic components of the soil. The overall reflectance level can be very low in the visible wavelengths due to the dark soil surface.

The components of the N1 crust were each measured separately (Fig. 31.8). All reflectance curves of the soil crust components show a small chlorophyll absorption feature at 0.68 μm. Spectra of the dark soil lichen *Collema* and dark rock lichens and varnish are very similar. As a result, areas with such

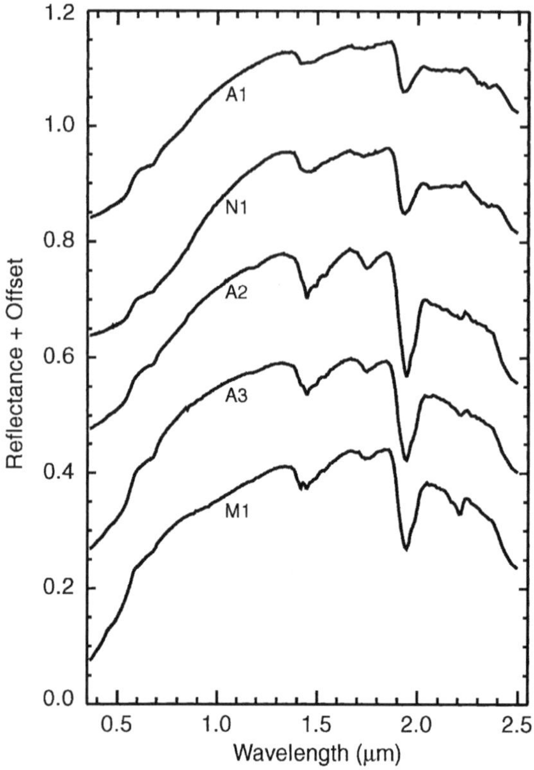

Fig. 31.7. Laboratory reflectance spectra of biological soil crusts from the Colorado Plateau region of the United States. Scale of vertical axis is given in relative units; offsets of 0.20 units were used in order to avoid overlapping of spectra (see text for explanation of symbols)

rocks may be indistinguishable from soil crusts in the 0.35–2.5 µm region. The top curve in Fig. 31.8 is the reflectance from "light" soil crusts. In general, this type of soil crust has a much greater resemblance to bare soil/sand and is the most difficult to identify in remote sensing data, especially with broad-band sensors.

Figure 31.9 shows the reflectance spectrum of extracted filaments of *Microcoleus*. Figure 31.10 shows the chlorophyll *a* absorption feature isolated using the continuum-removal process of Clark and Roush (1984). The chlorophyll absorption band center is estimated to be at 0.681 µm. The absorption feature to the left of the chlorophyll *a* center is due to phycobilins, unique pigments produced by cyanobacteria. One class of phycobilins, the phycocyanins, is blue, with a strong absorption feature at approximately 0.625 µm (Sirevag 1992). The continuum-removed reflectance spectrum in Fig. 31.10 does indeed show the presence of the phycocyanin absorption estimated at approximately 0.627 µm. In addition to these two pigments, cyanobacteria contain a large amount of carotenoids. Carotenoids, which include the fat-soluble hydrocarbon carotenes and the oxygen-containing xanthophylls, absorb violet and blue light at the shortwave end of the electromagnetic spectrum. In Fig. 31.9, the carotenoid absorption features

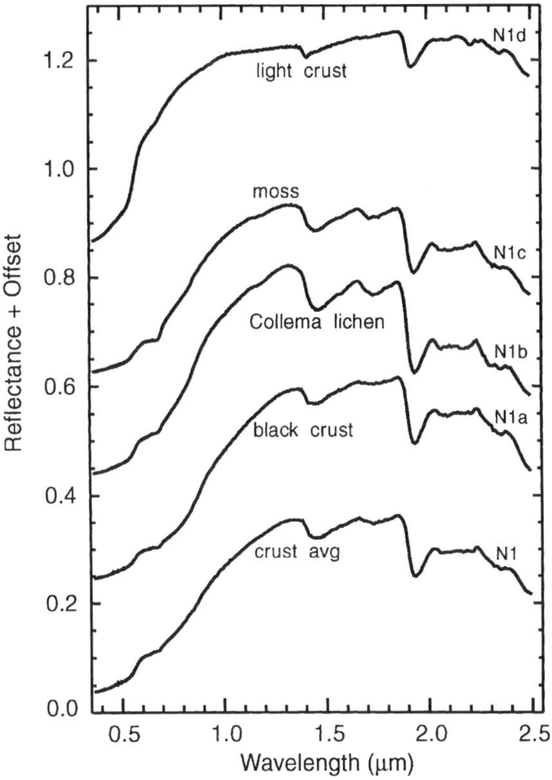

Fig. 31.8. Laboratory reflectance spectra of a biological soil crust N1 and its isolated components. Scale of vertical axis is given in relative units; offsets of 0.20 units were used in order to avoid overlapping of spectra (see text for explanation of symbols)

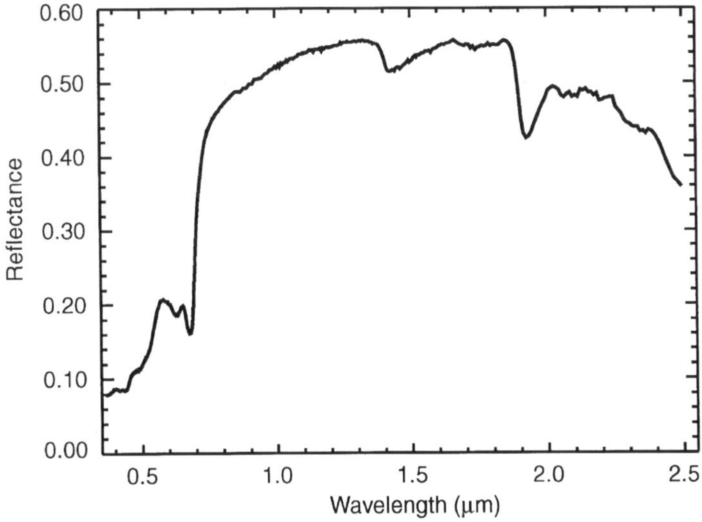

Fig. 31.9. Laboratory reflectance spectrum of cyanobacterial (*Microcoleus vaginatus*) filaments extracted from a biological soil crust

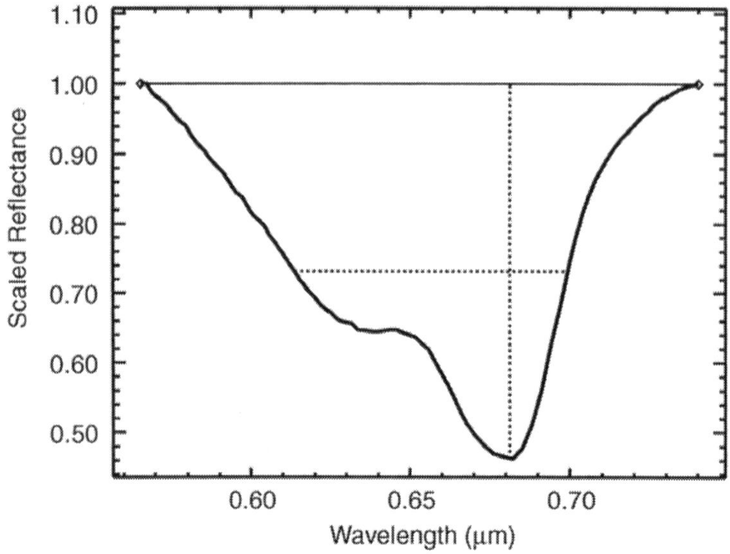

Fig. 31.10. Continuum-removed reflectance of the major pigment absorption feature of extracted cyanobacterial filaments (shown with the *continuum line* and *band center line*). Linear continuum analysis, using endpoints of 0.565 and 0.740 microns, revealed the following absorption feature parameters: band center at 0.681 microns, band depth of 0.538 (*vertical dotted line*), FWHM (full width at half maximum) of 0.0857 microns (*horizontal dotted line*), and a shoulder absorption centered at 0.627 microns

are likely the cause of the low reflectance level at wavelengths lower than 0.55 μm.

31.4.2 Field Reflectance Spectra of Biological Soil Crusts

Field reflectance spectra of soil crusts were measured with an ASD FR field spectrometer in southern Utah (Arches NP) and Arizona (Organ Pipe Cactus NM and Petrified Forest NP) (Fig. 31.11). The spectra in Fig. 31.11 have reflectance deleted for regions with strong absorption by atmospheric water at 1.4 and 1.9 μm. Also, the measurements had a low signal-to-noise ratio (S/N) at wavelengths greater then 2.4 μm, so reflectance in this region was deleted. Finally, the area from 0.9–1.0 μm was deleted due to decreasing instrument sensitivity. The upper curve (FS1) in Fig. 31.11 is for a "light" soil crust, and compares well with the laboratory measurement of "light" crust (N1d, Fig. 31.8). All crust types showed chlorophyll *a* absorption at approximately 0.68 μm. FS4 showed this absorption only weakly. Wet soil crust (FS5) showed chlorophyll *a* absorption more intensely than dry soil; however, the spectrum of wet soil crust was dominated by water absorption features beyond 1.0 μm.

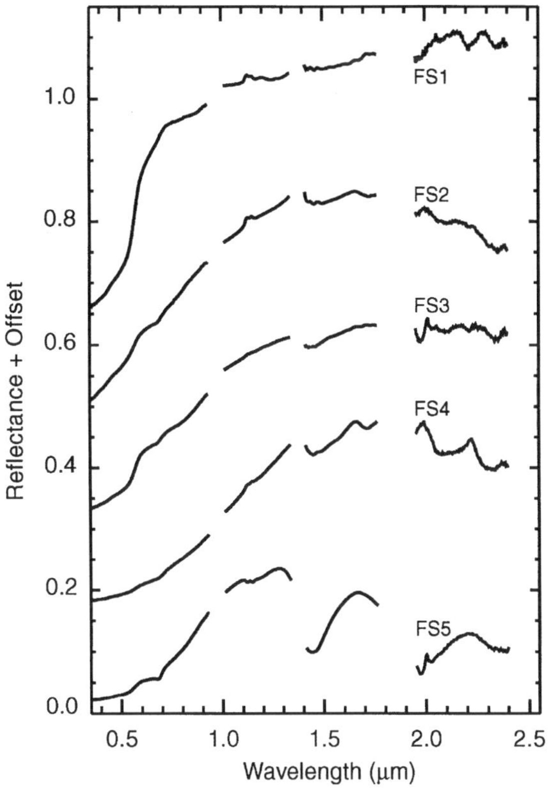

Fig. 31.11. Field reflectance spectra of biological soil crusts from locations throughout the desert southwestern United States. Reflectance has been deleted for wavelength regions with high atmospheric absorption and wavelength regions with instrument artifacts. Scale of vertical axis is given in relative units; offsets of 0.15 units were used in order to avoid overlapping of spectra (see text for explanation of symbols)

31.4.3 Mapping Soil Crusts Using Imaging Spectroscopy Data

In contrast to the broad-band remote sensing of crusts in the Negev Desert (presented in Sect. 31.3.1), which exploited the differential absorption properties of crusts at two wavelengths, soil crusts in the Colorado Plateau were identified and mapped using spectral feature analysis. Airborne Visible and Infra-Red Imaging Spectrometer (AVIRIS) data were collected on June 19, 1995. A reference spectral library for materials occurring in Arches NP was compiled that included crust (A1), blackbrush, dune sand, and linear combinations of these materials. The USGS Tetracorder algorithm was then used to compare the observed AVIRIS spectral features to the reference spectra and to select the best match (Clark et al. 1993b, 2002). The Tetracorder algorithm is an expert system that compares the absorption features in each AVIRIS pixel to the characteristic absorptions of more than 300 materials, including minerals, mineral mixtures, vegetation, water, snow, and man-made materials (Clark et al. 1993c).

From the Tetracorder analysis of AVIRIS data, a map of soil-crust cover was generated (Fig. 31.12). As discussed earlier, the chlorophyll absorption arising

Fig. 31.12. Results of mapping green plant cover and soil-crust/lichen cover using spectral feature analysis of AVIRIS data for Arches National Park, Utah, USA. (R.F. Kokaly, unpubl.)

from the soil crusts and lichen cover on rocks is very similar. Thus, the signatures in the remote sensing data that match crust cover may also be lichen-covered rocks and the results are labeled as such. Areas with significant crust cover are portrayed in the left-hand image. The center part of the scene shows a large area of 80 % soil crusts/lichen cover (indicated by the red pixels). The areas of 95 % soil crust/lichen cover (blue pixels) along the edges of the Salt Wash are most likely to be lichens growing on the exposed sandstone on the edge of this canyon. The maps portray a few well-defined patches of soil crust in the Salt Valley. Field studies confirmed the presence of several of the larger crust areas.

Vegetation cover is portrayed in the right-hand image of Fig. 31.12 and shows that vascular vegetation cover is very low in most areas. Since the AVIRIS data were collected after all annual and most perennial plants were brown, these results were expected. The riparian area along Salt Wash shows a high vegetative cover. There are also some areas of moderate green vascular plant cover in the upper portion of the map that are most likely due to the presence of evergreen juniper and pinyon pine. These results have not yet been field-checked.

In summary, laboratory and field reflectance spectra of biological soil crusts of the Colorado Plateau exhibit a similar weak absorption feature in the visible region due to biogenic pigments. This spectral feature of a soil crust was compared to the data collected by an airborne spectrometer (AVIRIS). This approach of spectral feature analysis applied to remotely-sensed spectra was used to map the distribution of soil crusts in Arches National Park. Spectra of soil crusts were most easily identified in areas of high crust cover and low vegetation cover. Laboratory measurements of black lichen cover on exposed rock and measurements of well-developed soil crusts had similar spectral properties. Thus, identification of soil crusts using remote sensing may be further complicated in areas with significant lichen cover.

31.5 The Southern Idaho Experience

Field spectroscopy and remote sensing studies of biological soil crusts were conducted in 1995 and 1996 at the US Sheep Station in southeastern Idaho. The Sheep Station (lat. 44°14'44"N, lat. 112°2'47"W, 1650 m elevation a.s.l.) is a US Department of Agriculture research facility consisting entirely of rangelands on windy, rolling plains dominated mostly by sagebrush steppe (West 1983). The climate is semiarid with cold winters and warm summers. Average annual precipitation is 325 mm, including 70 cm of snow, with an average annual temperature of 6.1 °C. Soils are mollisols derived from loess, residuum, or alluvium overlying basalt bedrock of varying age and, thus, of varying soil thickness (Bork et al. 1998b). Vascular plant cover and litter are greatest over the thicker soils. Moss, lichen, bare soil, and exposed bedrock are more abundant on the younger, less-weathered material.

This study was done in two phases. First, repeated close-range multispectral measurements were made of ten plots selected to represent each cover category as fully as possible. The plots were then revisited during June, July, and August of 1995. The second phase of the study involved random placement of ten 1.75-m^2 circular plots in each of six different fenced paddocks and exclosures with differing grazing history, albeit all still considered sagebrush steppe (Bork et al. 1998b). The spectral measurements were made at mid-day on cloud-free days in June, July, and August of 1996. An ASD Personal Spectrometer was suspended from a portable aluminum boom with the detector located 3.4 m over the plot. Cover of all vascular plant species and ground cover categories was determined within 24 h after the spectrometer data were collected. Data concerning soil crusts, bare soil, and exposed basalt rock are discussed in detail here (specific results dealing with the vascular species can be found in the cited papers).

31.5.1 Spectral Reflectance of Ground Cover

The mean spectral response curves (with 95 % confidence intervals) for June are shown in Fig. 31.13. The comparison of dry bare soil to rock and to microphytes was statistically significant throughout the entire spectrum (Bork et al. 1998a). All three curves are relatively flat, soil-like curves, with little deviation in the visible and NIR. The narrow-band response curve of microphytes was sinuous relative to that of dry bare soil and basalt rock, but much less so compared to vascular plants (Bork et al. 1998a). Microphytes (lichens) were unique in that they had a slight reflectance increase in the R to NIR region around 730 nm. Thus, biological soil crusts remained spectrally distinct in all wavebands compared to the other two curves discussed here. This suggests that multiple narrow bands could possibly be used to distinguish between these cover components. While the class "microphytes" was spectrally distinct, this class was a complex mixture of mosses, lichens, algae, and cyanobacteria of many different species.

Mean percent reflectance and their standard deviations all plotted closely to the brightness axis, with bare dry soil to the upper right, rock in the lower left, and crust organisms slightly above rock (Table 31.4). These components all have low simulated NDVI values, although biological soil crusts showed greater reflectance than bare soil or rock.

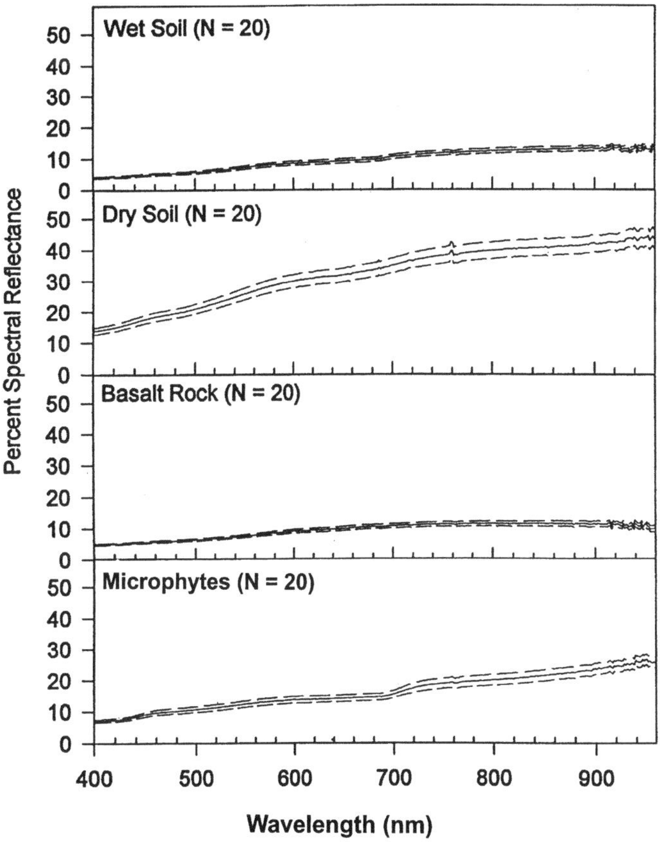

Fig. 31.13. Mean narrow-band spectral response curves (with 95 % confidence intervals) of microphytes and the nonvegetational ground cover components at the US Sheep Expt. Sta., Idaho, June 1995. (Bork et al. 1998a)

Table 31.4. Mean percent reflectance (and standard deviation) of ground cover components for the simulated TM values and resulting NDVI (listed in order of increasing NDVI). (Subset of Table 1 in Bork et al. 1998a)

Soil components	Blue-green (450–520 nm)		Green (520–600 nm)		Red (630–690 nm)		Near IR (760–900 nm)		NDVI (IR-R/IR+R)	
	Mean	SD	Mean	SD	Mean	SD	Mean	SD	Mean	SD
Dry soil	20.2	3.0	31.0	4.0	38.3	4.7	48.2	5.6	0.115	0.02
Exposed basalt rock	5.4	0.8	7.4	1.1	9.4	1.4	11.9	1.8	0.117	0.04
Microphytes	9.7	1.6	12.3	2.2	13.6	2.3	19.1	3.7	1.167	0.04

31.5.2 Regressions of Broad-band Reflectance and NDVI with Soil Cover

Calibrated predictive relationships were developed between the simulated broad-band NDVIs and the amounts of bare soil, basalt rock, crust cover, and vascular plant cover. The coefficient of determination of moss was 0.128, the lowest of any category of ground cover. Bare dry soil had the highest coefficient of determination (r^2=0.43). Blue band reflectance was best for lichens and rock. In fact, simple broad-band reflectances were often better predictors than the narrow-band data used in the companion studies (Bork et al. 1998a; 1999a,b,c). These results are not surprising, considering that quantitative changes in the soil-based components (Table 31.5) are more likely to change in the vertical position (i.e., magnitude) of the spectral response curves rather than their shapes.

The leading spectral variables related to moss and rock cover were simple regressions derived from July data. Lichen showed best with June data. Multiple dates were no more predictive than the single best date. This was not surprising because crust organisms are less likely to change phenologically, and thus spectrally, over the duration of the growing season than macrophytes. Multiple regression did not increase predictability over simple, one-variable models for lichen, moss or rock.

Table 31.5. The leading predictive equations obtained for each cover component using simple and multiple regression of the BB data, southern Idaho

Component	Simple regression[a]	Multiple regression
Moss	Y = 20.55-101.70 (Jy G/B-NDVI)	n/a
Rock	Y = -2.07 + 0.49 (Jy Blue)	Y = -2.07 + 0.49 (Jy Blue)
Soil	Y = 183.70-1150.35 (AugNDVI) + 1851.27 (x^2)*	Y = 32.35 + 8.91 (Aug Blue)

[a] Simple regressions marked with a * are 2nd-order, curvilinear functions.

31.5.3 Regressions of Narrow-Band Reflectance

The second phase of the study was done during the 1996 field season. Randomly located plots (ten in each of six different paddocks or exclosures) were evaluated for spectral reflectance and presence of various cover categories. Logarithmic transforms of the cover data produced equivalent-to-poorer fits than the raw data, as did nonlinear regressions. Narrow bands that appeared to be most unique were chosen from correlation diagrams (Bork et al. 1999a). The final set of narrow-band indices was derived by integrating areas under each raw spectral response curve. No smoothing was done on these data. Slopes were determined by calculating the change in reflectance (first derivative) between successive pairs of adjacent narrow bands.

Narrow-band variables (both simple and slope-based) had increased coefficients for soil and rock when a curvilinear function was used. Mosses had a more substantial coefficient of determination using either simple or slope-based narrow-band variables. The best fits for bare soil came from the August data, whereas lichen data from June had the best fit. Lichen and bare soils showed no increase in coefficients of determination by switching from broad-band to narrow-band variables. Mosses failed to gain an improved model of predictability when multiple regression was used. Lichen and bare soil had an increased r^2 using narrow-band models (from 0.222 to 0.674). Despite the statistical significance ($p<0.01$) of the simple regression coefficients for lichen, rock, and moss, these regressions explained relatively little of the variation between plots (40 % at most). This may reflect the limited quantities of these components at this study site (Table 31.6). Simple spectral variables such as broad-band blue, green, or red reflectance, or by narrow-band reflectances in those spectral regions, best represented all soil-based cover components.

Table 31.6. Mean cover (%) and standard deviations (SD) for nine rangeland indicator components ($n=60$); southern Idaho. (Bork et al. 1998b)

Component	Mean cover	±	SD
Grass	15.7	±	8.08
Herb	23.8	±	9.02
Shrub	23.0	±	10.15
Dead shrub	7.1	±	5.89
Lichen	4.9	±	11.46
Litter	19.0	±	5.89
Moss	2.6	±	3.28
Rock	0.8	±	1.28
Soil	18.5	±	12.31

Multiple regression predictability did not increase over simple models for lichen, moss, and rock. One exception was lichen. That model incorporated 5 June narrow-band data. Of these bands, four were from the blue-green and green regions, while only one was from longer wavelengths. This indicates that the R-NIR shift (680 to 730 nm) is not as important for identifying lichen as the other regions.

In summary, it was found that the broad-band reflectances were often better predictors of the cover of microphytes than narrow-band reflectances. This is probably a result of the quantitative changes in organisms on the soil surface being more closely related to the magnitudes of the spectral response curves than their shapes. All curves were relatively flat and soil-like with little deviation in the visible and NIR. Lichens were unique in that there was a slight increase in reflectance in the R to NIR region around 730 nm. Thus, narrow band detection was indicated to be more effective when biological crusts were dominated by lichens.

31.6 Conclusions

Remote sensing studies from spacecraft, aircraft, as well as in situ field and laboratory data have investigated the possibility of quantifying the amount of biological soil crusts, higher plants, and bare soil, based on their spectral reflectance. This chapter shows the ability of spectroscopic and image-processing analyses to differentiate biological soil crusts from other ground components and, thus, to successfully map the spatial distribution of soil crusts. Spectra of biological soil crusts from different locations (the Negev Desert in Israel, semiarid SW USA, and southern Idaho) were found to have similar signatures. However, spectral reflectance of biological soil crusts can be changed due to precipitation, thus altering the strength of the pigment absorption and the value of calculated NDVI. This relationship may be a complication in identifying crusts from a single image, but may be an advantage for multitemporal studies that could be timed to take advantage of rain events.

Additional spectral studies are needed to develop robust site-invariant mapping methods and to link biological soil crusts distribution maps to ecological studies and disturbance/response models. Such models would have significant importance in desertification/rehabilitation and climate change studies.

Acknowledgments. The study in the Israeli site was partially conducted at the Nizzana research site that is managed by the Hebrew University of Jerusalem Minerva Arid Ecosystems Research Center. Data belonging to the Israeli research were analyzed with the kind help of Dr. Eyal Ben-Dor (Tel Aviv University), Dr. Giora Kidron (Ben Gurion University of the Negev), Dr. Inka Dor, and Professor Avinoam Danin (Hebrew University of Jerusalem). The USGS study was aided by spectroscopic measurements conducted by Dr. Gregg A. Swayze and Dr. Jayne Belnap's expert knowledge of soil crusts of the Colorado Plateau region of the United States of America.

References

Allan JS (1977) The plant communities of Arches National Park. PhD Dissertation, Brigham Young University, Provo, UT

Bannari A, Morin D, Bonn F, Heute AR (1995) A review of vegetation indices. Remote Sensing Reviews 13:95–120

Belnap J (1990) Microbiotic crusts: their role in past and present ecosystems. Park Sci 10:3–4

Ben-Dor E, Inbar Y, Chen Y (1997) The reflectance spectra of organic matter in the visible near infrared short wave infrared region (400–2500 nm) during a controlled decomposition process. Remote Sens Environ 61:1–15

Bork EW, West NE, Price KA (1998a) In situ narrow-band reflectance characteristics of cover components in a sagebrush steppe rangeland. Geocarto Int 13:5–15

Bork EW, West NE, Walker JW (1998b) Three-tip sagebrush steppe response to long-term seasonal sheep grazing. J Range Manage 51:293–300

Bork EW, West NE, Price KP (1999a) Spectral variables for rangeland cover quantification. Int J Remote Sens 20:3641–3662

Bork EW, West NE, Price KP, Walker JW (1999b) Rangeland cover component quantification using calibrated broad (TM) and narrow-band (1.4 nm) spectrometry. J Range Manage 52:249–257

Bork EW, West NE, Price KP (1999c) Calibration of broad- and narrow-band spectral variables for rangeland cover component quantification. Int J Remote Sens 20:3641– 3662

Clark RN (1999) Spectroscopy of rocks and minerals and principles of spectroscopy. In: Rencz AN (ed) Remote sensing for the earth sciences: manual of remote sensing 3rd Edition, vol 3. John Wiley, New York, pp 3–58

Clark RN, Roush TL (1984) Reflectance spectroscopy – quantitative-analysis techniques for remote-sensing applications. J Geophys Res 89:6329–6340

Clark RN, Swayze GA, Heidebrecht, KB, Goetz, AFH, Green RO (1993a) Comparison of methods for calibrating AVIRIS data to ground reflectance. Summaries of the 4th Annual JPL Airborne Geoscience Workshop, vol 1: AVIRIS Workshop, Jet Propulsion Laboratory Publication 93–26, Pasadena, CA, pp 31–34

Clark RN, Swayze GA, Gallagher A (1993b) Mapping Minerals with Imaging Spectroscopy. US Geological Survey, Office of Mineral Resources Bull 2039, Denver, CO, pp 141–150

Clark RN, Swayze GA, Gallagher AJ, King TVV, Calvin WM (1993 c) The US Geological Survey, digital spectral library: version 1:0.2 to 3.0 microns., US Geological Survey Open File Report 93–592, Denver, CO

Clark RN, Swayze GA, Livo KE, Kokaly RF, Sutley SJ, Dalton JB, McDougal RR, Gent CA (2002) Imaging spectroscopy: earth and planetary remote sensing with the USGS tetracorder expert systems. J Geophys Res (in press)

Danin A, Bar-Or Y, Dor I, Yisraeli T (1989) The role of cyanobacteria in stabilization of sand dunes in southern Israel. Ecol Mediterr 15:55-64

Elvidge CD (1990) Visible and near infrared reflectance characteristics of dry plant material. Int J Remote Sens 11:1775-1795

Green RO, Eastwood ML, Sarture CM, Chrien TG, Aronsson M, Chippendale BJ, Faust, JA, Pavri BE, Chovit CJ, Solis M, Olah MR, Williams O (1998) Imaging spectroscopy and the airborne visible/infrared imaging spectrometer (AVIRIS). Remote Sens Environ 65:227-248

Houssa R, Pion JC, Yesou H (1996) Effects of granulometric and mineralogical composition on spectral reflectance of soils in the Sahelian area. Photogr Remote Sens 51: 284–298

Huete AR, Post DF, Jackson RD (1984) Soil spectral effects on 4-space vegetation discrimination. Remote Sens Environ 15:155-165

Jacobberger PA (1989) Reflectance characteristics and surface processes in stabilized dune environments. Remote Sens Environ 28:287-295

Karnieli A (1997) Development and implementation of spectral crust index over dune sands. Int J Remote Sens 18:1207–1220

Karnieli K, Sarafis V (1996) Reflectance spectrometry of cyanobacteria within soil crusts – a diagnostic tool. Int J Remote Sens 17:1609–1615

Karnieli A, Tsoar H (1995) Spectral reflectance of biogenic crust developed on desert dune sand along the Israel-Egypt border. Int J Remote Sens 16:369–374

Karnieli A, Shachak M, Tsoar H, Zaady E, Kaufman Y, Danin A, Porter W (1996) The effect of microphytes on the spectral reflectance of vegetation in semiarid regions. Remote Sens Environ 57:88–96

Karnieli A, Kidron GJ, Glaesser C, Ben-Dor E (1999) Spectral characteristics of cyanobacteria soil crust in semiarid environment. Remote Sens Environ 69:67–75

Kidron GJ (1995) The impact of microbial crust upon rainfall-runoff-sediment yield relationships on longitudinal dune slopes at Nizzana, western Negev Desert, Israel. Unpubl. PhD Thesis, The Hebrew University of Jerusalem, Jerusalem (Hebrew with English summary)

Kokaly RF, Clark RN (1999) Spectroscopic determination of leaf biochemistry using band-depth analysis of absorption features and stepwise multiple linear regression. Remote Sens Environ 67:267–287

Kokaly RF, Clark RN, Swayze GA (1994) Vegetation and cryptobiotic soils mapping in arid regions. Spectral Analysis Workshop, Desert Research Institute, Reno, NV. http://speclab.cr.usgs.gov/PAPERS.arches.crypto.94/arches.crypto.dri.html.

Kokaly RF, Clark RN, Livo KE (1998) Mapping the biology and mineralogy of Yellowstone National Park using imaging spectroscopy. In: Green RO (ed) Summaries of the 7th annual Jet Propulsion Laboratory airborne earth science workshop. Jet Propulsion Laboratory Publ 97–21, vol 1, pp 245–254

Lange OL, Kidron GJ, Büdel B, Meyer A, Kilian E, Abeliovich A (1992) Taxonomic composition and photosynthetic characteristics of the biological soil crusts covering sand dunes in the western Negev Desert. Funct Ecol 6:519–527

Martin ME, Newman SD, Aber JD, Congalton RG (1998) Determining forest species composition using high spectral resolution remote sensing data. Remote Sens Environ 65:249–254

O'Neill AL (1994) Reflectance spectra of microphytic soil crust in semi-arid Australia. Int J Remote Sens 15:675–681

Roberts DA, Gardner M, Church R, Ustin S, Scheer G, Green RO (1998) Mapping chaparral in the Santa Monica Mountains using multiple end-member spectral mixture models. Remote Sens Environ 65:267–279

Rouse JW, Haas RH, Schell JA, Deering DW, Harlan JC (1974) Monitoring the vernal advancements and retrogradation (greenwave effect) of natural vegetation. NASA/GSFC Final Report, NASA, Greenbelt, MD

Sirevag R (1992) Thermophilic phototrophs. In: Kristjannson JK (ed) Thermophilic bacteria. CRC Press, Ann Arbor, MI, pp 195–219

Stoner ER, Baumgardner MF (1981) Characteristic variations in reflectance of surface soils. Soil Sci Soc Am J 45:1161–1165

Swayze GA (1997) The hydrothermal and structural history of the Cuprite Mining District, Southwestern Nevada: an integrated geological and geophysical approach. Unpubl PhD Dissertation, University of Colorado, Boulder

Swayze GA, Smith KS, Clark RN, Sutley SJ, Pearson RM, Vance JS, Hageman PL, Briggs PH, Meier AL, Singleton MJ, Roth S (2000) Using imaging spectroscopy to map acidic mine waste. Environ Sci Technol 34:47–54

Tromp M, Steenis Z (1996) Deriving sub-pixel soil characteristics in northern Burkina Faso with spectral unmixing. ED Proc ISSS Int Symp (Working Groups RS and DM) on Monitoring Soils in the Environment with Remote Sensing and GIS. Ouagadougou, Burkina Faso, pp 269–284

Tsoar H, Karnieli A (1996) What determines the spectral reflectance of the Negev-Sinai sand dunes? Int J Remote Sens 17:513–525

Wessels DCJ, Van Vuuren DRJ (1986) Landsat imagery – its possible use in mapping the distribution of major lichen communities in the Namib Desert, South West Africa. Madoqua 14:369-373

West NE (1983) Western intermountain sagebrush steppe. In: West NE (ed) Temperate deserts and semideserts., Ecosystems of the world, vol 5. Elsevier, Amsterdam, pp 351–397

32 Monitoring and Management of Biological Soil Crusts

R. Rosentreter, D.J. Eldridge, and J.H. Kaltenecker

32.1 Introduction

Monitoring is the collection and analysis of repeated observations so that changes over time can be assessed. Typically, monitoring is used to evaluate changes in landscape condition in relation to defined management goals. The objectives of a monitoring program will determine the position in the landscape in which measurements will be made and the period over which data will be collected and assessed. Monitoring is often designed so that measurements can be made by more than one observer, and the level of change which is acceptable is usually determined before monitoring commences.

The principal aim of monitoring is to provide an objective basis for either changing or maintaining a current management practice. Thus, monitoring is intimately associated with management, with a feedback of information to the land manager and therefore the management process. In this chapter we discuss how soil crusts can be monitored, and the role of grazing, fire, and recreational use in the management of landscapes in which biological soil crusts are a major component. Monitoring using remotely sensed techniques are discussed in Chapter 31.

32.2 Crusts and Rangelands Monitoring

During the past century, rangeland managers grappled with methods to assess the health and trend of landscapes (see Tueller 1988). Techniques concentrated on the recording of vascular plant attributes such as cover, frequency, presence/absence, abundance, and biomass of various species, particularly perennial plants (Friedel and Bastin 1988; Friedel et al. 1988; Holechek et al. 1989; Milton et al. 1998).

While many scientists acknowledge the close links between biological soil crusts and rangeland condition assessment (see Klopatek 1992), soil crusts

Ecological Studies, Vol. 150
J. Belnap and O.L. Lange (eds.) Biological Soil Crusts: Structure, Function, and Management

and their component organisms have rarely been recorded during field-based assessment (West 1990). Early efforts to classify the surface of soils and to include biological soil crusts in these assessments were developed in the semiarid woodlands of eastern Australia (Tongway and Smith 1989). This system was later refined and extended to other landscape types (Tongway and Hindley 1995). In eastern Australia, the Department of Land and Water Conservation has been collecting data (including crust cover) on the condition and trend of rangelands since the mid-1980s (Green 1992), and workers in the western US (Pellant 1996; USDI 1997) have recently included crust cover as a component of monitoring programs.

32.2.1 Long-Term Monitoring Efforts

Biological soil crusts are frequently monitored as part of a larger study rather than being monitored for their own sake, and the length of time between successive monitoring events will depend on the aims of the study and the purpose for which the data are intended. Two examples of long-term monitoring of crust cover illustrate how the data differ in their intensity of measurement as well as in their level of input.

In eastern Australia the Department of Land and Water Conservation has been monitoring rangelands condition since 1990 at more than 350 sites (Rangelands Assessment Program; Green 1992). These sites comprise seven rangeland types with distinct vegetation and soil communities. Vegetation and soil characteristics are monitored at each 300 × 300-m site within 52 0.5-m^2 quadrats placed regularly over the site. The cover of biological soil crusts is assessed within each quadrat, along with assessment of plant cover, erosion, and litter.

Long-term data from the Rangelands Assessment Program indicate some interesting trends. Biological crust cover on three rangeland types (ATVC, NFPL, SAND) was very low (<10 %) and changed little over time (Fig. 32.1; see legends for interpretation of abbreviations). These landscapes are dominated by either clays or sands, and crust cover is intrinsically sparse. For the other four range types (GRAN, HRED, BLUE, ROBE) there was a general trend of increasing crust cover over time, with short-term increases and decreases in cover. Declining rainfall since 1990 resulted in reduced vascular plant biomass (Eldridge and Stafford 1999) so that biological soil crust was more apparent under conditions of sparse vegetation cover. In the absence of other data collected during the monitoring program, the results could easily have been interpreted as a real increase in crust cover brought about by some change in management or climate. The data also show that ground cover of vascular plants was higher than average in 1995 and lower than average in 1997 in the granite (GRAN)

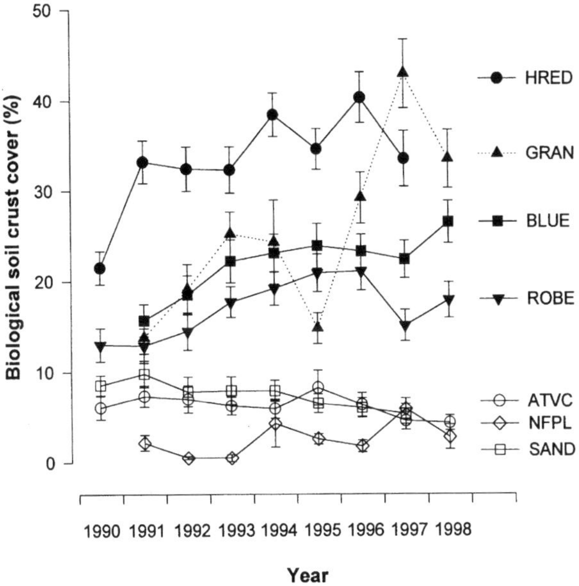

Fig. 32.1. Changes in mean (± standard error of the mean) cover (%) of biological soil crusts at 350 sites within seven rangeland plant communities in eastern Australia from 1990 to 1997. Plains sites: *BLUE Maireana* sp.; *GRAN* granitic soils, *Acacia aneura* and *Eucalyptus*; *HRED* hard red earth with *A. aneura*; *ROBE Alectryon oleifolius-Callitris glaucophylla*. *ATVC* plains and relict drainage lines with *Atriplex vesicaria*; *NFPL* relict floodplains with perennial grasses; *SAND* sandplains with dense woody shrubs

Table 32.1. Mean (± standard error of the mean) cover (%) for moss and *Collema tenax* in biological crusts of Virginia Park, Canyonlands National Park. Values are means of 60 0.1 m^2 frames, each containing 20 point hits, within three permanent 0.5-ha plots in two grassland communities. The 1967 values are from Kleiner and Harper (1977)

	Hilaria community		*Stipa/Sporobolus* community	
	Moss	*Collema*	Moss	*Collema*
1967	21.1	19.5	19.2	17.9
1996	18.5 ± 2.8	19.6 ± 2.0	17.6 ± 3.4	19.1 ± 3.0
1997	23.6 ± 2.4	15.1 ± 1.8	21.0 ± 2.4	9.1 ± 1.4
1998	27.8 ± 2.2	13.4 ± 1.4	24.1 ± 2.3	11.1 ± 1.1
1999	24.7 ± 2.2	14.0 ± 1.7	23.7 ± 2.2	7.0 ± 1.1

rangeland type, resulting in marked apparent fluctuations in cover of biological soil crusts.

In a study in southeastern Utah, biological soil crusts were sampled in two grassland types (dominated by *Hilaria jamesii* and *Stipa hymenoides/ Sporobolus* spp.) in 1964 (Kleiner and Harper 1977) and then annually from 1996 to 1999. Values for both moss and *Collema* lichens were similar in 1964 and 1996 (Table 32.1). However, between 1996 and 1999, both communities saw an increase in moss cover (>33 %) and a decrease in *Collema* cover (29 % in *Hilaria* and 42 % in *Stipa* community). The biggest change occurred between 1996 and 1997, and may have been partially a result of increased cool-season moisture in 1996, which would have favored mosses at the expense of *Collema*.

32.2.2 Monitoring Using Morphological Groups

Field identification of individual crust species can often be difficult, particularly for broad-scale monitoring programs. As strong relationships exist between the form (morphology) of biological soil-crust organisms and the way in which they function in relation to disturbance and ecological processes, morphological groups have been proposed as surrogates for species for monitoring of crusts in arid areas (Kaltenecker 1997; Ponzetti et al. 1998; Eldridge and Rosentreter 1999).

Morphological groups eliminate the need for exact determination of genera and species and avoid the problems associated with complex, often confusing changes in nomenclature. They allow easy identification that is independent of reproductive structures, and the ability to monitor more sites more quickly with less specialized staff. Measures of cover and abundances of morphological groups can be obtained more rapidly and simply than measuring individual species. Morphological groups may be too coarse to allow the detection of rare species but can provide substantial information about changes in biological soil-crust communities over time.

Kaltenecker (1997) used morphological groups to examine the recovery of biological soil crusts after fire in *Artemisia* shrublands in southern Idaho, USA (Table 32.2). Unburned (control) sites were compared with burned sites where perennial grasses had been sown as part of a postfire revegetation treatment, or where the site was left untreated and had become dominated by annual exotic weeds. Crusts were dominated by short mosses such as *Bryum lisae, B. argenteum, Ceratodon purpureus*, and *Didymodon rigidulus*, along with a variety of lichens. The results indicated substantial differences in cover of morphological groups between the three treatments (Table 32.2). Thus, although individual taxa were not measured, information was still gained regarding biological crust dynamics in conjunction with changes to vascular plant community structure following fire.

Table 32.2. Cover (%) of lichens and mosses, arranged by morphological group, in unburned *Artemisia tridentata* communities (control) and two postfire treatments (seeded and unseeded). Cover was assessed using 20-m line intercept transects. (Kaltenecker 1997)

	Control	Seeded	Unseeded
Lichens			
Gelatinous	8.1	0.8	0
Crustose	3.8	0.4	0
Squamulose	2.4	0.2	0
Fruticose/foliose	0.6	0	0
Mosses			
All	58.5	45.1	35.0
Short	23.2	43.1	34.0

32.2.3 Field Techniques for Monitoring

32.2.3.1 When and Where to Monitor

The length of time between successive monitoring events will depend on the aims of the study, the purpose to which the data are intended, climate of the study area, and the composition of the biological crusts in question. Some biological soil-crust components, e.g., phycolichens, grow very slowly, and differences are difficult to detect over short periods of time (<5 years). Others, such as bryophytes, cyanobacteria, and several cyanolichens, can respond more rapidly, particularly in moist, cool weather. Most monitoring efforts occur during the peak growing season to measure the full biological potential of the site. As some crust groups are difficult to detect when the soil surface is dry, soils can be sampled during naturally moist periods or by artificially moistening the surface prior to measurement (Rosentreter 1986; Kaltenecker 1997; Kaltenecker et al. 1999b).

Monitoring can be used to assess impacts of a specific land use, to measure recovery, or to ascertain normal background variation in crust populations, and location of sampling sites will vary depending on intent of the effort. In areas of active disturbance, soil crusts are likely to be best developed in areas protected from trampling such as under shrubs, or adjacent to obstacles such as fallen trees and rocks. Shallow rocky areas often provide refugia for crust organisms which have been destroyed by trampling. However, these refugia may not reflect the exact potential of the area in question. For instance, species composition and percent cover under shrub canopies and on shallow soils can be quite different than those found in interspaces or deeper soils.

Repeated sampling of relatively undisturbed reference sites is often critical in order to measure recovery rates and natural variation in crusts in the absence of disturbance.

32.2.3.2 How to Monitor

Biological soil crusts are typically measured using standard or slightly modified rangeland assessment techniques. The size of the sampling unit should be appropriate to the size, density and spatial distribution of the organisms being studied, their habitats and the nature of the impact being investigated. For example, the impacts of off-highway vehicles on crust communities may be best examined by using repeated photo-points, remote sensing, or aerial photography interpretation on large plots of up to several hectares in size. Where finer-scale assessments are needed, line-intercept or quadrat methods, stratified within relevant vegetation or soil zones, are more appropriate. The size and number of samples will depend on a compromise between statistical considerations and what is feasible.

Line-Intercept and Line-Point Methods. Line-intercept methods, where the proportion of the line comprising various species or groups is measured, are generally best for communities where the vegetation (including the biological soil crust) is strongly patterned and soil-crust organisms have distinct boundaries. Line-intercept is difficult in densely crusted and species-rich areas, or where individual crust species are intermingled (Kaltenecker 1997). As soil-crust organisms are very small, the line must be placed near the soil surface in exactly the same location at each measurement time. This can be achieved by placing permanent markers such as steel stakes along the route of the line. Line-point methods differ in that entities are recorded at predetermined or random points along a line. These methods are easier to use than line-intercept where vegetation is not strongly patterned, but a large number of points may be required for statistical efficiency. A variant of the line-intercept method is the short-focus telescope (Pickard and Seppelt 1984).

Quadrat-Based Methods. Quadrats are generally effective where soils are sparsely crusted, or where crusts occur in distinct patches. As in vascular plant studies, quadrats may be nested. Cover is relatively quick and easy to assess, and can be recorded by individual species or by morphological group. In Australia, frequency and cover of the soil crust is commonly assessed in 0.5-m^2 quadrats (Eldridge and Bradstock 1994; Eldridge 1996, 1999; Eldridge and Tozer 1996). Quadrat size may be reduced where crust cover is dense or greater detail is required. For example, Rogers and Lange (1971) used a quadrat size of 15 × 15 cm to examine changes in soil-crust floristics in relation to stock watering points in semiarid and subtropical Australia. Very small cores of soil crust (circular quadrats) can be extracted and assessed in the laboratory under a microscope (Eldridge et al. 2000). Rectangular plots of 20 × 50 cm are frequently used to monitor shrub-steppe vegetation, including crusts (Daubenmire 1959; Ponzetti et al. 1998) in the western United States.

Cover in quadrats can be estimated using either cover classes or point-quadrat methods, and whilst both methods are comparable (Ponzetti et al. 1998), there are tradeoffs in efficiency and effort. Trained recorders can quickly and consistently read many plots using ocular cover classes in a relatively short period of time. Point-quadrat methods can be less subjective, but uncommon species are often missed and species diversity may be underestimated.

32.3 Biological Soil Crusts and Rangeland Management

There are many factors to consider in the management of landscapes where biological soil-crust communities are a major component. Total protection from disturbance is often the easiest way to maintain or improve biological soil crusts, but it is not often feasible or desirable. Biological soil crusts can often be maintained by manipulating the type, intensity, timing, frequency, duration, or extent of disturbance. The impact of various management strategies on biological soil crusts can be assessed by comparing areas with protected relic sites or rangeland reference sites (National Research Council 1994) which provide an indication of the ecological potential of a site.

32.3.1 Grazing

Management of biological soil crusts should aim to reduce the impact of grazing when crusts are brittle or susceptible to compression. In high altitude or cool-season rangelands, crusts on all soil types are least vulnerable to disturbance when the soils are frozen or snow-covered (Chap. 27). On sandy or sandy loam soils, crusts are less susceptible to disturbance when moist or wet and on clay or clay loam soils when crusts are dry (Marble and Harper 1989; Memmott et al. 1998). In general, early to mid-wet season grazing and light to moderate stocking rates are recommended.

In the United States, implementation of rest-rotation strategies that minimize the frequency of surface disturbance during dry seasons, and maximize the period between disturbances, will reduce impacts to biological soil crusts.

Winter grazing is beneficial to vascular plant communities, and can substantially reduce the costs of supplementary feeding. Winter grazing also most closely imitates the grazing behaviour of native herbivores (Burkhardt 1996). Livestock should be removed before the end of the wet season to allow recovery of soil-crust organisms before the oncoming dry periods. Wet, muddy conditions should also be avoided in order to prevent crusts becoming buried (Kaltenecker and Wicklow-Howard 1994; Kaltenecker et al. 1999b). In Australia, practices such as "low risk" or conservative stocking are recommended to ensure that biological soil crusts are not destroyed by trampling. Conservative stocking involves the use of small groups of animals in a larger number of paddocks, at stocking rates which are generally below the district average (see Chap. 29).

In general, strategies which disperse livestock throughout the rangeland rather than concentrating them in one location will reduce damage to crusts (Eldridge and Greene 1994). Livestock can be dispersed by (1) locating water (or other supplements) on sites with low potential for biological soil-crust development such as on rocky areas, (2) providing more than one watering point per paddock, (3) using brush barriers or fencing to discourage livestock from moving across sensitive areas. Where stock are concentrated for feeding, such as during droughts, feeding should be confined to the one small area to minimize the damage to other areas. Livestock should be excluded from reference sites, and sites with highly erodible soils or sparse plant cover. In general, decisions on stocking levels and season of use should be made annually, jointly by managers and users, with optimal coverage of both vascular plants and biological soil crusts as the management objective (Kaltenecker and Wicklow-Howard 1994; Kaltenecker et al. 1999b).

32.3.2 Fire

Prescribed fire can be a useful tool for manipulating the cover, composition, or vigor of vegetation communities. For example, burning may be useful on more productive sites with low potential for exotic plant invasion to reduce high woody vegetation densities resulting from overgrazing. However, biological soil-crust organisms are generally killed by hot ground fires, resulting in loss of cover and species diversity (see Chaps. 28, 29).

Historically, wildfires in semiarid and arid landscapes produced a patchwork mosaic of burnt and unburnt patches. These were probably small, occurring at a scale ranging from individual plants to landscapes. This has also resulted in a mosaic of successional stages of plants and crusts, with propagules readily available to replenish burned sites. Unlike many controlled (manipulated) burns, wildfires were also relatively infrequent, allowing crust organisms to recolonize burned sites before another burn. In the semiarid

Eucalyptus woodlands in eastern Australia, researchers are using fire to alter the vegetation community to make it more suitable for the reintroduction of the mallee fowl, a large ground-dwelling bird. Whilst controlled burning has benefited the native vegetation, if the fires are too frequent (> once every 16 years), the biological soil crust is converted into one dominated by cyanobacteria at the expense of lichens and mosses (Eldridge and Bradstock 1994).

There is now much evidence to show the strong links between the lack of biological soil crusts and fire. In the western United States, overgrazing and disturbance of crusts in *Artemisia* shrublands results in removal of soil crusts and often invasion by annual grasses such as *Bromus tectorum* (West 1994). Large, unnaturally frequent fires have become commonplace in these annual grasslands. These fires often spread to adjacent intact shrub communities, further destroying remnant patches of crust, and converting the shrublands into grasslands (Whisenant 1990). Management of intact shrub communities aims to maintain healthy soil crusts to prevent invasion by annual grasses. This is being accomplished by grazing regimes to minimize disturbance of the crust and reduce competition from exotic species, selective use of fire to prevent fuel buildup, and restrictions on off-road vehicle use.

Management of burned areas aims to prevent reestablishment of exotic plants whilst maximizing the recolonization of soil-crust species. Burned sites are assessed to determine whether they will regenerate naturally or if they require rehabilitation. Many burned sites require revegetation to stop exotic plant invasion, and most techniques require some soil-surface disturbance. This may appear inconsistent with recovery of biological crusts, but failure to treat sites can result in irreversible dominance by annual species, which prevents the return of well-developed biological soil crusts (Kaltenecker 1997; Kaltenecker et al. 1999a). Similarly, it has been suggested that grazing can reduce the vigor and cover of annual weeds such as *Bromus tectorum*. However, the effective use of livestock in controlling exotic annual plants has not been demonstrated, and grazing late in the growing season may be detrimental to biological soil crusts as well as to native vegetation.

Once revegetated, protection from grazing and recreational use is often necessary for recovery of the biological soil crust and the vascular plant community. Restoration can be enhanced by sowing local ecotypes of native plants using seeding methods which minimize disturbance to the soil surface.

32.3.3 Recreational Use Management

Many recreational activities have similar impact as livestock, and therefore the same management principles apply. However, there are also major differences. People are often more difficult to control, their access is often unlimited, and they tend to go where they want, despite the use of barriers.

People also have a greater affinity for open vegetation, as it is easier to walk or drive through. Consequently, people are often harder to control than livestock.

Concentration of recreational use is generally desirable, and the use of designated campsites reduces the impact of haphazard placement of sites by individuals. Trails minimize the amount of biological soil crust that is disrupted by trampling, and in some high-value areas, walkways may be built over the tracks to prevent soil damage. Campers and hikers can be educated on where to place campsites, how to travel through country where there are no tracks or trails, and how to use the environment without causing damage.

There are a number of general management practices which reduce the impact on soils and soil crusts. These include: (1) restrict road locations to less sensitive areas; (2) design roads and drainage to minimize erosion and sedimentation; (3) promote extensive, low-density uses such as hiking and backpacking during late fall and winter periods; (4) restrict access during sensitive dry seasons; (5) restrict high-density, high-impact uses such as tree cutting and firewood collection to short periods during late fall and winter, and rotate these areas, ensuring long recovery periods of 5–10 years minimum before they are reused; (6) exclude sensitive sites; (7) provide designated trails, and restrict their use in high density recreational areas; (8) provide interpretive guides and information on the value of biological soil crusts at significant access points; (9) require analyses of the environmental impacts to biological soil crusts of all developments such as right-of-ways, oil, gas and exploration permits, and permits to drill.

References

Burkhardt JW (1996) Herbivory in the Intermountain West. An overview of evolutionary history, historic cultural impacts and lessons from the past. Station Bull 58. Idaho Forest, Wildlife and Range Experim Station, University of Idaho, Moscow

Daubenmire R (1959) A canopy-coverage method of vegetational analysis. Northwest Sci 33:43–64

Eldridge DJ (1996) Dispersal of microphytes by water erosion in an Australian semi-arid woodland. Lichenologist 28:97–100

Eldridge DJ (1999) Dynamics of moss- and lichen-dominated soil crusts in a patterned *Callitris glaucophylla* woodland in eastern Australia. Acta Ecol 20:159–170

Eldridge DJ, Bradstock RA (1994) The effect of time since fire on the cover and composition of cryptogamic soil crusts on a *Eucalyptus* shrubland soil. Cunninghamia 3:521–527

Eldridge DJ, Greene RSB (1994) Assessment of sediment yield from a semi-arid red earth soil with varying cryptogam cover. J Arid Environ 26:221–232

Eldridge DJ, Rosentreter RR (1999) Morphological groups: a framework for monitoring microphytic crusts in arid landscapes. J Arid Environ 41:11–25

Eldridge DJ, Stafford MR (1999) Rangeland health in the western Riverina: 1990–1997. Dept of Land and Water Conserv, Sydney

Eldridge DJ, Tozer ME (1996) Distribution and floristics of bryophytes in soil crusts in semiarid and arid eastern Australia. Aust J Bot 44:223–247

Eldridge DJ, Semple WS, Koen TB (2000) Dynamics of cryptogamic soil crusts in a derived grassland in south-eastern Australia. Aust J Ecol 25:232–240

Friedel MH, Bastin GN (1988) Photographic standards for estimating comparative yield in arid rangelands. Aust Rangel J 10:34–48

Friedel MH, Chewings VH, Basin GN (1988) The use of comparative yield and dry-weight-rank techniques for monitoring arid rangeland. J Range Manage 41:430–435

Green D (1992) Rangeland assessment in NSW. Proc 7th Bien Conf Aust Rangel Soc, Cobar NSW, pp 267–268

Holechek JL, Pieper RD, Herbel CH (1989) Range management principles and practices. Prentice Hall, Englewood-Cliffs, NJ

Kaltenecker JH (1997) The recovery of microbiotic crusts following post-fire rehabilitation on rangelands of the western Snake River Plain. Unpubl MS thesis, Boise State University, Idaho

Kaltenecker J, Wicklow-Howard M (1994) Microbiotic soil crusts in sagebrush habitats of southern Idaho. Contract Report. On file with: Interior Columbia Basin Ecosystem Management Project, 112 E. Poplar, Walla Walla, WA 99362

Kaltenecker JH, Wicklow-Howard M, Pellant M (1999a) Biological soil crusts: natural barriers to *Bromus tectorum* L. establishment in the northern Great Basin, USA. In: Eldridge D, Freudenberger D (eds) Proc VI Int Rangeland Congr, vol 1. VI Int Rangeland Congr, Aitkenvale, Queensland, Australia

Kaltenecker JH, Wicklow-Howard MC, Rosentreter R (1999b) Biological soil crusts in three sagebrush communities recovering from a century of livestock trampling. In: McArthur ED, Ostler WK, Wambolt CL (eds) Proceedings: Shrubland Ecotones. Proceedings RMRS-P-11, USDA Forest Service, Rocky Mountain Research Station, Ogden, Utah, pp 222–226

Kleiner EF, Harper KT (1977) Soil properties in relation to cryptogamic groundcover in Canyonlands National Park. J Range Manage 30:202–205

Klopatek JM (1992) Cryptogamic crusts as potential indicators of disturbance in semiarid landscapes. In: McKenzie DH, Hyatt DE, McDonald VJ (eds) Ecological indicators. Elsevier, New York, pp 773–786

Marble JR, Harper KT (1989) Effect of timing of grazing on soil-surface cryptogamic communities in Great Basin low-shrub desert: a preliminary report. Great Basin Nat 49:104–107

Memmott KL, Anderson VJ, Monsen SB (1998) Seasonal grazing impact on cryptogamic crusts in a cold desert ecosystem. J Range Manage 51:547–550

Milton SJ, Dean WRJ, Ellis RP (1998) Rangeland health assessment: a practical guide for ranchers in arid Karoo shrublands. J Arid Environ 39:253–265

National Research Council (1994) Rangeland health – new methods to classify, inventory, and monitor rangelands. National Academy Press, Washington, DC

Pellant M (1996) Use of indicators to qualitatively assess rangeland health. In: West NE (ed) Rangelands in a sustainable biosphere. Proc Vth Int Rangel Congr, Salt Lake City, Utah, pp 434–435

Pickard J, Seppelt RD (1984) Phytogeography of Antarctica. J Biogeogr 11:83–102

Ponzetti J, Youtie B, Salzer D, Kimes T (1998) The effects of fire and herbicide on microbiotic crust dynamics in high desert ecosystems. Final Rept Nature Conservancy of Oregon Forest and Rangeland Ecosystem Science Center, Biological Resources Division, US Geological Survey, Portland, Oregon

Rogers RW, Lange RT (1971) Lichen populations on arid soil crusts around sheep watering places in South Australia. Oikos 22:93–100

Rosentreter R (1986) Compositional patterns within a rabbitbrush (*Chrysothamnus*) community of the Idaho Snake River Plain. In: McArthur ED, Welch BL (eds) Proc Symposium: Biology of *Artemisia* and *Chrysothamnus*. Gen Tech Report INT-200, Ogden, Utah, pp 273–277

Tongway DJ, Hindley N (1995) Manual for assessment of soil condition of tropical grasslands. CSIRO, Australia

Tongway DJ, Smith EL (1989) Soil surface features as indicators of rangeland site productivity. Aust Rangel J 11:15–20

Tueller PT (1988) Application of plant sciences to rangeland management and inventory. Martinus Nijhoff/W Junk, Amsterdam

USDI (1997) National resource inventory: Colorado pilot handbook. Bureau of Land Management, Denver, CO

West NE (1990) Structure and function of microphytic soil crusts in wildland ecosystems of arid and semi-arid regions. Adv Ecol Res 20:179–223

West NE (1994) Effects of fire on salt-desert shrub rangelands. In: Monsen SB, Kitchen SG (eds) Proceedings, ecology and management of annual rangelands. General Techn Rept INT-GTR-313. US Dept Agric, Forest Serv, Intermountain Res Stn, Ogden, UT; pp 170–175

Whisenant SG (1990) Changing fire frequencies on Idaho's Snake River Plains: ecological and management implications. In: McArthur ED, Romney EM, Smith SD, Tueller PT (eds) Proc – Symp on Cheatgrass Invasion, Shrub Die-off, and Other Aspects of Shrub Biology and Management. General Technical Report INT-276. US Dept of Agric, Forest Serv, Intermountain Res Stn, Ogden, UT, pp 4–10

Part VIII

Conclusions

33 Structure and Functioning of Biological Soil Crusts: a Synthesis

J. BELNAP and O.L. LANGE

33.1 Introduction

Biological soil crusts play many important ecological roles worldwide, as amply demonstrated throughout this Volume. Given these roles, and the fact that soils in all semiarid and arid lands (representing over 30 % of the Earth's surface) have some degree of biological soil-crust development, these organisms are clearly a substantial force in shaping the structure and function of many ecosystems worldwide. Biological soil crusts are known to increase the stability of often easily eroded soils, influence local hydrological cycles in regions that receive limited precipitation, and increase soil fertility. Biological soil crusts occur in many diverse climates and soils. While some of the published information on soil crusts appears contradictory, chapters in this Volume show that most of the apparent controversy can be resolved by accounting for soil texture and climate (especially separating areas where soils do or do not freeze). Also, as can be seen in this Volume, new information has helped resolve what appeared to be contradictions in older studies.

33.2 Taxonomic Composition and Biogeography

Part I of this Volume showed that biological soil crusts occur in arid and semiarid regions throughout the world, as well as in locally suited sites in temperate, boreal, and arctic environments. While soil crusts exist in unconnected and seemingly dissimilar vegetation types and climates, they are taxonomically and structurally very similar. Most of the typical and often dominant cyanobacteria, lichen, and moss genera have a cosmopolitan distribution, even on the species level. Ecologically similar species, closely related taxa, and/or convergent life forms often replace each other both on neighboring sites and at sites separated by large distances (e.g., *Fulgensia fulgens* and *F. bracteata*; many species of vagrant lichens). On a regional basis, the

Ecological Studies, Vol. 150
J. Belnap and O.L. Lange (eds.) Biological Soil Crusts: Structure, Function, and Management

relative dominance of the different species varies with climate and soils. However, many species maintain similar habitat preferences regardless of where they occur worldwide (e.g., the preferential occurrence of certain lichen taxa on stabilized calcareous soils, the salt tolerance of some *Microcoleus* species, the dominance of filamentous cyanobacteria in sandy soils, or the similar suite of phycolichens on most gypsiferous soils), which can also structure crust communities within a given site or region.

Our present understanding of floristics and phytogeography of biological soil crusts on a worldwide basis is still very limited. We are far from being able to construct statistically sound phytosociological groupings or to make detailed comparisons of soil-crust communities on anything except a local basis. It is apparent that we lack even rudimentary knowledge about distribution of biological soil crusts and of their species composition in many regions on most continents. Because species composition of a given soil crust strongly influences the role it plays in a specific ecosystem, additional surveys are needed to better understand the importance of soil crusts in a given area. Systematic work with taxonomically difficult and often scarcely studied groups of soil-crust organisms is urgently needed so comparisons of soil crusts can be made on a worldwide basis. Ecologists should be encouraged to view soil-crust components as part of the vegetation community, and not to restrict their studies to vascular plants. In addition, all groups of soil-crust components should be addressed in field studies, rather than the commonly observed tendency to restrict studies to one group (e.g., lichens, mosses, or cyanobacteria).

An exciting and important goal should be the systematic creation of a comprehensive, worldwide monographic treatment of the phytosociology and biogeography of biological soil crusts. Communities need to be classified with respect to the vegetation zones and higher plant formations in which they occur, ranging from the arid and semiarid lands in the Americas through the Eurasian and African continent to Australia, and from the Arctic, boreal, and temperate belts to Antarctica. This knowledge can then serve as a foundation for a solid assessment and appreciation of the ecological role that these soil crusts play worldwide. In addition, such an assessment will enable land managers to know what crust species should occur in a given ecosystem, and what functional roles these species might play in a given setting. This, then, can provide management and restoration goals for specific regions.

33.3 Heterotrophic Associates

As discussed in *Part II*, soil crusts consist not only of photosynthetic organisms, but also contain significant fungal, microbial, and invertebrate populations. Little is known about individual heterotrophic crustal components or interactions between them and the autotrophic crust components. Microfungi have been the most extensively studied, but these studies have been mostly descriptive and limited to a few geographical locations. Microfungi are known to colonize early after disturbance and to increase aggregation of soil particles, thus increasing soil stability. However, microfungi may also profoundly influence the physiological functioning of other crust species. For instance, both vesicular-arbuscular mycorrhizae and dark-septate microfungi are known to secrete acids that break the chemical bonds that keep phosphorus (P) unavailable to plants in high-pH soils. Microfungi may similarly enhance P acquisition by crust species, and thus differentially influence the relative success of cyanobacteria, lichens, or mosses on a given soil. Also, soil microbes are likely to utilize carbon provided by autotrophic components of the soil crusts. This source of carbon may be an important determinant of decomposition rates (and thus nutrient availability to vascular plants) in deserts, especially in the large interspaces that occur between plants.

33.4 Soil-Crust Structure

Water is everything to the poikilohydric organisms that dominate soil crusts. Thus, while the major determinant of crust structure can vary with the scale being examined, moisture availability is always an important factor in determining crust structure, which is discussed in *Part III*.

On a regional scale, the structure of a given biological soil crust is mostly controlled by climate. In very hot and hyperarid areas, soil crusts are smooth and dominated by cyanobacteria. As rainfall increases, the diversity and cover of mosses and lichens increases as well, and rugose crusts can be found. In regions where frost-heaving occurs, the pinnacled and rolling crusts are dominated by mosses and lichens.

On the landscape scale, the structure of soil crusts is heavily influenced by soil characteristics. Fine-textured soils generally support greater cover of a more diverse lichen and moss community than immediately-adjacent coarse-textured soils. Soils with carbonates tend to support greater lichen diversity (especially phycolichens) than soils with low or no carbonates, while gypsiferous soils support the greatest lichen diversity of all. Excessive salt restricts crust species. Topographic position is important, as polar-facing slopes will support the greatest moss and lichen cover and diversity within a landscape, unless heavy vascular plant cover precludes their presence.

On the meter and centimeter scale, crust structure is influenced by microclimate factors. Greater lichen and moss cover and diversity are found in microhabitats where the combination of moisture, light, and temperature is the most favorable for photosynthesis. Soil chemistry may also be important in determining moss and lichen distributions on this scale.

On the millimeter (or smaller) scale, crust structure is mostly influenced by light: that is, by the balance between sufficient light for photosynthesis and protection from excessive light and excessive UV exposure. Species with no or low mobility (mosses, lichens, and small cyanobacteria) must reside at the soil surface in order to obtain sufficient light for photosynthesis. However, intense sun radiation (especially in the shorter wavelengths) at the soil surface requires these species to have protective mechanisms, such as heavy pigmentation. Species without such pigmentation (e.g., *Microcoleus*) must utilize behavioral adaptations to obtain sufficient light for photosynthesis, while avoiding excessive light and UV radiation.

33.5 Carbon and Nitrogen Fixation

Semiarid and arid ecosystems have limited vascular plant productivity; thus, in these regions, the carbon contribution of soil crusts can be significant. In some desert regions (e.g., the coastal fog zone of Namibia), the area-related chlorophyll content of the soil crusts is in the range of C3 plant leaf. Thus, the seemingly bare desert surface is actually covered by photosynthetic machinery somewhat like a giant leaf. The maximal photosynthetic capacity of this "leaf" is similar to that of phanerogamous plants growing in the same area. However, metabolic activity of the poikilohydric autotrophs is restricted to the short periods of time when the crusts become hydrated by high humidity, snow, fog, dew, or rain. Nevertheless, a substantial stream of carbon is delivered from the crust autotrophs to the ecosystem, contributing to soil and humus formation and carbon availability to the soil heterotrophs.

Our knowledge of CO_2 exchange and photosynthetic production of soil crusts in the different ecosystems of the world is still very limited. The few data available are reported in *Part IV* of this Volume. However, most observations are extrapolations from measurements under controlled conditions and many address only a single organism (e.g., lichen thalli), rather than soil crust as a complex community. While allowing us to make general estimates, these laboratory studies cannot facilitate regional comparisons. Long-term monitoring of soil crust carbon fixation in nature is needed to establish carbon balances under different environmental conditions. In addition, specific ecophysiological work is needed to understand how soil organisms adapt to the extreme microclimate of the soil surface, and to explain metabolic performance of the soil-crust community. With such information, we will then be able to model and quantify carbon exchange in soil crusts of different ecosystems.

There is another important aspect to understanding the carbon exchange of soil crusts. As the dominant ground cover in arid and semiarid areas, soil crusts cover a substantial proportion of the Earth's surface. Thus, they may play a substantial role in the CO_2 fluxes between the ground and the atmosphere. Discussion about the causes of the present global increase in atmospheric CO_2 concentration, and possible mitigation measures, need to include the role of biological soil crusts during their different successional stages. Thus, future measurements and modeling work need to include large-scale estimates of how biological soil crusts contribute to the global carbon budget.

Cyanobacteria and cyanolichens are able to fix nitrogen (N), and soil crusts can be the dominant source of nitrogen in some desert ecosystems. Because only a few habitats have been investigated, it is not yet known how important these crusts are to worldwide N budgets. N fixation rates increase with temperature until a thermal limit of about 30 °C is reached. Thus, in deserts N inputs are greatest during cooler seasons, while in temperate and polar regions, N inputs are greatest in the warmest season. Absolute estimates of N inputs are currently not known for many ecosystems, as ^{15}N calibration of the commonly used acetylene-reduction method has not been done. The lack of conversion ratios for other environments, and year-round data on fixation rates, is a critical problem that needs to be addressed.

Much of the N fixed is immediately leaked to the surrounding soil. It is not known why this occurs: it may be that these simple organisms are unable to stop leakage, that they excrete N to prevent toxicity, or that they excrete N to attract beneficial organisms. The remainder of the N is incorporated in the ecosystem upon decaying of the cyanobacteria and cyanolichens.

Only a handful of studies have addressed N losses in deserts, especially in relation to biological soil crusts. There are many unanswered questions regarding how crusts affect volatilization and denitrification rates. While the crust environment may appear conducive to increased N losses, direct measures have unexpectedly shown very low rates of both volatilization and denitrification during most of the year. Losses appear greater in hot deserts than cool deserts, and higher in summer than at other times of the year. For sandy soils in southeastern Utah, N inputs and losses appear decoupled in time: when inputs are high during fall, winter, and spring, losses are neglible. In summer, the opposite is true, as inputs are low and losses are high. Further studies are needed in other environments that account for time of year, substrate availability, and differences in crustal components and soil characteristics (e.g., texture, nutrient status, and moisture content). With such information, we can then model N budgets for different soil-crust types (including different successional stages) in different ecosystems.

33.6 Interactions with Vascular Plants

Much controversy surrounds the effect of biological soil crusts on the germination and establishment of vascular plants. Some authors claim that crusts inhibit plant establishment and that crusts need to be broken or removed if greater vascular plant cover is desired, while others claim quite the opposite. As discussed in *Part IV* of this Volume, separating results of studies in regions where soils do not freeze from regions with soil freezing resolves most, if not all, this controversy. Regions that lack freezing generally have flat soil surfaces that are further smoothed by cyanobacteria. In such areas, water and wind easily move seeds, organic matter, and water off crusted surfaces to nearby obstacles (e.g., vascular plants or rocks), thus indirectly inhibiting plant establishment in plant interspaces. However, most desert soils have some inherent physical crusting, and the contribution of the biological versus the physical crusts to increased seed movement is not known. In regions where frost-heaving crusts create a roughened surface, the opposite is true: movement of seeds, organic matter, and water is inhibited across crusted surfaces, and the presence of crusts has been found to enhance seedling establishment in plant interspaces. Similarly, in cool regions, the presence of crusts either increases or does not affect seed germination, while in hot regions, germination of native plants shows a full range of responses (positive, negative, and no response).

Plants growing on crusted soil often show higher concentrations and/or greater total accumulation of various essential nutrients when compared to plants growing in adjacent, uncrusted soils. There are many possible mechanisms to explain this finding: (1) crusts fix and secrete both N and C, and it has been shown that the fixed N is utilized by nearby plants; (2) C additions may stimulate microbial populations, increase decomposition rates of plant litter, and thus increase nutrient availability to living plants; (3) exopolymers secreted by crustal components bind metals on the polysaccharide surfaces, thus keeping them plant-available; (4) silts trapped by the sticky polysaccharide material hold positively charged plant macronutrients, keeping them from being leached downwards; (5) most crustal components secrete powerful metal chelators, keeping the metals in solution; (6) there is a strong positive correlation between the presence of crusts and plant mycorrhizal infections; such infections are known to enhance nutrient uptake of plants; and/or (7) the presence of biological soil crusts increases soil surface temperatures by up to 14 °C relative to adjacent uncrusted soils, thus greatly increasing rates of nutrient uptake by vascular plants.

33.7 Soil Stability and Hydrology

All studies reported in *Part V* show that soil crusts increase soil stability. However, as with vascular plants, there has been much disagreement in the literature about their effect on water infiltration. As summarized in Chapter 24, most differences in study results can be resolved if soil texture and site location are taken into account. Fine-textured soils, in regions both with and without frost-heaving, show an increase in water infiltration with the presence of biological soil crusts. Unfortunately, all studies on sandy soils have occurred in regions with no frost-heaving. Here, increased smoothing and sealing by cyanobacteria crusts decreases water infiltration. However, no studies have been done on frost-heaved sandy soils (e.g., Colorado Plateau, Great Basin, Asian steppes), where increased surface roughness from heaving is expected to increase water infiltration.

33.8 Disturbance to Biological Soil Crusts

Most arid and semiarid regions evolved with little soil-surface disturbance, as food and water limited the number and distribution of animals. These same ecosystems occur where biological soil crusts are most evident, and where human use is increasing rapidly. Extensive discussion in *Parts V, VI,* and *VII* shows that in most regions where biological crusts are a dominant feature, crusts are reduced or extirpated by intense or frequent soil-surface disturbances (e.g., grazing, recreation, and fire). This is especially true in soils with low aggregate stability such as sands. Disturbance to crusts generally results in reduced C and N inputs, and increased soil loss from wind and water erosion.

The effects of disturbance by livestock on soil crusts is currently controversial, as crusts can be found in both grazed and ungrazed regions. While most desert soil crusts show little resistance or resilience to grazing, there are some more temperate regions where soil crusts occur despite heavy use by hooved animals (e.g., the Serengeti Plain of east Africa, the Mongolian steppes, the Great Plains of the US, and Oman). While lichen and moss cover and diversity is generally lower in these areas compared to similar ungrazed areas, crusts still persist. Common features of these more resistant areas include (1) soils that are highly weathered, fine to loamy textured, and calcareous, with high water-holding capacity, high bulk density, and some physical crusting; (2) high water availability (due to rainfall of relatively high amounts, occurring during cold seasons, and/or rain ponds on the soil surface); (3) closely spaced plant material or rocks; (4) early-successional crust species (e.g., *Collema* and *Catapyrenium* species, surface *Nostoc commune*; and (5) intermittent animal use.

33.9 Recovery from Disturbance

Reported recovery rates of soil crusts after disturbance vary widely, as they depend on climate, soil type, and severity of disturbance. Factors that impede recovery include intense or frequent disturbance, large surface-volume ratios of disturbed areas, sandy soils, and high potential evapotranspiration (low rainfall and high temperatures). As the severity of one or more of these variables is lessened, recovery rates increase. Nevertheless, experiments and subsequent extrapolations show that a severely disturbed crust in most semiarid or arid landscapes usually requires many decades, or even centuries, for full recovery.

33.10 Crust Monitoring and Management

Because land uses can have such a profound impact on the structure and functioning of biological soil crusts, it is essential that the condition of soil crusts be monitored where anthropogenic activities are continually recurring. The principal aim of monitoring is to provide an objective basis for either changing or maintaining a current management practice. Thus, monitoring is intimately associated with management, providing information to the land manager and therefore the management process. During the past century, assessment of rangeland health has concentrated on perennial vascular plant attributes (e.g., cover, frequency). While many scientists acknowledge the close links between biological soil crusts and rangeland condition, soil-crust organisms have rarely been documented in field surveys. It is important that such measures be incorporated into rangeland health assessments, as the condition of soil crusts can tell the manager much about the stability and fertility of the site. Remote sensing is likely to be a critical tool in this assessment effort.

Studies show that concentration of use, both spatially and temporally, in desert landscapes is generally desirable. For recreational uses, this means (1) designated campsites and trails and (2) restricting roads to less sensitive areas, designing them to minimize erosion and sedimentation; and eliminating off-road travel. For grazing, protection of soil crusts requires that most low-elevation areas should be used only during wet periods, that livestock numbers be low, and grazing intermittent.

33.11 Summary

This Volume has shown that biological soil crusts occur in many ecosystems worldwide. Where they occur, soil crusts contribute to hydrological balance and erosion control, and are sources of nitrogen and carbon. Unlike vascular plant cover, crustal cover is not reduced in drought and, unlike rain crusts, these organic crusts are present year-round. They also contain extremely long-lived organisms. Consequently, biological soil crusts can offer ecosystem services continuously through time, in spite of conditions often limiting other soil-surface protectors. Unfortunately, the ever-increasing human population is demanding a greater land base for livelihood and recreation. As humans push further into the semiarid and arid lands previously considered unsuitable for such activity, the cover and diversity of crusts will be concomitantly reduced. Thus, it is a crucial task for present and future managers of these drier regions to find a balance between human needs and desires and sustainable use of these arid and semiarid landscapes.

We hope that this Volume, which summarizes our present knowledge on the structure, function, and management of biological soil crusts, will contribute to the appreciation and understanding of soil-crust ecosystems such that they become an object of concern and conservation for both the public and land managers. The well-being of our planet depends on maintaining the stability and fertility of our soils. With sufficient care, the small and inconspicuous soil-crust ecosystems that cover so much of our planet like a living skin will continue to provide this service long into the future.

Subject Index

Page numbers in *italics* denote figures, photos, or tables.

Taxonomic Index

Page numbers in *italics* denote figures, photos, or tables. Letters in parantheses indicate large taxonomic groups: algae (A), bacteria (Ba), bryophytes (B), cyanobacteria (C), fungi (F), lichens (L), microfauna (MF), phanerogams (P), and pteridophytes (Pt).

Ecological Studies

Volumes published since 1997

Volume 125
Ecology and Conservation of Great Plains Vertebrates (1997)
F.L. Knopf and F.B. Samson (Eds.)

Volume 126
The Central Amazon Floodplain: Ecology of a Pulsing System (1997)
W.J. Junk (Ed.)

Volume 127
Forest Decline and Ozone: A Comparison of Controlled Chamber and Field Experiments (1997)
H. Sandermann, A.R. Wellburn, and R.L. Heath (Eds.)

Volume 128
The Productivity and Sustainability of Southern Forest Ecosystems in a Changing Environment (1998)
R.A. Mickler and S. Fox (Eds.)

Volume 129
Pelagic Nutrient Cycles: Herbivores as Sources and Sinks (1997)
T. Andersen

Volume 130
Vertical Food Web Interactions: Evolutionary Patterns and Driving Forces (1997)
K. Dettner, G. Bauer, and W. Völkl (Eds.)

Volume 131
The Structuring Role of Submerged Macrophytes in Lakes (1998)
E. Jeppesen et al. (Eds.)

Volume 132
Vegetation of the Tropical Pacific Islands (1998)
D. Mueller-Dombois and F.R. Fosberg

Volume 133
Aquatic Humic Substances: Ecology and Biogeochemistry (1998)
D.O. Hessen and L.J. Tranvik (Eds.)

Volume 134
Oxidant Air Pollution Impacts in the Montane Forests of Southern California (1999)
P.R. Miller and J.R. McBride (Eds.)

Volume 135
Predation in Vertebrate Communities: The Białowieża Primeval Forest as a Case Study (1998)
B. Jędrzejewska and W. Jędrzejewski

Volume 136
Landscape Disturbance and Biodiversity in Mediterranean-Type Ecosystems (1998)
P.W. Rundel, G. Montenegro, and F.M. Jaksic (Eds.)

Volume 137
Ecology of Mediterranean Evergreen Oak Forests (1999)
F. Rodà et al. (Eds.)

Volume 138
Fire, Climate Change and Carbon Cycling in the North American Boreal Forest (2000)
E.S. Kasischke and B. Stocks (Eds.)

Volume 139
Responses of Northern U.S. Forests to Environmental Change (2000)
R. Mickler, R.A. Birdsey, and J. Hom (Eds.)

Volume 140
Rainforest Ecosystems of East Kalimantan: El Niño, Drought, Fire and Human Impacts (2000)
E. Guhardja et al. (Eds.)

Volume 141
Activity Patterns in Small Mammals: An Ecological Approach (2000)
S. Halle and N.C. Stenseth (Eds.)

Volume 142
Carbon and Nitrogen Cycling in European Forest Ecosystems (2000)
E.-D. Schulze (Ed.)

Volume 143
Global Climate Change and Human Impacts on Forest Ecosystems: Postglacial Development, Present Situation and Future Trends in Central Europe (2001)
J. Puhe and B. Ulrich

Volume 144
Coastal Marine Ecosystems of Latin America (2001)
U. Seeliger and B. Kjerfve (Eds.)

Volume 145
Ecology and Evolution of the Freshwater Mussels Unionoida (2001)
G. Bauer and K. Wächtler (Eds.)

Volume 146
Inselbergs: Biotic Diversity of Isolated Rock Outcrops in Tropical and Temperate Regions (2000)
S. Porembski and W. Barthlott (Eds.)

Volume 147
Ecosystem Approaches to Landscape Management in Central Europe (2001)
J.D. Tenhunen, R. Lenz, and R. Hantschel (Eds.)

Volume 148
A Systems Analysis of the Baltic Sea (2001)
F.V. Wulff, L.A. Rahm, and P. Larsson (Eds.)

Volume 149
Banded Vegetation Patterning in Arid and Semiarid Environments (2001)
D. Tongway and J. Seghieri (Eds.)

Volume 150
Biological Soil Crusts: Structure, Function, and Management (2001)
J. Belnap and O.L. Lange (Eds.)

Volume 151
Ecological Comparisons of Sedimentary Shores (2001)
K. Reise (Ed.)

Volume 152
Global Biodiversity in a Changing Environment: Scenarios for the 21st Century (2001)
F.S. Chapin, O. Sala, and E. Huber-Sannwald (Eds.)

Volume 153
UV Radiation and Arctic Ecosystems (2002)
D.O. Hessen (Ed.)

Volume 154
Geoecology of Antarctic Ice-Free Coastal Landscapes (2002)
L. Beyer and M. Bölter (Eds.)

Volume 155
Conserving Biological Diversity in East African Forests: A Study of the Eastern Arc Mountains (2002)
W.D. Newmark

Volume 156
Urban Air Pollution and Forests: Resources at Risk in the Mexico City Air Basin (2002)
M.E. Fenn, L. I. de Bauer, and T. Hernández-Tejeda (Eds.)

Volume 157
Mycorrhizal Ecology (2002)
M.G.A. van der Heijden and I.R. Sanders (Eds.)

Volume 158
Diversity and Interaction in a Temperate Forest Community: Ogawa Forest Reserve of Japan (2002)
T. Nakashizuka and Y. Matsumoto (Eds.)

Volume 159
Big-Leaf Mahogany: Genetic Resources, Ecology and Management (2003)
A. E. Lugo, J. C. Figueroa Colón, and M. Alayón (Eds.)

Volume 160
Fire and Climatic Change in Temperate Ecosystems of the Western Americas (2003)
T. T. Veblen et al. (Eds.)

Volume 161
Competition and Coexistence (2002)
U. Sommer and B. Worm (Eds.)

Volume 162
How Landscapes Change: Human Disturbance and Ecosystem Fragmentation in the Americas (2003)
G.A. Bradshaw and P.A. Marquet (Eds.)

Volume 163
Fluxes of Carbon, Water and Energy of European Forests (2003)
R. Valentini (Ed.)

Volume 164
Herbivory of Leaf-Cutting Ants: A Case Study on *Atta colombica* in the Tropical Rainforest of Panama (2003)
R. Wirth, H. Herz, R.J. Ryel, W. Beyschlag, B. Hölldobler

Made in the USA
Monee, IL
13 September 2026

62975067R00289